Global Cheesemaking Technology

Global Cheesemaking Technology

Cheese Quality and Characteristics

Edited by

Photis Papademas
Department of Agricultural Sciences, Biotechnology and Food Science,
Cyprus University of Technology, Cyprus

Thomas Bintsis
11 Parmenionos, 50200 Ptolemaida, Greece

WILEY

This edition first published 2018

Registered Offices
John Wiley & Sons, Inc., 111 River Street, Hoboken, NJ 07030, USA
John Wiley & Sons Ltd, The Atrium, Southern Gate, Chichester, West Sussex, PO19 8SQ, UK

Editorial Office
The Atrium, Southern Gate, Chichester, West Sussex, PO19 8SQ, UK

For details of our global editorial offices, customer services, and more information about Wiley products visit us at www.wiley.com.

Wiley also publishes its books in a variety of electronic formats and by print-on-demand. Some content that appears in standard print versions of this book may not be available in other formats.

Library of Congress Cataloging-in-Publication Data

9781119046158

Cover Design: Wiley
Cover Image: The cover photo is of Ragusano PDO cheese. For more details see Part II Section 8.8. Courtesy of Photis Papademas

Set in 10/12pt WarnockPro by SPi Global, Chennai, India

10 9 8 7 6 5 4 3 2 1

This book is dedicated to our families and to a great teacher, the late Dr R.K. Robinson.

Contents

List of Contributors

Noelia Aldai
Food Technology and Biochemistry and Molecular Biology
Faculty of Pharmacy – University of the Basque Country/EHU, Vitoria-Gasteiz
Spain

Efstathios Alichanidis
Department of Food Science and Technology, School of Agriculture
Aristotle University of Thessaloniki
Thessaloniki
Greece

Ylva Ardö
Department of Food Science
University of Copenhagen, Frederiksberg
Denmark

Everaldo Attard
Division of Rural Sciences and Food Systems, Institute of Earth Systems
University of Malta
Malta

Luiz Javier R. Barron
Food Technology and Biochemistry and Molecular Biology
Faculty of Pharmacy – University of the Basque Country/EHU, Vitoria-Gasteiz
Spain

Carina V. Bergamini
Facultad de Ingeniería Química
(Universidad Nacional del Litoral)
Santa Fe
Argentina

and

Instituto de Lactología Industrial (Universidad Nacional del Litoral – Consejo Nacional de Investigaciones Científicas y Técnicas)
Santa Fe
Argentina

Françoise Berthier
Unité de Recherches en Technologie et Analyses Laitières Rue de Versailles
France

Thomas Bintsis
11 Parmenionos
50200 Ptolemaida
Greece

Margherita Caccamo
CoRFiLaC
Ragusa
Italy

Stefania Carpino
CoRFiLaC – Consorzio Ricerca Filiera Lattiero Casearia, Ragusa
Italy

Darren R. Cooke
School of Food and Nutritional Sciences
University College Cork, Cork
Ireland

Eva-Maria Düsterhöft
NIZO Food Research
The Netherlands

Wim Engels
NIZO Food Research
The Netherlands

Elisabeth Eugster-Meier
Bern University of Applied Sciences
School of Agricultural, Forest and Food Sciences HAFL
Zollikofen, Switzerland

Marie-Therese Fröhlich-Wyder
Agroscope, Research Division Food Microbial Systems
Federal Department of Economic Affairs Education and Research EAER
Bern, Switzerland

Anthony Grupetta
Veterinary Regulations Directorate, Marsa
Malta

Katja Hartmann
Anton Paar GmbH
Germany

Karol Herian
Slovak Dairy Research Institute
Slovakia

Thom Huppertz
NIZO Food Research
The Netherlands

Erica R. Hynes
Facultad de Ingeniería Química
(Universidad Nacional del Litoral)
Santa Fe
Argentina

and

Instituto de Lactología Industrial
(Universidad Nacional del Litoral – Consejo Nacional de Investigaciones Científicas y Técnicas)
Santa Fe
Argentina

Ernst Jakob
Agroscope, Institute for Food Sciences IFS
Federal Department of Economic Affairs
Education and Research EAER, Bern
Switzerland

Paul Jelen
Department of Agricultural, Food and Nutritional Science, University of Alberta
Canada

Kimon-Andreas G. Karatzas
Department of Food and Nutrition Sciences
The University of Reading
United Kingdom

Kieran Kilcawley
Teagasc Food Research Centre
Moorepark, Fermoy, Co. Cork
Ireland

Paul S. Kindstedt
Department of Nutrition and Food Sciences
University of Vermont
United States

Giuseppe Licitra
Department of Agriculture, Nutrition and Environment
University of Catania, Catania
Italy

Sylvie Lortal
INRA, Agrocampus Ouest, Science et Technologie du lait et de l'oeuf
Rennes
France

F. Xavier Malcata
Laboratory of Engineering of Processes, Environment Biotechnology and Energy (LEPABE)
Portugal

and

Department of Chemical Engineering
University of Porto
Portugal

Paul L.H. McSweeney
School of Food and Nutritional Sciences
University College Cork, Cork
Ireland

Jelena Miocinovic
Department of Food Microbiology, Faculty of Agriculture, University of Belgrade
Serbia

Maria Belén López Morales
Food Science and Technology Department
International Excellence Campus for Higher Education and Research 'Campus Mare Nostrum', Veterinary Faculty
University of Murcia
Spain

Maurice O'Sullivan
School of Food and Nutritional Sciences
University College Cork, Cork
Ireland

Barbaros Özer
Ankara University
Faculty of Agriculture
Department of Dairy Technology
Ankara, Turkey

Photis Papademas
Department of Agricultural Sciences
Biotechnology and Food Science
Cyprus University of Technology, Limassol
Cyprus

Maria Cristina Perotti
Facultad de Ingeniería Química
(Universidad Nacional del Litoral)
Santa Fe
Argentina

and

Instituto de Lactología Industrial
(Universidad Nacional del Litoral – Consejo Nacional de Investigaciones Científicas y Técnicas)
Santa Fe
Argentina

Antonia Picon
Department of Food Technology
National Institute of Agricultural and Food Research and Technology (INIA)
Madrid
Spain

Zorica Radulovic
Department of Food Microbiology, Faculty of Agriculture, University of Belgrade
Serbia

Mertxe de Renobales
Biochemistry and Molecular Biology
Faculty of Pharmacy – University of the Basque Country/EHU, Vitoria-Gasteiz
Spain

Tânia G. Tavares
Laboratory of Engineering of
Processes, Environment Biotechnology and
Energy (LEPABE)
Portugal

and

REQUIMTE/Department of Chemical
Sciences
Faculty of Pharmacy
University of Porto
Portugal

İrem Uzunsoy
Bülent Ecevit University Caycuma
Vocational High School
Department of Food Technology
Zonguldak
Turkey

Florence Valence
INRA, Agrocampus Ouest, Science et
Technologie du lait et de l'oeuf
Rennes
France

Mailo Virto
Biochemistry and Molecular Biology
Faculty of Pharmacy – University of the
Basque Country/EHU, Vitoria-Gasteiz
Spain

Daniel Wechsler
Agroscope, Institute for Food Sciences IFS
Federal Department of Economic Affairs
Education and Research EAER, Bern
Switzerland

Preface

The history of cheese goes back to the Neolithic era, parallel to the origins of livestock domestication and dairying, and since then, more than 1000 cheese varieties have evolved. Although cheese is industrially produced in large quantities with a high degree of automation and totally controlled processes, the techniques are very similar to those produced with the traditional methods. Based on the same principles and following basic steps, cheesemakers blend science with 'art', producing a great variety of cheeses.

It is not clear whether cheesemaking is a simple or a complicated process. What is well known is that the impact of a number of different factors in each cheesemaking step is critical, and this is the main reason for the great variability in the characteristics of the final cheese. Thus, the regulation of each factor is vital for producing a cheese with the specific quality characteristics of its variety.

The purpose of this book is to describe (1) the manufacturing process of the most significant cheeses of the world and (2) the quality characteristics of the corresponding individual cheese. In addition, attention is paid to the scientific justification of the development of the final cheese characteristics, and the study of the impact of critical parameters on the development of cheese flavour and texture throughout maturation.

In Part I of the book, some fundamental topics are discussed in order to give a background for a better understanding of cheesemaking and the factors affecting cheese quality. Thus, the history of cheese is presented in Chapter 1; the behaviour of calcium in cheesemilk, during manufacture and during ripening and its impact on the rheological and functional properties of cheese in Chapter 2; cheese flavour development and sensory characteristics in Chapter 3; cheese microbial ecology and safety in Chapter 4; cheese with protected land- and tradition-related labels, traceability and authentication in Chapter 5; an overview of the cheesemaking process in Chapter 6 and traditional wooden equipment used for cheesemaking and their effect on quality in Chapter 7.

In Part II, the cheesemaking processes and the quality and sensory characteristics of 100 cheeses are described. Most of the cheeses presented are traditional products (50 of them with the PDO-Protected Designation of Origin designation). Experts on cheese science and technology gave a comprehensive description of cheese varieties that are important for their country. The cheeses are divided into 13 categories, and each is presented in a separate chapter. Relevant research on each cheese and extensive references to facilitate further studies and stimulate further research on specific aspects of cheesemaking are included.

We wish to express our sincere gratitude to all 43 contributors; for their high professionalism and cooperation.

Photis Papademas and Thomas Bintsis

Part I

1

The History of Cheese

Paul S. Kindstedt

Department of Nutrition and Food Sciences, University of Vermont, US

1.1 Introduction

The International Dairy Federation estimated that global cheese production in 2015 totalled approximately 23 million tonnes (IDF, 2016). This production was spread across six continents and included cheese made mainly from cow (20.7 million tonnes) milk. The remainder is composed of cheese from other species (buffalo, goat and sheep) as well as home-made and farmstead cheeses which do not appear in national statistics. How did this come about? More specifically, where, when and why did cheesemaking begin, how did it spread and evolve, and how did cheese attain such diversity, widespread distribution and prominence in our time? Although our understanding of the history of cheese remains very incomplete, various pieces of this vast puzzle can be fitted together to form a narrative that provides context for global cheesemaking in the twenty-first century.

1.2 Origins of Cheese

Until recently, the origins of cheese have remained mostly shrouded in the impenetrable fog of ancient prehistory. During the past two decades, however, groundbreaking advances in widely ranging fields of research and scholarship have yielded new insights into humanity's earliest experiences with cheese. Indeed, the convergence of multiple trains of research has pushed the likely beginnings of cheesemaking back to the Neolithic, perhaps nearly all the way back to the very origins of livestock domestication and dairying, which provided the context for the emergence of cheese.

Sheep and goats were first domesticated in the upper Euphrates and Tigris River valleys of Southwest Asia, as inferred from the study of archaeological skeletal remains. Advances in techniques to recover, evaluate and statistically analyse skeletal and dental remains for vital diagnostic characteristics such as size, sex and age of the animal at death, along with advances in interpretive frameworks based on ethnographic modelling of management strategies used by semi-nomadic shepherds in Southwest Asia today, have led to breakthroughs in the ability to detect the emergence, and track the spread, of livestock domestication (Vigne, 2011; Vigne & Helmer, 2007). Archaeozoological data clearly demonstrate the occurrence of drastic changes in the slaughtering profiles of sheep and goats, considered indicative of the onset of domestication, around the middle of the 9th millennium BC (Helmer, Gourichon & Vila, 2007; Vigne, 2011; Vigne et al., 2011). Similarly, cattle were also domesticated in the Middle Euphrates basin

Global Cheesemaking Technology: Cheese Quality and Characteristics, First Edition.
Edited by Photis Papademas and Thomas Bintsis.

slightly later, again based on archaeozoological analyses (Vigne, 2011). Furthermore, mitochondrial genetic studies of modern sheep, goats and cattle, along with analyses of mitochondrial DNA extracted from Neolithic skeletal remains, also support the conclusion that the earliest domestication of these livestock occurred in the Fertile Crescent region of Southwest Asia (Bollongino et al., 2012; Bonfiglio et al., 2012; Conolly et al., 2012; Edwards et al., 2007; Hiendleder et al., 2002; Meadows et al., 2007; Naderi et al., 2008). Thus, a considerable body of evidence indicates that goat, sheep and cattle domestication occurred for the first time in the same general region of the upper Fertile Crescent, aptly dubbed the 'cradle of agriculture', where the initial domestication of key founder grain crops such as wheat, barley, lentil, pea and chickpea also took place several centuries earlier (Weiss & Zohary, 2011).

It has been widely (though not universally) presumed that domesticated livestock in Southwest Asia were initially raised for their meat, hides and other products resulting from the animals' slaughter, and that the milking of goats, sheep and cattle did not commence until much later, for example, around the 4th millennium BC during the so-called 'secondary products revolution' (Sherratt, 1981, 1983). However, current archaeozoological and archaeochemical findings reveal that dairying was practised much earlier. For example, analyses of dental remains testify to the occurrence of sheep and goat slaughtering profiles, as early as the late 9th millennium BC, that are consistent with milk production (Helmer, Gourichon & Vila, 2007). Dairying practices appear to have then spread rapidly beyond their initial areas of origin, such that by the 8th millennium BC, Neolithic migrants from the northern Levantine mainland had transported domestic sheep and goats to Cyprus, where the animals were raised partly for milk production, as inferred from the early culling profiles observed there (Vigne, 2008; Vigne et al., 2011). Around the same period, archaeozoological remains of domestic cattle in the Northern Levant show similar evidence of culling strategies indicative of milking (Vigne & Helmer, 2007), which eventually spread to central and western Anatolia by the 7th millennium BC (Çakirlar, 2012; Evershed et al., 2008). Thus, ample indirect archaeozoological evidence points to dairying being practised almost from the beginning of the Neolithic when livestock were first domesticated. Indeed, it is not unreasonable to postulate that the harvesting of milk for human consumption may have been among the original reasons that inspired Neolithic farmers to domesticate ruminant livestock in the first place (Vigne, 2008; Vigne & Helmer, 2007).

The first direct evidence for dairying in the archaeological record, however, had to wait until the dawn of pottery making, during the 7th millennium BC. Recent advances in analytical techniques to recover lipid residues preserved within the fabric of ancient unglazed pottery sherds, and to identify the lipid sources based on stable carbon isotope (C^{12} and C^{13}) content, have enabled archaeochemists to reconstruct the contents of many ancient Neolithic pots at the time of their use (Dudd & Evershed, 1998; Mottram et al., 1999). Using this approach, Evershed et al. (2008) demonstrated definitively, and Thissen et al. (2010) corroborated, that milk production occurred as early as the 7th millennium BC in western Anatolia.

This same analytical approach has also made it possible to track the ancient practice of milk production through time and space by analysing pottery remains left behind by migrating Neolithic farmers. For example, a growing body of evidence in the field of archaeoclimatology strongly suggests that a substantial rise in sea level, followed by a major episode of climatic cooling, occurred during the late 7th millennium BC, which in turn precipitated social collapse among Neolithic farmers in Southwest Asia and triggered large-scale migrations out of Southwest Asia into Europe and elsewhere (Clare et al., 2008; Pross et al., 2009; Turney & Brown, 2007; Weninger et al., 2006). Among the evidence for Neolithic migration from Anatolia to Europe around this time are the analyses of potsherds recovered from the Balkan Peninsula that chronicle the spread of dairying as migrating Neolithic farmers transported their pottery-making technology and dairy subsistence strategy with them (Evershed et al., 2008). From there, Neolithic farmers continued their migration into Central, Eastern and Southern Europe by the 6th millennium BC (Craig et al., 2005; Salque et al., 2012; Spangenberg, Jacomet, & Schibler, 2006), the British Isles by the 5th millennium BC (Copley et al., 2003; Copley et al., 2005a,b), and the Western Baltic region, Scandinavia and Finland by the 5th/4th millennium BC (Craig et al., 2011; Cramp et al., 2014; Isaksson & Hallgren, 2012), leaving behind a trail of potsherds containing milk fat residues. Similar analyses have also confirmed the occurrence of dairying as early as the 5th millennium BC in Northern Africa (Dunne et al., 2012), and the 2nd millennium BC in the steppe zone of Central Asia (Outram et al., 2012). Thus, it appears that Neolithic farmers meticulously conserved dairying as a component of their subsistence strategy, even as they migrated vast distances, sometimes under conditions of great environmental stress.

The presence of milk fat residues in ancient potsherds does not necessarily indicate the occurrence of cheesemaking, only that the original pot contained milk in some form at the time of use. However, results from model studies of unglazed potsherds that were exposed to milk products and then buried to simulate conditions of archaeological pottery strongly suggest that the presence of milk fat residues in ancient potsherds constitutes telltale signs of concentrated dairy products such as butter and cheese. For example, unglazed potsherds that were deliberately exposed to liquid full fat milk only absorbed minute levels of milk fat within the pottery fabric, which rapidly degraded to undetectable levels upon burial of the sherds, probably due to microbial breakdown (Copley et al., 2005a; Dudd & Evershed, 1998). Therefore, it seems unlikely that ancient pots that contained only liquid raw milk at the time of use would have retained permanent measurable milk fat residues embedded within the pottery fabric. In contrast, model potsherds that were deliberately dosed with butter and then buried absorbed milk fat into the pottery fabric at 70 times the level observed for liquid milk, and the embedded milk fat underwent much less degradation during burial for up to one year, resulting in the abundant persistence of measurable milk fat residues (Copley et al., 2005a). It is evident, therefore, that concentrated dairy products such as butter and cheese, which contain high levels of milk fat and low levels of water and lactose, are much more likely than liquid milk to transfer abundant milk fat into the fabric of unglazed pottery in a stable form that may persist for immense periods of time under the right conditions; hence, the rationale for the use of milk fat residues as an indicator of concentrated dairy products such as butter or cheese.

Given this context, it is not surprising then, that milk fat residues have also been identified in sherds from Neolithic ceramic sieves recovered from Northeastern and Northwestern Europe, which have been dated to the 6th millennium BC (Salque et al., 2012, 2013). Remnants of Neolithic ceramic sieves have been observed widely in the archaeological material record throughout Central Europe, and similar ceramic sieves from the Bronze Age have been found in Central Italy, the Balkans, and the Indus River region of Pakistan (Barker, 1981; Bogucki, 1984; Gouin, 1997). It has long been suspected that these ancient pottery sieves were used to separate curds from whey during cheesemaking, on the basis of modern peasant ethnography that has documented the widespread use of similar sieves Central Italy, Central Europe, the Balkans and the Middle East (Barker et al., 1991; Gouin, 1997). The findings of Salque et al. (2012, 2013) confirm that Neolithic farmers used such sieves in cheesemaking some 7000 years ago in much the same way as is still practised today in some traditional societies.

In summary, the occurrence of milk fat residues in Neolithic potsherds and sherds, from ceramic sieves in particular, confirms with near certainty that cheesemaking was well under way in Southwest Asia and parts of Europe by the late Neolithic. However, a much earlier origin of cheesemaking, closer to the beginnings of dairying, is also possible. Genetic modelling based on modern human DNA sampling, combined with analyses of DNA recovered from Neolithic human skeletal remains, indicates that humans were universally adult lactose intolerant at the onset of dairying around the 9th millennium BC, due to the ubiquitous downregulation of the lactase enzyme (beta-galactosidase) that occurs after weaning in all mammals (Leonardi et al., 2012). Lacking the lactase production needed to break down lactose in the gut, early Neolithic adults were lactose intolerant, and it took several thousand years from the start of dairying before adult lactase persistence/lactose tolerance became widely established in the human population for the first time in Central Europe, sometime after the 6th millennium BC (Burger et al., 2007; Curry, 2013; Itan et al., 2009; Leonardi et al., 2012). This implies that the earliest harvesting of milk was intended exclusively for young children who were still suckling, to supplement the mothers' milk supply.

However, there is an additional possibility. The processing of milk into lactose-reduced products such as butter, and especially cheese, would have rendered a substantial fraction of the total nutrient portfolio of milk accessible to the Neolithic adult population. Dairying must have provided Neolithic farmers with very strong nutritional advantages for them to conserve milking practices over the many millennia and vast distances of migration that eventually enabled the successful genetic selection for the capacity to express lactase into adulthood. It is not unreasonable to postulate that cheesemaking may have commenced soon after the beginnings of dairying in the early Neolithic, which furnished the new farmers with a powerful nutritional incentive to culturally conserve their dairying practices through the long millennia that ebbed and flowed until adults, too, gained the capacity to benefit directly from consuming milk.

Unfortunately, there is no way to know for certain what Neolithic cheeses were like. Probably they were similar to the simplest cheeses still produced traditionally by semi-nomadic shepherds in Southwest Asia today: fresh, soft, acid coagulated and acid-heat coagulated types, which can be dried in the sun and preserved for later use (Gouin, 1997; Kindstedt, 2012). Alternatively, such types, when heavily salted, lend themselves to packing and preserving in sealed animal skins or clay pots, as is still practised today in Southwest Asia (Kamber, 2008), and which may account for some of the milk fat residues recovered from Neolithic potsherds discussed previously. Whether Neolithic cheesemakers perfected rennet-coagulated cheese is a matter for speculation. The culling of very young male livestock, practised from the beginning of dairying, afforded Neolithic farmers with ample opportunity to observe the milk clotting capacity of animal stomachs. It was only a matter of time before the connection between the clotted contents in the stomachs of the suckling lambs, kids and calves that were routinely culled, and the capacity of the stomach, and its curdled contents, to transform harvested liquid milk into a clotted state, inspired the birth of rennet-coagulated cheese.

From that point on, the basic technologies of acid, acid-heat and especially rennet-coagulated cheesemaking evolved in many different directions as cheesemakers in different places and at different times were confronted with new environmental, ecological, social and economic circumstances that caused them to adapt their practices and equipment to the world in which they found themselves. Great milestones in the circuitous evolution of cheesemaking were marked by the foundational technological advances that we take for granted today, such as the development of techniques and devices for cutting the coagulated mass of milk, for heating the cut mass of curd and whey and for separating whey from curds and applying pressure to the drained mass of curd, all of which facilitated the expulsion of whey from curds; the mastering of salt application levels and techniques; and the commandeering of local natural microenvironments for cheese storage and ripening. Taken collectively, these simple yet profound technical advances elegantly enabled cheesemakers to select for chemical characteristics and microbial populations in their cheeses that rendered positive outcomes that would otherwise be impossible (Kindstedt, 2014). The end result over the course of millennia has been the evolution of the major cheese families, each made up of seemingly endless variations on the family theme.

1.3 Cheese in Antiquity

It was not until several thousand years after start of cheesemaking, however, that descriptive information about cheeses and their manufacture began to be written down as humanity's first civilizations dawned. The earliest known examples of proto-writing, dating from the late 4th millennium BC, come from Uruk, the first great city-state of the Sumerian civilization of Southern Mesopotamia. These proto-cuneiform clay tablets represent the antecedents of humanity's first written language, and among the tablets recovered at Uruk are numerous administrative records that tabulate annual production figures for dairy products, primarily cheese and butteroil (ghee), produced from the milk of state-controlled herds of cattle and flocks of goats and sheep (Englund, 1991, 1995a; Green, 1980). The administrative complexity reflected in these clay records is astonishing and indicates that dairying and dairy processing had become very sophisticated. At the centre of Uruk's economic and political system stood two soaring temples dedicated to Inanna and An, the patron deities of Uruk whose cultic practices demanded a constant supply of agricultural products, including cheese and butter. These cultic practices not only underpinned the religious ideology of Uruk but also formed the basis of its centrally administered redistributive economy (Kindstedt, 2012).

The Inanna mythology of Uruk, and the Inanna-Demuzi cult that it inspired, institutionalised the routine cultic sacrifices of cheese and butter, which were subsequently replicated in various other Sumerian city-states during the 3rd millennium BC. Indeed, more than a thousand years after the initial rise of Uruk, sophisticated administrative oversight of cheese and butteroil production continued to be practised in Sumer, as is evident from abundant cuneiform records recovered from the massive city-state of Ur around the end of the 3rd millennium BC (Englund, 1995b; Gomi 1980). Other written accounts from Ur record the details of daily sacrifices of cheese and butter to Inanna and Ningal (Inanna's mythological mother), always in equal amounts ranging from about 29 to 54 litres of cheese and butteroil per day (Figulla, 1953). This strong linkage between cheese and religious expression is repeatedly observed in the Hittite, Greek and Roman civilizations that followed Sumer, the consequence of powerful currents of cultural influence that flowed northwards and westwards out of Mesopotamia from the Bronze Age forward (Kindstedt, 2012; McCormick, 2012).

Mesopotamia, however, was evidently not the only region where cheese was used as an element of religious expression during the 3rd millennium BC. Craig et al. (2015) uncovered striking evidence of the use of processed dairy products, most likely probably cheese, in religious practices in the vicinity of the Stonehenge megalithic complex in England, dating to around 2500 BC. Their findings, which were based on the identification of milk fat residues embedded in pottery sherds recovered at the site, raise intriguing questions as to whether these religious practices at Stonehenge originated independently of similar concurrent practices in Southern Mesopotamia (approximately 5000 km to the southeast of England), or whether they derived from a common pre-existing religious system that Neolithic migrants from the Levant and Anatolia brought with them to England and Southern Mesopotamia following the great migrations of the 7th millennium BC. Although direct evidence of the use of cheese in religious observances extending back to the 7th millennium BC is lacking, it is interesting to note that ceramic barrel-shaped vessels, believed to be butter churns, have been recovered from a 7th millennium BC Neolithic site in southwest Anatolia that seems to have been a cultic shrine; the churns may have been used for cultic ceremonies (Morris, 2014). Thus, a link between dairy products and religious practices in the early Neolithic seems possible. Unfortunately, detailed analyses of lipid residues in pottery sherds recovered from Neolithic Near East religious sites, which may help to elucidate this mystery, have yet to be reported. Returning to Southern Mesopotamia, a particularly noteworthy feature of Sumerian cuneiform literature from the standpoint of cheese history are modifiers that appear along with the term for cheese, which provide the first descriptive information about cheese in antiquity, and which indicate that cheeses were beginning to diversify. Modifiers that have been translated with reasonable certainty include terms for small and large cheese, herb-flavoured cheese, cheese with cereal grains added, milled or grated cheese, rich cheese, fresh cheese, sharp cheese, white cheese, stinking cheese, and dung cheese (Bottéro, 1985; Stol, 1993). None of these terms provide definitive insight into whether rennet-coagulated cheesemaking was practised in Sumer; however, a few terms have been noted among Sumerian cuneiform texts that could possibly be translated as animal rennet and plant rennet (Stol, 1993).

The first definitive evidence for rennet and rennet-coagulated cheesemaking does not appear in the archaeological record until the rise of the Hittite civilization in Anatolia during the late Bronze Age. Anatolia and Southern Mesopotamia maintained extensive trade networks and cultural exchanges during the Bronze Age; thus, the Hittites were profoundly influenced by Sumerian civilization. For example, they adapted the technique of cuneiform writing to the Hittite language and assimilated many Sumerian cultural features such as architectural forms and religious practices, including the use of cheese in various sacrificial rites (Kindstedt, 2012). Cuneiform texts from the mid-2nd millennia BC reveal that the Hittites performed sacrificial rites involving not only cheese but also rennet, suggesting that rennet had attained a revered status (Güterbock, 1968; Hoffner, 1995, 1998). Other Hittite texts clearly indicate that rennet-coagulated cheesemaking was firmly established in Hittite Anatolia by this time (Wainwright, 1959). Hittite modifiers for cheese that have been translated include terms for small cheese, large cheese, crumbled or grated cheese, scoured or finished cheese, and aged soldier cheese (Carter, 1985; Hoffner, 1966). The latter term suggests that the Hittites used cheese as a military ration, a practice that future armies and navies of Western civilization would often repeat, down to the present.

The Hittite Civilization collapsed around 1200 BC during a period of catastrophic upheaval throughout the Eastern Mediterranean that also triggered an abrupt end to the Greek Mycenaean Civilization, whom Homer referred to as the Achaeans. The cultural legacies of the Mycenaean, Hittite and Sumerian civilizations lived on, however, and helped shape the rise of classical Greek civilization a few hundred years later. The Greek world would come to embrace cheese in daily life and elevate its status to new heights in trade and gastronomic appreciation.

Cheese that the Greeks called 'fresh cheese' was a regular feature of the *opson*, or relish that accompanied the *sitos*, or main course of the Greek meal, which consisted of bread and cereal porridge (Neils, 2008; Wycherley, 1956). Fresh cheese mixed with honey also served as the filling for the beloved flaky cheesecake pastries known as *plakous* or *plakounta*. Fresh cheese probably was a simple rennet-coagulated, uncooked, unpressed or lightly pressed, surface-salted or brine-salted, rennet-coagulated type produced from sheep or goats milk, or blends of the two, much like the fresh white cheeses still produced throughout the Aegean and Eastern Mediterranean regions (Kamber, 2008). The term 'fresh cheese' in Greek literature also refers to the district of the Athens marketplace where cheese was bought and sold, and since every Greek polis (city-state) had a marketplace in the city-centre (agora), each also probably had its own fresh cheese district.

Beyond being common elements of the basic daily Greek meal, cheese and cheesecakes were enjoyed by the aristocracy during the symposium, or drinking party, which was the premier form of entertainment among the upper aristocratic classes (Grandjouan, Markson & Rotroff, 1989; Noussia, 2001). Exceptional local cheeses sometimes became items of maritime commerce, and some cheeses that acquired stellar reputations became identified by their place of origin, such as those from the islands of Cythnos and Chios in the Aegean Sea (Berlin, 1997; Casson, 1954; Migeotte, 2009). Many of the imported cheeses in Athens were probably variants of basic fresh cheese that, when stored and ripened in ceramic jars containing brine, were transformed into the flavourful Feta-type white brined cheeses that became ubiquitous throughout the Aegean and Balkan regions and have remained so to this day (Anifantakis & Moatsou, 2006; Kamber, 2008). Other imported cheeses that were highly esteemed in Athens came from the heavily Greek colonised island of Sicily, where hard, dry cheeses were crafted that were long-lasting yet flavourful enough to serve as condiments in cooking when grated. Sicilian grating cheeses probably consisted of small rennet-coagulated, uncooked, unpressed or lightly pressed, surface-salted sheep and goats milk cheeses similar to those produced in Sardinia and the Southern Italian peninsula today (Kindstedt, 2012). The use of such cheeses in cooking became so popular throughout the Greeks world that Archestratos, a renowned fourth-century BC chef and cookbook writer from Sicily, complained about the overuse of cheese sauces in cooked dishes of the time (Rapp, 1955). Thus, besides serving as a staple of peasant subsistence, cheese in the Greek world became a gourmet luxury food and a flavourful ingredient that added coveted gastronomic variety to an increasingly sophisticated food culture.

The Romans greatly admired Greek culture, and the Greek love of hard, dry pecorino grating cheeses captivated the Romans from the beginning. Indeed, the process began with Etruscans, forerunners of the Romans, whose aristocratic warriors left behind cheese graters, an essential feature of a Homeric feasting ritual that the Etruscans assimilated from the Greeks, in their seventh-century BC tombs (Ridgway, 1997; Sherratt, 2004). By the time of the Roman Empire, the bronze or iron cheese grater had become a standard utensil in the Roman kitchen. The Romans officially recognised two classes of cheese for tax purposes: *caseus mollis*, or soft cheese, and *caseus aridus*, or dry cheese. According to the first-century AD Roman agricultural writer Columella, both cheeses were made from sheep and/or goats milk by a common rennet-coagulated, uncooked, lightly pressed, surfaced-salted make procedure, but to produce the dry version, the salting and pressing steps were repeated, and the pressing pressure was increased (Forster & Heffner, 1954).

Conspicuously absent from Columella's instructional manual on cheesemaking, however, is any mention of one of the most ancient and beloved of cheeses of the Central Italian peninsula, the acid-heat-coagulated (Ricotta) types. The making of whole milk Ricotta seems to have dominated cheese production on the Italian peninsula during the 2nd millennium BC, as inferred from the abundant occurrence of ceramic devices referred to as 'milk boilers' in the archaeological record. Milk boilers, which were produced according to two different designs, were used throughout much of the Italian peninsula during the 2nd millennium BC to prevent heated milk from frothing and boiling over (Potter, 1979; Trump, 1965). Similar devices are still used today by shepherds in the Italian Apennines for the making of traditional Ricotta cheese (Barker, 1981; Barker et al., 1991). Milk boilers disappeared from the Italian archaeological record during the first millennium BC, however, which coincided with the rise of hard pecorino grating cheeses, suggesting that a shift from the making of whole milk Ricotta to whey Ricotta (which is less prone to frothing and boiling over, obviating the need for milk boilers) may have taken place in conjunction with the rise in hard pecorino grating cheese production (Kindstedt, 2012).

The Roman love affair with hard pecorino grating cheeses had not only culinary implications but also military implications as well. The vastness of the Roman Empire, with some 16,000 km of frontier to protect against the 'barbarians' beyond, presented daunting logistical challenges for Roman military planners that had to feed, clothe and otherwise provision a permanent force of nearly half a million soldiers to guard the Empire. To address these needs, Roman forts were endowed with agricultural lands that were used to produce wheat and to raise sheep and pigs for the provisioning of the legions (Bezeczky, 1996; Davies, 1971). Cheese was a basic ration of the Roman military, and the frequent occurrence of perforated heavy-duty ceramic press moulds in the archaeological material records from Roman forts throughout Europe indicate that the making of hard pecorino grating cheese often took place on site, perhaps by the soldiers themselves during times of peace (Davies, 1971; Niblett, Manning & Saunders, 2006). The widespread introduction of Roman cheesemaking technology to Europe north of the Alps left its mark on the future of European cheesemaking, particularly that of the conquering Anglo-Saxons in England, as discussed later.

The Romans were not the first to introduce cheesemaking to Europe north of the Alps, however. On the contrary, they frequently encountered vibrant cheesemaking activities among the Celtic peoples that they conquered, and many cheeses from the provinces to the north came to be imported to Rome, where they attained stellar reputations. Particularly noteworthy were the alpine cheeses that were made all along the arc of the Alps, and the cheeses from the Massif Central of France (Kindstedt, 2012). Thus, the Neolithic migration of dairy farmers from Southwest Asia to Central Europe that occurred thousands of years earlier ultimately gave rise to a very sophisticated and widely dispersed cheesemaking culture in Central Europe by the time of the Roman invasions.

1.4 Cheese in the Middle Ages and Renaissance

Virtually all aspects of medieval life in Europe were profoundly shaped by the two ubiquitous institutions that collectively formed the scaffolding for much of the economic, social, intellectual and spiritual infrastructure of medieval society: the manor and the Benedictine monastery. Cheesemaking in the Middle Ages was no exception. The manor and the monastery were fertile centres of cheesemaking activity, and the great proliferation of new varieties of cheese that came of age during this period is a testimony to the powerful influence that these institutions exercised over cheesemakers.

Because the continent of Europe encompasses extremely diverse physical environments (e.g. with respect to climate, topography, indigenous flora), manorial and monastic cheesemakers were confronted with a wide range of microenvironments, each with its own set of opportunities and constraints, depending on where they were situated. Furthermore, the social and economic structures of manorial and monastic communities differed at different times and in different regions across Europe, which imposed additional formative conditions and constraints on cheesemaking. All of this created incentives for European cheesemakers to develop novel practices and equipment to accommodate their diverse needs. On the other hand, in other regions, cheesemaking technology changed little from the basic methods used throughout the Mediterranean in antiquity. However, the radically different environmental, social and economic conditions of medieval Europe north of the Alps produced very different outcomes even though the same basic Mediterranean technology continued to be employed.

For example, manorial peasant families who made up the labour force of the large manors of Northwest Europe were typically allowed to raise a cow or two on common pastures, which furnished small but vital quantities of milk for the family needs. Peasant wives there employed a basic rennet-coagulated, uncooked, unpressed, surface-salted make procedure, using cow's milk, that was similar to that used by Greek shepherds to produce the ubiquitous 'fresh cheese', and that was used by Italian shepherds to produce the Roman *caseus mollis*, or soft cheese, which Columella described as 'cheese which is to be eaten within a few days while still fresh...' (Forster & Heffner, 1954). Manorial peasant wives probably often had to combine multiple milkings when making cheese because of the small quantity of milk available, which favoured high populations of lactic acid bacteria (and other bacterial species) in the cheesemilk. The end result was the production of small, high-moisture, low-pH (ca. pH 4.6) cheeses. In the warm climate of the Mediterranean, such cheeses spoil or dry out and become inedible within a few days. In the damp temperate climate of Northwest Europe, however, the environmental conditions present in damp cool cellars, or sometimes natural caves, that were used to store the cheeses selected for the prolific growth of surface of yeasts and moulds, especially the greyish-white mould *Penicillium camemberti,* which produced desirable transformations during storage instead of spoilage/rotting. The origins of the plethora of surface mould-ripened (e.g. bloomy rind) cheeses so beloved in Northwest Europe almost certainly had their earliest roots in the peasant manorial communities and, later, the peasant villages that emerged out of the breakup of the manors (Kindstedt, 2012).

In the same regions of Northwest Europe, Benedictine monastic cheesemakers practised the same basic rennet-coagulated, uncooked, unpressed, surface-salted make procedure as their manorial peasant neighbours but arrived at a very different outcome: the evolution of the bacterial smear-ripened cheeses, sometimes referred to as monastery cheeses. Monastic cheesemakers had the advantage of abundant fresh cow's milk from the monastic herd; there was no need to combine multiple milkings for cheesemaking. Warm fresh milk, used immediately after harvesting, ensured low populations of lactic acid bacteria, which resulted in high-moisture cheeses that were higher in initial pH than those of their manorial peasant neighbours. The high moisture, relatively high pH chemistry of the curd, combined with salting techniques that included surface smearing with brine and ready access to cool damp monastic cellars for storage provided the right combination of conditions for prolific yeast and coryneform bacterial growth on the cheese surface that pre-empted spoilage/rotting by transforming the cheese in new desirable ways during storage (Kindstedt, 2014).

In the Southern Massif Central of France, this same basic rennet-coagulated, uncooked, unpressed, surface-salted make procedure gave birth to another radically different cheese, Roquefort, which has become emblematic of the family of blue-veined cheeses. Although cheesemaking in the Roquefort region predated the Romans, it seems that important fine-tuning of the make procedure did not take place until around the eleventh century AD, when manorial sheep ranges and cheesemaking operations on the Larzac Plateau of the Southern Massif Central, and the ageing of cheeses in the famous Caves of Cambalou just below the Plateau, came under monastic control (Whittaker & Goody, 2001). The combination of high-moisture, low-pH sheep milk curd, along with intensive surface salting of the cheese (made possible courtesy of the Romans, who developed salt works along the Mediterranean coast of France and a system of roads ascending from the coast to Massif Central to transport the salt), and access to the Caves of Cambalou for ageing in a well-ventilated, near constant temperature (6–10°C) and humidity (95–98% relative humidity) environment, provided the right combination of conditions for prolific growth of *Penicillium roqueforti* mould growth that produced desirable transformations during storage in place of destructive spoilage/rotting (Kindstedt, 2012). In summary, the simple rennet-coagulated, uncooked, unpressed, surface-salted cheesemaking technology that became deeply embedded in the Mediterranean region in antiquity evolved into radically new families of cheese such as soft surface-ripened types (white mould-ripened and bacterial smear-ripened cheeses) and blue-veined types when practised in diverse European microenvironments.

In England, the conquering Anglo-Saxon aristocracies inherited Roman agricultural infrastructure along with the Roman technology for making small rennet-coagulated, uncooked, lightly pressed surface-salted dry pecorino cheeses of the type described by Columella. Evidently, the Anglo-Saxons continued to produce these small, hard pecorino cheeses on their demesnes for some 500 years until the Normans wrested control of England during the eleventh century AD. With the Normans came the blossoming of trade across the English Channel, including trade in cheeses, which coincided with noteworthy increases in the size of English demesne cheeses, as noted in monastic records of manorial holdings at the time. A change in cheese geometry almost certainly also occurred at this time, as the small cylindrical cheeses of the Anglo-Saxon period evolved into larger wheel-shaped cheeses by the end of the Middle

Ages (Kindstedt, 2012). In other words, English cheesemakers began to modify their practices in response to market opportunities/pressures brought on by trade. Indeed, as the Renaissance dawned and lucrative trade routes re-emerged across Europe after centuries of isolation that followed the collapse of the Roman Empire, cheesemakers in various regions responded to the new world of expanding trade networks with innovative new practices.

For example, cheesemakers in the highlands of Gruyère Switzerland began to produce increasingly larger cheeses during the Renaissance as the reputation of Gruyère cheese grew, and demand in lucrative distant markets soared (Birmingham, 2000). The production of large durable cheeses, which were tailored in size to be transported on foot (in head yokes) over steep mountain passes to Lake Geneva and then packed tightly in barrels for passage down the Rhone River to the Mediterranean, presented immense challenges for the alpine cheesemakers. Moisture control was particularly troublesome because large cheeses possess less surface area relative to their volume than do small cheeses, which slows down evaporative moisture loss outwards from the cheese centre to the surface, and diffusion of salt inwards from the surface to the centre, thereby elevating the risk of spoilage in the high-moisture, low-salt interior during ageing. To combat this, alpine cheesemakers went to great technical lengths to maximise whey expulsion during cheesemaking by cutting the curd into tiny rice-sized particles, cooking the curds to exceptionally high temperatures, and pressing the drained curds into thin wheel-shaped cheeses of immense diameters that maximised the surface area to volume ratio in the finished cheeses. By the end of the Middle Ages, new methods of cooking, pressing and salting developed in various regions of Europe had given birth to a new generation of larger cheeses, ranging from the more diminutive Gouda (ca. 7 kg) in Holland to the massive Parmesan (ca. 40 kg) in the Po River Valley of northern Italy and Cantal (ca. 40 kg) in the northern Massif Central of south-central France (Kindstedt, 2012).

1.5 Cheese in the Modern Era

The seventeenth century arguably marked a turning point in the history of cheese, which ushered in the modern era. The explosive growth of urban populations in rapidly expanding cities such as London, the establishment of truly global trade networks by major European powers as they competed to colonise east and west, and the onset of the Enlightenment, which gave rise to profound scientific advances that soon stimulated the scientific and industrial revolutions, collectively began to change the market forces that confronted modern cheesemakers, as well as the capacity of cheesemakers to respond to market forces with technical innovations. It is true, of course, that market forces affected cheese practices and inspired technical advances long before the seventeenth century, as in the aforementioned example of Gruyère cheese. However, the growing intensity of market forces, which increasingly emphasised efficiency and cost, began to affect cheesemakers in new ways that ultimately paved the way for the cheese factory and industrial cheesemaking.

The beginnings of the modern era are perhaps best illustrated by the transformation that took place in English cheesemaking during the seventeenth and eighteenth centuries, when London became England's foremost population centre. The sprawling metropolis of London created a mega-market that reshaped much of English agriculture, including English cheesemaking. Access to the cheese and butter markets of London was controlled by the London cheesemongers, a cartel of buyers and distributors, who began to apply intense pressure on their suppliers in East Anglia during the early seventeenth century to produce more butter along with their cheese or risk losing their contracts, butter being more profitable to sell in London than cheese. As the demand for butter grew, cheesemakers were forced to skim more cream from their milk before cheesemaking, resulting in cheese with progressively lower fat content. East Anglian cheesemakers lacked the technical expertise to develop high-quality reduced-fat cheeses (a challenge that cheesemakers still wrestle with today), and consequently their product quality deteriorated. The situation reached crisis proportions when the cheesemongers then began to source full fat cheese from Cheshire, effectively forcing East Anglia out of the London cheese market and relegating dairy farmers there to the production of butter. Thus, by the early eighteenth century, East Anglia, which had been London's premier cheese supplier for more than a century, essentially stopped producing cheese, and the Cheshire region became London's foremost supplier (Stern, 1973).

Cheesemakers in Cheshire then quickly came under pressure from the cheesemongers to produce ever-larger cheeses, which were more efficient to transport and distribute, and more profitable because they experienced less moisture loss (and therefore less yield loss) during storage than small cheeses due to their lower surface area relative to volume. However, the move to larger cheeses necessitated innovations in cheesemaking practices and equipment to produce cheese with lower moisture and higher salt contents in the centre that would withstand internal rotting during storage. Cheesemakers in Cheshire responded by phasing in a high-pressure pressing step, using newly developed heavy-duty presses and perforated press moulds, along with a new salting technique that replaced surface salting of the pressed cheese with dry salting of milled curd particles before pressing into cheese (Cheke, 1959; Fussell, 1966). Cheshire cheesemakers then had to develop an alternative protective coating and vapour barrier at the cheese surface to replace the dense rind produced by surface salting, which had previously served as a natural packaging that protected the surface from physical harm and prevented surface cracking. This was accomplished, imperfectly, by smearing inexpensive whey butter on the cheese surface (Kindstedt, 2012).

Despite the impressive, rapid-fire technical innovations developed by the cheesemakers of Cheshire, the region lost its pre-eminence in the London market by the mid-nineteenth century, displaced by cheese produced in the West Country to the south. Cheesemakers there combined a mild cooking or scalding step with the salting of milled curd before high-pressure pressing to render a new cheese variety that eventually came to be called Cheddar. Soon after, English Cheddar cheesemakers found themselves in a technological race for survival as lower-cost Cheddar-style cheese from America, and later Canada, New Zealand and Australia, flooded the London market. Ultimately, the English dairy industry was forced to reorient away from cheesemaking in favour of fresh liquid milk production for the burgeoning urban population of London and other major cities (Blundel & Tregear, 2006). By this time, the modern era of cheesemaking had reached a tipping point, with global market forces and technological innovations firmly in control of the fate of much that would come during the twentieth century and beyond.

Cheesemakers in America, who produced mostly English style cheeses during the seventeenth and eighteenth centuries and who closely emulated the technical innovations coming out of England, rendered this tipping point irreversible during the mid-nineteenth century with the introduction of factory cheesemaking. The factory system, supported by rapid advancements in the field of dairy science and a plethora of new mass-produced labour-saving equipment and utensils, enabled cheese to be made on ever-larger scales with ever-greater efficiency and consistency. By the end of the nineteenth century, the cheese factory had virtually eliminated traditional on-farm artisanal cheesemaking in America while generating astonishing increases in annual US cheese production (Kindstedt, 2012). Highly efficient, large-scale, technology-intense industrial cheesemaking eventually became the norm for many of the world's cheesemaking regions during the twentieth century, including the United States, western Canada, Australia, New Zealand, Ireland, Holland, Denmark and many other regions to varying degrees.

However, a sharp dichotomy also characterised the modern era of cheese from the beginning because many other cheesemaking regions tenaciously continued to produce hand-crafted artisanal cheeses on small scales using traditional practices, even as the factory gained ground elsewhere. Traditional artisanal cheesemaking often persisted in geographically isolated regions of Europe and Southwest Asia, and in regions with marginal lands that are poorly suited for agricultural purposes other than sheep and goat herding. Traditional cheesemaking also persisted in more accessible and fertile regions of Europe and beyond, where strong cultural conservatism prevailed and where traditional cheesemaking formed an integral component of the working landscape, such as in many parts of France and in Quebec, Canada.

As the twentieth century progressed, however, increasingly intense competition from lower-cost industrial cheeses, spurred on by global trade, posed grave challenges to the economic survival of these bastions of traditional cheesemaking. Artisanal cheeses by nature are very labour intensive to produce and not amenable to the cost savings that accompany economies of scale, rendering them much more expensive to the consumer than industrial cheeses (Bouma, Durham & Meunier-Goddik, 2014; Nicholson & Stephenson, 2007). Traditional cheeses also often utilise practices and equipment that conflict with the rapidly evolving global standards for hygiene and safety, posing further threats to their continued existence (Licitra, 2010). Thus, in the twentieth-first century, the long-term sustainability of traditional artisanal cheeses seems unlikely unless (1) modern safety regulations and traditional cheesemaking practices can be reconciled in ways that preserve the essence of traditional cheeses while satisfying the appropriate level of public health protection, and (2) the public can be convinced to pay much more for traditional cheeses than industrial cheeses, either in the form of higher prices or through public subsidies of some sort.

One encouraging model for how this might be accomplished emerged during the past few decades in the United States and several other developed countries, where a new public appreciation for traditional artisanal cheeses has arisen (Kindstedt, 2005). Traditional cheeses collectively offer a rich diversity of intrinsic physical and sensory characteristics that, arguably, are unmatched in industrial cheeses (Licitra, 2010). This diversity, contrasted with the perception of a growing segment of the public that industrial cheeses are bland and uninspiring, has helped to stimulate consumer interest in, and willingness to pay for, a new generation of artisanal cheeses, produced in traditional ways on small scales, but which often employ advanced practices and technologies that satisfy public health regulations while preserving traditional cheese character. Furthermore, the public's willingness to pay more for artisanal cheeses is also being encouraged by extrinsic attributes related to values that consumers hold, such as sustainability and stewardship of the environment, animal welfare, closeness to nature, and so on, which they associate with traditional cheesemaking (Wang et al., 2015). Consequently, small-scale artisanal cheesemakers have at their disposal powerful intrinsic and extrinsic drivers of the public's willingness to pay, which has enabled the new generation of traditional cheesemakers to experience remarkable growth during the last two decades. Effective management of these drivers of consumer willingness to pay, coupled with targeted adoption of technologies to satisfy public health regulations, will undoubtedly be among the keys to future sustainability of traditional cheesemaking worldwide.

References

Anifantakis, E. M. & Moatsou, G. (2006). Feta and other Balkan Cheeses. In Tamime, A. (ed.), *Brined Cheeses*. Blackwell Publishing Ltd, Oxford.

Barker, G. (1981). *Landscape and Society. Prehistoric Central Italy*. Academic Press, London.

Barker, G., Grant, A., Beavitt, P., Christie, N., Giorgi, J., Hoare, P., Leggio, T. & Migliavacca, M. (1991). *Ancient and modern pastoralism in Central Italy: An interdisciplinary study in the Cicolano mountains*. Papers of the British School at Rome, 59, 15–88.

Berlin, A. M. (1997). Archaeological sources for the history of Palestine: Between large forces: Palestine in the Hellenistic Period. *The Biblical Archaeologist*, 60 (1), 2–51.

Bezeczky, Dr. (1996). Amphora inscriptions – legionary supply? *Britannia*, 27, 329–336.

Birmingham, D. (2000). *Switzerland: A Village History*. St. Martin's Press, New York.

Blundel, R. & Tregear, A. (2006). From artisans to "factories": The interpretation of craft and industry in English cheese-making, 1650–1950. *Enterprise & Society*, 7 (4), 705–739.

Bogucki, P. (1984). Ceramic sieves of the linear pottery culture and their economic implications. *Oxford Journal of Archaeology*, 3 (1), 15–30.

Bollongino, R., Burger, J., Powell, A., Mashkour, M., Vigne, J.-D. & Thomas, M. G. (2012). Modern taurine cattle descended from small number of Near-Eastern founders. *Molecular Biology and Evolution*, 29 (9), 2101–2104.

Bonfiglio, S., Ginja, C., De Gaetano, A., Achilli, A., Olivieri, A., Colli, L., Tesfaye, K., Hassan Agha, S., Gama, L. T., Cattonaro, F., Penedo, M. C. T., Ajmone-Marsan, P., Torroni, A. & Ferretti, L. (2012). Origin and spread of *Bos Taurus*. New clues from mitochondrial genomes belonging to haplogroup T1. *PLoS One*, 7 (6), e38601.

Bouma, A., Durham, C. A., and Meunier-Goddik, L. (2014). Start-up and operating costs for artisan cheese companies. *Journal of Dairy Science*, 97, 3964–3972.

Bottéro, J. (1985). The cuisine of ancient Mesopotamia. *Biblical Archaeologist*, 48 (1), 36–47.

Burger, J., Kirchner, M., Bramanti, B., Haak, W. & Thomas, M. G. (2007). Absence of the lactase-persistence-associated allele in early Neolithic Europeans. *PNAS*, 104 (10), 3736–3741.

Çakirlar, C. (2012). Neolithic dairy technology at the European-Anatolian frontier: Implications of archaeozoological evidence from Ulucak Höyük, İzmir, Turkey, ca. 7000–5700 cal. BC. *Anthropozoologica*, 47 (2), 78–98.

Carter, C. (1985). Hittite *Hashas*. *Journal of Near Eastern Studies*, 44 (2), 139–141.

Casson, L. (1954). The grain trade of the Hellenistic world. *Transactions and Proceedings of the American Philological Association*, 85, 168–187.

Cheke, V. (1959). *The Story of Cheese-Making in Britain*. Routledge & Kegan Paul, London.

Clare, L., Rohling, E. J., Weninger, B. & Hilpert, J. (2008). Warfare in Late Neolithic/Early Chalcolithic Pisidia, Southwest Turkey. Climate induced social unrest in the late 7th millennium cal bc. *Documenta Praehistorica*, 35, 65–92.

Conolly, J., Manning, K., Colledge, S., Dobney, K. & Shennan, S. (2012). Species distribution modelling of ancient cattle from early Neolithic sites in SW Asia and Europe. *The Holocene*, 22 (9), 997–1010.

Copley, M. S., Berstan, R., Dudd, S. N., Docherty, G., Mukherjee, A. J., Straker, V., Payne, S. & Evershed, R. P. (2003). Direct chemical evidence for widespread dairying in prehistoric Britain. *PNAS*, 100 (4), 1524–1529.

Copley, M. S., Berstan, R., Dudd, S. N., Aillaud, S., Mukherjee, A. J., Straker, V., Payne, S. & Evershed, R. P. (2005a). Processing of milk products in pottery vessels through British prehistory. *Antiquity*, 79, 895–908.

Copley, M. S., Berstan, R., Mukherjee, A. J., Dudd, S. N., Straker, V., Payne, S. & Evershed, R. P. (2005b). Dairying in antiquity. III. Evidence from absorbed lipid residues dating to the British Neolithic. *Journal of Archaeological Science*, 32, 523–546.

Craig, O. E., Chapman, J., Heron, C., Willis, L. H., Bartosiewicz, L., Taylor, G., Whittle, A. & Collins, M. (2005). Did the first farmers of central and Eastern Europe produce dairy foods? *Antiquity*, 79, 882–894.

Craig, O. E., Shillito, L.-M., Albarella, U., Viner-Daniels, S., Chan, B., Cleal, R., Ixer, R., Jay, M., Marshall, P., Simmons, E., Wright, E. & Parker Pearson, M. (2015). Feeding Stonehenge: Cuisine and consumption at the Late Neolithic site of Durrington Walls. *Antiquity*, 89, 1096–1109, doi: 10.15.15184/aqy.2015.110.

Craig, O. E., Steele, V. J., Fischer, A., Hartz, S., Andersen, S. H., Donohoe, P., Glykou, A., Saul, H., Jones, D. M., Kock, E. & Heron, C. P. (2011). Ancient lipids reveal continuity in culinary practices across the transition to agriculture in Northern Europe. *PNAS*, 108 (44), 17190–17195.

Cramp, L. J. E., Evershed, R. P., Lavento, M., Halinen, P., Mannermaa, K., Oinonen, M., Kettunen, J., Perola, M., Onkamo, P. & Heyd, V. (2014). Neolithic dairy farming at the extreme of agriculture in Northern Europe. *Proceedings of the Royal Society B*, 281, 2014.0819.

Curry, A. (2013). The milk revolution. *Nature*, 500 (7460), 20–22.

Davies, R. W. (1971). The Roman military diet. *Britannia*, 2, 122–142.

Dudd, S. & Evershed, R. P. (1998). Direct demonstration of milk as an element of archaeological economies. *Science*, 282 (5393), 1478–1481.

Dunne, J., Evershed, R. P., Salque, M., Cramp, L., Bruni, S., Ryan, K., Biagetti, S. & di Lernia, S. (2012). First dairying in green Saharan Africa in the fifth millennium BC. *Nature*, 486 (7403), 390–394.

Edwards, C. J., Bollongino, R., Scheu, A., Chamberlain, A., Tresset, A., Vigne, J.-D., Daird, J. F., Larson, G., Ho, S. Y. W., Heupink, T., Shapiro, B., Freeman, A. R., Thomas, M. G., Arbogast, R.-M., Arndt, B., Bartosiewicz, L., Benecke, N., Budja, M., Chaix, L., Choyke, A. M., Mashkour, M., Özdogan, M., Schulting, R. J., Stephan, E., Uerpmann, H.-P., Vörös, I., Voytek, B., Bradley, D. G. & Burger, J. (2007). Mitochondrial DNS analysis shows a Near Eastern Neolithic origin for domestic cattle and no indication of domestication of European aurochs. *Proceedings of the Royal Society B*, 274, 1377–1385.

Englund, R. K. (1991). Archaic dairy metrology. *Iraq*, 53, 101–104.

Englund, R. K. (1995a). Late Uruk period cattle and dairy products: Evidence from proto-cuneiform sources. *Bulletin on Sumerian Agriculture*, 8, 33–50.

Englund, R. K. (1995b). Regulating dairy productivity in the Ur III period. *Orientalia*, 64, 377–429.

Evershed, R. P., Payne, S., Sherratt, A. G., Copley, M. S., Coolidge, J., Urem-Kotsu, D., Kotsakis, K., Özdogan, M., Özdogan, A. E., Nieuwenhuyse, O., Akkermans, P. M. M. G., Bailey, D., Andeescu, R., Campbell, S., Farid, S., Hodder, I., Yalman, N., Özbasaran M., Bicakci, E., Garfinkel, Y., Levy, T. & Burton, M. M. (2008). Earliest date for milk use in the Near East and Southeastern Europe linked to cattle-herding. *Nature*, 455 (7212), 528–531.

Figulla, H. H. (1953). Accounts concerning allocations of provisions for offerings in the Ningal-Temple at Ur. *Iraq*, 15 (2), 171–192.

Forster, E. S. & Heffner, E. H. (1954). *Lucius Junius Moderatus Columella on Agriculture*. Harvard University Press, Cambridge.

Fussell, G. E. (1966). *The English Dairy Farmer*. A.M. Kelley, New York.

Gomi, T. (1980). On dairy productivity at Ur in the late Ur III period. *Journal of the Economic and Social History of the Orient*, 23 (1/2), 1–42.

Gouin, P. (1997). Ancient oriental dairy techniques derived from archaeological evidence. *Food and Foodways*, 7 (3), 157–188.

Grandjouan, C., Markson, E. & Rotroff, S. I. (1989). *Hellenistic relief molds form the Athenian agora*. Hesperia Suppl., Vol. 23. American School of Classical Studies at Athens, Princeton.

Green, M. W. (1980). Animal husbandry at Uruk in the Archaic period. *Journal of Near Eastern Studies*, 39 (1), 1–35.

Güterbock, H. G. (1968). Oil plants in Hittite Anatolia. *Journal of the American Oriental Society*, 88 (1), 66–71.

Helmer, D., Gourichon, L. & Vila, E. (2007). The development of the exploitation of products from *Capra*and *Ovis* (meat, milk and fleece) from the PPNB to the early Bronze in the Northern Near East (8700 to 2000 bc cal.). *Anthropozoologica*, 42 (2), 41–69.

Hiendleder, S., Kaupe, B., Wassmuth, R. & Janke, A. (2002). Molecular analysis of wild and domestic sheep questions current nomenclature and provides evidence for domestication from two different subspecies. *Proceedings of the Royal Society B: Biological Sciences*, 269 (1494), 893–904.

International Dairy Federation (2016). Bulletin of the International Dairy Federation 485/2016 – The World Dairy Federation, Brussels, Belgium.

Isaksson, S. & Hallgren, F. (2012). Lipid residue analyses of Early Neolithic funnel-beaker pottery from Skogsmossen, eastern Central Sweden, and the earliest evidence of dairying in Sweden. *Journal of Archaeological Science*, 39, 3600–3609.

Hoffner, H. A. (1995). Oil in Hittite texts. *Biblical Archaeologist*, 58 (2), 108–114.

Hoffner, H. A. (1998). *Hittite Myths*, 2nd edition. Society of Biblical Literature. Scholars Press, Atlanta.

Hoffner, H. A. (1966). A native cognate to West Semitic *GBN "Cheese"? *Journal of the American Oriental Society*, 86 (1), 27–31.

Itan, Y., Powell, A., Beaumont, M. A., Burger, J. & Thomas, M. G. (2009). The origins of lactase persistence in Europe. *PLOS Computational Biology*, 5 (8), 1–13, e1000491.

Kamber, U. (2008). The traditional cheeses of Turkey: Cheeses common to all regions. *Food Reviews International*, 24, 1–38.

Kindstedt, P. S., with the Vermont Cheese Council. (2005). *American Farmstead Cheese*. Chelsea Green Publ., White River Jct.

Kindstedt, P. S. (2012). *Cheese and Culture. A History of Cheese and Its Place in Western Civilization.* Chelsea Green Publ., White River Jct.

Kindstedt, P. S. (2014). The basics of cheesemaking. In Donnelly, C. W. (ed.), *Cheese and Microbes*. ASM Press, Washington, DC.

Leonardi, M., Gerbault, P., Thomas, M. G. & Burger, J. (2012). *International Dairy Journal*, 22, 88–97.

Licitra, G. (2010). World wide traditional cheeses: Banned for business? *Dairy Science & Technology*, 90, 357–374.

Meadows, J. R. S., Cemal, I., Karaca, O., Gootwine, E. & Kijas, J. W. (2007). Five ovine mitochondrial lineages identified from sheep breeds of the Near East. *Genetics*, 175, 1371–1379.

McCormick, F. (2012). Cows, milk and religion: The use of dairy produce in early societies. *Anthropozoologica*, 47 (2), 99–111.

Migeotte, L., translated by J. Lloyd. (2009). *The Economy of the Greek cities. From the Archaic Period to the Early Roman Empire*. University of California Press, Berkeley.

Morris, S. (2014). Dairy Queen. Churns and milk products in the Aegean Bronze-Age. *Opuscula*, 7, 205–222.

Mottram, H. R., Dudd, S. N., Lawrence, G. J., Stott, A. W. & Evershed, R. P. (1999). New chromatographic, mass spectromic and stable isotope approaches to the classification of degraded animal fats preserved in archaeological pottery. *Journal of Chromatography A*, 833, 209–221.

Naderi, S., Rezaei, H.-R., Pompanon, F., Blum, M. G. B., Negrini, R., Naghash, H.-R., Balkiz, Ö., Mashkour, M., Gaggiotti, O. E., Ajmone-Marsan, P., Kence, A., Vigne, J.-D. & Taberlet, P. (2008). The goat domestication process inferred from large-scale mitochondrial DNA analysis of wild and domestic individuals. *PNAS*, 105 (46), 17659–17664.

Niblett, R., Manning, W. & Saunders, C. (2006). Verulamium: Excavations within the Roman town 1986–88. *Britannia*, 37, 53–188.

Nicholson, C. & Stephenson, M. W. (2007). *Financial performance value-added dairy operations in New York*, Vermont and Wisconsin. http://purl.umn.edu/9732. Accessed Mar. 17, 2014.

Neils, J. (2008). *The British Museum Concise Introduction to Ancient Greece*. Univ. Michigan Press, Ann Arbor.

Noussia, M. (2001). Solon's symposium. *The Classical Quarterly*, New Series, 51 (2), 353–359.

Outram, A. K., Kasparov, A., Stear, N. A., Varfolomeev, V., Usmanova, E. & Evershed, R. P. (2012). Patterns of pastoralism in later Bronze Age Kazakhstan: New evidence from faunal and lipid residue analyses. *Journal of Archeological Science*, 39, 2424–2435.
Potter, T. W. 1979. *The Changing Landscape of South Etruria*. St. Martins Press, New York.
Pross, J., Kotthoff, U., Müller, U. C., Peyron, O., Dormoy, I., Schmiedl, G., Kalaitzidis, S. & Smith, A. M. (2009). Massive perturbation in terrestrial ecosystems of the Eastern Mediterranean region associated with the 8.2 kyr B.P. climatic event. *Geology*, 37 (10), 887–890.
Rapp, A. (1955). The father of western gastronomy. *Classical Journal*, 51 (1), 43–48.
Ridgway, D. (1997). Nestor's Cup and the Etruscans. *Oxford Journal of Archaeology*, 16 (3), 325–344.
Salque, M., Radi, G., Tagliacozzo, A., Pino Uria, B., Wolfram, S., Hohhle, I., Stäuble, H., Whittle, A., Hoffmann, D., Pechtl, J., Schade-Lindig, S., Eisenhauer, U., and Evershed, R. P. (2012). New insights into the Early Neolithic economy and management of animals in Southern and Central Europe revealed using lipid residue analyses of pottery vessels. *Anthropozoologica*, 47 (2), 45–61.
Salque, M., Bogucki, P. I., Pyzel, J., Sobkowiak-Tabaka, I., Grygiel, R., Szmyt, M. & Evershed, R. P. (2013). Earliest evidence for cheese making in the sixth millennium BC in northern Europe. *Nature*, 493 (7433), 522–525.
Sherratt, A. G. (1981). Plough and pastoralism: Aspects of the secondary products revolution. In I. Hodder, G. Isaac & N. Hammond (eds.), *Pattern of the Past: Studies in Honour of David Clarke*. Cambridge University Press, Cambridge, UK, pp. 261–305.
Sherratt, A. (1983). The secondary exploitation of animals in the Old World. *World Archaeology*, 15 (1), 90–104.
Sherratt, S. (2004). Feasting in Homeric epic. *Hesperia*, 73 (2), 301–337.
Spangenberg, J. E., Jacomet, S. & Schibler, J. (2006). Chemical analyses of organic residues in archaeological pottery from Arbon Bleiche 3, Switzerland – evidence for dairying in the late Neolithic. *Journal of Archaeological Science*, 33, 1–13.
Stoll, M. (1993). Mik, butter and cheese. *Bulletin on Sumerian Agriculture*. 7, 99–113.
Stern, W. M. (1973). Cheese shipped coastwise to London towards the middle of the eighteenth century. *Guildhall Miscellany*, 4 (4), 207–221.
Thissen, L., Özbal, H., Türkekul Biyik, A., Gerritsen, F. & Özbal, R. (2010). The land of milk? Approaching dietary preferences of Late Neolithic communities in NW Anatolia. *Leiden Journal of Pottery Studies*, 26, 157–172.
Trump, D. (1965). *Central and Southern Italy before Rome*. Frederick A. Praeger, New York.
Turney, C. S. M. & Brown, H. (2007). Catastrophic early Holocene sea level rise, human migration and the Neolithic transition in Europe. *Quaternary Science Reviews*, 26, 2036–2041.
Vigne, J.-D. (2008). Zooarchaeological aspects of the Neolithic diet transition in the Near East and Europe, and their putative relationships with the Neolithic demographic transition. In Bocquet-Appel, J.-P. & Bar-Yosef, O. (eds.), *The Neolithic Demographic Transition and Its Consequences*. Springer Sci. and Bus. Media B.V.
Vigne, J.-D. (2011). The origins of animal domestication and husbandry: A major change in the history of humanity and the biosphere. *Comptes Rendus Biologies*, 334, 171–181.
Vigne, J.-D. & Helmer, D. (2007). Was milk a "secondary product" in the Old World Neolithisation process? Its role in the domestication of cattle, sheep and goats. *Anthropozoologica*, 42(2), 9–40.
Vigne, J.-D., Carrère, I., Briois, F. & Guilaine, J. (2011). The early process of mammal domestication in the Near East: New evidence from the pre-Neolithic and pre-pottery Neolithic in Cyprus. In The Origins of Agriculture: New data, new ideas, *Current Anthropology*, 52 (Suppl. 4), 255–271.
Wainwright, G. A. (1959). The Teresh, the Etruscans and Asia Minor. *Anatolian Studies*, 9, 197–213.

Wang, Q., Thompson, E. & Parsons, R. (2015). Preferences for farmstead, artisan, and other cheese attributes from a conjoint study in the Northeast United States. *International Food and Agribusiness Management Review*, 18 (2), 17–36.

Weiss, E. & Zohary, D. (2011). The Neolithic Southwest Asian founder crops: Their biology and archaeobotany. In The Origins of Agriculture: New data, new ideas, *Current Anthropology*, 52 (Suppl. 4), 237–254.

Weniinger, B., Alram-Stern, E., Bauer, E., Clare, L., Danzeglocke, U., Jöris, O., Kubatzki, C., Rollefson, G., Todorova, H. & van Andel, T. (2006). Climate forcing due to the 8200 cal yr BP event observed at Early Neolithic sites in eastern Mediterranean. *Quaternary Research*, 66, 401–420.

Whittaker, D. & Goody, J. (2001). Rural manufacturing in the Rouergue from antiquity to the present: The examples of pottery and cheese. *Comparative Studies in Society and History*, 43 (2), 225–245.

Wycherley, R. E. (1956). The market of Athens: Topography and monuments. *Greece & Rome, Second Series*, 3 (1), 2–23.

2

From Micelle to Melt: The Influence of Calcium on Physico-chemical Properties of Cheese

Darren R. Cooke and Paul L.H. McSweeney

School of Food and Nutritional Sciences, University College Cork, Cork, Ireland

2.1 Introduction

The calcium content of cheese has a major influence on a number of its physicochemical properties. Calcium influences the rheological and functional properties of cheese due to calcium-dependent interactions between casein proteins (Lucey, Johnson & Horne, 2003). Functional properties of cheese such as melting and stretch are of critical importance when the cheese is used as a food ingredient, that is, as pizza toppings, lasagne layers, slices for hamburgers, and so on (Lucey, 2008). Most textural, rheological and functional properties are dependent on molecular interactions involving calcium in the *para*-casein matrix, the origin of which can be traced back to the behaviour of calcium in the cheesemilk, during manufacture and throughout ripening. It should be noted that the only cheeses discussed in this chapter are those made from bovine milk.

The natural calcium content of bovine milk depends on numerous factors such as breed, stage of lactation, geography, mastitis and diet (Holt, 1985). Bovine milk typically contains 26–32 mmol Ca/kg (Gaucheron, 2005). About two-thirds of the total calcium in milk exists in insoluble complexes associated with casein micelles known as colloidal calcium phosphate (CCP). The calcium in milk exists in a dynamic equilibrium between the insoluble form (CCP) and the soluble forms (free calcium ions and soluble undissociated calcium complexes with phosphate and citrate) in the aqueous phase (Holt, 1985). The majority of insoluble calcium exists in CCP nanoclusters, which are of critical importance to the structure of the casein micelle as they can crosslink numerous casein molecules and reduce electrostatic repulsion, allowing formation of the casein micelle (Horne, 1998). The exact form of calcium phosphate in CCP nanoclusters remains a controversial topic, with many different models suggested and revised (Holt et al., 1989, 1998; Little & Holt, 2004; McGann et al., 1983). The precipitation of calcium phosphate to form CCP nanoclusters is largely influenced by pH and temperature, which is significant in the formation of mineral precipitates on heat exchanger surfaces used for thermal processing of milk (De Jong, 2008; Lucey & Horne, 2009).

Rennet coagulation is the primary manufacturing step involved in the production of most cheese varieties, and calcium has a major influence on this process. Calcium has no influence on the enzymatic phase of rennet coagulation if the pH is kept constant, but the aggregation phase is highly dependent on ionic calcium (Ca^{2+}) concentration (Van Hooydonk, Hagedoorn & Boerrigter, 1986). Cleavage of the glycomacropeptide from κ-casein reduces the net negative

Global Cheesemaking Technology: Cheese Quality and Characteristics, First Edition.
Edited by Photis Papademas and Thomas Bintsis.

charge and steric repulsion of micelles, allowing micelles to come into close contact with each other and aggregate in the presence of Ca^{2+} (Fox et al., 2000). Sufficient Ca^{2+} activity is required for proper rennet coagulation (Udabage, McKinnon & Augustin, 2001). At constant Ca^{2+} activity, a lower CCP content increases the rennet coagulation time (RCT) (Zoon, Van Vliet & Walstra, 1988). At a constant pH, a lower CCP also increases the RCT (Choi, Horne & Lucey, 2007). It is well known that addition of calcium to milk can reduce the RCT and improve gel properties (Udabage, McKinnon & Augustin, 2001; Zoon, Van Vliet & Walstra, 1988).

Throughout the cheese manufacturing process, decreases in pH cause partial solubilisation of CCP, and the residual CCP concentration in the finished cheese has a major influence on the rheological properties of the cheese (Lucey & Fox, 1993; Lucey et al., 2005; O'Mahony, Lucey & McSweeney, 2005). In Cheddar cheese, the residual CCP remaining in the cheese after manufacture partially solubilises during the first month of ripening (Hassan, Johnson & Lucey, 2004; Lucey et al., 2005; O'Mahony, Lucey & McSweeney, 2005), during which time a pseudo-equilibrium between soluble and insoluble calcium phosphate is reached (Hassan, Johnson & Lucey, 2004). This equilibrium is commonly termed the 'calcium equilibrium' of cheese, and alteration of manufacturing steps, that is, pH alterations, acid development and addition of calcium salts and calcium-binding salts, can alter the calcium equilibrium of cheese (Brickley, Lucey & McSweeney, 2009; Choi et al., 2008; Lee, Johnson & Lucey, 2005). Addition of calcium at sufficient levels can alter the microstructure of cheese, increasing the density of the *para*-casein matrix (Ong et al., 2013). The decrease in the CCP content of cheese during early ripening is principally responsible for the initial softening of cheese and an increase in its meltability (Lucey et al., 2005; O'Mahony, McSweeney & Lucey, 2006). A number of studies have reported a decrease in firmness and increased meltability in cheeses with reduced CCP concentrations due to alterations in manufacturing steps (Choi et al., 2008; Joshi, Muthukumarappan & Dave, 2002; Mizuno & Lucey, 2005). Softening, meltability and stretch are among the most important functional properties of heated cheese. Understanding the relationship between the rheological behaviour of cheese and its calcium content at the molecular level is of great importance when studying functionality improvements in cheese.

2.2 Calcium Equilibrium in Bovine Milk

2.2.1 Forms of Calcium in Milk

Bovine milk contains 26–32 mM Ca, with ~69% associated with casein micelles in an insoluble form (CCP) or directly bound Ca^{2+}, and ~31% present as soluble forms in the aqueous phase (Lucey & Horne, 2009). The majority of soluble calcium exists in undissociated complexes formed mainly with citrate (as Cit^{3-}) and also to a lesser extent with inorganic phosphate (as a mixture of $H_2PO_4^-$ and HPO_4^{2-}) (Gaucheron, 2005). Only about ~2 mM of the soluble calcium exists as free ionic calcium ions (Ca^{2+}) (Tsioulpas, Lewis & Grandison, 2007; Van Hooydonk, Hagedoorn & Boerrigter, 1986). About 90% of the total citrate and 50% of the total inorganic phosphate in milk is soluble (Holt, 2004). The behaviour of calcium and phosphate in milk dictate the so-called calcium pseudo-equilibrium in milk, that is, the distribution of Ca between the soluble phase and insoluble casein-bound (colloidal) phase (Lucey & Horne, 2009). Of the major forms of inorganic phosphate in the aqueous phase of milk ($H_2PO_4^-$ and HPO_4^{2-}), the $H_2PO_4^-$ form has a low affinity for calcium, whereas the HPO_4^{2-} form has a relatively high affinity for this metal (Mekmene, Le Graet & Gaucheron, 2009); however, the low concentration of the $CaHPO_4$ complex in the aqueous phase of milk (~0.6 mM) is due to its low solubility (Mekmene, Le Graet & Gaucheron, 2009). It is noteworthy that a very small proportion of the insoluble calcium in milk is bound to whey proteins (Holt, 1985), principally to α-lactalbumin,

Table 2.1 Mineral composition of bovine milk (Gaucheron, 2005).

Mineral	Concentration (mg/kg)	Concentration (mmol/kg)
Calcium	1043–1283	26–32
Magnesium	97–146	4–6
Inorganic phosphate	1805–2185	19–23
Total phosphate	930–992	30–32
Citrate	1323–2079	7–11
Sodium	391–644	17–28
Potassium	1212–1681	31–43
Chloride	772–1207	22–34

which binds ~0.5 mmol/kg (Lucey & Horne, 2009); however, this can be considered negligible. The calcium phosphate pseudo-equilibrium is the most important aspect of the milk salts system in terms of the casein micelle structure; however, the other salts and their ions in the system also have a major effect. Sodium, potassium and chloride are present at high concentrations in milk (Table 2.1), with ~95% of each being present in the soluble phase. Potassium has the highest concentration of all the ions in milk (31–43 mmol/kg) and so has a great influence on milk ionic strength. In addition to inorganic phosphate, milk also contains various forms of organic phosphate which can be found in phosphoseryl residues of caseins, phospholipids, nucleotides, nucleic acids and ATP.

2.2.2 Colloidal Calcium Phosphate

The salts associated with casein micelles in milk are collectively referred to as colloidal calcium phosphate (Holt, 1985). This term encompasses the crosslinking calcium phosphate in nanoclusters (also known as micellar or casein-bound calcium phosphate) and also calcium directly bound to casein molecules (Ca caseinate).

It is generally accepted that CCP nanoclusters are particles of amorphous hydrated calcium phosphate linked to casein phosphoseryl clusters, measuring ~2.5 nm in diameter, and are distributed throughout the casein micelle (Holt, 2004; McGann et al., 1983). The spacing between nanoclusters has been estimated to be 18 nm (De Kruif & Holt, 2003). There may be up to ~800 of these nanoclusters in a casein micelle with a radius of 100 nm (Holt, 2004). Along with calcium phosphate, McGann et al. (1983) reported that CCP contains citrate, magnesium (Mg) and zinc (Zn) at molar ratios to Ca averaging 0.05, 0.03 and 0.003, respectively.

These CCP nanoclusters can be viewed as having two primary roles in casein micelles, namely, neonatal nutrition and micellar structural integrity. CCP nanoclusters allow milk to contain concentrations of calcium and phosphate well in excess of saturation levels. The more important role of CCP in the context of this chapter is that CCP nanoclusters are proposed to be one of the main crosslinking pathways in the formation of casein micelles. In the dual-binding model of the casein micelle proposed by Horne (1998) (Figure 2.1), polymerization of casein molecules proceeds via two possible pathways: (1) hydrophobic interactions between non-polar residues on adjacent casein molecules, with more than two molecules possibly interacting at these junctions and (2) CCP nanocluster crosslinks between hydrophilic regions of certain caseins. The CCP crosslinks act as bridges between two or more casein molecules that contain phosphoseryl cluster sequences. The interaction of the positively charged CCP nanoclusters with negatively charged phosphoseryl clusters reduces electrostatic repulsion between casein molecules, allowing attractive hydrophobic interactions to dominate.

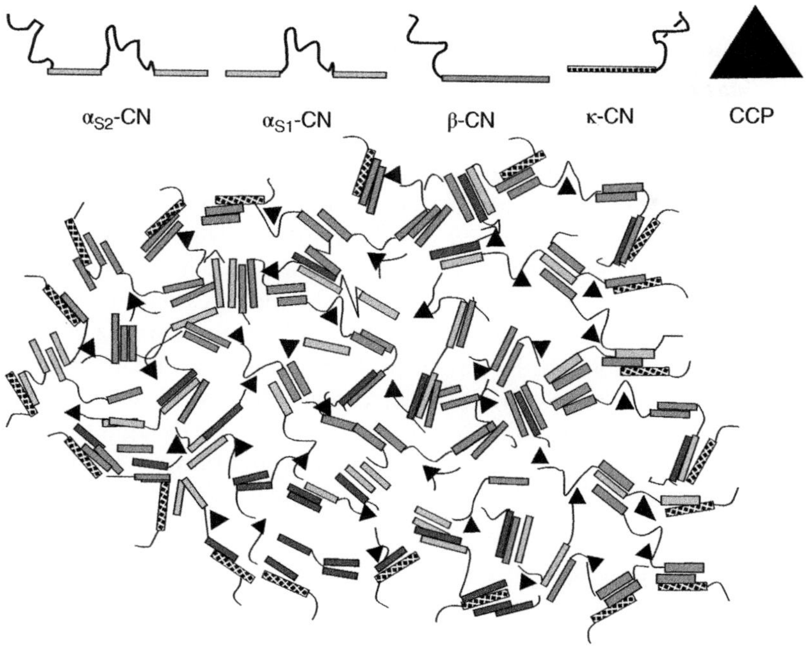

Figure 2.1 Schematic representation of the 'dual-binding model' of the casein micelle (Horne, 1998) revised by Lucey and Horne (2009). CN is casein, and CCP is colloidal calcium phosphate nanoclusters.

α_{S1}-, α_{S2}- and β-casein are multi-phosphorylated proteins, and all contain at least one phosphoseryl cluster, whereas κ-casein lacks a phosphoseryl cluster (Figure 2.2). Phosphoseryl clusters are essential for the nucleation and stabilization of the calcium phosphate salts that comprise the core of CCP nanoclusters. Aoki, Umeda and Kako (1992) proposed that at least three phosphoseryl residues are required for a casein molecule to be crosslinked by CCP nanoclusters. These sequences on casein molecules have the specific motif Ser(P)$_3$-Glu$_2$ (Holt & Sawyer, 1988). Thus, according to this model, only sequences that have three consecutive phosphoseryl residues are involved in the stabilization of CCP nanoclusters; that is, each of α_{S1}- and β-casein has one of these sequences, and α_{s2}-casein has two. Holt (2004) defined phosphate

α_{S1}-casein (8P)

---- - SerP$_{46}$-Glu-SerP$_{48}$- ---- - SerP$_{64}$-Ile-SerP$_{66}$-SerP$_{67}$-SerP$_{68}$- -------- -SerP$_{75}$- ------
---SerP$_{115}$- -------

α_{S2}-casein (11 or 12P)

--- - SerP$_8$-SerP$_9$-SerP$_{10}$- --- -SerP$_{16}$- --------- -SerP$_{56}$-SerP$_{57}$-SerP$_{58}$-Glu-Glu-SerP$_{61}$- -
----------- -SerP$_{129}$- Thr-SerP$_{131}$- ----- -SerP$_{143}$- -----

β-casein (5P)

-------- -SerP$_{15}$-Leu-SerP$_{17}$-SerP$_{18}$-SerP$_{19}$- ------- -SerP$_{35}$- --------------

κ-casein (1P)

---------------- -SerP$_{149}$- ------------

Figure 2.2 Phosphoseryl residue positions on bovine casein molecules (Horne, 2006).

centres (PC) as at least two phosphorylated residues in a short sequence, which would give α_{S1}-, α_{S2}- and β-casein two, three and one of these phosphate centres, respectively. Such discrepancies in defining phosphoseryl clusters/centres can lead to different models of nanoclusters. These negatively charged sequences interact with and stabilise the positively charged CCP nanocluster core. Essentially, casein phosphoseryl clusters convert an intrinsically unstable milk system into a thermodynamically stable system (Holt, 2004).

CCP nanoclusters are thought to exist in a metastable state where growth into a macroscopic phase (leading to tissue calcification) is prevented by the rheomorphic structure of caseins along with their phosphoseryl clusters (Holt, 2004). The exact structure of CCP is still controversial. A proportion of colloidal calcium in milk is also directly bound to casein molecules. α_{S1}-Casein has a higher Ca^{2+}-binding capacity than β-casein, which is thought to be a consequence of the higher phosphoseryl content of the former protein (Dickson & Perkins, 1971). However, most of the phosphoseryl clusters that exist in casein micelles are involved in stabilizing CCP nanoclusters. Apart from phosphoseryl residues, Ca^{2+} is also thought to bind to carboxyl groups of glutamic and aspartic acid residues, phenolic groups of tyrosyl residues, sulfhydryl groups of cystedyl residues and imidazole groups of histidyl residues (Dickson & Perkins, 1971; Gaucheron et al., 1997).

2.2.3 Modification of Calcium Equilibrium in Bovine Milk

Changes in milk environmental conditions and solution properties, that is, alteration of pH, temperature, ionic strength and addition of various mineral salts, can have a major effect of the distribution and form of calcium in milk. Upon addition of calcium to milk, for example, in the form of $CaCl_2$, an increase in both casein-bound calcium and inorganic phosphate is observed (Philippe et al., 2003; Udabage, McKinnon & Augustin, 2000, 2001; Van Hooydonk, Hagedoorn & Boerrigter, 1986). This co-precipitation of calcium and inorganic phosphate to the colloidal phase is indicative of the formation of CCP nanoclusters. It is estimated that ~10% of the total amount of phosphoseryl clusters in the casein micelle is unreacted (Holt, 2004). Philippe et al. (2003) suggested that new CCP nanoclusters formed at unreacted phosphoseryl clusters after Ca^{2+} addition to milk may differ from the natural form of CCP. Addition of calcium to milk also increases the level of soluble calcium. Addition of strong calcium sequestering agents such as trisodium citrate and EDTA can decrease both the CCP content of casein micelles and Ca^{2+} activity in milk (Choi, Horne & Lucey, 2007; Udabage, McKinnon & Augustin, 2000, 2001). Addition of sodium chloride to milk increases the ionic strength and can displace Ca^{2+} directly bound to casein by Na^+, thus increasing the soluble calcium; however, CCP nanoclusters are not thought to be affected (Van Hooydonk, Hagedoorn & Boerrigter, 1986).

Decreasing the pH of milk results in solubilisation of CCP, with all of the CCP being completely soluble at ~pH 5.0 (Lucey & Horne, 2009). The buffering capacity of milk is reliant on this solubilisation of CCP, which results in the formation of phosphate ions that combine with H^+, causing buffering (Lucey et al., 1993). During acidification, milk exhibits the maximum buffering capacity at ~pH 5 (Hassan, Johnson & Lucey, 2004; Lucey et al., 1993). Increasing milk pH results in the formation of additional CCP (Lucey and Horne, 2009) as the amount of calcium and phosphate bound by phosphoseryl sequences increases with pH (Cross et al., 2005). The solubility of calcium phosphates decreases at high temperatures, which results in the formation of heat-induced CCP, which re-solubilises when milk is allowed to cool (Lucey & Horne, 2009). Alteration of the calcium equilibrium of milk by any of the aforementioned mechanisms can have a major influence on the various processing steps involved in cheese manufacture, cheese composition and physicochemical properties of the finished cheese.

Bovine milk has been the primary focus of detailed studies of milk calcium equilibrium; however, the findings from these studies can be tentatively extrapolated to the milk of other species.

In the case of caprine milk, differences in casein proteins between individuals, large variation in mineral content and casein genetic polymorphism can influence calcium equilibrium. Genetic polymorphism of α_{S1}-casein impacts the number of available phosphoseryl centres; for example, milks with F-type α_{S1}-casein genetic variants have less available phosphoseryl clusters than average goat milk, and milk with O-type have no α_{S1}-casein (Pierre, Michel & Le Graet, 1995; Tziboula-Clarke., 2002), which impacts casein micelle properties Pierre, Michel and Le Graet (1995) found that O-type milk had a higher soluble Ca content than A-type milk (high α_{S1}-casein). This indicates that less CCP was formed due to lower availability of phosphoseryl centres to sequester Ca^{2+}. Cheese with O-type milk would likely have lower CCP content and have more meltability than an equivalent cheese with A-type milk.

2.3 Calcium Equilibrium in Cheese

2.3.1 Changes in the Calcium Equilibrium of Cheese during Ripening

The process of cheese ripening involves numerous microbiological, biochemical and physicochemical changes, many of which are interrelated. Collectively, these changes are responsible for the conversion of the rubbery, bland young cheese into a mature cheese with characteristic flavour, texture and aroma (Fox & McSweeney, 1998; Lucey, Johnson & Horne, 2003). One of the major factors governing changes in the structure and texture of cheese is its calcium content. Calcium content varies between cheese varieties due to their unique manufacturing procedures; in particular, the pH at whey drainage has a major influence on the final calcium content of cheese (Lucey & Fox, 1993). Typical calcium contents of Camembert, Cheddar and Emmental are 350, 720 and 970 mg/100 g cheese, respectively (O'Brien & O'Connor, 2004), which are linked to the differences in texture between these varieties. However, the insoluble calcium content of cheese is much more important than total calcium in regard to cheese structure and inherent textural properties (Lucey & Fox, 1993). As mentioned in Section 2.1, a dynamic equilibrium between insoluble calcium bound to the casein micelles (CCP) and soluble calcium in the aqueous phase exists in milk. A similar situation is thought to occur in cheese, where calcium solubilises from the residual CCP in the *para*-casein matrix during ripening to become part of the aqueous phase of cheese in order to attain a so-called pseudo-equilibrium of calcium phosphate between the soluble and insoluble phases of cheese (Hassan, Johnson & Lucey, 2004). This is commonly termed the 'Ca-equilibrium' of cheese. In Cheddar cheese, the proportion of insoluble calcium decreases during ripening from an initial level of ~72% to ~58% during the first three months of ripening (Hassan, Johnson & Lucey, 2004; Lucey et al., 2005) with very little change in insoluble calcium observed beyond this time, even when ripened for up to nine months (Lucey et al., 2005). Most studies have reported that the majority of the changes in calcium equilibrium actually occur within the first month of ripening (Hassan, Johnson & Lucey, 2004; Lucey et al., 2005; O'Mahony, Lucey & McSweeney, 2005).

2.3.2 Methods of Calcium Equilibrium Determination in Cheese

Two efficient methods of determining the insoluble calcium content of cheese have been developed and used successfully during the past two decades. One approach is the cheese juice method (Hassan, Johnson & Lucey, 2004; Lee, Johnson & Lucey, 2005; Morris et al., 1988), in which the serum phase is extracted from cheese by applying hydraulic pressure to grated cheese. The extracted serum or 'juice' is assumed to be compositionally equal to the aqueous phase of cheese (Morris et al., 1988), and so contains soluble calcium at the same concentration as the aqueous phase of the cheese. The insoluble calcium content of the cheese can be

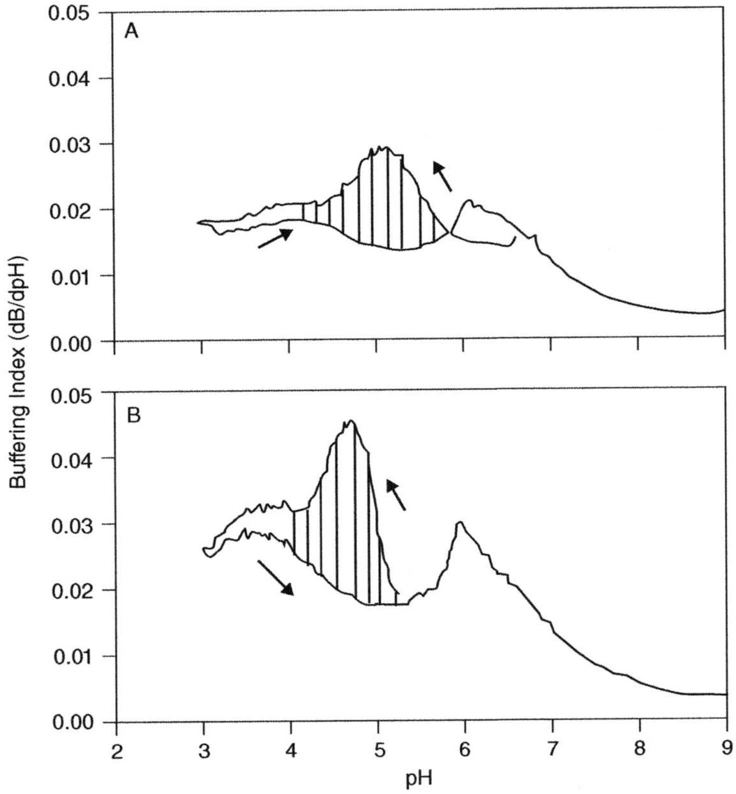

Figure 2.3 Buffering curves of milk (A) and Cheddar cheese (B) titrated from initial pH to pH 3.0 with 0.5 N HCl and then back-titrated to pH 9.0 with 0.5 N NaOH. Hatched area represents the buffering due to colloidal calcium phosphate. Arrows indicate the direction of the titration (Hassan et al., 2004).

estimated by comparing the total calcium content of the cheese to that of the juice. The second method is the acid-base titration method (Hassan, Johnson & Lucey, 2004; Lucey, Gorry & Fox, 1993; Lucey et al., 2005), which relates the buffering capacity of cheese to its residual CCP content. In this method, the buffering capacity of the cheese and the milk it was made from are determined. The buffering peaks observed in milk between ~pH 5.8 to 4.1 and in cheese from ~pH 5.1 to 4.0 are an index of their CCP content (Hassan, Johnson & Lucey, 2004; Lucey, Gorry & Fox, 1993; Lucey et al., 1993) (Figure 2.3). Both the buffering capacities and calcium contents of the milk and cheese are used to calculate the insoluble calcium content. Hassan, Johnson and Lucey (2004) reported no statistical difference between these two methods for accuracy in their determination of the percentage of insoluble calcium in cheese (Figure 2.4).

2.3.3 Manipulation of Calcium Equilibrium in Cheese

A typical manufacturing protocol for Cheddar cheese is shown in Figure 2.5. A decreased total calcium content in cheese is accompanied by a decreased insoluble calcium content (Choi et al., 2008). By altering acid development during the manufacture of Cheddar cheese, Lee, Johnson and Lucey (2005) produced cheeses with very low pH values (<4.9) that contained lower total calcium and insoluble calcium than cheeses with higher pH values. This study showed that the concomitant decrease in insoluble calcium with decrease in pH that is observed in milk can also be found in cheese. However, in the same study, Lee, Johnson and Lucey (2005)

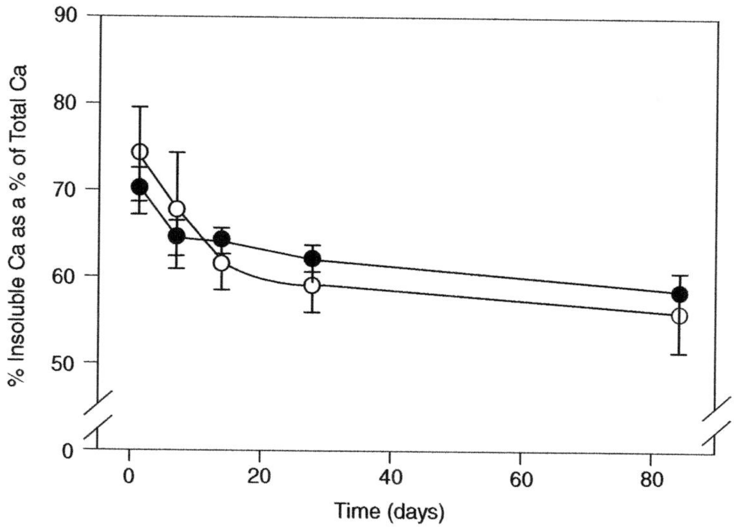

Figure 2.4 Changes in the percent insoluble calcium content (expressed as a percentage of total cheese calcium) as a function of ripening time in Cheddar cheese determined by acid-base titration (o) and cheese juice (•) methods (Hassan et al., 2004).

reported that the insoluble calcium content did not decrease below ~41% of total calcium during ripening in a cheese with a pH value of ~4.7 (Figure 2.6). This observation highlights the difference in mineral–casein interactions between cheese and milk; that is, CCP is completely solubilised below pH 5.0 in milk (Lucey & Horne, 2009). Lee et al. (2010) also observed a decrease in the insoluble calcium content in Colby cheese with a low pH induced by alteration of the manufacturing pH. Choi et al. (2008) produced directly acidified cheeses with reduced insoluble calcium levels by lowering the pH of the cheesemilk and adding EDTA to it. Pastorino, Hansen and McMahon (2003a) decreased the pH of Cheddar cheese after manufacture by injecting 20% (w/w) glucono-δ-lactone into cheese during early ripening, which decreases the insoluble calcium content.

Mizuno and Lucey (2005) supplemented non-fat *pasta filata* cheese with trisodium citrate (TSC) via addition at the dry-salting step of manufacture. Addition of TSC decreased the insoluble calcium level. In a similar study, Brickley, Lucey and McSweeney (2009) added TSC to Cheddar cheese at the salting stage and observed a decrease in insoluble calcium during the first month of ripening. In both studies, the authors attributed the decrease in insoluble calcium to the calcium-sequestering ability of TSC, which causes solubilisation of a proportion of CCP in the cheeses. Pastorino, Hansen and McMahon (2003b) was unsuccessful in altering the insoluble calcium level of Cheddar cheese by injecting the cheese with 40% TSC solution during early ripening. Immersing slices of four-month-old Cheddar cheese in synthetic Cheddar cheese aqueous phase (SCCAP) solutions of varying calcium concentration, O'Mahony, McSweeney and Lucey (2006) observed an increase in the CCP concentration of cheese when the calcium concentration of the SCCAP solution exceeded that of the cheese serum. The opposite effect was observed with SCCAP solutions containing lower calcium concentrations than the cheese serum, where solubilisation of cheese CCP occurred. These observations are explained by calcium equilibration between the cheese and solutions in which it was immersed.

2.3.4 Mechanisms of Calcium Equilibrium Changes during Cheese Ripening

Comparing the environment surrounding CCP in milk to the CCP in cheese, milk contains ~2.8% casein and ~87% water and has a pH of ~6.6, whereas Cheddar cheese, for example,

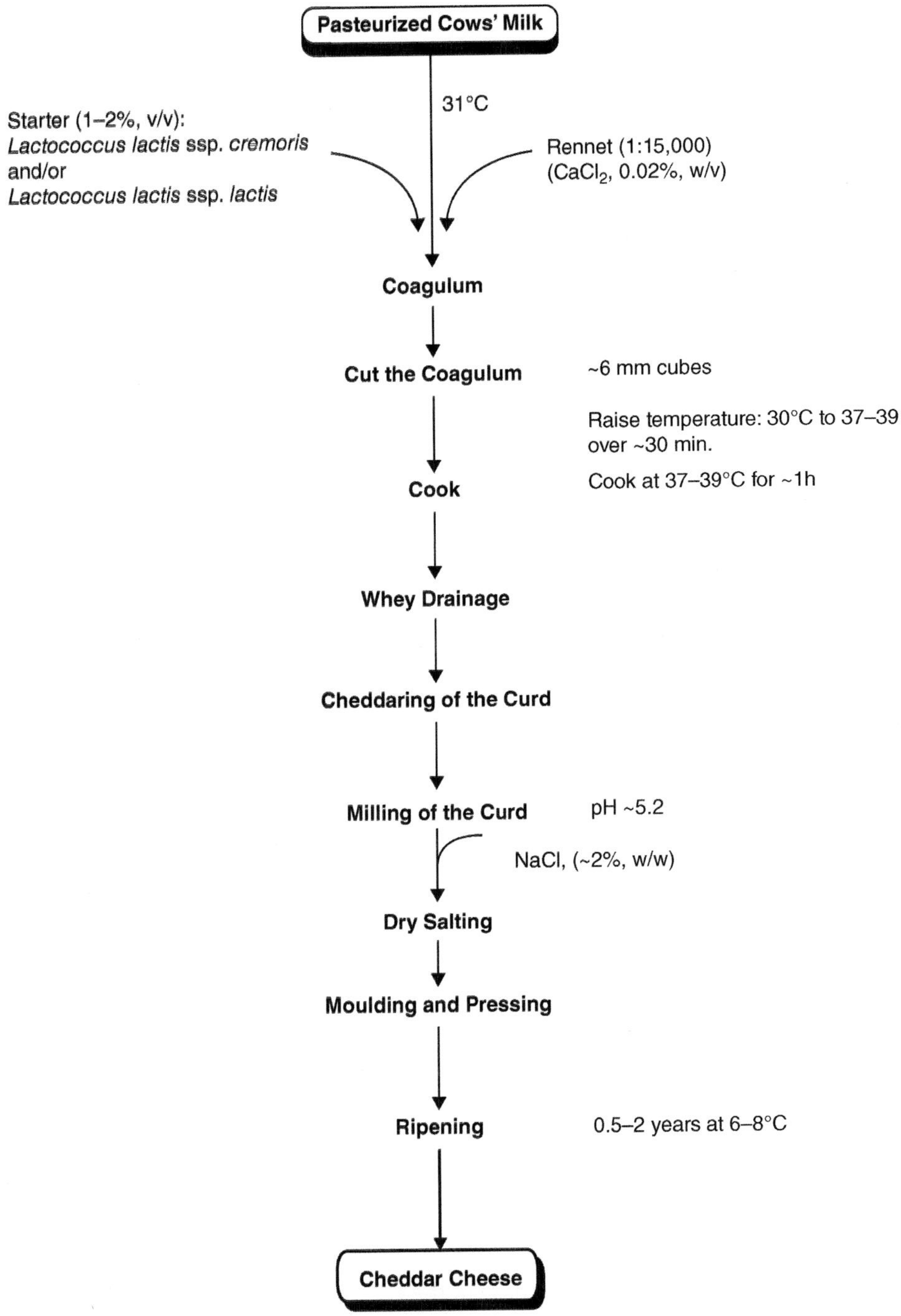

Figure 2.5 Typical manufacturing protocol for Cheddar cheese (Fox et al., 2000).

initially contains ~25% casein and 35% to 39% water and a pH of 5.1 to 5.3 (Fox & McSweeney, 1998; Lawrence et al., 2004). The low pH and reduced water content of cheese may have a major effect on the saturation state and form of calcium phosphate. The lower pH in cheese and changes in pH during ripening may affect the form of P_i that is present, that is, conversion of HPO_4^{2-} to $H_2PO_4^-$ and vice versa, depending on the ripening age. As each of these P_i anionic

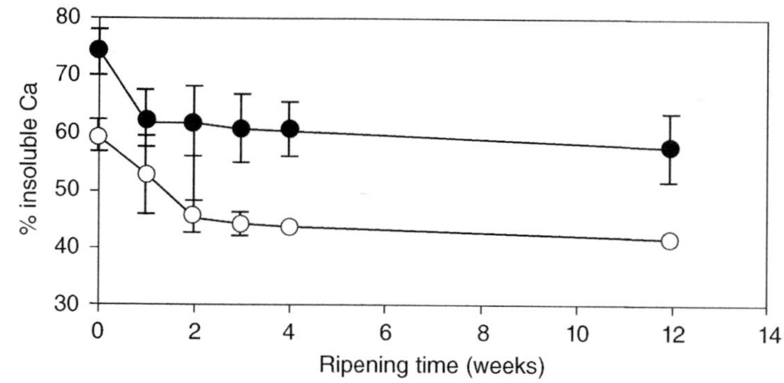

Figure 2.6 Changes in the percent insoluble Ca content (percentage of total calcium in cheese) in Cheddar cheeses with pH decrease from 5.06 to 4.91 (•) and 4.91 to 4.77 (o) during ripening (Lee et al., 2005).

species has different affinities for calcium (Mekmene, Le Graet & Gaucheron, 2009), this conversion may alter the composition of CCP nanoclusters. During the first few weeks of Cheddar cheese ripening, starter lactic acid bacteria convert residual lactose to lactic acid (Fox et al., 2000), which can slightly decrease cheese pH depending on the salt-in-moisture level that controls microbial growth. The buffering effect of CCP in cheese likely prevents any excessive decrease in pH as acidification causes solubilisation of CCP, which liberates PO_4^- ions that combine with H^+, producing buffering (Hassan, Johnson & Lucey, 2004; Lucey, Gorry & Fox, 1993a). It is likely that solubilisation of CCP by this mechanism contributes to the decrease in insoluble calcium levels during early ripening. Cheeses with very low pH values (<4.9) may likely have an acidic calcium phosphate form in CCP nanoclusters as basic calcium phosphate complexes form over a pH range from ~5.0 to 9.0 (Cross et al., 2005).

The composition of cheese juice from a one-month-old Cheddar cheese was studied in detail by Morris et al. (1988). They reported that approximately 57% of calcium, 89% of P_i and 55% of the citrate was insoluble at this stage of ripening. These levels of insoluble P_i and citrate are much higher than their levels in milk, that is, ~46 and ~11% for P_i and citrate, respectively (Gaucheron, 2005). These high values were suggested by Morris et al. (1988) to be the consequence of the formation of calcium phosphate or calcium citrate crystals in the aqueous phase of the cheese due to supersaturation of these salts. The crystalline forms were suggested to be tricalcium citrate and brushite, as a more acidic calcium phosphate form would be expected to form at the pH of cheese juice. Morris et al. (1988) proposed that areas with high local pH near or within bacterial colonies may act as nucleation sites for precipitation of calcium phosphate and crystal formation as the degree of supersaturation of calcium phosphate salts would increase in these areas. Lee, Johnson and Lucey (2005) observed that calcium lactate crystals in Cheddar cheese ripened for ~3 months and proposed that crystallisation of calcium salts would effectively decrease the soluble calcium level and possibly promote further loss of calcium from the CCP to balance calcium equilibrium.

Another consideration is that proteolysis in cheese may play some role in changes in calcium equilibrium during ripening. α_{S1}- and β-casein are the principal structural components of the *para*-casein matrix of cheese, and both of these proteins are hydrolysed to varying degrees during ripening by certain enzymes. As both of these casein molecules can interact with CCP nanoclusters via their phosphoseryl cluster sequences, the location of these sequences and the sites of enzymatic cleavage on these proteins are of interest as they may have a role to play in calcium equilibrium at some point during ripening. The phosphoseryl cluster sequence with three consecutive phosphoseryl residues in α_{S1}-casein occurs between residues 64 and 68 and in β-casein between residues 15 and 19 (Horne, 2006). α_{S1}- and β-casein are hydrolysed mainly by chymosin and plasmin, respectively. Hydrolysis products of α_{S1}-and β-casein formed during Cheddar cheese ripening include α_{S1}-CN(f24-98), α_{S1}-CN(f24-101), α_{S1}-CN(f24-109), β-CN(f1-28) and β-CN(f1-105/107) (Sousa, Ardo & McSweeney, 2001), all of which contain a phosphoseryl cluster. Loss of stabilizing phosphoseryl clusters may destabilize CCP nanoclusters. O'Mahony, Lucey and McSweeney (2005) provided evidence that the solubilisation of CCP during the first few weeks of ripening is not influenced by proteolysis of α_{S1}-casein. In this study, inhibition of residual chymosin activity by pepstatin greatly reduced the cleavage of the Phe_{23}-Phe_{24} bond in α_{S1}-casein, allowing ~91% of α_{S1}-casein to remain intact after 180 days of ripening. On the basis of the findings of this study, it is likely that the rapid decrease in insoluble calcium in cheese within the first few weeks of ripening is not directly related to proteolysis. Gagnaire et al. (2001) reported that 17 peptides out of 91 identified in mature Emmental cheese were phosphopeptides derived from α_{S2}- and β-casein. After extensive hydrolysis of caseins associated with CCP nanoclusters in long ripened cheese, solubilisation of phosphopeptides may induce a lack of CCP stabilisation, and so calcium phosphate may grow into macrocrystals. As cheese ripens, the estimated spacing between CCP nanoclusters increases from ~17 nm to ~40 nm by 6 weeks of ripening (Lucey, Johnson & Horne, 2003; Tunick et al., 1997). Lucey, Johnson and Horne (2003) suggested that a type of Ostwald ripening of nanoclusters may occur during ripening. Destabilisation of nanoclusters by proteolysis may facilitate such a mechanism.

As calcium can bind directly with caseins, especially at phosphoseryl residues (Dickson & Perkins, 1971), it is likely that hydrolysed casein fragments that incorporate a phosphoseryl cluster (casein phosphopeptides) contain bound calcium and possibly affect calcium equilibrium. At the same time, casein phosphopeptides that were already associated with CCP nanoclusters may remain associated, as casein hydrolysates containing phosphoseryl clusters have the ability to stabilize CCP (Cross et al., 2005; Holt et al., 1998). In the core-shell nanocluster model proposed by Holt et al. (1998), it is proposed that the nanocluster core is surrounded by a peptide 'shell' containing ~49 casein molecules. Presumably, proteolytic breakdown of these associated caseins may interfere with the structure of nanoclusters, destabilising the metastable amorphous calcium phosphate core and inducing crystal formation. In an attempt to isolate CCP nanoclusters, Choi, Horne and Lucey (2011) digested casein micelles in milk with trypsin and papain, producing CCP nanoclusters stabilized by phosphopeptide fragments of caseins. The authors reported that a small proportion of nanoclusters did not survive the digestion/dialysis processes. This finding supports the hypothesis that hydrolysis of caseins may affect calcium equilibrium in cheese. Proteolysis also reduces the 'free' water content of maturing cheese, as cleavage of peptide bonds liberates a NH_2^+ group and a COO^- group that both compete for water (Creamer & Olson, 1982; Lucey, Johnson & Horne, 2003). This reduction in free water may concentrate the soluble calcium phosphate in the aqueous phase of cheese.

Formation of calcium lactate crystals in Cheddar cheese is also related to calcium equilibrium. The presence of calcium lactate crystals in Cheddar cheese is a quality defect which occurs as white crystals on the cheese surface. Racemization of L-lactate to D-lactate by non-starter lactic acid bacteria occurs in Cheddar cheese (McSweeney and Fox, 2004). Ca-D-lactate has a lower solubility than Ca-L-lactate, and so a higher incidence of calcium lactate precipitation occurs with higher conversion to D-lactate (Chou et al., 2003). As the ratio of D/L lactate increases, cheese becomes increasing susceptible to calcium lactate precipitation (Chou et al., 2003; Johnson et al., 1990). Higher soluble calcium levels and higher levels of residual lactose make cheese more susceptible to the formation of calcium lactate crystals (Agarwal et al., 2006; McSweeney & Fox, 2004).

An interesting case of calcium equilibrium in a cheese system other than Cheddar is the change of calcium distribution during ripening of surface-ripened cheeses such as Brie and Camembert (Figure 2.7). In these cheeses, moulds grow on the cheese surface that metabolize

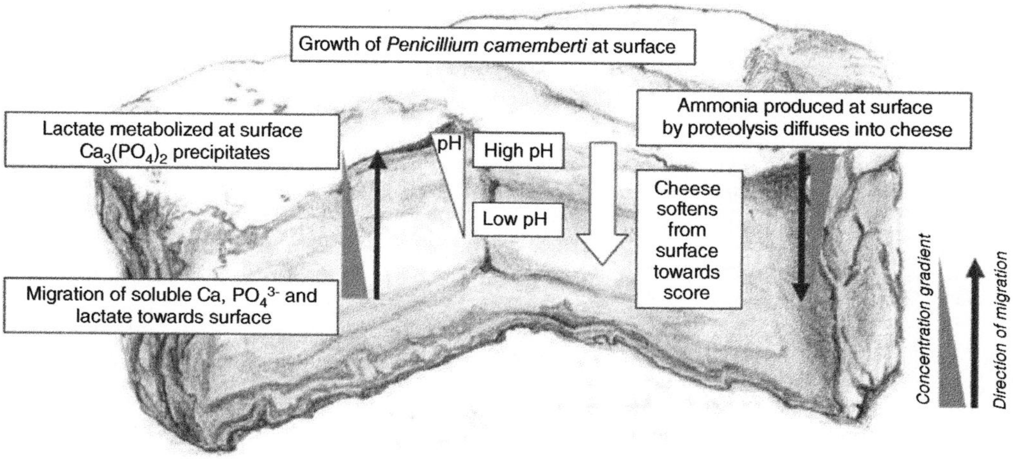

Figure 2.7 Schematic representation of the changes that occur in calcium, phosphate, lactate, ammonia and pH gradients in Camembert-type cheese during ripening (McSweeney & Fox, 2004).

lactate and produce ammonia, which produces a pH gradient in the cheese, with a high pH at the surface and a low pH in the centre of the cheese (Abraham et al., 2007; Karahadian & Lindsay, 1987). The high pH at the surface causes calcium phosphate to precipitate, as a result of which soluble calcium phosphate migrates from the centre to balance the calcium equilibrium at the surface during ripening (Le Graet et al., 1983). The precipitated forms of calcium phosphate at the surface are suggested to be dicalcium or tricalcium phosphate, as the $Ca:P_i$ ratio at the surface is ~1.87:1.00 (Le Graet et al., 1983). This process essentially solubilizes CCP at the centre of the cheese, reducing the attraction between caseins, which decreases the structural integrity of the cheese matrix. Considerable amounts of calcium should also migrate to the surface as calcium lactate, with the lactate being metabolized by the surface flora (Karahadian & Lindsay, 1987). Boutrou et al. (1999) reported a decrease in Ca and P_i concentrations and the $Ca:P_i$ ratio of juice extracted from Camembert during ripening, and attributed this to the precipitation of calcium phosphate that occurs near the cheese surface.

2.4 The Influence of Calcium on Cheese Rheology and Functionality

Cheese has grown in commercial importance to the food industry over the past few decades as it is being used increasingly as an ingredient in a wide variety of prepared foods (Figure 2.8), for example, pizza toppings and cheese slices on hamburgers (Lucey, 2008). For cheese to be used as a food ingredient, it must have suitable functional properties such as softening, meltability, stretchability and browning. In particular, Cheddar and Mozzarella have received much attention during the past two decades in the context of modulation of their functional properties relating to calcium, for example, improvements in meltability. The functional properties of cheese are governed by their rheological properties, and so an intimate understanding of cheese rheology is essential when selecting or designing a cheese for use as a food ingredient.

2.4.1 The Influence of Calcium Equilibrium on Cheese Microstructure

The microstructure of cheese has a major influence on its textural and functional properties. Physicochemical changes occur in the cheese matrix during ripening, which lead to increased

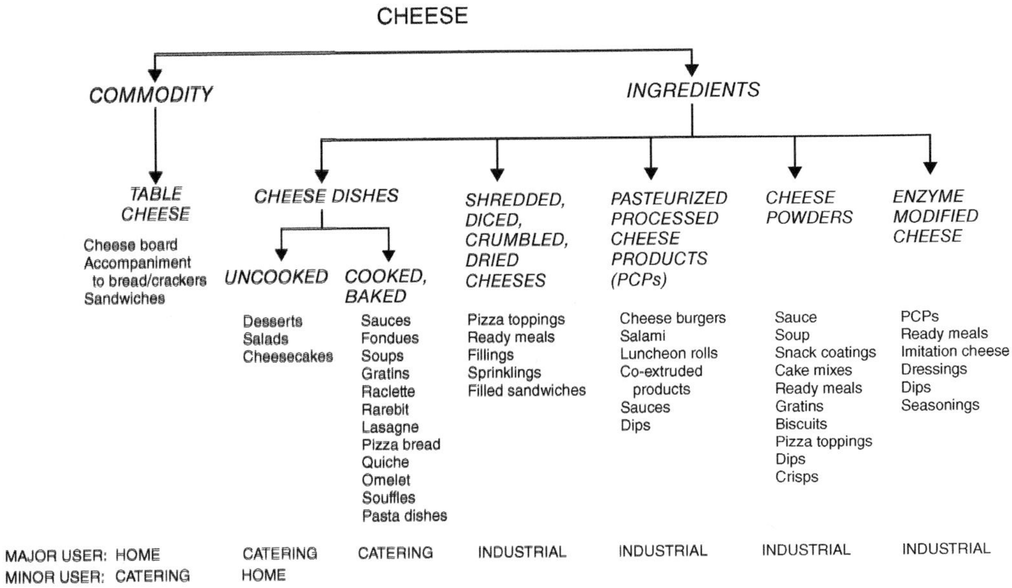

Figure 2.8 Uses of cheese as a food ingredient (Fox et al., 2000).

hydration of the matrix. Hydration increases due to casein proteolysis, pH changes and solubilisation of CCP (Fox et al., 2000). Increased hydration leads to physical expansion or swelling of the matrix, which pushes the fat globules closer together. They coalesce into fat pools, which increases the free fat upon melt (Fox et al., 2000). Two of the most popular methods to visualise cheese microstructure are (1) light microscopy, for example, confocal scanning laser microscopy (CSLM) and (2) electron microscopy, for example, scanning electron microscopy (SEM) (El-Bakry & Sheehan, 2014). Casein mineralization has a major influence on the microstructure, as formation of CCP increases the matrix density. Ong et al. (2013) observed that supplementing cheesemilk with calcium at 300 and 600 mg/L produced cheeses with a greater number of micron-sized pores in the structure and attributed this to reduced micellar fusion. Cheese hardness also increased with >100 mg/L addition levels. This is due to an increased number of CCP crosslinks in the matrix. The influence of the higher calcium level on the cheese microstructure can be seen in Figure 2.9, where higher calcium content results in contraction of the *para*-casein matrix. In the study of Pastorino et al. (2003), injection of 40% $CaCl_2$ solution into cheese post manufacture resulted in contraction of the matrix and release of water into the pores of the matrix. This was attributed to increased casein–casein bonding, leading to decreased matrix hydration. Reducing the calcium content of the cheesemilk decreases the matrix density by allowing more hydration and swelling of the matrix, creating more voids in the microstructure (Joshi, Muthukumarappan & Dave, 2003). From these studies, it can be concluded that the CCP content of cheese has a major influence on the microstructure.

2.4.2 Determination of the Rheological Properties of Cheese

Rheology is the study of the flow and deformation of matter when subjected to stress or strain. The rheological behaviour of cheese can be categorized as viscoelastic. A viscoelastic material exhibits aspects of both solid (elastic) and liquid (viscous) rheological behaviours under a wide range of conditions. In the relationship between stress and strain, there are three regions of importance for cheese (Figure 2.10). The first region is the linear viscoelastic region, where the

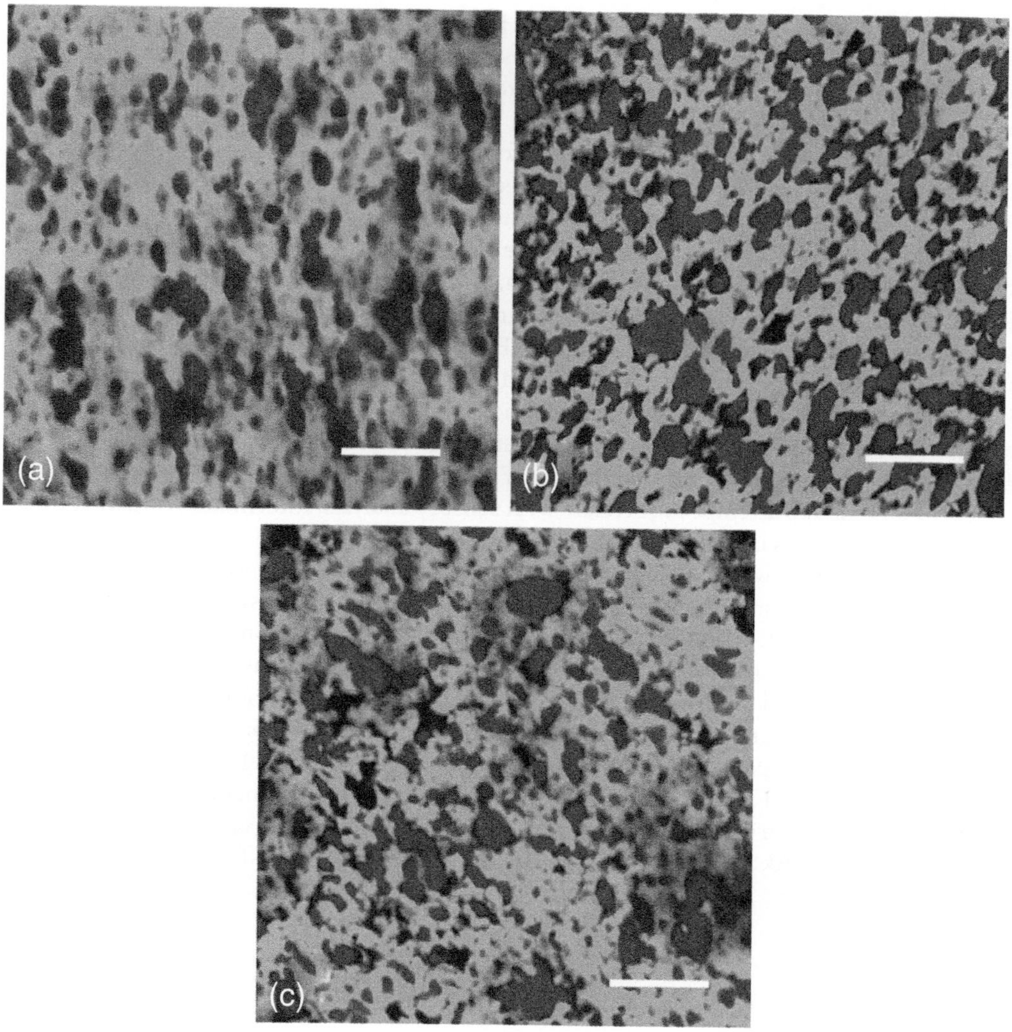

Figure 2.9 The microstructure of Cheddar cheese prepared using cheesemilk with the addition of (a) 0, (b) 300 or (c) 600 mg $CaCl_2$ per litre of milk. The Nile Red stained fat appears red, and the Fast Green FCF stained protein appears green in these images. The scale bars are 10 mm in length (Ong et al., 2013).

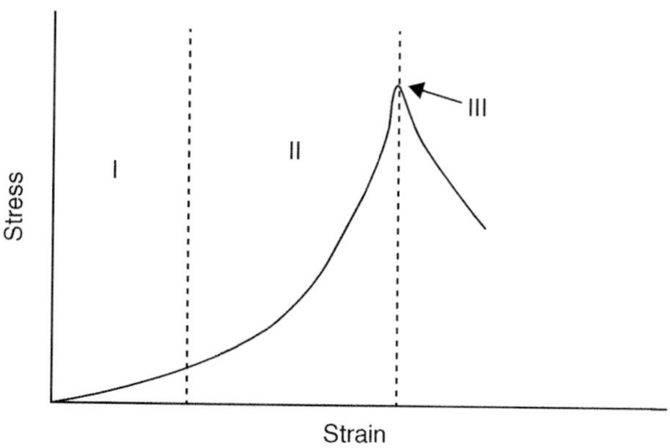

Figure 2.10 Rheological regimes for the viscoelastic behaviour of cheese (Foegeding et al., 2003).

strain and stress are directly proportional, and Hooke's law is obeyed. The second region is non-linear, displaying a stress increase that is less proportional to strain, which increases the slope, and the curve becomes increasingly concave until the third region is reached, where the cheese fractures (Foegeding et al., 2003).

Force compression tests are commonly used for cheese rheology studies. In these tests, a cheese sample with a defined geometry is subjected to varying stress and strain over time between two parallel plates, with the top plate causing deformation of the sample. Depending on the rheological information required, a cheese rheologist may carry out large deformation or small deformation tests. In large deformation, non-linear deformation is achieved where structural bonds are broken and do not reform during the experiment; that is, the stress–strain curve passes through regions 1, 2 and possibly 3 (Figure 2.10). This type of deformation occurs in uniaxial compression tests (fracture tests) and texture profile analysis (TPA). In TPA, the cheese sample is subjected to a double compression cycle, with the sample being compressed by a certain percentage (e.g., 70%) of its initial height for each compression. A typical texture profile of cheese obtained from TPA is shown in Figure 2.11. TPA can be referred to as an imitative test as it is used to attempt to correlate with sensory analysis. Information obtained from TPA compression curves include fracturability, hardness, adhesiveness, cohesiveness, springiness, gumminess and chewiness (Tunick, 2000). Large deformation properties strongly depend on the size of the largest inhomogeneities or 'weak spots' in the cheese matrix (Luyten & Van Vliet, 1996), which may be curd junctions, cracks and eyes, depending on the cheese variety.

The most important small deformation tests are termed dynamic rheological tests, where cheese samples are placed between two parallel plates and subjected to small-amplitude oscillatory shear and a strain that must be within the linear viscoelastic region (Figure 2.10) for the cheese. In this type of experiment, there exists a type of equilibrium between bond breakage and reformation. Typical frequency and strain regimes applied to Cheddar cheese during dynamic small-amplitude oscillatory rheology tests (DSAOR) are $\leq$1 Hz and $\leq$0.5%, respectively (Gunasekaran & Ak, 2003; Lucey et al., 2005). Parameters derived from DSAOR include

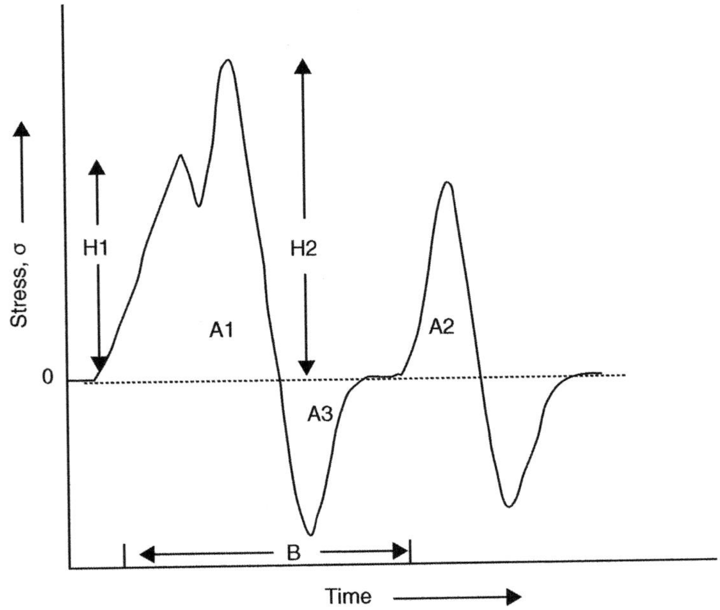

Figure 2.11 A typical texture profile of cheese obtained from texture profile analysis. Areas (A) and heights (H) of the curve used to calculate TPA parameters, for example, H2 = hardness (Fox et al., 2000).

the storage modulus (G′), the loss modulus (G′′), the complex modulus G* and the loss tangent (LT), also known as tan δ, which is the ratio of viscous to elastic moduli (G′′/G′) (Gunasekaran & Ak, 2003). The information obtained from DSAOR can be related to the molecular interactions occurring in the cheese (Lucey, Johnson & Horne, 2003). A common DSAOR test used in cheese rheology studies is the temperature sweep, where the temperature of the cheese sample is usually increased to >70°C at a fixed frequency and strain.

Cheese is thermorheomorphic, and its rheological behaviour changes considerably during heating. Typical changes in G′ and LT as a function of temperature during Cheddar cheese ripening are shown in Figure 2.12. Heating cheese from refrigeration temperatures to 40°C causes a decrease in the G′ of cheese, which is partially attributed to melting of milk fat, but above 40°C all of the fat in cheese is thought to be liquid (Walstra and Jenness, 1984), and further decrease in G′ > 40°C is the result of the interactions between caseins (Lucey, Johnson &

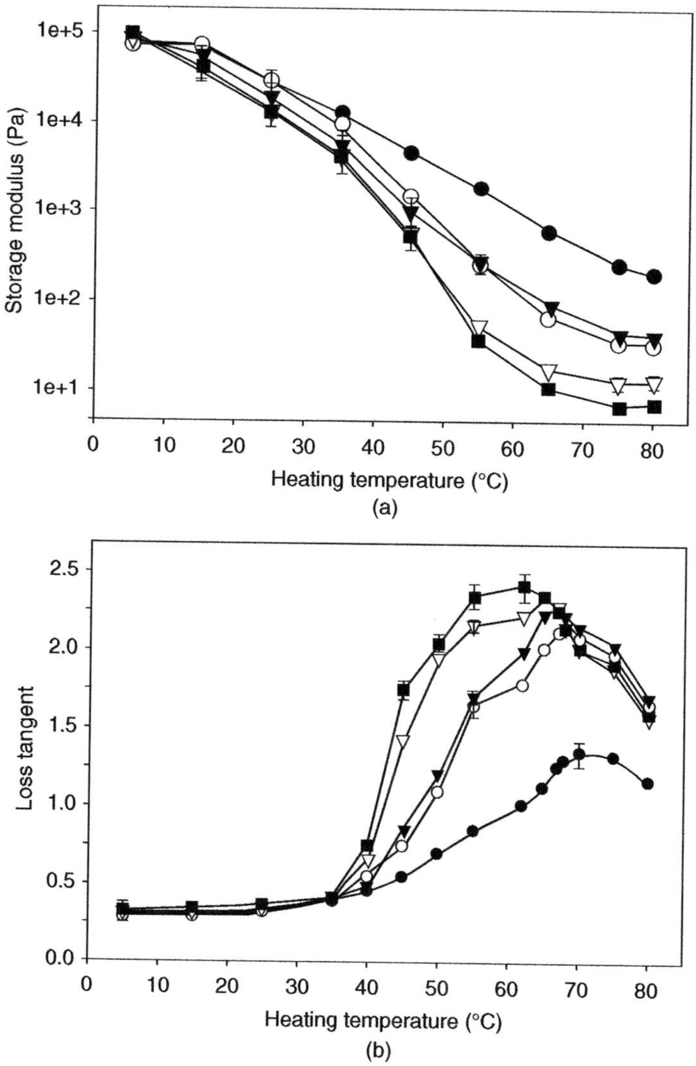

Figure 2.12 Changes in the storage modulus (a) and the loss tangent (b) as a function of temperature for Cheddar cheese ripened for 3 days (•), 1 month (o), 2 months (▾), 3 months (▿) and 9 months (▪), obtained from dynamic small amplitude oscillatory rheology (Lucey et al., 2005).

Horne, 2003). As the temperature increases, so do hydrophobic interactions between protein molecules (Bryant & McClements, 1998), which is thought to result in shrinking of casein particles and a decrease in contact area, causing a decrease in overall gel strength that leads to decreased G′ values in cheese (Lucey, Johnson & Horne, 2003). Other interesting parameters that can be derived from temperature sweeps include the temperature at LT = 1, which can be taken as the melting point of cheese (Gunasekaran & Ak, 2003), and the maximum loss tangent (LT_{max}), which indicates the point of highest bond mobility and can be used as an index of the maximum meltability of cheese (Lucey et al., 2005). As insoluble calcium is involved in the structural integrity of the *para*-casein matrix of cheese via CCP crosslinks, the level of insoluble calcium in cheese has a major influence on its rheological properties, primarily due to electrostatic interactions (Lucey, Johnson & Horne, 2003).

2.4.3 Influence of Calcium on Rheological Properties of Unmelted Cheese

Most cheese varieties do not exhibit flow at the refrigeration temperature to ~25°C, and so large deformation and fracture properties are appropriately measured in this temperature range. The machinability of cheese, that is, the ability to be cut, sliced or shredded by machines, is also better when the cheese displays moderate to high hardness in this lower-temperature range (Lucey, 2008). At low temperatures, hydrophobic interactions are weak, which allows casein molecules to exist in an expanded form. This increases the contact area between caseins, leading to more casein–casein interactions that account for the firmness observed (Lucey et al., 2005; Lucey, 2008). As calcium phosphate is more soluble at low temperatures (Lucey & Horne, 2009), solubilisation of CCP may increase electrostatic repulsion and also contribute to the increase in contact area in the lower-temperature range. Originally, it was thought that chymosin-mediated cleavage of the Phe_{23}–Phe_{24} bond of α_{S1}-casein was responsible for the softening observed during the early ripening stages of Cheddar cheese (Creamer & Olson, 1982). By inhibiting residual chymosin activity in Cheddar cheese, O'Mahony, Lucey and McSweeney (2005) still observed a significant softening of Cheddar cheese during early ripening and correlated this with the decrease in insoluble calcium content in cheese within the first month of ripening. Changes in Cheddar cheese texture beyond the first month of ripening are principally attributed the proteolytic degradation of the *para*-casein network (Lucey et al., 2005).

Reducoffing the total calcium content by manufacturing cheeses with lower pH values can lead to softer cheeses (Sheehan and Guinee, 2004); however, the low pH increases the rate of hydrolysis of α_{S1}-casein during ripening, reducing cheese firmness (Watkinson et al., 2001), so an indirect effect of lower insoluble calcium could be greater susceptibility to hydrolysis due to the reduced attractive forces between caseins (Fox, 1970). Watkinson et al. (2001) reported an increase in cheese firmness with increase in pH for directly acidified cheeses with initial pH values ranging from ~5.2 to 6.2, where the compositions of the cheeses were relatively similar. Presumably, as calcium phosphate complexes are less soluble at higher pH values, the increased firmness in this study was attributable to increased CCP crosslinking in the cheeses as the pH increased. Ong et al. (2013) observed a significant increase in the TPA hardness values of Cheddar cheeses made with 100 to 600 mg/L $CaCl_2$ added to the cheesemilk. Supplementing Cheddar cheese with $CaCl_2$ at the salting stage, Brickley, Lucey and McSweeney (2009) observed an increase in TPA hardness values in a calcium-supplemented cheese containing ~1160 mg Ca/100 g cheese compared to a control cheese containing ~828 mg Ca/100 g cheese during ripening. The authors attributed this increase in hardness to the increased CCP content in the calcium-supplemented cheese. Addition of TSC at the salting stage of cheesemaking can decrease the TPA hardness values of cheese (Brickley, Lucey & McSweeney, 2009; Mizuno &

Lucey, 2005), due to reduction of CCP crosslinks via calcium sequestration by TSC forming soluble calcium citrate complexes.

2.4.4 Influence of Calcium on Cheese Melt and High Temperature Cheese Rheology

When cheese is heated above 40°C, there is a dramatic reduction in its dynamic moduli (G′ and G′′) and an increase in LT, indicating the prevalence of a more viscous-like behaviour in the cheese. Moisture and liquid fat are the plasticizing agents during cheese melt (Muthukumarappan, Wang & Gunasekaran, 1999), facilitating the flow of casein particles past each other when heated. Hydrophobic interactions increase with temperature (Bryant & McClements, 1998), and it is proposed that the decreased elasticity of heated cheese is the result of casein particles contracting on themselves, causing individual casein particles to shrink, which reduces the contact area and in turn the intermolecular/particle bonding in the *para*-casein network (Lucey, Johnson & Horne, 2003). Hydrophobic interactions increase to a maximum strength up to 60°C–70°C and lose strength thereafter (Bryant & McClements, 1998). As electrostatic repulsion also increases with temperature (Bryant & McClements, 1998), Lucey, Johnson and Horne (2003) proposed that cheese melt occurs when electrostatic repulsion becomes the dominant casein interaction. Increased protein hydration is also thought to increase cheese melt (McMahon & Oberg, 1999; Sheehan & Guinee, 2004) as protein hydration increases repulsion between molecules (Bryant and McClements, 1998). The presence of CCP crosslinks in casein micelles greatly reduces the localized electrostatic repulsion (Horne, 1998), so the insoluble calcium content in cheese plays a key role in determining the melting behaviour of heated cheese (Choi et al., 2008). Halloumi has less fat and more moisture than varieties such as Cheddar. The high-temperature treatment of curd (90°C–95°C × 30 min) likely at least partially inactivates chymosin and substantially decreases microbial populations (Abd El-Salam & Alichanidis, 2004). This ensures very little proteolysis in early ripening. As it is eaten fresh, the *para*-casein network is sufficiently strong to inhibit melt of Halloumi at high temperatures (see Part II, Section 7.5).

The melting of cheese occurs in two stages: softening and flow. As cheese is heated, it always softens before it flows, but some varieties do not flow after softening (Lucey, Johnson & Horne, 2003). Softening of cheese refers to a loss of elasticity when cheese is heated. Flowability of cheese on heating may be defined as the displacement of contiguous molten planes of the *para*-casein matrix as mediated by heat-induced stress (Guinee, Auty & Fenelon, 2000). Flowability and meltability are interchangeable terms in the context of heated cheese. As mentioned previously, analysis of cheese using DSAOR temperature sweeps can provide invaluable information about cheese melting behaviour. In cheese rheology studies, it is useful to use more than one type of test to evaluate cheese melt. Numerous methods exist for measuring cheese meltability, such as the Schreiber test, where cheese cylinders are subjected to oven temperatures (e.g., 232°C for 5 min) (Altan, Turhan & Gunasekaran, 2005) or the 'squeeze-flow' approach using the UW Meltmeter, which carries out 'melt profile analysis' on the cheese, yielding information such as softening temperature and degree of flow (Lucey et al., 2005; Muthukumarappan, Wang & Gunasekaran, 1999). However, a lack of correlation between melting tests can occur (Park, Rosenau & Peleg, 1984). It is known that the type of heating system, the rate of heating and sample geometry greatly influence the results of melting analyses (Lucey, Johnson & Horne, 2003).

Cheeses with reduced calcium content may soften and melt at lower temperatures and exhibit higher meltability (Joshi, Muthukumarappan & Dave, 2003). Lucey et al. (2005) found that the increase in melt (as indicated by LT_{max}) during the first few weeks of Cheddar cheese ripening is more significantly correlated with the decrease in levels of insoluble calcium than with the extent of proteolysis. Solubilisation of CCP crosslinks during early ripening increases the electrostatic repulsion between casein particles, weakening the *para*-casein matrix sufficiently to increase meltability. Subsequent increases in cheese meltability after the first month of ripening can be attributed to proteolytic breakdown of the matrix (Lucey et al., 2005). When α_{S1}-casein is degraded, the remaining larger peptides no longer interact with other caseins, weakening the matrix and thus increasing melting during ripening (Joshi, Muthukumarappan & Dave, 2003). A resultant decrease in the temperature at which LT_{max} occurs during ripening suggests that less thermal energy is required to achieve maximum meltability as ripening proceeds (Lucey et al., 2005). Selecting cheese at the appropriate ripening age is essential for its use as a functional ingredient in foods and also for its use as an ingredient in the manufacture of processed cheese; in particular, the insoluble calcium content of natural cheese can have a major influence on the rheological properties of the processed cheese made from it (Guinee & O'Kennedy, 2009).

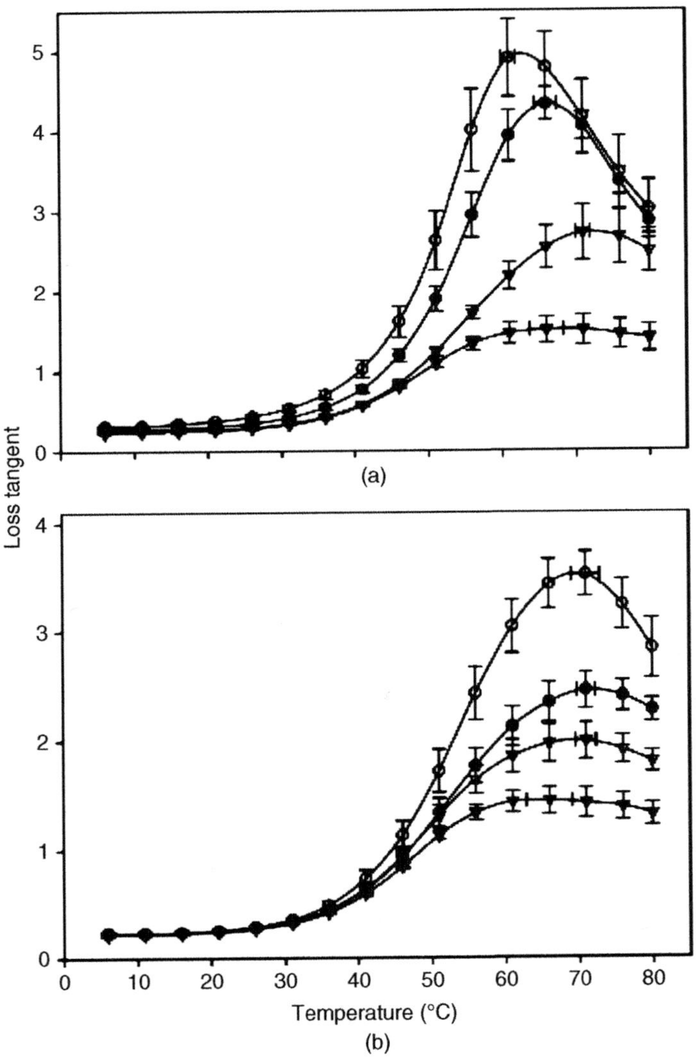

Figure 2.13 Loss tangent as a function of temperature for cheese (a) made from milk acidified to pH 6.0 (▾), 5.8 (▿), 5.6 (•), or 5.4 (o) and cheese (b) made from milk that had 0 (▾), 2 (▿), 4 (•) or 6 mM (o) EDTA added before cheesemaking (Choi et al., 2008).

As discussed in Section 2.3, manufacturing protocols can be altered to produce cheeses with lower pH values that result in lower levels of insoluble calcium. Reducing the insoluble calcium content of cheese can lead to increased LT values in cheese heated at pH values above 5.0 (Choi et al., 2008; Lee, Johnson & Lucey, 2005; Lee et al., 2010; Pastorino, Hansen & McMahon, 2003a). However, decreasing cheese pH below 4.9 inhibits the meltability of cheese even in cheeses with reduced levels of insoluble calcium (Lee, Johnson & Lucey, 2005; Lee et al., 2010) as electrostatic attraction becomes the dominant casein interaction due to the proximity of the pH to the isoelectric point of casein (pH 4.6) (Lucey, Johnson & Horne, 2003). Cheese varieties with pH values less than 4.9 such as cottage cheese or Feta exhibit very little flow as a result of the attractive interactions in the cheese matrix (Lucey, 2008). A decrease in cheese pH not only results in solubilisation of CCP but also in a decrease in the net charge of caseins. Choi et al. (2008) separated the effects of charge and insoluble calcium level on the rheological properties of cheese by producing cheeses with the same pH values but varying insoluble calcium levels by addition of EDTA to cheesemilk. It was found that cheeses with the same pH (~5.7) and composition exhibited increased LT values (Figure 2.13) and decreased G′ values at high temperatures as insoluble calcium level decreased, indicating a decrease in elastic-like properties as the CCP level reduces.

By immersing Cheddar cheese slices in SCCAP with varying calcium concentrations, O'Mahony, McSweeney and Lucey (2006) reported a significant decrease in LT (Figure 2.14) and increase in G′ values at 70°C with increasing cheese CCP concentration, and the opposite effect was observed with decreasing CCP concentration. The holding time of melted cheese at certain temperatures can also influence its flow properties. Kuo et al. (2001) reported that the extent of cheese flow decreased with increasing holding time of melted cheese at 60°C. The authors attributed this finding to increased aggregation in the cheese mass due to a greater number of hydrophobic interactions at this temperature, which is in the temperature range, where hydrophobic interactions are strongest (Bryant and McClements, 1998). It is evident that the types of bonding involved in cheese melting are very sensitive to temperature and pH.

Cheese stretch is an important functional property at high temperatures in Mozzarella, which is commonly used as a pizza topping. Stretchability is the ability of melted cheese to form fibrous strands that elongate without breaking under tension during ripening (Kindstedt, 1993). For cheese to stretch correctly, a low insoluble calcium content is required (Lucey, 2008), as stretch is impaired when casein–casein interactions are too strong (Lucey, Johnson & Horne, 2003).

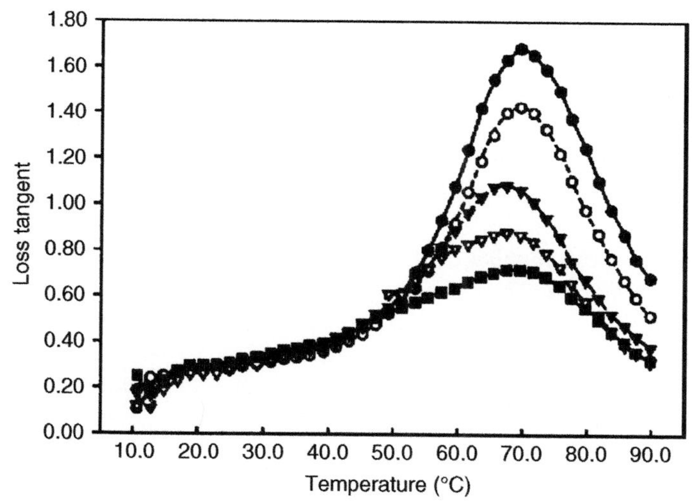

Figure 2.14 Loss tangent as a function of temperature from DSAOR for Cheddar cheese slices incubated in synthetic Cheddar cheese aqueous phase solutions containing 1.39 (•), 2.78 (o), 5.56 (▾), 6.95 (▿) or 8.34 (▪) g of calcium/L (O'Mahony, McSweeney & Lucey, 2006).

2.5 Conclusions

The concentration and form of calcium in cheesemilk and cheese is of critical importance to the many textural, rheological and functional properties of cheese. By manipulating the calcium equilibrium of cheese, the cheesemaker can modulate the rheological properties of cheese and enhance its functionality. Accelerating the development of cheese texture during ripening may also be useful as mature cheese is more expensive. A detailed understanding of how calcium influences all aspects of cheesemaking is essential for a food technologist developing cheese with tailored functional properties for ingredient food applications.

Further characterisation and understanding of the structure of CCP nanoclusters in the casein micelles of milk are of great importance for dairy chemistry. Most of the existing knowledge about calcium equilibrium in cheese simply encompasses the quantity of soluble and insoluble calcium, but there is a lack of information on the soluble forms of calcium, that is, calcium lactate, calcium phosphate, calcium citrate, and so on, and also the form of insoluble calcium in cheese. Studies on the form of CCP nanoclusters in cheese and changes in their structure during ripening are also of interest as they will provide more information about the mechanisms controlling rheological properties in ripened cheese, which is important for manipulating the functional properties of cheese.

The development of novel methods for altering the calcium equilibrium in cheese should be pursued in order to determine which methods are the most efficient in terms of cost and flexibility in the extent of alteration. While most of the studies on calcium equilibrium have focused on Cheddar and Mozzarella, it would be interesting if calcium equilibrium studies were carried out on a wider range of cheese varieties that are currently used for or have potential as food ingredients. A more precise understanding of the mechanisms involved in cheese melting, that is, the relationship between hydrophobic and electrostatic interactions, is required in order to enhance the accuracy and efficiency in modulation of functional properties.

References

Abd El-Salam, M. H. & Alichanidis, E. (2004). Cheese varieties ripened in brine. In Fox, P. F., McSweeney, P. L. H., Cogan, T. M. & Guinee, T. P. (eds.), *Cheese Chemistry, Physics and Microbiology: Volume 2. Major Cheese Groups*, 3rd edition. Elsevier Academic Press, London, UK, pp. 227–249.

Abraham, S., Cachon, R., Colas, B., Feron, G. & De Coninck, J. (2007). Eh and pH gradients in Camembert cheese during ripening: Measurements using microelectrodes and correlations with texture. *International Dairy Journal*, 17, 954–960.

Agarwal, S., Powers, J. R., Swanson, B. G., Chen, S. & Clark, S. (2006). Cheese pH, protein concentration, and formation of calcium lactate crystals. *Journal of Dairy Science*, 89, 4144–4155.

Altan, A., Turhan, M. & Gunasekaran, S. (2005). Short communication: Comparison of covered and uncovered Schreiber test for cheese meltability evaluation. *Journal of Dairy Science*, 88, 857–861.

Aoki, T., Umeda, T. & Kako, Y. (1992). The least number of phosphate groups for crosslinking of casein by colloidal calcium phosphate. *Journal of Dairy Science*, 75, 971–975.

Boutrou, R., Gaucheron, F., Piot, M., Michel, F., Maubois, J. L. & Leonil, J. (1999). Changes in the composition of juice expressed from Camembert cheese during ripening. *Lait*, 79, 503–513.

Brickley, C. A., Lucey, J. A. & McSweeney, P. L. H. (2009). Effect of the addition of trisodium citrate and calcium chloride during salting on the rheological and textural properties of Cheddar-style cheese during ripening. *International Journal of Dairy Technology*, 62, 527–534.

Bryant, C. M. & McClements, D. J. (1998). Molecular basis of protein functionality with special consideration of cold-set gels derived from heat-denatured whey. *Trends in Food Science and Technology*, 9, 143–151.

Choi, J., Horne, D. S. & Lucey, J. A. (2007). Effect of insoluble calcium concentration on rennet coagulation properties of milk. *Journal of Dairy Science*, 90, 2612–2623.

Choi, J., Horne, D. S., Johnson, M. E. & Lucey, J. A. (2008). Effects of the concentration of insoluble calcium phosphate associated with casein micelles on the functionality of directly acidified cheese. *Journal of Dairy Science*, 91, 513–522.

Choi, J., Horne, D. S. & Lucey, J. A. (2011). Determination of molecular weight of a purified fraction of colloidal calcium phosphate derived from the casein micelles of bovine milk. *Journal of Dairy Science*, 94, 3250–3261.

Chou, Y. E., Edwards, C. G., Luedecke, L. O., Bates, M. P. & Clark, S. (2003). Nonstarter lactic acid bacteria and aging temperature affect calcium lactate crystallization in Cheddar cheese. *Journal of Dairy Science*, 86, 2516–2524.

Creamer, L. K. & Olson, N. F. (1982). Rheological evaluation of maturing Cheddar cheese. *Journal of Food Science*, 47, 631–636.

Cross, K. J., Huq, N. L., Palamara, J. E., Perich, J. W. & Reynolds, E. C. (2005). Physicochemical characterization of casein phosphopeptide-amorphous calcium phosphate nanocomplexes. *Journal of Biological Chemistry*, 280, 15362–15369.

De Jong, P. (2008). Thermal processing of milk. In Britz, T. J.,& Robinson, R. K. (eds.), *Advanced Dairy Science and Technology*. Blackwell Publishing Ltd., Oxford, UK, pp. 1–34.

De Kruif, C. G. & Holt, C. (2003). Casein micelle structure, functions and interactions. In Fox, P. F. & McSweeney, P. L. H. (eds.), *Advanced Dairy Chemistry: Volume 1. Proteins*, 3rd edition. Kluwer Academic/Plenum Publishers New York, New York, USA, pp. 233–276.

Dickson, I. R. & Perkins, D. J. (1971). Studies on interactions between purified bovine caseins and alkaline-earth-metal ions. *Biochemical Journal*, 124, 235–240.

El-Bakry, M. & Sheehan, J. (2014). Analysing cheese microstructure: A review of recent developments. *Journal of Food Engineering*, 125, 84–96.

Foegeding, E. A., Brown, J., Drake, M. & Daubert, C. R. (2003). Sensory and mechanical aspects of cheese texture. *International Dairy Journal*, 13, 585–591.

Fox, P. F. (1970). Influence of aggregation on the susceptibility of casein to proteolysis. *Journal of Dairy Research*, 37, 173–180.

Fox, P. F. & McSweeney, P. L. H. (1998). *Dairy Chemistry and Biochemistry*. Blackie Academic & Professional, London, UK.

Fox, P. F., Guinee, T. P., Cogan, T. M. & McSweeney, P. L. H. (2000). *Fundamentals of Cheese Science*. Aspen Publishers Inc., Gaithersburg, Maryland, USA.

Gagnaire, V., Molle, D., Herrouin, M. & Leonil, J. (2001). Peptides identified during Emmental cheese ripening: Origin and proteolytic systems involved. *Journal of Agricultural and Food Chemistry*, 49, 4402–4413.

Gaucheron, F., Le Graet, Y., Boyaval, E. & Piot, M. (1997). Binding of cations to casein molecules: Importance of physicochemical conditions. *Milchwissenschaft*, 52, 322–327.

Gaucheron, F. (2005). The minerals of milk. *Reproduction Nutrition Development*, 45, 473–483.

Guinee, T. P., Auty, M. A. E. & Fenelon, M. A. (2000). The effect of fat content on the rheology, microstructure and heat-induced functional characteristics of Cheddar cheese. *International Dairy Journal*, 10, 277–288.

Guinee, T. P. & O'Kennedy, B. T. (2009). The effect of calcium content of Cheddar-style cheese on the biochemical and rheological properties of processed cheese. *Dairy Science and Technology*, 89, 317–333.

Gunasekaran, S. & Ak, M. M. (2003). *Cheese Rheology and Texture*. CRC Press LLC., Boca Raton, Florida, USA.

Hassan, A., Johnson, M. E. & Lucey, J. A. (2004). Changes in the proportions of soluble and insoluble calcium during the ripening of Cheddar cheese. *Journal of Dairy Science*, 87, 854–862.
Holt, C. (1985). The milk salts: Their secretion, concentrations and physical chemistry. In Fox, P. F. (ed.), *Developments in Dairy Chemistry: 3. Lactose and Minor Constituents*. Elsevier Applied Science Publishers, London, UK, pp. 143–181.
Holt, C. & Sawyer, L. (1988). Primary and predicted secondary structures of the caseins in relation to their biological functions. *Protein Engineering*, 2, 251–259.
Holt, C., Van Kemenade, M. J. J. M., Nelson, L. S., Sawyer, L., Harries, J. E., Bailey, R. T. & Hukins, D. W. L. (1989). Composition and structure of micellar calcium phosphate. *Journal of Dairy Research*, 56, 411–416.
Holt, C., Timmins, P. A., Errington, N. & Leaver, J. (1998). A core-shell model of calcium phosphate nanoclusters stabilized by β-casein phosphopeptides, derived from sedimentation equilibrium and small-angle X-ray and neutron-scattering measurements. *European Journal of Biochemistry*, 252, 73–78.
Holt, C. (2004). An equilibrium thermodynamic model of the sequestration of calcium phosphate by casein micelles and its application to the calculation of the partition of salts in milk. *European Biophysics Journal*, 33, 421–434.
Horne, D. S. (1998). Casein interactions: Casting light on the black boxes, the structure in dairy products. *International Dairy Journal*, 8, 171–177.
Horne, D. S. (2006). Casein micelle structure: Models and muddles. *Current Opinion in Colloid and Interface Science*, 11, 148–153.
Johnson, M. E., Riesterer, B. A. & Olson, N. F. (1990). Influence of nonstarter bacteria on calcium lactate crystallization on the surface of Cheddar cheese. *Journal of Dairy Science*, 73, 1145–1149.
Joshi, N. S., Muthukumarappan, K. & Dave, R. (2002). Role of soluble and colloidal calcium contents on functionality of salted and unsalted part-skim Mozzarella cheese. *Australian Journal of Dairy Technology*, 57, 203–210.
Joshi, N. S., Muthukumarappan, K. & Dave, R. I. (2003). Understanding the role of calcium in functionality of part skim Mozzarella cheese. *Journal of Dairy Science*, 86, 1918–1926.
Karahadian, C. & Lindsay, R. C. (1987). Integrated roles of lactate, ammonia, and calcium in texture development of mold surface-ripened cheese. *Journal of Dairy Science*, 70, 909–918.
Kindstedt, P. (1993). Effect of manufacturing factors, composition, and proteolysis on the functional characteristics of Mozzarella cheese. *Critical Reviews in Food Science and Nutrition*, 33, 167–187.
Kuo, M. I., Wang, Y. C., Gunasekaran, S. & Olson, N. F. (2001). Effect of heat treatments on the meltability of cheeses. *Journal of Dairy Science*, 84, 1937–1943.
Lawrence, R. C., Gilles, J., Creamer, L. K., Crow, V. L., Heap, H. A., Honore, C. G., Johnston, K. A. & Samal, P. K. (2004). Cheddar cheese and related dry-salted cheese varieties. In Fox, P. F., McSweeney, P. L. H., Cogan, T. M. & Guinee, T. P. (eds.), *Cheese Chemistry, Physics and Microbiology: Volume 2. Major Cheese Groups*, 3rd edition. Elsevier Academic Press, London, UK, pp. 71–102.
Le Graet, Y., Lepienne, A., Brule, G. & Ducruet, P. (1983). Migration du calcium et des phosphates inorganiques dans les fromages a pate molle de type Camembert au cours de l'affinage. *Lait*, 63, 317–332.
Lee, M. R., Johnson, M. E. & Lucey, J. A. (2005). Impact of modifications in acid development on the insoluble calcium content and rheological properties of Cheddar cheese. *Journal of Dairy Science*, 88, 3798–3809.
Lee, M. R., Johnson, M. E., Govindasamy-Lucey, S., Jaeggi, J. J. & Lucey, J. A. (2010). Insoluble calcium content and rheological properties of Colby cheese during ripening. *Journal of Dairy Science*, 93, 1844–1853.

Little, E. M. & Holt, C. (2004). An equilibrium thermodynamic model of the sequestration of calcium phosphate by casein phosphopeptides. *European Biophysics Journal*, 33, 435–447.

Lucey, J. A., Gorry, C. & Fox, P. F. (1993). Changes in the acid-base buffering curves during the ripening of Emmental cheese. *Milchwissenschaft*, 48, 183–186.

Lucey, J. A., Hauth, B., Gorry, C. & Fox, P. F. (1993). The acid-base buffering properties of milk. *Milchwissenschaft*, 48, 268–272.

Lucey, J. A. & Fox, P. F. (1993). Importance of calcium and phosphate in cheese manufacture – A review. *Journal of Dairy Science*, 76, 1714–1724.

Lucey, J. A., Johnson, M. E. & Horne, D. S. (2003). Invited review: Perspectives on the basis of the rheology and texture properties of cheese. *Journal of Dairy Science*, 86, 2725–2743.

Lucey, J. A., Mishra, R., Hassan, A. & Johnson, M. E. (2005). Rheological and calcium equilibrium changes during the ripening of Cheddar cheese. *International Dairy Journal*, 15, 645–653.

Lucey, J. A. (2008). Some perspectives on the use of cheese as a food ingredient. *Dairy Science and Technology*, 88, 573–594.

Lucey, J. A. & Horne, D. S. (2009). Milk salts: technological significance. In McSweeney, P. L. H. & Fox, P. F. (eds.), *Advanced Dairy Chemistry: Volume. 3. Lactose, Water, Salts and Minor Constituents*, 3rd edition. Springer, New York, New York, USA, pp. 351–389.

Luyten, H. & Van Vliet, T. (1996). Effect of maturation on large deformation and fracture properties of (semi-)hard cheeses. *Netherlands Milk and Dairy Journal*, 50, 295–307.

McGann, T. C. A., Buchheim, W., Kearney, R. D. & Richardson, T. (1983). Composition and ultrastructure of calcium phosphate-citrate complexes in bovine milk systems. *Biochimica et Biophysica Acta*, 760, 415–420.

McMahon, D. & Oberg, C. (1999). Deconstructing Mozzarella. *Dairy Industries International*, 64, 23–26.

McSweeney, P. L. H. & Fox, P. F. (2004). Metabolism of residual lactose and of lactate and citrate. In Fox, P. F., McSweeney, P. L. H., Cogan, T. M. & Guinee, T. P. (eds.), *Cheese Chemistry, Physics and Microbiology: Volume 1. General Aspects*, 3rd edition. Elsevier Academic Press, London, UK, pp. 361–371.

Mekmene, O., Le Graet, Y. & Gaucheron, F. (2009). A model for predicting salt equilibria in milk and mineral-enriched milks. *Food Chemistry*, 116, 233–239.

Mizuno, R. & Lucey, J. A. (2005). Effects of two types of emulsifying salts on the functionality of nonfat *pasta filata* cheese. *Journal of Dairy Science*, 88, 3411–3425.

Morris, H. A., Holt, C., Brooker, B. E., Banks, J. M. & Manson, W. (1988). Inorganic constituents of cheese: analysis of juice from a one-month-old Cheddar cheese and the use of light and electron microscopy to characterize the crystalline phases. *Journal of Dairy Research*, 55, 255–268.

Muthukumarappan, K., Wang, Y. C. & Gunasekaran, S. (1999). Estimating softening point of cheeses. *Journal of Dairy Science*, 82, 2280–2286.

O'Brien, N. M. & O'Connor, T. P. (2004). Nutritional aspects of cheese. In Fox, P. F., McSweeney, P. L. H., Cogan, T. M., & Guinee, T. P. (eds.), *Cheese Chemistry, Physics and Microbiology: Volume 1. General Aspects*, 3rd edition. Elsevier Academic Press, London, UK, pp. 573–581.

O'Mahony, J. A., Lucey, J. A. & McSweeney, P. L. H. (2005). Chymosin-mediated proteolysis, calcium solubilization, and texture development during the ripening of Cheddar cheese. *Journal of Dairy Science*, 88, 3101–3114.

O'Mahony, J. A., McSweeney, P. L. H. & Lucey, J. A. (2006). A model system for studying the effects of colloidal calcium phosphate concentration on the rheological properties of Cheddar cheese. *Journal of Dairy Science*, 89, 892–904.

Ong, L., Dagastine, R. R., Kentish, S. E. & Gras, S. L. (2013). The effect of calcium chloride addition on the microstructure and composition of Cheddar cheese. *International Dairy Journal*, 33, 135–141.

Park, J., Rosenau, J. R. & Peleg, M. (1984). Comparison of four procedures of cheese meltability evaluation. *Journal of Food Science*, 49, 1158–1170.

Pastorino, A. J., Hansen, C. L. & McMahon, D. J. (2003a). Effect of pH on the chemical composition and structure-function relationships of Cheddar cheese. *Journal of Dairy Science*, 86, 2751–2760.
Pastorino, J., Hansen, C. L. & McMahon, D. J. (2003b). Effect of sodium citrate on structure-function relationships of Cheddar cheese. *Journal of Dairy Science*, 86, 3113–3121.
Pastorino, J., Ricks, N. P., Hansen, C. L. & McMahon, D. J. (2003). Effect of calcium and water injection on structure-function relationships of cheese. *Journal of Dairy Science*, 86, 105–113.
Philippe, M., Gaucheron, F., Le Graet, Y., Michel, F. & Garem, A. (2003). Physicochemical characterization of calcium-supplemented skim milk. *Lait*, 83, 45–59.
Pierre, A., Michel, F. & Le Graet, Y. (1995). Variation in size of goat milk casein micelles related to casein genotype. *Lait*, 75, 489–502.
Sheehan, J. J. & Guinee, T. P. (2004). Effect of pH and calcium level on the biochemical, textural and functional properties of reduced-fat Mozzarella cheese. *International Dairy Journal*, 14, 161–172.
Sousa, M. J., Ardo, Y. & McSweeney, P. L. H. (2001). Advances in the study of proteolysis during cheese ripening. *International Dairy Journal*, 11, 327–345.
Tsioulpas, A., Lewis, M. J. & Grandison, A. (2007). Effect of minerals on casein micelle stability of cows' milk. *Journal of Dairy Research*, 74, 167–173.
Tunick, M. H., Cooke, P. H., Malin, E. L., Smith, P. W. & Holsinger, V. H. (1997). Reorganization of casein submicelles in Mozzarella cheese during storage. *International Dairy Journal*, 7, 149–155.
Tunick, M. H. (2000). Rheology of dairy foods that gel, stretch, and fracture. *Journal of Dairy Science*, 83, 1892–1898.
Tziboula-Clarke, A. (2002). Goat milk. In Fuquay, J. W. & Fox, P. F. (eds.), *Encyclopedia of Dairy Sciences*. Elsevier Academic Press, London, UK, pp. 1270–1279.
Udabage, P., McKinnon, I. R. & Augustin, M. (2000). Mineral and casein equilibria in milk: Effects of added salts and calcium-chelating agents. *Journal of Dairy Research*, 67, 361–370.
Udabage, P., McKinnon, I. R. & Augustin, M. A. (2001). Effects of mineral salts and calcium chelating agents on the gelation of renneted skim milk. *Journal of Dairy Science*, 84, 1569–1575.
Van Hooydonk, A. C. M., Hagedoorn, H. G. & Boerrigter, I. J. (1986). The effect of various cations on the renneting of milk. *Netherlands Milk and Dairy Journal*, 40, 369–390.
Walstra, P. & Jenness, R. (1984). *Dairy Chemistry and Physics*. John Wiley & Sons, New York, New York, USA.
Watkinson, P., Coker, C., Crawford, R., Dodds, C., Johnston, K., McKenna, A. & White, N. (2001). Effect of cheese pH and ripening time on model cheese textural properties and proteolysis. *International Dairy Journal*, 11, 455–464.
Zoon, P., Van Vliet, T. & Walstra, P. (1988). Rheological properties of rennet induced skim milk gels. 3. The effect of calcium and phosphate. *Netherlands Milk and Dairy Journal*, 42, 295–312.

3

Cheese Flavour Development and Sensory Characteristics

Kieran Kilcawley[1] *and Maurice O'Sullivan*[2]

[1] *Teagasc Food Research Centre, Moorepark, Fermoy, Co. Cork, Ireland*
[2] *School of Food and Nutritional Sciences, University College Cork, Cork, Ireland*

3.1 Introduction

Cheese is one of the most diverse food products with in excess of 500 different varieties commercially available. The extent of diversity encompasses a wide range of visual, physical and flavour attributes. The flavour of cheese is governed by three main biochemical pathways; glycolysis, lipolysis and proteolysis (Figure 3.1). In general terms, the extent of each of these processes is characteristic of the individual cheese variety. Cheese flavour comprises texture, taste and aroma (arguably also colour). In fact, it has long been postulated, by the 'component balance theory', that cheese flavour is the result of the correct balance and concentration of a wide variety of volatile flavour compounds (Mulder, 1952). An updated statement of this theory might be that the flavour of each cheese variety is the result of the correct balance and concentration of both non-volatile (taste) and volatile (aromatic) compounds influenced by cheese composition (Delahunty & Piggott, 1995; Overington et al., 2010; Subramanian, Harper & Rodriguez-Saona, 2009; Urbach, 1993).

It is well established that only low-molecular-weight water-soluble compounds are responsible for cheese taste (Aston & Creamer, 1986; Kubickova & Grosch, 1998; Rychlik, Warmke, & Grosch, 1997; Warmke, Belitz & Grosch, 1996). There are only five basic accepted taste sensations: 'sour', 'sweet', 'salt', 'bitter' and 'umami'; however, other sensations and interactions exist that increase the complexity of taste, such as acid, 'hot' and 'cooling', astringency and mouthcoating. Salt (NaCl) is of particular importance as it directly impacts taste and acts as a flavour enhancer, and influences the structure and rheological properties of cheese. The extent of the impact of salt on cheese flavour depends upon its concentration, cheese composition and the age of the cheese.

The biochemical reactions in cheese are primarily initiated by the addition of microbial populations and/or exogenous enzymes during production. Cheese is a dynamic product, with many varieties having up to 100 billion bacteria per gram, all of which metabolise carbohydrates, lipids and/or proteins to create a myriad of aromatic and sapid compounds that contribute to cheese flavour. These biochemical reactions are in turn influenced by the milk (type, quality and treatment), production equipment/processes, indigenous/exogenous microbial populations (selection and concentration), indigenous/exogenous enzymes (selection and concentration), salting (dry and brine) and production processes and ripening regimes (time, temperature and humidity), all of which help contribute to the wide variety of cheeses available.

Global Cheesemaking Technology: Cheese Quality and Characteristics, First Edition.
Edited by Photis Papademas and Thomas Bintsis.

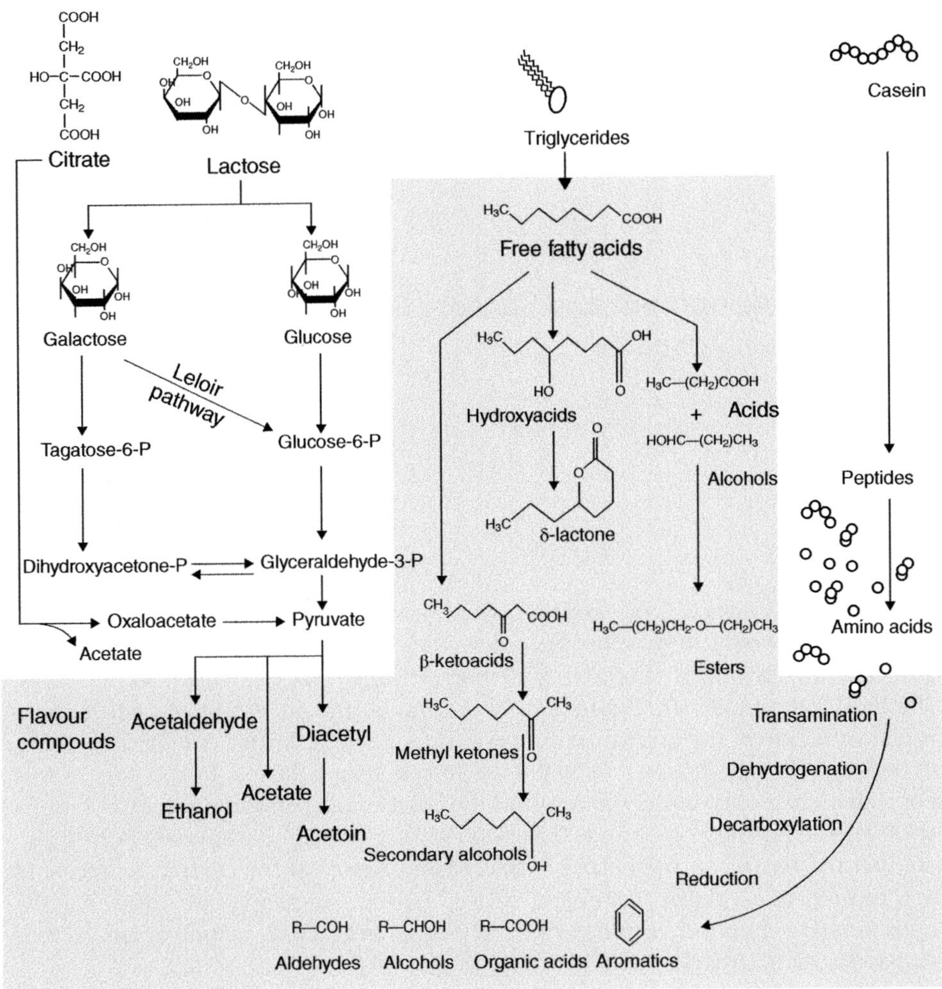

Figure 3.1 Biochemical pathways involved in the formation of flavour compounds in cheese. Published with permission from Marilley & Casey (2004).

A range of different approaches exist to gain information on the sensory character of cheese, which can be broadly segmented into four different areas: (1) expert grading, (2) difference methods, (3) descriptive methods and (4) affective methods. The choice of sensory approach depends upon factors such as the complexity of information required, application, cost and time.

3.2 Biochemical Pathways Involved in Cheese Flavour

3.2.1 Glycolysis

The metabolism of lactose to lactate is often the first step in cheese manufacture and is typically initiated by the addition of a selected starter culture(s) (defined or undefined). Lactic acid bacteria are Gram-positive, non-motile and non-spore forming, grow anaerobically and obligately ferment carbohydrates (Walstra, Wouters & Geurts, 2006). They require a carbon source for energy and obtain energy in the form of adenosine triphosphate (ATP). In rennet-coagulated cheeses, most of the lactose is actually lost in the whey as un-metabolised lactose or as lactic acid (lactate). Starter cultures metabolise lactose readily as it has a higher energy value than other substrates present and thus aids growth and survival in the milk (during production), and in the curd (post whey drainage) and in the early stages of cheese ripening.

Most lactose is utilised within two weeks of cheese curd formation (Fox et al., 1993; McSweeney, 2004), and some residual lactose remains in some varieties, depending upon the make procedure and also due to the fact that it is likely compartmentalised away from the starter bacteria. The rate of lactose metabolism during cheese production is influenced by the salt in moisture (S/M) content. In semi-hard cheeses, such as Cheddar, if the S/M level is too high (>6%), the growth of starter bacteria may be inhibited, resulting in lower levels of lactose metabolism and a higher pH (Lawrence & Gilles, 1987). This can encourage growth of advantageous non-starter lactic acid bacteria (NSLAB) early in production/ripening, which can result in uncontrolled flavour development. Also, low S/M% levels can result in rapid growth of starter bacteria and excessive proteolysis and also subsequent enhanced metabolism of amino acids, again leading to unbalanced or off-flavour development (Lawrence & Gilles, 1987). The lactose content is also controlled in the manufacture of many Dutch and Swiss cheeses, where whey is replaced with warm water (curd washing) during production to control the pH and moisture content and limit NSLAB growth. In these cheeses, lactose is utilised rapidly by the starter culture, which also helps control starter activity. In brine-salted cheeses, the net movement of sodium and chlorine ions from the brine into the curd results in an osmotic pressure difference between the moisture in the curd and the brine. The amount of salt retained and moisture expelled from the curd depends upon the brine concentration and brining time (Floury et al., 2010). Thus, the S/M percentage changes over this period and diffusion of salt into the curd continues after the brining process during ripening, which can impact starter bacteria and subsequent metabolic activity (Daly, McSweeney & Sheehan, 2010; Jeanson et al., 2011). In surface mould-ripened cheeses, the lactose is metabolised to L-lactate by mesophilic starter bacteria, and the subsequent growth of bacteria and yeasts on the surface metabolises the lactate to CO_2 and water (McSweeney & Sousa, 2000). In some brine-salted cheeses, clostridium species can metabolise lactate to CO_2 to H_2O and butyric acid, causing major defects, but this is less common in large-scale cheese production.

Homofermentation involves the glycolytic pathway. Strains such as *Lactococcus lactis* utilise the glycolytic pathway to generate L-lactate, while *Lactobacillus helveticus* or *Streptococcus thermophilus* generates both L- and DL-lactate. Heterofermenters such as *Leuconostoc* species and some *Lactobacillus* species convert lactose to glucose or galactose-6-phosphate using the Embden–Myerhof pathway. The glucose may be oxidised to pyruvate, which can be converted to the important cheese flavour compounds lactate, diacetyl, acetoin, acetaldehyde and ethanol (Cogan & Hill, 1993). Lactic acid has a slightly tart taste, but also more importantly influences the final pH of cheese, thus also influencing overall cheese flavour perception. Acetaldehyde has a characteristic yoghurt aroma, while diacetyl (2,3-butanedione) and acetoin (3-hydroxy-2-butanone) impart a very characteristic creamy buttery aroma, with acetic acid providing a sharp vinegar note (Curioni & Bosset, 2002; Smit G., Smit B.A. & Engles, 2005). Ethanol likely has a minimal direct contribution to cheese flavour as it is described as having a dry dust aroma (Curioni & Bosset, 2002), but it is much more important in relation to the formation of ethyl esters with free fatty acids, via esterification or alcoholysis as these compounds have a very high odour activity and are responsible for fruity flavours in many cheeses.

Lactic Acid Bacteria (LAB) also metabolise pyruvate by lactate dehydrogenase to lactate and in doing so produce NAD+. It can be converted to either L-lactate or D-lactate or both depending upon the types of lactate dehydrogenase present (Walstra, Wouters & Geurts, 2006). The metabolism of pyruvate can result in the formation of formic acid, acetic acid, ethanol and acetoin and acetaldehyde. Formic acid is very volatile and typically only found intermittently in cheese at low levels, and it has a similar odour to acetic acid. Acetaldehyde is generally produced by LAB that do not contain alcohol dehydrogenase activity (Singh, Drake & Cadwallader, 2003; Smit G., Smit B.A. & Engles, 2005) and can also be produced from amino acid metabolism.

Some LAB can metabolise citric acid, but citrate is only a minor constituent of milk, about 8 mmol L^{-1} (Wilkinson & Kilcawley, 2007). However, citrate metabolism can also result in the formation of acetic acid, diacetyl, acetoin, 2,3-butanediol and CO_2, but its importance in cheese is likely limited to specific short-ripened cheese types (Monnet et al., 1995) and some Dutch varieties (Sable & Cottenceau, 1999). Diacetyl, acetoin and 2,3-butanediol are commonly found in cheese, but their production is primarily due to pyruvate metabolism. 2,3-Butanediol has a characteristic fruity aroma, and CO_2 is very important in Dutch and Swiss cheese as it is responsible for eye (holes) formation (McSweeney, 2004). Diacetyl is an important compound in other cheese varieties, such as Quarg and Cottage cheese.

In cheeses such as Cheddar lactic acid can be racemised to D-lactate by NSLAB or oxidised to acetic acid. Mesophilic NSLAB can also metabolise residual lactose to lactate, ethanol and CO_2. Residual lactose fermentation is important in Swiss and Dutch cheese. In Dutch cheese, lactose is converted to L-lactate within 24 hours. In Swiss cheese, lactose is metabolised to L-lactate by *S. thermophilus*, but this strain is unable to metabolise galactose, which is converted by lactobacilli to L- and D-lactate. If galactose is not fully metabolised, it may act as a substrate for NSLAB growth, resulting in uncontrolled flavour development. In Swiss-type cheeses, the hot room stage promotes growth of propionibacteria that metabolise L-lactate to propionic acid, acetic acid and CO_2, which contribute to Swiss cheese flavour, texture and visual characteristics. Propionic acid directly contributes to the sweet flavour; however, in some Swiss cheeses, facultative heterofermentative lactobacilli are also added to retard propionic acid production and metabolise citrate to formic acid, acetic acid and CO_2 (Frohlich-Wyder, 2003). In Mozzarella cheese, galactose-negative cultures are often used as galactose is important in the Maillard reaction and thus helps provide the brown surface characteristics when used as a pizza topping cheese.

3.2.2 Lipolysis

Milk fat consists of triglycerides (~98%), the vast majority of which are even-numbered saturated fatty acids esterified on glycerol. A major characteristic of bovine milk fat is the presence of water-soluble short-chain fatty acids (SCFAs) with eight or fewer carbons (Jensen, Ferris & Lammi-Keefe, 1991; Jensen, Gander & Sampugna, 1962), which are volatile and highly odour active. Lipolysis is the hydrolysis of free fatty acids from tri-, di- and mono-acylglycerides and is carried out by two hydrolytic enzymes: esterases and lipases. Both enzymes catalyse the same reaction, the hydrolysis of the ester bond of a glyceride yielding a fatty acid and an alcohol (glycerol), but operate in different environments and have different specificities. The hydrolysis of SCFA from tri-, di- or mono-acylglycerides is catalysed by esterases, which act on soluble substrates (fatty acids with eight or fewer carbons) in an aqueous environment (Collins, McSweeney & Wilkinson, 2003). The hydrolysis of longer-chain water-insoluble fatty acids (fatty acids with more than eight carbons) is carried out by lipases, which act on emulsified fat in a non-polar environment and work on the interface between the aqueous and fat phases (Arpigny & Jaeger, 1999). The distribution of fatty acids in milk tri-acylglycerides is non-random (Angers et al., 1998). Table 3.1 provides an approximate distribution of fatty acids in bovine milk fat; however, the distribution and concentration are influenced by breed, stage of lactation and diet. Bovine milk fat is present as emulsified globules surrounded by a thin membrane called the milk fat globule membrane (MFGM). The MFGM consists of a complex mixture of proteins, phospholipids, glycoproteins, tri-acylglycerides, cholesterol and other

Table 3.1 Approximate fatty acid distribution and level in bovine milk fat.

Fatty Acid	Position%			Triacylglyceride %
	Sn-1	*Sn-2*	*Sn-3*	
$C_{4:0}$	0	0	35.4	11.8
$C_{6:0}$	0	0.9	12.9	4.6
$C_{8:0}$	1.4	0.7	3.6	1.9
$C_{10:0}$	1.9	3.0	6.2	3.7
$C_{12:0}$	4.9	6.2	0.6	3.9
$C_{14:0}$	9.7	17.5	6.4	11.2
$C_{15:0}$	2.0	2.9	1.4	2.1
$C_{16:0}$	34.0	32.3	5.4	23.9
$C_{16:1}$	2.6	2.8	3.6	2.6
$C_{17:0}$	0.8	1.3	1.0	0.8
$C_{18:0}$	7.0	10.3	9.5	7.0
$C_{18:1}$	24.0	30.0	18.9	24.0
$C_{18:2}$	2.5	1.7	3.5	2.5
Total (%)	100	100	100	100

Table modified from that shown in Christie and Clapperton (1982).

minor components and acts as a natural emulsifying agent, enabling the fat to remain dispersed in the aqueous phase of milk (Wilkinson, 2007).

The sources of lipolytic enzymes in cheese are relatively widespread, and they can originate from a number of different sources: (1) the indigenous milk lipase (lipoprotein lipase), (2) pregastric esterases/rennet pastes, (3) starter bacteria, (4) adjunct starter bacteria, (5) NSLAB, (6) yeasts and moulds and (7) the addition as exogenous lipases (Deeth & Fitz-Gerald, 1995; Fox & Wallace, 1997; McSweeney & Sousa, 2000). Lipoprotein lipase is inactivated at high-temperature, short-time (HTST) pasteurisation conditions (Hickey et al., 2007), but plays an important role in raw milk cheeses or cheeses produced with thermised milk. Pregastric esterases are produced from glands at the base of the tongue and are added to some Italian/Greek cheeses (McSweeney, 2004). In addition, some animal rennet pastes which are manufactured from calves' stomachs can also contain pregastric esterase activity as the enzyme is transferred into the stomach as the animal suckles. Starter bacteria contain significant amounts of esterase activity and influence the extent of lipolysis over ripening (Alewijn, 2006; Hickey et al., 2006; Lopez et al., 2006).

The extent of lipolysis varies considerably between cheese types. Figure 3.2 is a histogram of approximate fatty acid levels in a range of cheeses. However, the extent of lipolysis will also vary within each cheese variety, due to differences in manufacture, milk and ripening times. The aroma threshold of free fatty acids is influenced substantially by the pH and the composition of the cheese. There are considerable variations in polarity among the fatty acids, which is reflected by the taste threshold differences between aqueous and fat components of some cheese types. The pH has an influence on the taste, as the dissociated fatty acid anions at high pH are less flavour active and less volatile, and are perceived as soaps. Compared to Dutch-type and Cheddar cheeses, Camembert and blue-veined cheeses contain extremely high levels of free fatty acids. In Camembert and some blue-veined cheeses, the flavour perception is reduced, and fatty acids are not immediately associated with rancid off-flavours, because of their higher

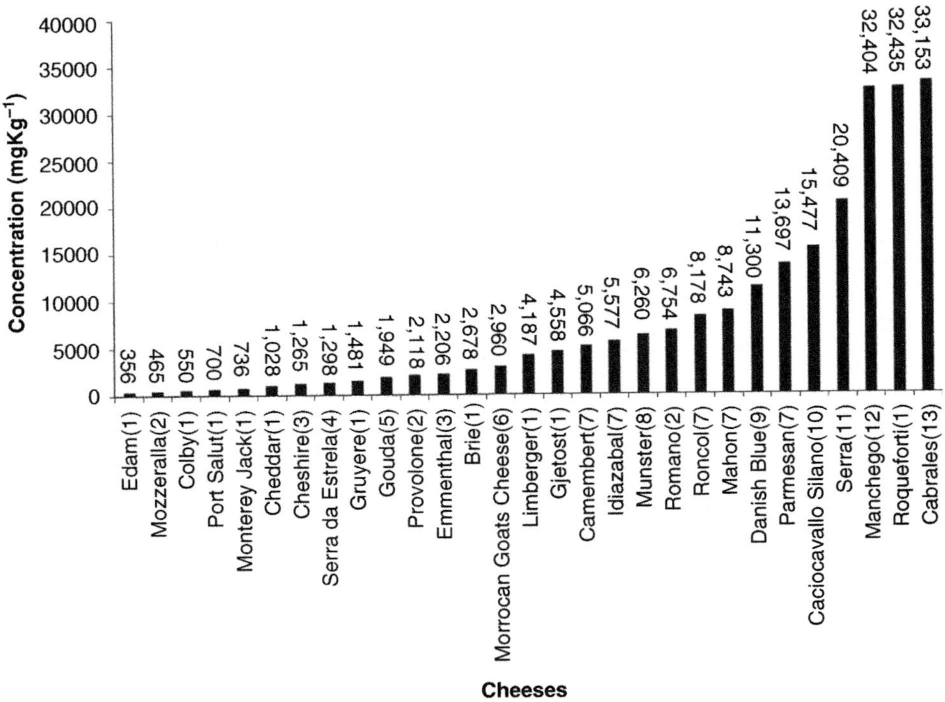

Figure 3.2 Examples of the extent of lipolysis in different cheeses. [1]Woo, Kollodges & Lindsay, 1984; [2]Woo & Lindsay, 1984; [3]McNeill & Connolly, 1989; [4]Partidario, 1999; [5]Iyer et al., 1967; [6]El Galiou et al., 2013; [7]De la Fuente et al., 1993; [8]De Leon-Gonzalez et al., 2000; [9]Alewijn, 2006; [10]Corsetti et al., 2001; [11]Macedo & Malcata, 1996; [12] Proveda, Pérez-Coello & Cabezas, 1999; [13]Alonso et al., 1987

pH, although it can be a problem if the pH is lower (Alewijn, 2006). The flavour thresholds of fatty acids also differ considerably, but are also dependent upon the cheese composition (matrix effect).

In terms of directly influencing cheese flavour, the SCFFA resulting from hydrolysis of dairy fat; butyric ($C_{4:0}$), caproic ($C_{6:0}$) and caprylic ($C_{8:0}$) are very important in many cheese varieties, partly because they are relatively easily hydrolysed due to their position on the glyceride backbone (mainly Sn-3) and their water solubility. These SCFFA are described as having a cheesey aroma (Qian & Burback, 2007), but are also described as rancid, sweaty and goaty (caproic and caprylic), which can be perceived as positive or negative depending upon the cheese variety. Levels of caproic and caprylic acids are higher in caprine milk, and caprylic is higher in ovine milk than in bovine milk (Markiewicz-Keszycka et al., 2013) and are thus important in goats and sheep milk cheeses. Some other acids such as formic ($C_{1:0}$), acetic ($C_{2:0}$) and propionic ($C_{3:0}$) acids are chemically similar to SCFFA, but arise from primarily carbohydrate metabolism rather than from fat hydrolysis. Valeric ($C_{5:0}$), iso-valeric (3-methyl butanoic acid) and iso-butyric (2-methyl propionic) are also prominent acids in cheese, but these are primarily present in cheese as products of amino acid metabolism.

The metabolism of individual free fatty acids is very important as methyl ketones, secondary alcohols, esters (methyl esters, ethyl esters and propyl esters) and lactones are all odour active and contribute to cheese flavour. Methyl ketones (2-ketones) are major contributors to blue-veined cheese flavour and are formed by β-oxidation of free fatty acids. The β-oxidation pathway produces methyl ketones with one carbon less than the parent free fatty acid plus CO_2. In blue-veined cheeses, the free fatty acids are intracellularly toxic to the mould, and therefore the mould produces enzymes that β-oxidise and decarboxylate these fatty acids to methyl ketones, to reduce toxicity and provide a source of energy. Both 2-heptanone and 2-nonanone are important methyl ketones in blue-veined and hard Italian cheeses and are described as having a blue cheese, fruity, sweet and fruity, musty, rose, tea-like aroma, respectively (Qian & Burbank, 2007). Other ketones such as acetophenone (also a product of amino acid metabolism) and 4-methylacetopheone are produced by the microflora of the surface smear of cheeses (Urbach, 1997). These ketones are described as having strong musty, floral notes reminiscent of orange blossoms (Qian & Burbank, 2007). The ethyl ketone 1-octen-3-one is thought to be an important component in Parmesan, Grana Padano and Pecorino cheeses and has a mushroom-like earthy aroma (Kubickova & Grosch, 1998; Qian & Burbank, 2007). Ketones are found in a wide variety of non-mould cheeses, and alternate production pathways or non-enzymatic mechanisms may exist. Forss (1979) identified that heating of milk can induce ketone formation, and Alewijn (2006) postulated that some methyl ketones in cheeses like Gouda may arise directly from esterified β-keto acids. Urbach (1997) also suggested that some ketones may be formed during analytical extraction processes used to concentrate volatiles from cheese directly from β-ketoacids during heating. Ketones such as 2-butanone and diacetyl are produced from carbohydrate metabolism; 2-butanone has a buttery, sour milk aroma (Curioni & Bosset, 2002) and is found in a wide variety of cheeses. Acetone is another commonly found ketone in cheese, has a wood pulp, hay aroma (Curioni & Bosset, 2002) and is derived from the β-oxidation of butyric acid (Walstra et al., 2005), but also from carbohydrate and amino acid metabolism.

Secondary alcohols in cheese are formed by the reduction of their corresponding methyl ketones (Engels & Visser, 1997). In blue-veined cheese, Penicillium species are directly responsible for the production of 2-pentanol, 2-heptanol and 2-nonanol from acetone, 2-heptanone and 2-nonanone, respectively (Collins, McSweeney & Wilkinson, 2003). The aroma attributes of these secondary alcohols have been described by Qian and Burbank (2007), Curioni and Bosset, (2002) and Kubickova and Grosch (1998) – 2-pentanol (green, fruity, fresh), 2-heptanol (fruity, earthy, green, sweet) and 2-nonanol (fatty, green) – but are thought not to be as important as their corresponding methyl ketones because of their low aroma activity (Singh, Drake & Cadwallader, 2003). 2-Butanol is produced from 2-butanone, has a fruity aroma (Qian & Burbank, 2007) and is present in many cheeses. Also, 1-octen-3-ol has a mushroom-like aroma (Qian & Burbank, 2007) and is derived from 1-octen-3-one. It may have a role in the flavour of Camembert, Parmesan, Grana Padano and Pecorino cheeses, but may also provide a metallic aroma at high concentrations (Moio & Addeo, 1998).

Liu, Holland and Crow (2004) highlighted the two main enzymatic mechanisms for the biosynthesis of esters, esterification and alcoholysis. Esterification is the formation of esters from alcohols and carboxylic acids, whereas alcoholysis is the production of esters from alcohols and acylglycerols or from alcohols and fatty acyl-CoAs derived from the metabolism of fatty acids, amino acids and/or carbohydrates. In alcoholysis, fatty acyl groups from acylglycerols and acyl-CoA derivatives are directly transferred to alcohols and are the major mechanism of ester biosynthesis by dairy lactic acid bacteria. Ethyl, methyl, propyl, butyl and isobutyl esters have all been reported in cheese. Ethyl esters tend to predominate due to the presence of ethanol from carbohydrate metabolism and also possibly from amino acid catabolism. As SCFFA are relatively abundant in cheese, ethyl esters of these acids are more common. Alewijn, Sliwinski and Wouters (2005) found correlations between ethyl esters and the presence of their associated SCFFA in Gouda cheese. Ethyl acetate, ethyl butyrate, ethyl hexanoate, ethyl octanoate and ethyl decanoate are found in most cheese varieties. In general they all have fruity aromas, but distinct aroma differences exist. Liu, Holland and Crow (2004) summarised the aroma of each of these esters as follows: ethyl acetate (solvent, fruity, pineapple), ethyl butanoate (apple, banana, sweet, fruity, fragrant), ethyl hexanoate (banana, pineapple, sweet, fruity, wine-like, brandy, powerful), ethyl octanoate (pear, sweet, fruity, banana, pineapple, apricot, wine, floral) and ethyl decanoate (apple, brandy, grape-like, fruity, oily). Ethyl acetate may also be formed from esterification of ethanol with acetyl coenzyme A (Collins, McSweeney & Wilkinson, 2003). Esters can also be formed by yeasts (Jollivet et al., 1994). Engels and Visser (1997) were able to directly associate fruity notes in Gruyere, Parmesan and Proosdij cheese with ethyl butanoate, as did Lawlor et al. (2003) with blue-type cheeses, but McSweeney and Sousa (2000) also highlighted that this could be a defect in Cheddar cheese. In general, ester formation is minimal in cheeses such as Gouda and Cheddar in comparison to Italian-type cheeses, where they are important characteristic aroma compounds. The likely route of ester formation in semi-hard cheeses is alcoholysis (Liu, Holland & Crow, 2004). In Swiss-type cheeses, ethyl esters of propionic acid, such as ethyl propionate, which has a pineapple, fruity, solvent-like aroma (Barron et al., 2005; Qian & Burbank, 2007), are present. In addition, other esters such as butyl propanoate and 1-methyl propyl propionate have also been identified in Swiss-type cheeses (Engels & Visser, 1997; Preininger & Grosch, 1994; Thierry, Salvat-Brunaud & Maubois, 1999). Esters, methyl ketones and fatty acids are the characteristic components of blue-veined cheese (Qian, Nelson & Bloomer, 2002). In general, limitations in ester formation tend to be the availability of an alcohol, especially in hard and semi-hard cheeses, rather than the presence of a fatty acid or esterase activity. Esters are also formed by carbohydrate and amino acid metabolism (Smit G., Smit, B.A. & Engles, 2005).

Aldehydes derived from autoxidation of free fatty acids also have a role in cheese flavour development, although due to their relatively low odour thresholds and due to the fact that they are easily reduced to alcohols or acids, their role in cheese flavour may not be as significant as other lipid metabolites. Many aldehydes confer green, grass-like aromas or malty nuances (Qian & Burbank, 2007). Most saturated and unsaturated aldehydes are derived from the oxidation of saturated and unsaturated free fatty acids, respectively, but this is not a prevalent mechanism in many cheeses due to a low redox potential. Other aromatic and branched-chain aldehydes result from the metabolism of amino acids and carbohydrates and tend to be more prevalent in cheese.

Lactones are cyclic compounds formed by the intramolecular esterification of hydroxyacids through the loss of water (McSweeney, 2004) and are widely found in milk (Urbach, 1997). Many studies on the volatile profiles of milk or cheese report significant levels of lactones, mainly because of their low volatility rather than their absence. Lactones in general are described as having a buttery-type character (Wilkinson, 2007). Lactones can be derived by heat in the presence of water and hydroxyacids, with γ- and δ-lactones derived from 4- and 5-hydroxy fatty acids, respectively (Alewijn, 2006) being more stable, and thus more common in cheeses (Collins, McSweeney & Wilkinson, 2003). This is one suggestion why lactones are at higher concentrations in cheeses such as Parmesan and Grana Padano as these have high cooking temperatures during production (Qian & Burbank, 2007). δ-Dodecalactone and δ-Tetradecalactone are known to be important constituents of blue-veined cheese (Jolly & Kosikowski, 1975) and are described as having a fresh, fruity, peach aroma (Curioni & Bosset, 2002: Qian & Burbank, 2007). Alewijn, Sliwinski and Wouters (2005) reported that lactones contribute to the flavour of Gouda cheese, although their relatively high flavour threshold may limit their overall contribution to cheese flavour (Qian & Burbank, 2007). Eriksen (1976) and Alewijn et al. (2007) have suggested that lactones may be formed by a one-step non-enzymatic reaction in which a hydroxyl fatty acid esterified in a triacylglycerol undergoes an intramolecular transesterification to release a lactone directly. It also reported that lactones may be derived directly from keto acids after they are reduced to hydroxyacids (Wong, Ellis & LaCroix, 1975). The formation of lactones in bovine milk appears to be influenced by diet, season, stage of lactation and breed (Fox et al., 2000).

3.2.3 Proteolysis

Proteolysis contributes to the softening of cheese texture during ripening due to the hydrolysis of caseins within the curd and through a decrease in water activity (McSweeney, 2004), and directly and indirectly to cheese flavour through the hydrolysis of caseins to small peptides and amino acids, and via the metabolism of amino acids by starter bacteria, secondary cultures, NSLAB, yeasts and moulds. Direct associations between proteolysis and the development of mouth-feel, texture, taste and aroma of maturing cheese have been reported (Pripp et al., 2006). Proteolysis is the most complex of the three biochemical pathways involved in cheese ripening, mainly due to the diversity of the potential enzymatic and chemical reactions involved. In simple terms, proteolysis is initiated primarily by the coagulant's action on casein and subsequently by the action of cell wall proteinases of lactic acid bacteria and intracellular enzymatic activity. The specificity of coagulants varies, depending upon their source and purity, although the main role is to cleave κ-casein to produce para-κ-casein and glycomacropeptide, which are essential for milk coagulation in the cheesemaking process (McSweeney, 2004). However, some also have activity towards other caesins, mainly α_{S1}-casein and β-casein. Milk also contains indigenous proteolytic enzymes, namely, plasmin and cathepsin D, that can also play an active role in primary proteolysis. The rate of activity of these indigenous proteinases and the coagulant is highly pH dependent. The large peptides generated during primary proteolysis have no direct effect on cheese flavour, but are the primary substrates for cell wall proteinase, aminopeptidase and intracellular metabolic activity that produce essential components for most cheese flavours. Cheese flavour intensity has been correlated with small peptides and free amino acids and with off-flavours, such as bitterness (Pripp et al., 2006). Bitterness in cheese is due to the accumulation of small peptides that have hydrophobic end sequences due to the presence of certain amino acids at the carboxy or amino terminal. The main bitter amino acids are isoleucine, phenylalanine, leucine, methionine, proline and valine, although lysine, tryptophan, tyrosine, histidine and arginine are also considered bitter (Kilcawley, 2017). The intensity of bitterness is directly related to the number of hydrophobic amino acids, the sequence of amino acids and the size of the peptide. It has also been suggested that bitterness is also influenced by calcium chloride, magnesium chloride and bitter amino acids (Salles et al., 2000; Toelstede & Hofmann, 2008a, 2008b). Some peptides have been associated with other potential flavour attributes. Toelstede and Hofmann (2009) and Toelstede, Dunkel and Hofmann (2009) identified γ-glutamyl (γ-G) dipeptides to be associated with umami flavour, and other γ-G peptides with mouthfulness and thickness termed 'kokumi' in Gouda cheese. They have recently identified more γ-G peptides in Parmesan cheese and highlighted their significance to Parmesan flavour (Hillmann & Hofmann, 2016; Hillmann et al., 2016), as well as identifying that key γ-glutamyltransferase enzymes involved in their production originate from raw milk rather than the LAB used in Parmesan cheese production. Roudot-Algaron et al. (1994) have also identified γ-G peptides in Comté cheese and associated them with salty, sour, brothy, metallic and sour flavours. Amino acids are also associated with sweetness (proline, lysine, alanine, glycine, serine and threonine), sourness (histidiine, aspargine and glutamic acid) and umami flavour (leucine, tyrosine, asparagine and glutamic acid) (McSweeney, 1997).

Many cheese aroma compounds are formed from the metabolism of amino acids (Figure 3.3), a major pathway in cheese flavour development. Amino acid transamination results in the formation of α-keto acids, which are key intermediates in the process (Ganesan et al., 2007). α-Ketoglutaric acid is the most widely studied α-keto acid acceptor for amino acid transamination and has been identified as a rate limiting factor in flavour development in different cheeses (Banks et al., 2001; Yvon, Berthelot & Gripon, 1998; Yvon & Rijnen, 2001). Ganesan et al. (2004) identified different keto acids, including pyruvate used as ketoacid receptors by lactococci or lactobacilli. However, as pyruvate is utilised by other enzymes in carbohydrate metabolism, it

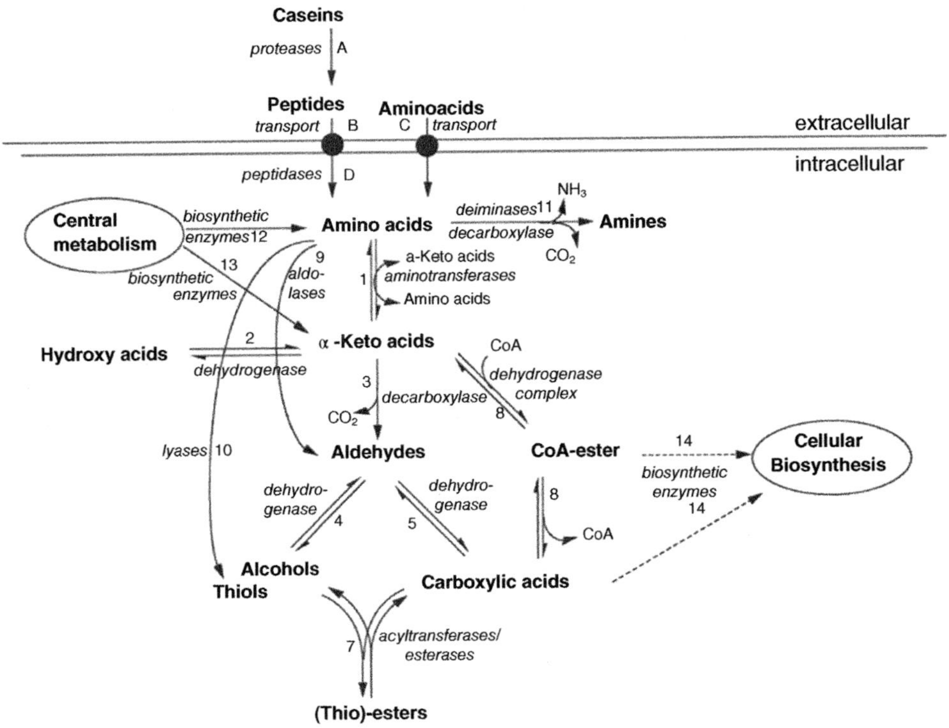

Figure 3.3 Overview of general protein conversion pathways for flavour formation in cheese. Adapted from Smit G., Smit B.A. & Engles (2005).

may not be as widely available as other keto acids. α-ketoglutaric acid is produced by the glutamate dehydrogenase pathway by oxidative deamination of glutamate (Helinck et al., 2004; Tanous et al., 2002), and this is likely the main mechanism of formation. However, Tanous et al. (2005) also demonstrated that some strains of *Lc. lactis* could produce α-ketoglutaric acid from citrate and glutamate by the action of citrate permease, citrate lyase and aspartate aminotransferase activity. Possibly the most important aroma compounds resulting from amino acid metabolism are aldehydes, alcohols, carboxylic acids, thiols and thio-esters. Table 3.2 lists some of the main metabolic components derived from amino acids in cheese. The associated aroma attributes of the compounds are provided (where available), and it is apparent that some components have negative or positive attributes. Obviously the concentration of each compound, odour activity and cheese composition (matrix effect) influence sensory perception. Thus, some compounds with apparent negative aroma attributes may actually play a positive role at low concentrations (such as sulphur-derived aromas), but others resulting from the metabolism of aromatic amino acids (skatole, indole, phenol, p-cresol) are generally less desirable in most cheese varieties.

Thioesters are formed by the reaction of fatty acids with free sulphydryl groups, often methanethiol (Molimard & Spinnler, 1996) and produce S-methylthioesters, and thus effectively arise from amino acid catabolism (Smit G., Smit B.A. & Engles, 2005). Some studies have highlighted the importance of specific amino acid metabolic components as characteristic flavour compounds for some cheese types. Smit G., Smit B.A. and Engles (2005) identified 3-methylbutanal, 3-methylbutanol, 2-methylpropanol, methanethiol, dimethylsulphide (DMS), and dimethyltrisulphide (DMTS) as key aroma compounds in Gouda cheese. Singh, Drake and Cadwallader (2003) identified 3-methylthiopropanal, 3-methylbutanoic acid, pentanoic acid,

Table 3.2 Important amino acids involved in amino acid metabolism in cheese and the potential aromatic compounds.

	Amino Acid	Ketoacid	Aldehyde	Acid	Alcohol	Ester	Ketone	Other
BCAA	Leucine	α-Ketoisocaproic acid	3-Methylbutanal (*malty, powerful, cheese*)	3-Methyl butanoic acid (*rancid, sweat, cheese, putrid*)	3-Methyl butanol (*fresh cheese, breathtaking, alcoholic*)	Ethyl-3-methylbutanoate (*fresh cheese*)		
			2-Methylpropanal (*banana, malty, chocolate-like*)	2-Methyl propanoic acid (*rancid, sweaty, sweet, apple-like*)	2-Methyl propanol (*penetrating, alcohol, wine-like*)	Ethyl isobutanoate (*unripe fruit, green*)		
	Valine	α-Ketoisovaleric acid	2-Methyl propanal (*banana, malty, chocolate-like*)	2-Methyl propanoic acid (*rancid, sweaty, sweet, apple-like*)	2-Methyl propanol (*penetrating, alcohol, wine-like*)	Ethyl isobutanoate (*unripe fruit, green*)		
	Isoleucine	α-Keto-3-methyl-pentanoic acid	2-Methylbutanal (*dark, chocolate, malt*)	2-Methyl butanoic acid (*fruity, waxy, sweaty- fatty acid*)	2-Methyl butanol (*wine*)			
ArAA	Phenylalanine	Phenyl pyruvate	Benzaldehyde (*Bitter almond oil, sweet cherry*)	Benzoic acid (*balasmic*)	Phenylmethanol (*floral, rose, phenolic, balsamic*)	Ethyl benzoate (*floral*)	Acetophenone (*almond, musty, glue*)	
			Phenylacetaldehyde (*rosy, violet-like*)	Phenylacetic acid (*Lily, jasmine, metallic, medicinal*)	Phenylethanol (*unclean, rose, violet-like, honey*)	Phenylethyl acetate (*floral, rose-like*)		
	Tryptophan	Indole-3-pyruvate	Indole-3-acetaldehyde	Indole-3-acetic acid	Tryptophol			Skatole (*faecal*) Indole (*putrid*) Tryptamine
	Tyrosine		p-OH-Phenyl aldehyde	p-OH-Phenyl lactate	OH-Phenylethanol			p-cresol (*medicinal*)
			OH-Benzaldehyde	p-OH-Phenyl acetate				Phenol (*medicinal*)
	Histidine							Histamine

(Continued)

Table 3.2 (Continued)

	Amino Acid	Ketoacid	Aldehyde	Acid	Alcohol	Ester	Ketone	Other
Sulphur	Methionine	α-Keto methylthio butyrate	Methional (*cooked potato, meat-like, sulphur*)	Methylthiobutyric acid (*chives*)	Methionol (*sweet, sulphurous, vegetable soup*)	Ethyl-3-methylthio propionate		Dimethyldisulphide (*cabbage*)
			Methylthio-acetaldehyde		Methanethiol (*rotting cabbage, cheese, vegetative, sulphur*)	S-Methyl thioacetate (*cooking cauliflower*)		Dimethyltrisulphide (*cabbage, garlic, sulphurous*)
			3-(Methylthio) propanal (*cooked/boiled potato*)	Methylthiopropionic acid (*cheesey*)	3-(Methylthio) propanol (*cooked/boiled potato*)	S-Methyl thioacetate (*cooking cauliflower*)		Dimethylsulphide (*cabbage, garlic, sulphur*)
		α-Keto-γ-methylthio butyrate				S-Methyl thiopropionate (*cheesey*)		
		α-Ketobutyric acid		Acetic acid (*vinegar, sour, acid*)				
				Propionic acid (*fruity, pungent, sweetish*)				
				Isobutyric acid (*cheese, rancid, butter*)				
				Butyric acid (*faecal, cheesy, sweet, sweat, rancid*)				
				Isovaleric acid (*Swiss cheese, cheese rind, sweetish, sweaty*)				
				Caproic acid (*bad breath, goaty*)				
		Indole-3-pyruvate	Indole-3-acetaldehyde	Indole-3-acetic acid				Skatole (*faecal*)

	Cysteine					S-Methyl thiopropionate (*cheesey*)		Hydrogen sulphide (*rotten eggs*)
						S-Methyl thiobutyrate (*cheesey*)		Carbonyl sulphide (*cooked cabbage-like*)
Asp	Aspartic acid	Pyruvate	Acetaldehyde	Propionic acid (*fruity, pungent, sweetish*)	2,3-Butanediol (*fruity*)		Acetone (*wood pulp, hay*)	
		Oxaloacetate		Acetic acid (*vinegar, sour, acid*)			2,3-butanedione (*buttery, nuts*)	
				Isobutyric acid (*cheese, rancid, butter*)			Acetoin (*buttery, acid milk*)	
				Caproic acid (*bad breath, goaty*)				

phenylacetic acid, 2-methylbutanal, 3-methylbutanal, ρ-cresol, skatole, 2-phenyl ethanol and DMTS as important contributors to Cheddar cheese flavour; with Smit G., Smit B.A. and Engles (2005) also identifying 3-methylbutanal, 3-methylbutanoic acid, methional, methanethiol, dimethyldisulphide (DMDS) and DMTS as key components of Cheddar flavour. Qian and Burbank (2007) highlighted that 2-methylpropanoic acid, 2-methylbutanoic acid, pentanoic acid, 2-methyl propanal, 3-methylbutanal, 2-methylbutanal, acetophenone, phenylacetaldehyde, 3-methylbutanol, methional, methanethiol, DMS, DMDS and DMTS contribute to Parmesan cheese flavour. Smit G., Smit B.A. and Engles (2005) identified 3-methylbutanoic acid, 3-methylbutanal, benzaldehyde, phenylacetaldehyde, methional, methanethiol and DMS in Camembert, and Roger, Degas and Gripon (1988) identified 2-phenylethyl acetate as an important ester in Camembert cheese. S-methylthioesters are very potent esters and are thought to be very important in mould- and surface mould-ripened cheeses and in some Swiss-type cheeses (Cuer et al., 1979; Lecanu et al., 2002; Sable & Cottenceau 1999). Smit G., Smit B.A. and Engles (2005) identified methional, 3-methylbutanal and skatole as important components in Swiss-type cheese. Proveda et al. (2008) identified 3-methylbutanoic acid, pentanoic acid, 2-phenylethanol and 3-methylthiopropanal in goat's cheese.

Some caution must be exercised with regard to relating volatile compounds to cheese flavour, mainly due to the fact that the combined odour effect of these compounds may differ considerably from that of individual compounds in pure form. Also, many studies using advanced gas chromatographic mass spectrometric techniques fail to incorporate odour activity (gas chromatography olfactometry) analysis or descriptive sensory analysis (nor take into account limitations or inherent bias of extraction/concentration techniques, column phases, detector sensitivity), which is important when working with complex fermented foods such as cheese. Most studies on the volatile profiles of cheese have an inherent bias built into the methodology that results in a failure to capture the complete or true volatile profile of the cheese. This bias generally results from the choice of extraction technique, choice of column phase, the sensitivity of mass spectrometer and data processing limitations, which are also related to cost and time factors. Thus, it is possible or even likely that the influence of some compounds may have been overestimated, while others have been overlooked.

3.3 Sensory Methods

3.3.1 Grading Methods

Grading schemes are defect-based judgements for cheese which have been developed over the years as a tool to determine the reliability and reproducibility of processes for quality control, awards, as a guide to optimum storage and for marketing strategy. For the latter, the strategy determines what cheeses are selected for specific markets (Muir, 2010). During grading, large numbers of cheeses can be rapidly scored for overall flavour and texture quality on the basis of an idealised concept of the perfect cheese or graded for a specific market. Graders are typically looking for negative attributes and often have a predetermined list of defects that have been developed for that cheese through experience, which can be either formal or informal (Delahunty & Drake, 2004; Delahunty & Murray, 1997; Drake, 2007; Kilcawley, 2017). A grader can assess over a hundred cheeses in a session and does not always provide a score, just an overall comment as an assessment of quality. For example, in the case of Cheddar cheese, the approach typically involves the grader making rapid visual assessments of cheese blocks, followed by further assessments, such as taking a sample cheese plug using a cheese trier. The grader assesses how cleanly the plug emerges, along with its appearance, colour and adhesiveness to the trier. A small sample of the plug is repeatedly manipulated between the index finger and thumb. The aroma of this warm cheese is inhaled, and finally the cheese may be rolled within the mouth for aroma and taste (Kilcawley, 2017).

Some countries have defined courses and accredited training for graders, while other countries source graders by simply assessing individuals for their ability to perceive key flavour and aroma attributes. In the latter case, training occurs through mentoring with an experienced grader or graders for a period of time (Kilcawley, 2017). For cheeses with long potential maturation times, the grader plays a vital role in monitoring ongoing quality development. Experienced factory graders can diagnose problems and provide feedback to those in production to aid improvement and consistency of manufacture for future reference (Muir, 2010). Some experienced graders also use additional compositional data to get a better overall assessment of quality, and also potentially gain a better understanding of predicted ripening characteristics (Kilcawley, 2017).

3.3.2 Difference Methods

In general, difference testing involves determining the difference between two (paired comparison), three (triangle) or four (tetrad) cheeses. These tests can be categorised into overall difference tests and attribute-specific directional difference tests. Difference tests are methods which determine if there is a detectable sensory difference between samples, whereas attribute difference testing determines whether there is a perceived specified attribute difference between samples. The most common difference tests are the 'duo-trio test', the 'triangle test', the 'simple same-difference test' and the 'A–not A' test (Lawless & Heymann, 1998a; Piggott, Simpson & Williams, 1998). Attribute difference tests also include the 'simple ranking test' and the 'alternative forced choice (AFC) test', which are more sensitive in the detection of sensory differences between samples. A difference test can become an AFC test when specific differences are asked of assessors. For example, in a triangle test three samples are given to the assessor, two are identical and the assessor is asked to pick the odd one out. This could become a 3 AFC test if, for example, the assessor was asked to pick out the sweetest sample. Similarly, a paired comparison test becomes a 2 AFC test when assessors are given a criterion to differentiate between samples, that is, sweetness, bitterness and so on. The panellist has to choose a sample in an AFC test even if they cannot differentiate between the samples. The tetrad test is more powerful than the triangle test, but the AFC tests are more powerful than both (Xia et al., 2015). The tetrad test is a difference test involving four samples where the assessor is presented with blind-coded samples, with two samples of one product and two samples of another product. The assessors must then group the products into two groups according to their similarity. Note that these instructions are different from asking the subjects to identify the two most similar samples (Ennis & Rousseau, 2012). The probability of guessing the right answer is similar to the triangle test (33%). The tetrad test has also attracted much interest due to its potential to provide increased power without specifying an attribute. This greater power means that for the same sample size, an existing difference is less likely to be missed (Ennis & Jesionka, 2011; Ennis & Rousseau, 2012). For preference testing where more than two samples are used, consumers can rank their preference; this is termed 'ranked preference testing', but a major limitation of this type of study is that no information on liking or disliking is captured (Drake, 2007).

3.3.3 Affective Sensory Testing

Affective sensory testing methods use hedonics (liking) to capture the emotive sensory response of naïve assessors, who can be either consumers or assessors analogous to the consumer. Products are scored on preference or on their 'liking' of attributes such as appearance, flavour or texture and ultimately their overall impression of a product or 'overall acceptability'. The assessors should also be regular consumers of the cheese (Delahunty & Drake, 2004; Delahunty & Murray, 1997).

Ideally the numbers of assessors required for affective testing, with the exception of focus groups, is much greater than with descriptive tests as the sample size must be representative of a larger consumer population. It is a sensory science convention that hedonic tests and more analytical or descriptive tests should not be undertaken with the same respondents. This is certainly true of trained sensory panels for descriptive profiling, where hedonic elements should never be included. The rationale is that as the panellists are trained to respond to the defined sensory attributes, any hedonic response they may have can be biased and therefore unreliable.

Affective sensory tests can be qualitative (focus groups) or quantitative (preference, sensory acceptance tests, consumer tests). Qualitative and quantitative affective types of tests usually are at opposite ends of the research and development spectrum. Moderated focus groups generally use relatively small numbers of suitable screened individuals (typically 8–12) and are good for testing product and packaging concepts; they are also not very expensive. Understanding consumer perception of cheese flavour is crucial for effective marketing and product development. However, as the numbers of participants are low, results need to be interpreted with caution (Drake, 2007).

Sensory acceptance testing can be employed during the development and optimisation processes as a means of assessing variant suitability in a hedonic fashion. It involves anywhere from 25 to 75 individuals and should be performed in duplicate (Stone, Bleibaum & Thomas, 2012ab; Stone & Sidel, 2004). Consumer tests are generally undertaken after the completion of product development as a means of final validation before product launch and involve a large number of consumers (>100). Overall, the greater the number of assessors used for affective analysis, the greater the statistical reliability of the data obtained.

3.3.4 Descriptive Sensory Profiling

Descriptive methods involve the training of panellists to quantitatively determine the sensory attributes in a cheese or more usually a selection of cheeses. Descriptive analysis is the most powerful sensory tool in cheese flavour research. It can be used to differentiate cheeses on the basis of a full complement of sensory characteristics/attributes/lexicons and to obtain a quantitative description of all the sensory aspects that can be identified (Singh, Drake & Cadwallader, 2003). Assessors are trained to measure the attributes associated with the relevant sensory modalities of 'appearance', 'aroma', 'flavour', 'texture', 'taste' and 'aftertaste'. The language is descriptive and non-hedonic, in that assessors are not asked how much they rate or like the cheese. The different methods for descriptive profiling include the 'flavour profile', 'texture profile', 'free choice profiling', 'spectrum descriptive analysis' and 'quantitative descriptive analysis (QDA). Some of these are methods are widely used for product development and for research purposes, while others appear to be of less practical use.

Free choice profiling (FCP) involves panellists developing their own descriptive terms (Delahunty et al., 1997; Williams & Arnold, 1985). The problem with this method is the subjective correlation of terms derived by different assessors may not, in reality be related, and for this reason it has not been widely adopted. However, descriptive analysis is a method in which defined sensory terms are quantified by sensory panellists. Detailed descriptions of sensory terminology and procedural guidelines for the identification and selection of descriptors for establishing a sensory profile by a multidimensional approach have been described in ISO (1992) and ISO (1994). A list of descriptive terms, determined initially and referred to as a lexicon or descriptive vocabulary, describe the specific sensory characteristics of the cheese(s) and can be used to evaluate the changes. The two most commonly used methods are the QDA and the spectrum method. The spectrum method was developed in the 1970s (Civille & Szczesniak, 1973) and is a descriptive profiling method which prescribes the use of a strict technical sensory vocabulary using reference materials. This method is pragmatic in that it provides the tools to design a descriptive procedure for a given product category. Its principal characteristic is that the panellist scores the perceived intensities with reference to pre-learned 'absolute' intensity scales. The purpose is to make the resulting profiles universally understandable. The method provides an array of standard attribute names ('lexicons'), each with a set of standards that define a scale of intensity (Meilgaard, Civille & Carr, 1999; Muñoz & Civille, 1992). These descriptive terms have been developed and employed for the sensory evaluation of cheese, for example, Van Hekken et al. (2006) for Mexican cheese. With the spectrum method, the scales are anchored using extensive reference points which may include a range of foods corresponding to food reference samples, which apparently reduces panel variability. Panellists develop their list of attributes by evaluating a large array of products within the category. Products may be described in terms of only one attribute (e.g. 'appearance' or 'aroma') or, they may be trained to evaluate all attributes (Murray, Delahunty & Baxter, 2001). One of the drawbacks is that extensive training (of panellists) is required when using the spectrum method. Other potential drawbacks include cultural differences of panels and the difficulty of quantifying an attribute over a range of different products (Murray, Delahunty & Baxter, 2001).

In the QDA method, experts with cheese knowledge can evaluate cheeses and suggest descriptive terms that specifically describe the cheese and the sensory dimension to be examined to produce an initial 'meta' sensory list. This vocabulary can be based on terms suggested by the panellists themselves in discussions under supervision of the panel leader. Sensory lexicons can also be provided which consist of lists of sensory terms describing 'appearance', 'aroma', 'flavour', 'texture', 'taste' and 'aftertaste' attributes. These attributes/lexicons are available for various cheese products: Cheddar cheese (Drake et al., 2005) and French cheese (Rétiveau, Chambers & Esteve, 2005). The QDA method was first proposed by Stone et al. (1974) and relies heavily on statistical analysis to determine the appropriate terms, procedures and panellists to be used for the analysis of a specific product. Each descriptive test has three stages: (1) selecting a panel, (2) establishing the language or vocabulary to describe the cheese and (3) quantifying the sensory results (Delahunty & Drake, 2004). A panel of typically 8–12 individuals is used to assess products in the controlled environment. Upwards of 50 or more individuals may need to be screened before a trained group is achieved, a trained group consisting of individuals who are able to discern each required attribute. They should have normal colour vision and be screened for ageusia, which is the inability to taste, and anosmia, the inability to detect odours (ISO, 1991, 1992). Screening tests could include difference testing, sensory threshold testing or ranking tests which are described in detail in ISO, 1993. Assessors are trained to identify and quantify a wide range of specific sensory attributes. Such attributes relate to flavour, aroma and texture and in some cases appearance. Initially, this statistical analysis is used in the sensory term reduction process during training, and this training of the QDA panel requires the use of product references to stimulate the generation of sensory terms. These references help the panellist define and quantify the attribute they are assessing and greatly assist in the training process. Sample presentation order must be unbiased with samples presented in a randomised order to prevent first-order and carry-over effects (MacFie et al., 1989). The panel leader acts as moderator and facilitator, without directly influencing the group. Panellists cannot discuss data, terminology or samples after each taste session, but must rely on the discretion of the panel leader for any information on their performance. Feedback is provided by the facilitator on the basis of a statistical analysis of the taste session data (Lawless & Heymann, 1998b; Meilgaard, Civille & Carr, 1999). Training continues to reduce the number of sensory lexicons, finally followed by sensory profiling. Panels act as a highly trained instrument and require regular 'calibration' or training to ensure accuracy. These panels are costly to operate mainly because panels operate best when regularly active.

Descriptive sensory methods can be expensive and time consuming because of the necessity to train and profile individual panellists over extended periods of time (O'Sullivan, Kerry & Byrne, 2011). On the opposite end of the sensory spectrum, affective methods are restricted to hedonic assessments, including acceptability or preference, but do not describe the product. Traditionally, this issue was solved by combining descriptive data and hedonic data using data statistical tools involving predominantly chemometrics in a method called 'preference mapping'. In addition the 'intensive' use of consumers for sensory tests is not accepted by everybody in the sensory community (Worch et al., 2014). Consumers can only tell you what they 'like' or 'dislike' (Lawless & Heymann, 1998b). Some sensory scientists consider that using consumers for sensory descriptive tasks is not appropriate as consumers lack consensus and repeatability, or comprehension of the meaning of the sensory attributes (Lawless & Heymann, 1998b; Stone & Sidel, 2004). In contrast, other sensory scientists have shown through different studies that consumers can describe the sensory characteristics of products with a precision comparable to experts (Worch et al., 2014). However, only 'simple' sensory attributes/terms can be used (cannot use technical or chemical terms) (Worch, Lê & Punter, 2010; Worch et al., 2014), and larger numbers of consumers are required to make up for a lack of appropriate training. Rapid sensory methods have been designed to provide more cost-effective solutions to these problems and to close the divide between the rigid rules of classic descriptive profiling and the emotional responses involved with affective sensory methods.

3.3.5 Rapid Sensory Methods

Rapid sensory evaluation methods can provide quick results with respect to the end user, and it requires less resources. These methodologies are also more flexible and can be used with semi-trained assessors or consumers, providing sensory maps similar to that operated in classic descriptive analysis where highly trained panels are used (Varela & Ares, 2012). A wide number of different rapid techniques exist, but perhaps flash profiling and ranking descriptive analysis (RDA) have the most potential for cheese. With flash profiling, assessors develop their own attributes to describe the products, with their own vocabulary, limited only by their sensory skills, and then quantify personal attributes using line scales; the method is based on the assumption that panellists do not differ in their perceptions, but solely in their ability to describe them (Murray, Delahunty & Baxter, 2001; Richter et al., 2010). Assessors are introduced to the samples and after a short instructional presentation are told to generate their own vocabulary free of choice on the basis of their own sensory perception, but to attempt to cover the sensory variations in the samples. After generating relevant attributes, they are allowed to see other assessors' vocabularies and to add or substitute attributes in their own list as they so wish. For each attribute, samples are ranked according to their intensity on an ordinal scale anchored from 'lower' to 'higher'. Unlike the ranking test proposed by Rodrigue et al. (2000), where ties (same score) are not allowed, flash profiling allows the assessors to apply the same rank to two or more samples if no difference is perceived (Dehlholm et al., 2012). This method offers a compromise over conventional descriptive methods and is thus a very rapid sensory profiling technique. Flash profiling takes into account the diversity of the point of view of the assessors. The combination of each assessor description reflects different points of view depending on the importance given by assessor to each sensory modality. This combination enriches the description (Dairou & Sieffermann, 2002). All the samples are presented to the assessors at the same time, without the requirement of a familiarisation phase. From the start, assessors can discriminate any relevant attributes using very intuitive ordinal scales. Disadvantages of flash profiling include the fact that the number of samples that can be profiled is limited, and large sample sets may confuse the assessors. However, Tarea, Cuvelier and Siefffermann (2007) demonstrated that up to 49 samples could be assessed in a flash profile in one session (2–5 h), but a significant contributing factor to this was the fact that the assessors were highly motivated, experienced and also trained and could take breaks during the profiling session. Other potential drawbacks are that the individual panel lexicons may be quite varied and are thus open to semantic interpretation (Dairou & Sieffermann, 2002), but the core attributes should have some level of consensus. Flash profiling has also been compared with classical descriptive methods (Dairou & Sieffermann 2002; Delarue & Sieffermann, 2004; Loescher et al., 2001).

RDA is a modification of flash profiling (Richter et al., 2010). Unlike flash profiling, the sensory lexicon is not developed in a free-choice-type manner with the same attributes for both RDA and the traditional QDA method, so there was no issue with differences in semantic consensus as described for flash profiling. RDA uses a greater number of assessors, and these untrained assessors come to a consensus on the quantity of the sample to be evaluated as well as the procedure. Samples are then ranked using ordinal scales for each of the defined (qualitative) lexicons, whereas interval scales were used in the traditional descriptive techniques. Samples are presented simultaneously and ranked for appearance and aroma attributes followed by another session for texture and flavour attributes. This method allows for the discrimination of samples with an efficiency similar to that displayed by the descriptive methods of the QDA and FCP. Richter et al. (2010) also suggested that when insufficient time is available to train a panel, the use of an untrained panel and a ranking test should be considered. Richter et al. (2010) observed that it was important to train a panel in order to obtain good descriptor conceptualisation and greater panel consensus with RDA and suggested that a more intense qualitative training for RDA would potentially allow more consistent results, primarily for the complex attributes of texture.

3.4 Data Analysis, Chemometrics and Preference Mapping

Data analysis is the key to unlocking the valuable data gathered from sensory, flavour chemical analysis and from associated compositional, microbial, rheological or other methods. Preference analyses techniques enable us to relate external information about perceived product characteristics to consumer preference ratings in order to understand what attributes of a product are driving preferences (Meilgaard, Civille & Carr, 2007; Van Kleef, van Trijp & Luning, 2006). Modern multivariate data analytical software readily facilitates the analysis of different data streams, for example, the subjective hedonic determinations and the more objective measurements such as sensory profiling data (trained descriptive panel data) and physicochemical analysis (rheological, compositional, flavour chemistry data). Principal component analysis (PCA) enables a multidimensional matrix to be simplified and described graphically.

3.5 Conclusion

Considerable progress has been made towards fully elucidating the metabolic (enzymatic and biochemical) reactions that contribute to cheese flavour in lactic acid bacteria, yeasts and moulds. Advances in biochemical screening, analytical methodology and sensory techniques continue to enhance our understanding of the complexity of flavour development within cheese varieties. The use of chemometric approaches to ascertain associations between different data streams (volatile and non-volatile data, sensory data and compositional parameters) will continue to provide researchers and industry with tools to make better, more consistent cheese.

References

Alewijn, M. (2006). *The Formation of Fat-Derived Flavour Compounds During Ripening of Gouda-Type Cheese.* PhD thesis, Wageningen University, the Netherlands.

Alewijn, M., Sliwinski, E. L. & Wouters, J. T. M. (2005). Production of fat-derived (flavour) compounds during the ripening of Gouda cheese. *International Dairy Journal*, 15, 733–740.

Alewijn, M., Smit, B. A., Sliwinski, E. L. & Wouters, J. T. M. (2007). Formation mechanism of lactones in Gouda cheese. *International Dairy Journal*, 17, 59–66.

Alonso, L., Juárez, M., Ramos, M. & Martín-Alvarez, P. J. (1987). Effects of changes during ripening and frozen storage on the physiochemical and sensory characteristics of Cabrales cheese. *International Journal of Food Science and Technology*, 22, 525–534.

Angers, P., Tousignant, E., Boudreau, A. & Arul, J. (1998). Regiospecific analysis of fractions of bovine milk fat triacylglycerols with the same partition number. *Lipids*, 33 (12), 1195–1201.

Arpigny, J. L. & Jaeger, K. E. (1999). Bacterial lipolytic enzymes, classification and properties. *Biochemistry Journal*, Oct. 1, 343 (Pt. 1), 177–183.

Aston, J. W. & Creamer, L. K. (1986). Contribution of the components of the water-soluble fraction to the flavour of Cheddar cheese. *New Zealand Journal of Dairy Science and Technology*, 21, 229–248.

Banks, J. M., Yvon, M., Gripon, J. C., Fuente, M. A. D. L., Brechany, E. Y., Williams, A. G. & Muir, D. D (2001). Enhancement of amino acid catbolims in Cheddar cheese using α-ketoglutarate: Amino acid degradation in relation to volatile compounds and other aroma character. *International Dairy Journal*, 11, 235–243.

Barron, L. J. R., Redondo, Y., Aramburu, M., Pérez-Elortondo, F. J., Aibisu, M., Nájera, A. I. & De Renobales, M. (2005). Variations in volatile compounds and flavour in Idiazabal cheese manufactured from ewe's milk in farmhouse and factory. *Journal of the Science of Food and Agriculture*, 85, 1660–1671.

Christie, W. W. & Clapperton, J. L. (1982). Structures of the triglycerides of cow's milk, fortified milks (including infant formulae), and human milk. *Journal of the Society of Dairy Technology*, 35, 22–24.

Civille, G. V. & Szczesniak, A. S. (1973). Guidelines to training a texture profile panel. *Journal of Texture Studies*, 4, 204–223.

Cogan, T. M. & Hill, C. (1993). Cheese starter cultures. In Fox, P. F. (eds.), *Cheese: Chemistry, Physics and Microbiology.* 2nd edition. Chapman and Hall, London, pp. 193–257.

Collins, Y. F., McSweeney, P. L. H. & Wilkinson, M. G. (2003). Lipolysis and free fatty acid catabolism in cheese: A review of current knowledge. *International Dairy Journal*, 13, 841–866.

Corsetti, A., Carbo, M. R., Albenzio, M., Di Cagno, R., Gobbetti, M. & Fox, P. F. (2001). Microbiology and biochemistry of Caciocavallo cheese. *International Journal of Food Microbiology*, 13, 297–309.

Cuer, A. G., Dauphin, A., Kergomard, J. P. & Adda, J. (1979). Production of S-methylthioacetate by Micrococcus cheese strains. *Agricultural Biological Chemistry*, 43, 1783–1784.

Curioni, P. M. G. & Bosset, J. O. (2002). Key odorants in various cheese types as determined by gas chromatography-olfactometry. *International Dairy Journal*, 12, 959–984.

Dairou, V. & Sieffermann, J. M. (2002). A comparison of 14 jams characterized by conventional profile and a quick original method, the Flash profile. *Journal of Food Science*, 67, 826–834.

Daly, D. F., McSweeney, P. L. H. & Sheehan, J. J. (2010). Investigation of underlying factors in split and secondary fermentation defects in rindless block Swiss-type cheese. *Australian Journal of Dairy Science*, 63, 214–216.

De la Fuente, M. A. & Juárez, M. (1993). Review: Determination of free fatty acids in dairy products. *Revista Espanola de Cienca y Technologia Alimentos*, 33, 247–269.

De Leon-Gonzalez, L. P., Wendorff, W. L., Ingham, B. H., Jaeggi, J. J. & Houck, K. B. (2000). Influence of salting procedure on the composition of Muenster-type cheese. *Journal of Dairy Science*, 83, 1396–1401.

Deeth, H. C. & Fitz-Gerald, C. H. (1995). Lipolytic enzymes and hydrolytic rancidity in milk and milk products. In Fox, P. F. (eds.), *Advanced Dairy Chemistry Volume 2, Lipids.* Chapman & Hall, London, pp. 247–308.

Delahunty, C. M. & Drake, M. A. (2004). Sensory character of cheese and its evaluation. In Fox, P. F., McSweeney, P. L. H., Cogan, T. M. & Guinee, T. P. (eds.), *Cheese: Chemistry, Physics and Microbiology, Volume 1. General Aspects.* Elsevier, Oxford, pp. 455–487.

Delahunty, C. M., McCord, A., O'Neill, E. E. & Morrissey, P. A. (1997). Sensory characterisation of cooked hams by untrained consumers using free-choice profiling. *Food Quality and Preference*, 8, 381–388.

Delahunty, C. M. & Murray, J. M. (1997). Organoleptic evaluation of cheese. In Cogan, T. M., Fox, P. F. & Ross, R. P. (eds.), *Proceedings of the 5th Cheese Symposium.* Teagasc, Dublin, pp. 90–97.

Delahunty, C. M. & Piggott, J. R. (1995). Current methods to evaluate contribution and interactions of components to flavour of solid foods using hard cheese as an example. *International Journal of Food Science & Technology*, 30, 555–570.

Dehlholm, C., Brockhoff, P. B., Meinert, L., Aaslyng, M. D. & Bredie, W. L. P. (2012). Rapid descriptive sensory methods – comparison of free multiple sorting, partial napping, napping, flash profiling and conventional profiling. *Food Quality and Preference*, 26, 267–277.

Delarue, J. & Sieffermann, J. M. (2004). Sensory mapping using Flash profile. Comparison with a conventional descriptive method for the evaluation of the flavour of fruit dairy products. *Food Quality and Preference*, 15, 383–392.

Drake, M. A. (2007). Invited Review: Sensory analysis of dairy foods. *Journal of Dairy Science*, 90, 4925–4937.

Drake, M. A., Yates, M. D., Gerard, P. D., Delahunty, C. M., Sheehan, E. M., Turnbull, R. P. & Dodds, T. M. (2005). Comparison of differences between lexicons for descriptive analysis of Cheddar cheese flavour in Ireland, New Zealand, and the United States of America. *International Dairy Journal*, 15, 473–483.

Engels, W. J. M. & Visser, S. (1997). A comparative study of volatile compounds in the water-soluble fraction of various types of ripened cheese. *International Dairy Journal*, 48, 127–140.

El Galiou, O., Zantar, S., Bakkali, M. & Laglaoui, A. (2013). Lipolysis and proteolysis during the ripening of Fresch Moroccan Goats milk cheese. *World Journal of Dairy & Food Sciences*, 8, 201–206.

Ennis, J. M. & Jesionka, V. (2011). The power of sensory discrimination methods revisited. *Journal of Sensory Studies*, 26, 371–382.

Ennis, J. M. & Rousseau, B. (2012). *Reducing Costs with Tetrad Testing.* The Institute for Perception, IFPress, 15, 3–4. (http://www.ifpress.com/wp-content/uploads/2012/04/15-1-Reducing-Costs-with-Tetrad-Testing.pdf.

Eriksen, S. (1976). Flavours of milk and milk products. I. Role of lactones. *Milchwissenschaft*, 31, 549–550.

Floury, J., Jeanson, S., Aly, S. & Lortal, S. (2010). Determination of the diffusion coefficients of small solutes in cheese: A review. *Dairy Science & Technology*, 90, 477–508.

Forss, D. A. (1979). Review of the progress of dairy science: Mechanisms of formation of aroma compounds in milk and milk products. *Journal of Dairy Research*, 46, 691–706.

Fox, P. F. & Wallace, J. M. (1997). Formation of flavour compounds in cheese. *Advances in Applied Microbiology*, 45, 17–85.

Fox, P. F., Law, J., McSweeney, P. L. H. & Wallace, J. (1993). Biochemistry of cheese ripening. In Fox, P. F., McSweeney, P. L. H., Cogan, T. M. & Guinee, T. P. (eds.), *Cheese: Chemistry, Physics and Microbiology*, 2nd edition, Vol. 1. Chapman and Hall, London, UK, pp. 389–438.

Fox, P. F., Guinee, T. P., Cogan, T. M. & McSweeney, P. L. H. (2000). Biochemistry of cheese ripening. In Fox, P. F., Guinee, T. P., Cogan, T. M. & McSweeney, P. L. H. (eds.), *Fundamentals of Cheese Science*. Aspen Publishers, USA, pp. 236–278.

Frohlich-Wyder, M-T. (2003). Yeasts in dairy products. In Boekhout, T. & Robert, B. (eds.), *Yeasts in Food, Beneficial and Detrimental Aspects*. Berh's Verlag, Hamburg, pp. 209–237.

Ganesan, B., Seefeldt, K., Koka, R. C., Dias, B. & Weimer, B. (2004). Monocarboxylic acid production by lactococci and lactobacilli. *International Dairy Journal*, 14, 237–246.

Ganesan, B., Weimer, B. C., Qian, M. C. & Burbank, H. M. (2007). Compounds associated with cheese flavor. In Weimer, B. C. (eds.), *Improving the Flavour of Cheese*. Woodhead Publishing Ltd, Cambridge England, pp. 26–51.

Helinck, S., Le Bars, D., Moreau, D. & Yvon, M. (2004). Ability of thermophilic lactic acid bacteria to produce aroma compounds from amino acids. *Applied Environmental Microbiology*, 70, 3855–3861.

Hickey, D. K., Kilcawley, K. N., Beresford, T. P. & Wilkinson, M. G. (2006). Starter bacteria are the prime agents of lipolysis in Cheddar cheese. *Journal of Agricultural and Food Chemistry*, 54, 8229–8235.

Hickey, D. K., Kilcawley, K. N., Beresford, T. P. & Wilkinson, M. G. (2007). Lipolysis in Cheddar cheese made from raw, thermized and pasteurised milks. *Journal of Dairy Science*, 90, 47–56.

Hillmann, H. & Hofmann, T. (2016). Quantitation of key tastants and re-engineering the taste of Parmesan cheese. *Journal of Agricultural and Food Chemistry*, 64, 1794–1805.

Hillmann, H., Behr, J., Ehrmann, M. A., Vogel, R. F. & Hofmann, T. (2016). Formation of kokumi-enhancing γ-glutamyl dipeptides in Parmesan cheese by means of γ-glutamyltransferase activity and stable isotope double-labeling studies. *Journal of Agricultural and Food Chemistry*, 64, 1784–1793.

ISO. (1991). *International Standard 3972*. Sensory Analysis Methodology, Method of investigating sensitivity of taste. Ref. No. ISO 3972:1991 (E). *International Organization for Standardization*, Genève, Switzerland.

ISO. (1992). *International Standard 5496*. Sensory Analysis Methodology, Initiation and training of assessors in the detection and recognition of odours. Ref. No. ISO 5496:1992 (E). *International Organization for Standardization*, Genève, Switzerland.

ISO. (1993). *International Standard 8586-1. Sensory Analysis Methodology, General guidance for the selection, training and monitoring of assessors*. Ref. No. ISO 8586–1:1993 (E). *International Organization for Standardization*, Genève.

ISO (1994). *International Standard. 11035. Sensory analysis-identification and selection of descriptors establishing a sensory profile by a multidimensional approach*. Ref. no. ISO 11035:1994 (E). *International Organization for Standardization*, Genève.

Iyer, M., Richardson, T., Amundson, C. H. & Boudreau, A. (1967). Improved technique for analysis of free fatty acids in butteroil and Provolone cheese. *Journal of Dairy Science*, 50, 285–291.

Jeanson, S., Chadœuf, J., Madec, M. N., Aly, S., Floury, J., Brocklehurst, T. F. & Lortal, S. (2011). Spatial distribution of bacterial colonies in a model cheese. *Applied Environmental Microbiology*, 77, 1493–1500.

Jensen, R. G., Gander, G. W. & Sampugna, J. (1962). Fatty acid composition of the lipids from pooled, raw milk. *Journal of Dairy Science*, 45, 329–331.

Jensen, R. G., Ferris, A. M. & Lammi-Keefe, C. J. (1991). The composition of milk fat. *Journal of Dairy Science*, 74, 3228–3243.

Jollivet, N., Chateaud, J., Vayssier, Y., Bensoussan, M. & Belin, J. M. (1994). Production of volatile compounds in model milk and cheese media by eight strains of *Geotrichum candidum*. *Journal of Dairy Research*, 61, 241–248.

Jolly, R. C. & Kosikowski, F. V. (1975). Flavour development in pasteurized milk blue cheese by animal and microbial lipase preparations. *Journal of Dairy Science*, 58, 846–921.

Kilcawley, K. N. (2016). Cheese Flavour. In Fox, P. F., Guinee, T. P., Cogan, T. M. & McSweeney, P. L. H (eds.), *Fundamentals of Cheese Science*, 2nd edition. Springer, New York, pp. 443–474.

Kubickova, J. & Grosch, W. (1998). Evaluation of flavour compounds of Camembert cheese. *International Dairy Journal*, 8, 11–16.

Lawless, H. T. & Heymann, H. (1998a). Sensory evaluation in quality control. In Lawless, H. T. & Heymann, H. (eds.), *Sensory Evaluation of Food, Principles and Practices*. Chapman and Hall, New York, pp. 548–584.

Lawless, H. T. & Heymann, H. (1998b). Descriptive analysis. In Lawless, H. T. & Heymann, H. (eds.), *Sensory Evaluation of Food, Principles and Practices*. Kluwer Academic/Plenum Publishers, New York, London, Dordrecht, Boston, pp. 341–378.

Lawlor, J. B., Delahunty, C. M., Sheehan, J. & Wilkinson, M. G. (2003). Relationships between sensory attributes and the volatile compounds, non-volatile and gross compositional constituents of six blue-type cheeses. *International Dairy Journal*, 13, 481–494.

Lawrence, R. C. & Gilles, J. (1987). Cheese composition and quality. In *Milk: The Vital force*. Proceedings of the XXII International Dairy Congress. The Hague, Dordrecht. D. Reidel Publishing, the Netherlands, pp. 111–121.

Lecanu, L., Ducruet, V., Jouquand, C., Gratadoux, J. J. & Feigenbaum, A. (2002). Optimization of headspace solid-phase microextraction (SPME) for the odor analysis of surface ripened cheese. *Journal of Agricultural and Food Chemistry*, 50, 3810–3817.

Liu, S. Q., Holland, R. & Crow, V. (2004). Esters and their biosynthesis in fermented dairy products: A review. *International Dairy Journal*, 14, 923–945.

Loescher, E., Sieffermann, J. M., Pinguet, C., Kesteloot, R. & Cuvlier, G. (2001). *Development of a list of textural attributes on pear/apple puree and fresh cheese: Adaptation of the Quantitative Descriptive Analysis method and use of Flash profiling*. Presented at the 4th Pangborn Sensory Science Symposium, Dijon, France.

Lopez, C., Briard-Bion, V., Camier, B. & Gassi, J-Y. (2006). Milkfat thermal properties and solid fat content in Emmenthal cheese: A differential scanning calorimetry study. *Journal of Dairy Science*, 89, 2894–2910.

Macedo, A. C. & Malcata, F. X. (1996). Changes in the major free fatty acids in Serra cheese throughout ripening. *International Dairy Journal*, 6, 1087–1097.

MacFie, H. J., Bratchell, N., Greenhoff, K. & Vallis, L. V. (1989). Designs to balance the effect of order of presentation and first-order carry-over effects in hall tests. *Journal of Sensory Studies*, 4, 129–148.

Marilley, L. & Casey, M. G. (2004). Flavours of cheese products: metabolic pathways, analytical tools and identification of producing strains. *International Journal of Food Microbiology*, 90, 139–159.

Markiewicz-Keszycka, M., Czyżak-Runowska, C., Lipińska, P. & Wójtowski, J. (2013). Fatty acid profile of milk – a review. *Bulletin-Veterinary Institute in Pulawy*, 57, 135–139.

McNeill, G. P. & Connolly, J. F. (1998). A method for the quantification of individual free fatty acids in cheese: Application to ripening of Cheddar-type cheese. *Irish Journal of Food Science and Technology*, 13, 119–128.

McSweeney, P. L. H. (1997). The flavour of milk and dairy products: III. Cheese: taste. *International Journal of Dairy Technology*, 50, 123–128.

McSweeney, P. L. H (2004). Biochemistry of cheese ripening. *International Journal of Dairy Technology*, 57 (2–3), 127–144.

McSweeney, P. L. H. & Sousa, M. J. (2000). Biochemical pathways for the production of flavour compounds in cheese during ripening. *Le Lait*, 80, 293–324.

Meilgaard, M. C., Civille, G. V. & Carr, B. T. (1999). *Sensory Evaluation Techniques*, 3rd edition. CRC Press, Boca Raton, New York.

Meilgaard, M. C., Civille, G. V. & Carr, B. T. (2007). *Sensory Evaluation Techniques*, 4th edition. CRC Press, Taylor & Francis Group, Boca Raton, USA.

Moio, L. & Addeo, F. (1998). Grana Padano cheese aroma. *Journal of Dairy Research*, 65, 317–333.
Molimard, P. & Spinnler, H. E. (1996). Review: Compounds involved in the flavour of surface mould-ripened cheeses: Origins and properties. *Journal of Dairy Science*, 79, 169–184.
Monnet, V., Condon, S., Cogan, T. M. & Gripon, J-C. (1995). Metabolism of starter cultures. In Cogan, T. M. & Accolas, J. P. (eds.), *Dairy Starter Cultures*. Wiley VCH New York, pp. 47–99.
Muir, D. D. (2010). The grading and sensory profiling of cheese. In Law, B. A. & Tamine, A. Y. (eds.), *Technology of Cheesemaking*, 2nd edition. Wiley-Blackwell Publication, Oxford, pp. 440–474.
Mulder, H. (1952). Taste and flavour-forming substances in cheese. *Netherlands Milk and Dairy Journal*, 6, 157–167.
Muñoz, A. M. & Civille, G. V. (1992). The spectrum descriptive analysis method. In Hootman, R. C. (eds.), *ASTM Manual on Descriptive Analysis*. Pennsylvania: American Society for Testing and Materials.
Murray, J., Delahunty, C. & Baxter, I. (2001). Descriptive sensory analysis: Past, present and future. *Food Research International*, 34, 461–471.
O'Sullivan, M. G., Kerry, J. P. & Byrne, D. V. (2011). Use of sensory science as a practical commercial tool in the development of consumer-led processed meat products. In Kerry, J. P. & Kerry, J. F. (eds.), *Processed Meats: Improving, Safety and Nutritional Quality*. Woodhead Publishing Ltd, UK, pp. 156–182.
Overington, A. R., Eyres, G. T., Delahunty, C. M., Silcock, P., Niimi, J., Holland, R. & Coolbear, T. (2010). Flavour release and perception in cheese bases. *Australian Journal of Dairy Technology*, 65, 162–164.
Partidario, A. (1999). Comparison of free fatty acids acids, volatile compounds and sensory characteristics of Serra da Estrela cheese at different ripening stages. *Milchwissenschaft*, 54, 381–385.
Piggott, J. R., Simpson, S. J. & Williams, S. A. R. (1998). Sensory analysis. *International Journal of Food Science & Technology*, 33, 7–12.
Preininger, M. & Grosch, W. (1994). Evaluation of key odorants of the neutral volatiles of Emmenthaler cheese by the calculation of odour activity values. *LWT-Food Science and Technology*, 27, 237–244.
Pripp, A. H., Skeie, S., Isaksson, T., Borge, G. I. & Sorhaug, T. (2006). Multivariate modelling of relationships between proteolysis and sensory quality of Prast cheese. *International Dairy Journal*, 16, 225–235.
Proveda, J. M., Pérez-Coello, M. S. & Cabezas, L. (1999). Evolution of the free fatty acid fraction in Manchego cheese during ripening. *Milchwissenschaft*, 54, 685–687.
Proveda, J. M., Sánchez-Palomo, E., Pérez-Coello, M. S. & Cabezas, L. (2008). Volatile composition, olfactometry profile and sensory evaluation of semi-hard Spanish goats cheeses. *Dairy Science & Technology*, 88, 355–367.
Qian, M. C. & Burbank, H. M. (2007). Hard Italian cheeses: Parmigiano-Reggiano and Grana Padano. In Weimer, B. C. (eds.), *Improving the Flavour of Cheese*. Woodhead Publishing Ltd., Cambridge England, pp. 421–443.
Qian, M. C., Nelson, C. & Bloomer, S. (2002). Evaluation of fat-derived aroma compounds in blue cheese by dynamic headspace GC/olfactometry-MS. *Journal of American Oil Chemists' Society*, 79, 663–667.
Rétiveau, A., Chambers, D. H. & Esteve, E. (2005). Developing a lexicon for the flavor description of French cheeses. *Food Quality and Preference*, 16, 517–527.
Richter, V., Almeida, T., Prudencio, S. & Benassi, M. (2010). Proposing a ranking descriptive sensory method. *Food Quality and Preference*, 21, 611–620.
Rodrigue, N., Guillet, M., Fortin, J. & Martin, J. F. (2000). Comparing information obtained from ranking and descriptive tests of four sweet corn products. *Food Quality and Preference*, 11, 47–54.

Roger, S., Degas, C. & Gripon, J. C. (1988). Production of phenyl ethyl alcohol and its esters during ripening of traditional Camembert. *Journal of Food Chemistry*, 28, 129–140.

Roudot-Algaron, F., Kerhoas, L., Le Bars, D., Einhorn, J. & Gripon, J. C. (1994). Isolation of γ-Gutamyl peptides from Comté cheese. *Journal of Dairy Science*, 77, 1161–1166.

Rychlik, M., Warmke, R. & Grosch, W. (1997). Ripening of Emmental cheese wrapped in foil with and without addition of *Lactobacillus casei* subsp. *casei*. III. Analysis of character impact flavour compounds. *LWT-Food Science and Technology*, 30, 471–478.

Sable, S. & Cottenceau, G. (1999). Current knowledge of soft cheeses flavor and related compounds. *Journal of Argicultural Food Chemistry*, 47, 4825–4836.

Salles, C., Herve, C., Septier, C., Demaizieres, D., Lesschaeve, I., Issanchou, S. & Le Quere, J. L. (2000). Evaluation of taste compounds in water-soluble extract of goat cheeses. *Food Chemistry*, 68, 429–435.

Singh, T. K., Drake, M. A. & Cadwallader, K. R. (2003). Flavor of Cheddar cheese: A chemical and sensory perspective. *Comprehensive Reviews in Food Science and Food Safety*, 2, 139–162.

Smit, G., Smit, B. A. & Engles, W. J. M. (2005). Flavour formation by lactic acid bacteria and biochemical flavour profiling of cheese products. *FEMS Microbiology Reviews*, 29, 591–610.

Stone, H., Bleibaum, R. N. & Thomas, H. A. (2012a). Affective testing. In Stone, H., Bleibaum, R. N. & Thomas, H. A. (eds.), *Sensory Evaluation Practices*, 4th Volume, 4th edition. Elsevier Academic Press, USA, pp. 291–325.

Stone, H., Bleibaum, R. N. & Thomas, H. A. (2012b). Test strategy and design of experiments. In Stone, H., Bleibaum, R. N. & Thomas, H. A. (eds.), *Sensory Evaluation Practices*, 4th edition, 4th Volume. Elsevier Academic Press, USA, pp. 117–157.

Stone, H. & Sidel, J. L. (2004). Introduction to sensory evaluation. In Stone, H. & Sidel, J. L. (eds.), *Sensory Evaluation Practices*, 3rd edition. Elsevier Academic Press, Oxford, UK, pp. 1–20.

Stone, H., Sidel, J., Oliver, S., Woolsey, A. & Singleton, R. C. (1974). Sensory evaluation by quantitative descriptive analysis. *Food Technology*, 28, 24–34.

Subramanian, A., Harper, W. J. & Rodriguez-Saona, L. E. (2009). Rapid prediction of composition and flavor quality of Cheddar cheese using ATR-FTIR spectroscopy. *Journal of Food Science*, 74, 292–297.

Tanous, C., Gori, A., Rijnen, L., Chambellon, E. & Yvon, M. (2005). Pathways for α-ketoglutarate formation by *Lactococcus lactis* and their role in amino acid catabolism. *International Dairy Journal*, 15, 759–770.

Tanous, C., Kieronczyk, A., Helinck, S., Chambellon, E. & Yvon, M. (2002). Glutamate dehydrogenase activity: A major criterion of the selection of flavour-producing lactic acid bacteria strains. *Antonie van Leeuwenhoek*, 82, 271–278.

Tarea, S., Cuvelier, G. & Siefffermann, J-M. (2007). Sensory evaluation of the texture of 49 commercial apple and pear purees. *Journal of Food Quality*, 30, 1121–1131.

Thierry, A., Salvat-Brunaud, D. & Maubois, J-L. (1999). Influence of thermophilic lactic acid bacteria strains on propionibacteria growth and lactate consumption in an Emmenthal juice-like medium. *Journal of Dairy Research*, 66, 105–113.

Toelstede, S., Dunkel, A. & Hofmann, T. (2009). A series of kokumi peptides impart the long-lasting mouthfulness of matured Gouda cheese. *Journal of Agricultural Food Chemistry*, 57, 1440–1448.

Toelstede, S. & Hofmann, T. (2008a). Sensomics mapping and identification of the key bitter metabolites in Gouda cheese. *Journal of Agricultural Food Chemistry*, 56, 2795–2804.

Toelstede, S. & Hofmann, T. (2008b). Quantitative studies and taste re-engineering experiments towards the decoding of the nonvolatile sensometabolome of Gouda cheese. *Journal of Agricultural Food Chemistry*, 56, 5299–5307.

Toelstede, S. & Hofmann, T. (2009). Kokumi-active glutamyl peptides in cheeses and their biogeneration by *Penicillium roqueforti*. *Journal of Agricultural Food Chemistry*, 57, 3738–3748.

Urbach, G. (1993). Relations between cheese flavour and chemical composition. *International Dairy Journal*, 3, 389–422.
Urbach, G. (1997). The flavour of milk and dairy products: II. Cheese: Contribution of volatile compounds. *International Journal of Dairy Technology*, 50, 79–89.
Varela, P. & Ares, G. (2012). Sensory profiling, the blurred line between sensory and consumer science. A review of novel methods for product characterization. *Food Research International*, 48, 893–908.
Van Hekken, D. L., Drake, M. A., Corral, F. J. M., Prieto, V. M. G. & Gardea, A. A. (2006). Mexican Chihuahua cheese: Sensory profiles of young cheese. *Journal of Dairy Science*, 89), 3729–3738.
Van Kleef, E., van Trijp, H. C. M. & Luning, P. (2006). Internal versus external preference analysis: An exploratory study on end-user evaluation. *Food Quality and Preference*, 17, 387–399.
Walstra, P., Geurts, T. J., Noomen, A., Jellema, A. & Van Boekek, M. A. J. S. (2005). Cheese ripening and properties. In Walstra, P., Geurts, T. J., Noomen, A., Jellema, A. & Van Boekek, M. A. J. S. (eds.), *Dairy Technology, Principles of Milk Properties and Processes*. Marcel Dekker Inc, New York and Basel, pp. 357–397.
Walstra, P., Wouters, J. T. M. & Geurts, T. J. (2006). Lactic fermentations. In Walstra, P., Wouters, J. T. M., and Geurts, T. J. (eds.), *Dairy Science and Technology*, 2nd edition. Talyor & Francis Group, Boca Raton, London, New York, pp. 357–397.
Warmke, R., Belitz, H. D. & Grosch, W. (1996). Evaluation of taste compounds of Swiss cheese (Emmentaler). *Zeitschrift für Lebensmittel-Untersuchung und –Forschung*, 203, 230–235.
Wilkinson, M. G. (2007). Lipolysis and cheese flavour development. In Weimer, B. (eds.), *Improving the Flavour of Cheese*. Woodhead Publishing Ltd, Cambridge, UK, pp. 102–120.
Wilkinson, M. G. & Kilcawley, K. N. (2007). Carbohydrate metabolism and cheese flavour development. In Weimer, B. (eds.), *Improving the Flavour of Cheese*. Woodhead Publishing Ltd., Cambridge, UK, pp. 55–69.
Williams, A. A. & Arnold, G. M. (1985). A comparison of the aroma of six coffees characterised by conventional profiling, free-choice profiling and similarity scaling methods. *Journal of the Science of Food and Agriculture*, 36, 204–214.
Wong, N. P., Ellis, R. & LaCroix, D. E. (1975). Quantitative determination of lactones in Cheddar cheese. *Journal of Dairy Science*, 58, 1437–1441.
Woo, A. H., Kollodges, S. & Lindsay, R. C. (1984). Modified techniques for improved gas chromatographic quantification of free fatty acids in dairy foods. *Journal of Dairy Science*, 67, 1517–1520.
Woo, A. H. & Lindsay, R. C. (1984). Concentrations of major free fatty acids and flavour development in Italian cheese varieties. *Journal of Dairy Science*, 67, 960–968.
Worch, T., Crine, A., Gruel, A. & Lê, S. (2014). Analysis and validation of the ideal profile method: Application to a skin cream study. *Food Quality and Preference*, 32, 132–144.
Worch, T., Lê, S. & Punter, P. (2010). How reliable are the consumers? Comparison of sensory profiles from consumers and experts. *Food Quality and Preference*, 21, 309–318.
Xia, Y., Zhang, J., Zhang, X., Ishii, R., Zhong, F. & O'Mahony, M. (2015). Tetrads, triads and pairs: Experiments in self-specification. *Food Quality and Preference*, 40, 97–105.
Yvon, M., Berthelot, S. & Gripon, J. C. (1998). Adding α-ketoglutarate to semi-hard cheese curd highly enhances the conversion of amino acids to aroma compounds. *International Dairy Journal*, 8, 889–898.
Yvon, M. & Rijnen, L. (2001). Cheese flavour formation by amino acid catabolism. *International Dairy Journal*, 11, 185–201.

4

Cheese Microbial Ecology and Safety

Antonia Picon

Department of Food Technology, National Institute of Agricultural and Food Research and Technology (INIA), Madrid, Spain

4.1 Introduction

Dairying constituted a major innovation in prehistoric societies. It allowed the preservation of milk and its transformation into a more digestible commodity. The presence of milk fat in specialised pottery vessels provided evidence for cheesemaking in the sixth millennium BC in northern Europe (Salque et al., 2013).

Many microorganisms are able to thrive in milk (which is rich in proteins, fats, carbohydrates, vitamins and minerals, has a high water content and a near-neutral pH). Lactic acid bacteria (LAB), a bacterial group including the genera *Lactococcus, Lactobacillus, Leuconostoc, Streptococcus* and *Enterococcus*, are the dominant population in raw milk. However, psychrotrophic populations (frequently including genera as *Pseudomonas* and *Acinetobacter*) become major components of milk microbiota during cold storage. Other strains of non-LAB genera, yeasts and moulds are also present in milk microbiota (Quigley et al., 2013). Milk microbiota can have a positive contribution, improving the organoleptic and textural properties of dairy products (Montel et al., 2014), or a negative effect, shortening milk shelf life during refrigerated storage (Capodifoglio et al., 2016; Uceda et al., 1994). Both negative and positive impacts on consumer's health have been claimed, for example, illness caused by consumption of pathogen-contaminated raw milk (Oliver et al., 2009), and stimulation of the immune system by raw milk microorganisms (Fernández et al., 2015).

4.2 Source of Microorganisms in Cheese

Cheese microorganisms are either associated with the ingredients used in cheese manufacture or components of the starter culture.

Although raw milk microbial load has decreased to levels of 5×10^3 to 10^4 colony forming units per mL (CFU/mL) after implementing procedures to improve its hygienic quality, raw milk still exhibits a significant microbial diversity. The teat surface (Verdier-Metz et al., 2012) and biofilms formed on different materials of the milking equipment can be direct sources of microorganisms (Marchand et al., 2012). The feed (Verdier-Metz et al., 2012) and the milking parlour and stable may also be indirect sources of microorganisms. Milk storage at refrigerated temperatures before cheesemaking shifts raw milk microbiota predominance from Gram-positive to Gram-negative bacteria (Rasolofo et al., 2010) and increases psychrotrophic bacteria

Global Cheesemaking Technology: Cheese Quality and Characteristics, First Edition.
Edited by Photis Papademas and Thomas Bintsis.

levels up to 10^5 CFU/mL (Ercolini et al., 2009). In total, more than 100 genera and 400 microbial species have been detected in raw milk. They are mainly Gram-negative bacteria (>90 species), Gram-positive and catalase-positive bacteria (>90 species), LAB (>60 species), yeasts (>70 species) and moulds (>40 species) (reviewed by Montel et al., 2014).

The contribution of rennet to the microbiota of cheese is considered very limited. The salt used in the preparation of traditional rennet inhibits microbial growth, and there is very little information on the microbiota of genetically engineered chymosin (Beresford et al., 2004).

Although the contribution of dry salt to the microbiota of cheese is considered limited, and the relative high salt content of brine inhibits the growth of many microorganisms, several studies indicate that the microbiota of commercial brines in white-brine cheese contribute to its volatile profile (Bintsis et al., 2000).

4.3 Factors Influencing the Growth of Microorganisms in Cheese

The cheesemaking process determines cheese composition and the environmental conditions that microorganisms will have to face during the process. Water and salt content, pH, redox potential, organic acid content and ripening conditions will control the growth of microorganisms in cheese.

During the first stages of cheese manufacture, water activity (a_w) is close to 1, supporting the growth of most microorganisms. However, after whey drainage, salting and during ripening, a_w levels drop to 0.988–0.917, values which are significantly lower than the optimal requirements of many microorganisms (Beresford et al., 2001). Salt concentration, which ranges from 0.7 to 7 g/100 g depending on cheese variety, would result in a reduction of a_w to values of 0.99–0.95, respectively. Since many microorganisms could grow under such conditions, other interacting factors also contribute to the inhibition of microbial growth (Beresford et al., 2001).

In cheese curd, the low pH values (4.5–5.3) inhibit the growth of acid-sensitive species. The real inhibitor is thought to be the undissociated form of the organic acids present in cheese (Beuchat & Golden, 1989).

In most cheese varieties, coagulation usually takes place at 30°C to 37°C, allowing the growth of most microorganisms. If a cooking step (at 37°C to 54°C) follows coagulation, some microorganisms could be inhibited. Cheese ripening temperatures are a compromise between the needs to promote ripening reactions and control the growth of the desirable secondary flora, and to prevent the propagation of potential spoilage and pathogenic bacteria. Higher ripening temperatures, which were used to shorten the maturation period (reviewed by Fox et al., 1996), promoted the expression of genes related to proteolysis, lipolysis and amino acid/lipid catabolism, increasing the maturation rate (de Filippis et al., 2016).

Cheese redox potential, around −250 mV, is one of the major factors determining the type of microorganisms that will grow in cheese. In the anaerobic cheese interior, only obligatory or facultatively anaerobic microbes grow, whereas on the cheese surface, predominantly obligate aerobes develop (Beresford et al., 2001).

4.4 Cheese Microbiota

4.4.1 Starter Bacteria

A starter culture is a microbial preparation of large numbers of cells of at least one microorganism, its main function being to cause a rapid acidification (Leroy & de Vuyst, 2004). The addition of selected starter cultures to pasteurised milk allows the fermentation process to be controlled and a standardised product to be obtained. Starter cultures, added to milk in large numbers, become predominant in cheese, especially at the early stages of ripening. In cheese curd, they already exceed 10^8 CFU/g. During cheese ripening, many starters lose viability and release their intracellular enzymes. This process, known as autolysis, has been correlated with increased proteolysis and is used to accelerate cheese ripening (Hannon et al., 2003).

LAB have a long and safe history of use in the production of fermented foods. They enhance shelf life and microbial safety, improve texture and contribute to the sensory profile of the end product. The LAB most often used as starter cultures are members of the genera *Lactococcus, Lactobacillus, Streptococcus, Leuconostoc* and *Enterococcus. Lactococcus lactis* and *Leuconostoc* spp. among mesophilic species (optimal growth temperature at 30°C) and *Streptococcus thermophilus, Lactobacillus delbrueckii ssp. bulgaricus* and *Lactobacillus helveticus* among thermophilic species (optimal growth temperature at 42°C) are all able to ferment lactose (Fox & McSweeney, 2004).

Lactococcus lactis is the main constituent of mesophilic dairy starter cultures used worldwide for the production of numerous fermented dairy products. Humans ingest up to 10^{18} lactococcal cells per year through the consumption of fermented dairy products (Mills et al., 2010). The small number of *Lc. lactis* strains used in fermented food production was mainly chosen for their acidification activity and resistance to bacteriophage infection. Nowadays the natural biodiversity within the *Lactococcus* genus is examined in an attempt to identify novel starter cultures that could fulfil consumer demands for more diverse flavour (Mills et al., 2010). Members of this genus can be isolated from raw milk, raw milk cheeses and non-milk environments and are collectively referred to as 'wild-type' *Lactococcus.*

Fifteen *Lc. lactis* genomes have been completed up to now and there are 70 more genomes in progress according to the data retrieved from NCBI (http://www.ncbi.nlm.nih.gov/genome/?term=Lactococcus+lactis; accessed August 18, 2016). Their genomes range in size from ~2.3 to 2.7 Mb. The availability of these complete lists of genes allows drawing full metabolic pathways (Oliveira, Nielsen & Forster, 2005) and exploiting some interesting characteristics, as flavour formation from amino acids (van Kranenburg et al., 2002).

Many of the traits that make these lactococcal strains suitable for dairy fermentations are encoded on plasmids: lactose utilisation, casein breakdown, bacteriophage resistance, bacteriocin production, antibiotic resistance, resistance to and transport of metal ions, and exopolysaccharide (EPS) production (Mills et al., 2010).

S. thermophilus is considered the second most important industrial starter. It has been traditionally used in combination with *Lb. delbrueckii* ssp. *bulgaricus* or *Lb. helveticus* for the manufacture of yogurt and high-temperature cooked hard cheeses. It is also used alone or in combination with lactobacilli for the production of Mozzarella and Cheddar cheeses (Mills et al., 2010).

According to the data retrieved from NCBI, 21 *S. thermophilus* genomes have been completed and eight more are nowadays in progress (http://www.ncbi.nlm.nih.gov/genome/?term=Streptococcus+thermophilus; accessed August 18, 2016). Their genome size is close to 1.8 Mb in most strains. It has lost the most important pathogenic determinants in its adaptation to the milk environment (Bolotin et al., 2004). In contrast with lactococcal plasmids, *S. thermophilus* plasmids are thought to play a relatively insignificant role.

The *Lactobacillus* species commonly used in dairy starters include the thermophilic *Lb. delbreuckii* ssp. *bulgaricus, Lactobacillus delbreuckii* ssp. *lactis, L. helveticus* and *Lactobacillus acidophilus.*

Five genomes of *Lb. delbrueckii,* all of them ssp. *bulgaricus,* have been completed up to now and there are 27 more in progress, including members of the ssp. *bulgaricus, delbrueckii* and *lactis* (http://www.ncbi.nlm.nih.gov/genome/?term=Lactobacillus+'especies'; accessed 18 August, 2016). With respect to *Lactobacillus helveticus* and *Lactobacillus acidophilus,* nine and three genomes have been already completed, and 13 more are in progress for each one. Their genome sizes are close to 2.0 Mb in both cases. This trend towards reductive evolution with incomplete metabolic pathways and few regulatory functions appears to be related to the transition to a nutritionally rich environment (van de Guchte et al., 2006).

4.4.2 Non-Starter LAB

Some LAB, known as non-starter LAB (NSLAB), are not deliberately added as part of primary or secondary starter cultures. They belong to the autochthonous milk microbiota or gain access to the cheese from environmental or technological sources (Montel et al., 2014). They become a significant proportion of the cheese microbiota of almost all cheese varieties and contribute to flavour formation during the ripening process. NSLAB are a highly heterogeneous group. Although non-starter lactobacilli constitute the majority of this population, lactococci, pediococci, enterococci, *Leuconostoc* sp. and thermophilic LAB are also part of it (Beresford et al., 2001; Settanni & Moschetti, 2010). NSLAB have the opposite kinetic of growth than primary starters, low levels (10^2–10^3 CFU/g) in curd and reaching a plateau at 10^7–10^9 CFU/g within a few months.

The *Lactobacillus* genus is a very heterogeneous microbial group containing 221 different species and 29 subspecies (www.bacterio.net), whose classification is constantly being reshuffled (Bernardeau et al., 2008). Among the non-starter lactobacilli frequently recovered from cheese are the following: *Lactobacillus farciminis* among the obligately homofermentative species; *Lactobacillus casei, Lactobacillus paracasei, Lactobacillus plantarum, Lactobacillus pentosus, Lactobacillus curvatus* and *Lactobacillus rhamnosus* among the facultatively heterofermentative species and *Lactobacillus fermentum, Lactobacillus buchneri, Lactobacillus parabuchneri* and *Lactobacillus brevis* among the obligately heterofermentative species (Settanni & Moschetti, 2010). The core microbiota of the NSLAB population is formed by *Lb. paracasei* and *Lb. plantarum*, together with *Lb. curvatus, Lb. rhamnosus* and *Lb. casei* (Gobbetti et al., 2015).

The genus *Enterococcus* consists of 55 species and 2 subspecies (www.bacterio.net). They are able to grow during refrigeration, survive pasteurisation, multiply during the fermentation process and contaminate finished products (Giraffa, 2003). Their wide range of growth temperatures (10°C–45°C) and their high tolerance to heat, acid (pH 4.0–9.6) and salt (up to 6.5% NaCl) allow them to persist during cheese ripening. *Enterococcus faecalis, Enterococcus faecium* and, to a lesser extent, *Enterococcus durans* are the most frequently isolated species from many cheese varieties (Giraffa, 2003; Settanni & Moschetti, 2010). Although their presence in fresh or soft cheese is related to poor hygienic conditions during cheesemaking, they can actively contribute to flavour development in ripened cheese due to their proteolytic, lipolytic and citrate breakdown activities (Giraffa, 2003). However, their association with the mammalian gastrointestinal tract and human infections, their antibiotic resistance, presence of virulence factors and production of biogenic amines require careful evaluation (Foulquié Moreno et al., 2006).

The genus *Leuconostoc* consists of 24 species and 7 subspecies (www.bacterio.net). They are associated with plant material and also found in milk and dairy products (Hemme & Foucaud-Scheunemann, 2004). In spite of their poor growth in milk, their ability to co-metabolise lactose and citrate producing lactate, acetate, CO_2, ethanol, acetaldehyde, diacetyl, acetoin and 2,3-butanediol, they contribute to the organoleptic properties of fresh and semi-hard (Edam and Gouda) cheese varieties, buttermilk and sour cream (McSweeney & Sousa, 2000). They are also able to synthesise dextrans from sucrose, or α-glucooligosaccharides (GOS) from maltose or isomaltose, which can be used as thickeners or texturisers in cultured milks or stabilisers in ice cream (Vedamuthu, 1994).

The genus *Pediococcus* consists of 15 species (www.bacterio.net). *Pediococcus acidilactici* and *Pediococcus pentosaceus* are frequently isolated from raw milk ripened cheeses (Settanni & Moschetti, 2010). They enhance flavour development, accelerating cheese ripening (Tzanetakis, Litopoulou-Tzanetaki & Vafopoulou-Mastrogiannaki, 1991). Most strains of *P. pentosaceus* produced diacetyl from serine (Irmler et al., 2013).

4.4.3 Propionibacteria

The dairy group of genus *Propionibacterium* comprises of four species: *P. freudenreichii, P. acidipropionici, P. jensenii* and *P. thoenii* (www.bacterio.net). They are the characteristic microbiota of Swiss-type cheeses. In raw milk, enough propionibacteria are present, but they are added to pasteurised milk to ensure levels of 10^3 CFU/g of cheese. They survive the high cooking temperature (~54°C) during cheesemaking and reach levels of 10^8 to 10^9 CFU/g of cheese after the warm ripening period (Beresford et al., 2001). They metabolise lactic acid to acetic and propionic acids and CO_2, and play a key role in the formation of free fatty acids and isovaleric acid, contributing to the characteristic flavour and appearance (eye formation) of Swiss-type cheeses (Thierry et al., 2004).

P. freudenreichii, used as secondary starter, is the best-studied species. Its metabolic activity was maximal at the end of the cold ripening period and was stable during the first two weeks of the warm ripening period, while lactate was still present (Falentin et al., 2010b). Whole-genome sequencing of *P. freudenreichii* CIRM-BIA1^T revealed its ability to cope with stress, resist phage attack and synthesise most vitamins and amino acids, and also several genes encoding surface proteins potentially involved in adhesion and immunoregulatory activity (Falentin et al., 2010a).

4.4.4 Micrococci and Staphylococci

Micrococci, staphylococci and coryneform bacteria (*Arthrobacter, Brachybacterium, Brevibacterium, Corynebacterium, Microbacterium* and *Rhodococcus* spp.) are aerobic bacteria that form part of the surface microbiota of many cheese types. They are present in large numbers on the surface of bacterial smear surface-ripened cheeses, after deacidification of the cheese surface by moulds and yeasts (Corsetti, Rossi & Gobbetti, 2001), and play a significant role in determining the final characteristics of these varieties (Irlinger & Bergère, 1999).

The genus *Micrococcus*, consisting of 17 species (www.bacterio.net), is very heterogeneous and more closely related phylogenetically to *Arthrobacter* than to *Staphylococcus*. *Micrococcus luteus* and *Micrococcus lylae* are two of the most frequently isolated species from milk and cheeses (Irlinger & Bergère, 1999). The beneficial roles of *Micrococcus* in cheese ripening are well documented. Although they are present at lower numbers than other microbial groups, being obligate aerobes and with optimum growth temperatures of 25°C–37°C, their enzymatic activities contribute to cheese ripening (Bhowmik & Marth, 1990). Cheese made with a culture of *Micrococcus* sp. INIA 528 (isolated from Manchego cheese) added to milk had higher levels of branched-chain aldehydes and alcohols and lower levels of diacetyl and acetoin than control cheese (Morales et al., 2010).

Staphylococci are facultatively anaerobes, which grow better under aerobic conditions. *Staphylococcus equorum, Staphylococcus xylosus* and *Staphylococcus carnosus* are the species typically isolated from milk and cheese. Most strains are coagulase-negative staphylococci (CNS) and can grow in the presence of 15% NaCl and between 18°C and 40°C. Their contribution to flavour development in cheese has not been investigated (Beresford, 2004). A few cases of nosocomial infection by CNS have been reported in immune-depressed patients, and transferable antibiotic resistance genes and haemolytic activity have been observed in some dairy isolates (Quigley et al., 2013).

4.4.5 Moulds and Yeasts

Moulds have an important role in the ripening of mould-ripened cheeses, where a complex consortium of yeasts, bacteria and filamentous fungi is formed during the maturation stage (Addis et al., 2001). Their contribution to proteolysis and lipolysis leads to texture, flavour and nutritional cheese quality improvements (Fox & McSweeney, 2004). Mould-ripened cheeses are divided into two groups: core mould-ripened and surface mould-ripened cheeses. In the first group (Roquefort, Gorgonzola, Stilton, Danish blue, Cabrales), *Penicillium roqueforti* grows and forms blue veins within the cheese, whereas in the second group (Camembert and Brie), *Penicillium camemberti* grows on the cheese surface (Beresford et al., 2001).

The spatial distribution of the bacterial community in Stilton has been reported to be highly heterogeneous (Ercolini, Hill & Dodd, 2003). On surface mould-ripening cheeses, the *Penicillium* layer gives them unique aroma and flavour characteristics due to the intense lipolytic and proteolytic action that occurred during maturation. Cheese elaborated with an O-type mesophilic starter culture and *Penicillium candidum* resulted in a more typical taste than those with a DL-type mesophilic starter and *P. camemberti* (Galli et al., 2016).

In other cheese varieties, the mould impact on ripening is not well understood. They can also cause appearance and flavour defects and synthesise toxic compounds (Banjara, Suhr & Hallen-Adams, 2015; Himery et al., 2014).

Recently, culture-independent DNA-based methods have led to the detection of fungi not previously reported (Delavenne et al., 2011) and allowed evaluation of fungal community dynamics in cheese, although most of their members are not very likely to survive pasteurisation and persist or grow in cheese (Himery et al., 2014). The most common moulds found in a study on 44 commercial cheeses were *Penicillium* (approximately 50% of isolates, and present in 45% of cheeses overall) and *Aspergillus* (15% of isolates, in 11% of cheeses). At the species level, *P. roqueforti* was the predominant mould, present in all blue cheese samples and in 23% of samples overall (Banjara, Suhr & Hallen-Adams, 2015).

Yeasts occur in many cheeses, and particularly in those made from raw milk. They play a major role in dairy fermentations due to their abilities to utilise lactose or galactose and to grow at low temperatures and high salt concentrations. Their high proteolytic and/or lipolytic activities make vital contributions to cheese texture and aroma characteristics. Flavour defects described as fruity and bitter off-flavours have also been attributed to yeast activity (Beresford, 2004).

Yeast species detected in raw milk include *Kluyveromyces marxianus, Kluyveromyces lactis, Rhodotorula mucilaginosa, Debaryomyces hansenii, Geotrichum candidum, G. catenulate, Pichia fermentans, Candida sake, Candida parapsilosis, Candida inconspicua, Trichosporon cutaneum, Trichosporon lactis, Cryptococcus curvatus, Cryptococcus carnescens* and *Cryptococcus victoriae* (Delavenne et al., 2011). Yeasts frequently found in cheese belong to the genera *Candida, Geotrichum, Kluyveromyces, Pichia, Rhodotorula, Saccharomyces, Trichosporon, Torulospora, Yarrowia* and *Zygosaccharomyces* spp. (Beresford, 2004).

In smear-ripened cheese varieties, lactate-metabolising yeasts aid in deacidifying the cheese surface and enable the outgrowth of the less acid-tolerant and aerobic bacteria of the smear consortium (Beresford et al., 2001; Corsetti, Rossi & Gobbetti, 2001). The interactions between yeast and bacteria may generate more complex volatile profiles, including S-methyl thioesters (Arfi et al., 2005).

In mould-ripened cheeses too, yeasts have a prominent role. *Debaryomyces hansenii*, tolerant to high salt concentrations, grew to maximum populations of 10^8–10^9 CFU/g within the first week of maturation and remained at this level throughout the process. *Yarrowia lipolytica* tended to develop later in the maturation process and reached maximum populations of 10^5 CFU/g (Addis et al., 2001). A study on the yeast community in Stilton cheese employing culture-independent and culture-dependent analyses showed that *Y. lipolytica* was present in the white core; *K. lactis* dominated in the blue veins, was less present in the white core, and had limited presence in the outer crust and *D. hansenii* was more present in the white core and outer crust than in the blue veins. A strong synergistic activity of *Y. lipolytica* and *P. roqueforti* in enhancing the production of ketone aroma compounds, characteristic of blue cheeses, was also observed (Gkatzionis et al., 2014).

Yeast species greatly changed during semi-hard ewe and goat raw milk cheese ripening. Although six to eight yeast species were found in one-day old cheese, most of them had vanished after six weeks. *D. hansenii* dominated at the end of ripening in all cheeses. *K. lactis*, which is able to ferment lactose and grow in the cheese interior, where other yeasts are scarce, was the second most frequently isolated species. *Y. lipolytica* was also present, although it was not a dominant yeast in the ripening process (Padilla, Manzanares & Belloch, 2014). *D. hansenii* and *Y. lipolytica* have been proposed as agents for accelerating Cheddar ripening (Ferreira & Viljoen, 2003).

4.4.6 Probiotics in Cheese

Probiotics are defined as live microorganisms which, when administered in adequate amounts, confer a health benefit on the host (FAO/WHO, 2001). Microorganisms from species of the genera *Lactobacillus* and *Bifidobacterium* are the main Gram-positive bacteria currently characterised as probiotics (FAO/WHO 2001), although several others such as *Propionibacterium, Streptococcus, Bacillus, Enterococcus, Escherichia coli* and yeasts are also used as probiotics. Each strain must have undergone controlled evaluation to document health benefits in the target host and their safety (FAO/WHO, 2002).

To exert their effects, probiotics should be present in a dairy food to a minimum level of 10^6 CFU/g, and the daily intake should be about 10^8 CFU/g (Shah, 2007). Probiotics have to be able to maintain their viability during industrial processes and storage, survive the passage through the gastrointestinal tract and adhere and colonise the gut mucosa, promoting immunostimulation without inflammatory effects. They should not be pathogenic, toxic, mutagenic or carcinogenic in the host organism, must be antagonistic to pathogens and genetically stable without a plasmid transfer mechanism, especially concerning antibiotic resistance (Saarela et al., 2000). Probiotics may exert their beneficial health effects by normalisation of the host's microbiota, by inhibition of pathogens, by interaction with the immune system of the host and through their own metabolic activity (Fernández et al., 2015).

Consumer interest in foods that confer health benefits has triggered the development of probiotic products. Mainly fermented milk or yogurt-like products containing specific strains of *Lb. acidophilus, Lb. casei, Bifidobacterium longum, Lb. fermentum, Lb. rhamnosus, Lactobacillus reuteri, Lactobacillus crispatus, Lb. plantarum, Bifidobacterium animalis* and *Bifidobacterium lactis* are available in the market, but many probiotic cheeses have also been developed worldwide (reviewed by Granato et al., 2010).

4.5 Cheese Pathogens

Although milk and dairy products are associated with a healthy diet, milk can be a major source of foodborne pathogens of human health significance (Oliver, Jayarao & Almeida, 2005). Pathogenic microorganisms can access raw milk by a direct transfer from blood (systemic infection) or udder infection (mastitis), or during or after milking from faeces, the animal skin, the environment, and so on (Verraes et al., 2015). Pasteurisation should effectively eliminate pathogenic microorganisms from milk. However, contact with pathogen-contaminated equipment or infected workers during cheesemaking can cause cheese contamination (Oliver, Jayarao & Almeida, 2005).

Campylobacter spp., verotoxigenic *Escherichia coli* (VTEC), *Salmonella* spp., and *Listeria monocytogenes* are the organisms that have been most frequently associated with foodborne illness outbreaks (Oliver et al., 2009).

Several control measures can be taken to limit the presence of pathogens in dairy products. At the farm level, the infection pressure in the stable and milk temperatures should be kept as low as possible to prevent pathogen presence and growth. During milk processing, efficient starter cultures should be used to restrict the pathogen growth. After cheese production, low temperatures from retail to consumer are important to ensure safety. Cleaning and disinfection of the material used are of utmost importance from milking to product commercialization (Verraes et al., 2015).

The development of a disease after consumption of contaminated dairy products depends on several factors such as the pathogenicity of the strain, the number of ingested microorganisms, the health condition of the consumer at the moment of ingestion, and so on. Persons belonging to the YOPI group (young, old, pregnant, immunodeficient) have a higher risk of infection for certain pathogens such as *L. monocytogenes* than healthy people. According to EFSA BIOHAZ Panel, *Salmonella, Campylobacter* and VTEC can cause illness in low numbers and are allocated a score of 3; *L. monocytogenes, Brucella, Mycobacterium bovis* and pathogenic species of *Streptococcus* have to grow to cause illness and have a score of 2 and *Staphylococcus aureus* has to grow to high numbers (often higher than 5 Log CFU/g) to be able to produce sufficient toxins to cause illness, and have a score of 1 (EFSA, 2013).

Dairy products were not the main causative agents of foodborne outbreaks in the European Union in 2013, but they were implicated in a few cases (Table 4.1). It was confirmed that one *Listeria* outbreak in Poland, one *Salmonella* outbreak in Slovakia and several VTEC outbreaks were caused by cheese, and several *Campylobacter* outbreaks were caused by milk. No monitoring on staphylococcal outbreaks is carried out nowadays.

The Rapid Alert System for Food and Feed (RASFF), a network for the immediate transmission of serious risks to human health derived from food or feed, involves all European Member States, Switzerland, Iceland, Liechtenstein and Norway, EFSA and the European Commission (EC). Several notifications caused by dairy products (Table 4.2) were included in the 2015 RASFF annual report (EC, 2016).

Foodborne multidrug resistance (MDR) outbreaks have been reported and are a matter of great concern (Doyle, 2015).

Table 4.1 Data on foodborne outbreaks in the European Union in 2013 and dairy foods implicated in some of these outbreaks (EFSA-ECDC, 2015).

Pathogen	Confirmed human cases	No. MS[1]	Fatality rate (n°)		Outbreaks, %		No. MS	Dairy food
Listeria	1,736	27	15.60%	(191)	12	0.2%	7	Cheese
VTEC	6,043	28	0.36%	(13)	73	1.4%	11	Cheese
Salmonella	85,268	27	0.14%	(59)	1,168	22.5%	22	Cheese
Campylobacter	214,779	27	0.05%	(56)	414	8.0%	16	Milk

1) MS = member states

Table 4.2 Data on selected food poisoning alerts involving foodborne pathogens in dairy products generated by RASFF in 2015 (EC, 2016).

Pathogen	Level	Food product	Origin	Notified by	Confirmed human cases
Listeria monocytogenes	6 ×10³ CFU / g	Raw cow milk cheese	France	France	1
VTEC (O26)		Raw cow milk cheese	Ireland	Ireland	2
Salmonella Enteritidis	Presence in 25 g	Raw milk Reblochon cheese	France	France	15
Staphylococcal enterotoxin		Raw milk cheese	France	France	114

4.5.1 Listeria monocytogenes

Listeria is a Gram-positive, facultative anaerobic, psychrotrophic, catalase-positive rod-shaped bacterium widely distributed in the environment (Warriner & Namvar, 2009).

The genus *Listeria* is composed of 19 species and 6 subspecies (www.bacterio.net), but only *L. monocytogenes* and *L. ivanovii* are pathogenic. *L. monocytogenes* can cause two forms of diseases: a febrile gastroenteritis and an invasive systematic disease (Warriner & Namvar, 2009). Gastroenteritis occurs in healthy individuals after one to seven days of consuming *L. monocytogenes* (>8 Log CFU) contaminated foods. Although symptoms disappear over a few days, *L. monocytogenes* may be shed for several weeks (Drevets & Bronze, 2008). The invasive disease mainly affects the YOPI group. The dose required to cause illness in susceptible individuals is thought to be of the order of 100–1000 cells. *L. monocytogenes* can invade the gastro-intestinal epithelium and becomes associated with monocytes, then subsequently the liver, spleen and lymphatic system. It can cross over to the nervous system and through the placental barrier, leading to septicaemia, encephalitis, meningitis and stillbirth (Drevets & Bronze, 2008). Although listeriosis incidence is low, the high morbidity and mortality rates in vulnerable populations make *L. monocytogenes* one of the most significant foodborne pathogens (Warriner & Namvar, 2009).

In raw cow milk, a frequency of *L. monocytogenes* between 0% and 10% has been reported, and its growth is limited by the microbiota present. Outbreaks linked to raw milk consumption were due to deficiencies in pasteurisation or post-pasteurisation contamination (Claeys et al., 2013).

Foods implicated in listeriosis outbreaks are ready-to eat (RTE) meats, soft cheese and seafood, due to their high protein, moderate a_w and low background microflora. Cheese was the causative agent of one listeriosis outbreak in 2013 in Poland (EFSA-ECDC, 2015) and one *L. monocytogenes* food poisoning alert included in the RASFF 2015 annual report (Table 4.2). The ability of *L. monocytogenes* to grow at low temperatures, form biofilms and resist sanitisers makes its elimination from food processing environments very difficult (Marchand et al., 2012).

4.5.2 Escherichia coli

Escherichia coli is a Gram-negative, facultative anaerobic rod-shaped bacterium that colonises the intestinal tract of mammals. Several highly adapted virulent *E. coli* clones are able to colonise a mucosal site, evade host defences, multiply and cause host damage (Karper, Nataro & Mobley, 2004). *E. coli* pathotypes are characterised by shared O (lipopolysaccharide, LPS) and H (flagellar) antigens that define serogroups (O antigen only) or serotypes (O and H antigens) (Nataro & Karper, 1998). Enterohaemorrhagic *E. coli* (EHEC) outbreaks are caused by toxin-producing strains (also known as STEC or VTEC) and have been associated with several foods. The systemic absorption of the toxin can lead to haemorrhagic colitis, non-bloody diarrhoea, haemolytic uremic syndrome and even death. Strains of the O157:H7 serotype are the most important VTEC pathogens, but other serotypes, as O26 and O111 serogroups, can also cause disease (Karper, Nataro & Mobley, 2004).

Cattle have been identified as the principal reservoir for VTEC, and the estimated frequencies for VTEC and *E. coli* serotype O157:H7 in milk in Europe are between 0% and 5.7% and between 0% and 2%, respectively (Claeys et al., 2013). Small ruminants may also harbour VTEC. Pasteurisation destroyed VTEC, and the pathogen is unable to grow in pasteurised refrigerated milk, although growth can occur under temperature abuse conditions (Claeys et al., 2013).

Foods implicated in VTEC outbreaks are bovine meat and products thereof, followed by vegetables, juices and other products thereof and cheese. Cheese was the causative agent of several VTEC outbreaks in 2013 (EFSA-ECDC, 2015) and one VTEC food poisoning alert included in the RASFF 2015 annual report (Table 4.2).

4.5.3 Salmonella enterica

Salmonella is a Gram-negative, facultative anaerobic, rod-shaped bacterium, usually motile by flagella, which produce gas from glucose, utilise citrate as a sole carbon source and is generally not able to use lactose (El-Gazzar & Marth, 1992). *Salmonella* is associated with the intestinal tract of mammals, is widespread in the environment and appears in a wide variety of foods.

Salmonella enterica ssp. *enterica* serovars Enteritidis and Typhimurium are one of the most commonly reported causative agents of human foodborne diseases. After consuming food that contains the pathogen, it enters the digestive tract, grows in the small intestine, and causes inflammation resulting in enterocolitis. The symptoms (abdominal pain, diarrhoea, nausea, vomiting, chills and fever) usually appear 12 to 36 h after eating contaminated food and last from 2 to 6 days (Gravani, 1984).

Most studies report a frequency of *Salmonella* spp. in bulk tank milk below 1% (Claeys et al., 2013). Salmonellosis outbreaks from milk and dairy products have been linked to inadequate pasteurisation or post-process contamination. Several MDR *Salmonella* outbreaks caused by cheese have been reported in the United States and France (Doyle, 2015).

Salmonella remained the most commonly detected causative agent in foodborne outbreaks (22.5% of outbreaks) reported in the EU in 2013. Eggs and egg products were the most frequently identified food vehicles, followed by sweets, chocolates, pork and pork products. One *Salmonella* outbreak for cheese consumption was reported in Slovakia in 2013 (EFSA-ECDC, 2015), and one *Salmonella* food poisoning alert was included in the RASFF 2015 annual report (Table 4.2).

4.5.4 Campylobacter spp.

Campylobacter is a Gram-negative, microaerophilic, rod-shaped bacterium with a spiral appearance and motile by one or two flagella which obtain energy from amino acids, or tricarboxylic acid cycle intermediates. It is a common commensal in the gastrointestinal tracts of wild and farm animals and widespread in the environment. *Campylobacter* is an infrequent cause of mastitis in dairy cattle, but faecal shedding by asymptomatic animals is considered to be the main source of bulk tank milk contamination (Ruegg, 2003). Most studies report a *Campylobacter* frequency in bulk tank milk of between 0% and 6% (Claeys et al., 2013).

Consumption of *Campylobacter*-contaminated milk causes a foodborne illness characterised by sporadic cases of chronic gastritis, enterocolitis and septicaemia. It can result in sequelae like *Campylobacter*-associated Guillain-Barré syndrome (Oliver, Jayarao & Almeida, 2005). Thermophilic (able to grow at 42°C) *Campylobacter* strains (*C. jejuni, C. coli, C. lari* and *C. upsaliensis*) are some of the most frequent etiological agents of bacterial gastroenteritis in humans (Kwan et al., 2008). The increasing evidence of strains of animal origin resistant to tetracycline and fluoroquinolones is recognised as an emerging public health problem (Doyle, 2015).

Campylobacter has been the most commonly reported gastrointestinal bacterial pathogen in humans in the EU since 2005. Broiler meat was the most frequently identified food vehicle, followed by other, mixed or unspecified poultry meat and products thereof and by milk and mixed food in the EU in 2013 (EFSA-ECDC, 2015).

4.5.5 Staphylococcus aureus

Staphylococcus is a Gram-positive, facultative anaerobic coccoid-shaped bacterium able to grow at low a_w, low pH and high salt concentration. *S. aureus* is one the most common causative bacterial agents of mastitis (Le Maréchal et al., 2011).

S. aureus food poisoning is caused by ingestion of food containing staphylococcal enterotoxins (SEs). SEs are small (MW = 24–28 kDa), water-soluble and highly stable proteins secreted into the culture medium by some *S. aureus* strains (Rosengren et al., 2010). At least 21 SE types have been described, being some of them (SEA to SEI, SER, SES and SET) emetic (Cretenet, Even & Le Loir, 2011). SEs withstand pasteurisation and are not inactivated by curd heating or low pH (Le Loir, Baron & Gautier, 2003). Symptoms include sudden onset of nausea, vomiting, abdominal cramps and diarrhoea (Rosengren et al., 2010).

S. aureus is the most frequent pathogen associated with raw milk cheeses (De Buyser et al., 2001). Several outbreaks caused by cheeses made of raw or pasteurised milk have been documented (Cretenet, Even & Le Loir, 2011). Storage of milk at temperatures above 10°C and/or poor starter culture activity during cheesemaking are the main factors responsible for *S. aureus* growth and staphylococcal intoxication. *S. aureus* levels tend to decrease during cheese ripening and might be absent in fully mature cheese, but SEs might be present (Cretenet, Even & Le Loir, 2011). One SE foodborne outbreak was reported in the RASFF 2015 annual report (Table 4.2).

MDR staphylococcal isolates, for example, methicillin-resistant *S. aureus* (MRSA), have been isolated from both humans and animals (Morgan, 2008).

4.6 Other Risks of Microbial Origin

Cheeses, as well as other fermented foods, could contain high levels of biogenic amines (BAs), which are the causative agent of food poisoning episodes (Halász et al., 1994).

The most important BAs in dairy foods are tyramine, histamine (produced by enzymatic decarboxylation of tyrosine and histidine, respectively), putrescine (synthesised via ornithine decarboxylation or agmatine deamination) and to a minor extent, cadaverine (originated by lysine decarboxylation). Gram-negative bacteria (mainly Enterobacteriaceae) present in milk are able to produce histamine, putrescine and cadaverine. However, the main BA producers in cheese are mostly LAB (Fernández et al., 2015). The use of non-producing BA starter cultures (Crow, Curry & Hayes, 2001), pasteurised or high-pressure-treated milk in cheesemaking (Novella-Rodríguez et al., 2002) or a high-pressure processing treatment in raw milk cheeses (Calzada et al., 2013) have been proposed as strategies to reduce BA build-up in cheese.

Mycotoxins are secondary metabolites synthesised by certain filamentous fungi that are reported to be carcinogenic, tremorogenic, haemorrhagic, teratogenic and dermatitic (Sweeney

& Dobson, 1998). Their presence in dairy products can be the result of indirect contamination, through ingestion of contaminated feed by lactating animals, or of direct contamination, through growth of moulds on dairy products. Although species of the genera *Penicillium* and *Aspergillus* isolated from various cheeses have been shown to produce a wide range of mycotoxins, the ripening temperatures and carbohydrate content in cheese are not well suited for mycotoxin production. However, studies reporting the presence of mycotoxins in cheese have been published (reviewed by Himery et al., 2014).

4.7 Growth and Survival of Bacterial Pathogens in Cheese

The behaviour of *L. monocytogenes* in cheese depends mainly upon pH and temperature conditions during manufacture, ripening and storage. During storage of Colby, *L. monocytogenes* levels gradually decreased from 3.5 to 1.5 Log CFU/g (Yousef and Marth, 1988). Feta cheese does not support *L. monocytogenes* growth (Papageorgiou & Marth, 1989a). In Mozzarella, stretching the curd (even containing 6.2×10^4 CFU/g) in water at 77°C for 3–4 min caused complete demise of the pathogen (Buazzi, Johnson & Marth, 1992). *L. monocytogenes* levels increased 10-fold during Camembert manufacture, decreased during the first 18 days of ripening and increased as the pH rose (Ryser & Marth, 1987b). In blue cheese, after a slight increase during cheesemaking, *L. monocytogenes* levels decreased 2.6 Log units during the first 50 days of ripening (Papageorgiou & Marth, 1989b). In Cheddar, growth of *L. monocytogenes* during manufacture was inhibited by proper acid development (Ryser & Marth, 1987a). Even post-process contamination of low-salt Cheddar did not support the growth of *L. monocytogenes* (Shrestha et al., 2011a). In naturally contaminated raw milk farmhouse Cheddar, *L. monocytogenes* never exceeded 20 CFU/g and could not be detected after five months of ripening (Dalmasso & Jordan, 2014). During Parmesan ripening, *L. monocytogenes* levels decreased almost linearly and faster than reported for other hard cheeses and was not detected after 2–16 weeks of ripening (Yousef & Marth, 1990).

Studies on the fate of *Escherichia coli* O157:H7 in different types of cheese are numerous. In Cottage cheese, *E. coli* O157:H7 levels increased 100-fold during the manufacturing process, but death occurred during cooking of the curd and whey (Arocha et al., 1992). In 'Queso Fresco' made from milk inoculated with two O157:H7 strains, no growth of either strain was observed during a two-month storage period at 8°C (Kasrazadeh & Genigeorgis, 1995). Studies on post-manufacture contamination of *E. coli* O157:H7 in three Greek whey cheeses inoculated the day after production at 10^6 CFU/g showed that the pathogen was able to grow (1.3 Log CFU/g increase) in cheeses stored at 12°C and survived in cheeses stored at 2°C (Govaris, Koidis & Papatheodorou, 2001). *E. coli* O157:H7, inoculated at 10^4 CFU/mL in milk, survived the manufacturing process of Camembert and Feta and grew during the storage (65 and 75 days, respectively) period (Ramsaran et al., 1998). In Mozzarella made from unpasteurised milk inoculated with *E. coli* O157:H7 at 10^5 CFU/mL, stretching at 80°C completely destroyed the pathogen but stretching at 70°C only achieved a 10-fold reduction (Spano et al., 2003). In Cheddar and Gouda cheeses manufactured from unpasteurised milk contaminated with one of three strains of *E. coli* O157:H7 at 20 CFU/mL, pathogen levels increased to approximately 145 CFU/g in one-day-old cheeses. Counts dropped significantly to mean levels of 25 and 5 CFU/g after 60 days in Cheddar and Gouda, respectively, but remained detectable after selective enrichment for more than 270 days in both cheese types. Results suggest that the 60-day ageing requirement alone is insufficient to completely eliminate levels of viable *E. coli* O157:H7 in stirred-curd Cheddar or Gouda cheese manufactured from raw milk contaminated with low levels of this pathogen (D'Amico, Druart & Donnelly, 2010). In a study on the survival of three pathogens in Danish raw milk cheese by pyrosequencing and quantitative RT-PCR, the highest percentage of *E. coli* sequence reads were found at 7 days of ripening and decreased in the later ripening stages. *E. coli* growth appeared to be affected by the cooking temperature and the rate of acidification but not by the starter culture used or the indigenous microbiota of raw milk (Masoud et al., 2012).

Salmonella survival in several cheese varieties has also been studied. In Colby and Cheddar cheeses, the rate and amount of acid production during cheesemaking, the pH of cheese, and the type and size of the starter inoculum were important factors in suppressing its growth and survival, while salt, moisture, chemical additives and milk pasteurisation (before artificial contamination) had little or no effect. Addition of large numbers of *Propionibacterium* and *Leuconostoc* seemed to favour *Salmonella* survival, whereas lactobacilli and enterococci had no effect (Hargrove, McDonough & Mattingly, 1969). In Cottage cheese, *Salmonella* survived cooking temperatures of 43.3°C and 46.1°C but not those ranging from 51.7°C to 54.4°C. No decrease in *Salmonella* levels were observed in Creamed Cottage stored at 4.4°C (McDonough, Hargrove & Tittsler, 1967). Milk-inoculated *Salmonella* Javiana grew and survived the acid-ripening phase of Mozzarella-type cheese but did not survive the 60°C stretching and moulding (Eckner et al., 1990). *Salmonella* levels in low-salt Cheddar contaminated with a five-strain *Salmonella* cocktail at 4 Log CFU/g were much higher than in standard-salt Cheddar, irrespective of their low or high pH. Survival of *Salmonella* during storage for up to 90 days at 4°C or 10°C and for up to 30 days at 21°C was also stated (Shrestha et al., 2011b). In Domiati cheese manufactured from pasteurised buffalo milk with 5% or 10% NaCl and inoculated with 4 to 6 Log CFU/mL *Salmonella* Typhi, survival times of *S.* Typhi were 34 and 16 days in cheeses with low or high NaCl, respectively (Naguib, Sabbour & Nour, 1979).

Studies on fresh farm-made cheeses found that enterotoxigenic *S. aureus* were frequently found in raw milk fresh cheeses at levels of concern (Jakobsen et al., 2011; Rosengren et al., 2010). The use of starter cultures significantly lowered the levels of *S. aureus* in raw milk fresh cheeses (Rosengren et al., 2010). In soft-curd cheeses elaborated with milk containing approximately 7.3 Log CFU/mL *S. aureus*, pathogen levels reached values of approximately 8.5 Log CFU/g after manufacture and showed no significant decreases during storage (López-Pedemonte et al., 2006). In Camembert, a 1–3 Log units increase during cheesemaking and stable *S. aureus* populations during ripening were reported. SE were also produced in cheeses with initial *S. aureus* populations of 3–6 Log CFU/mL (Meyrand et al., 1998). Numerous studies on the growth of *S. aureus* in Cheddar were performed after several staphylococcal food poisoning outbreaks in the United States in the 1960s (Johnson, Nelson & Johnson, 1990). The *S. aureus* population grew until pressing and then decreased during ripening (Ibrahim et al., 1981a). The absence of an active starter favoured the growth of *S. aureus* (Ibrahim et al., 1981b). Salting and a temperature decrease induced an increase in the population of *S. aureus*, probably due to a lower starter activity in those conditions (Ibrahim et al., 1981a). SE A production depended on the size of the inoculum and the activity of the starter and was indirectly influenced by the salt concentration (Ibrahim et al., 1981a). In Manchego cheese, the *S. aureus* population grew until pressing and then decreased during ripening (Nuñez et al., 1988). An active starter culture lowered *S. aureus* counts from day 1 to the end of ripening (Gaya et al., 1988). In Saint-Nectaire and Salers cheeses made with raw milk whose coagulase-positive staphylococci (SC+) counts ranged from undetectable (<10 CFU/mL) to 3.03 Log CFU/mL, levels on day 1 ranged from 2.82 to 6.84 Log CFU/g, and SE were detected in two Salers cheeses whose SC+ counts on day 1 were above 5 Log CFU/g and whose pH values at 6 h were 6.5–6.6 (Delbes et al., 2006). In cheeses where the curd is scalded at high temperatures (e.g. 52°C–55°C for a maximum of 60 min for Emmental or Gruyère), the risk of *S. aureus* growth is greatly reduced. *S. aureus*–LAB interactions in a cheese matrix studied by DNA microarrays found that *L. lactis* affected the carbohydrate and nitrogen metabolisms and the stress response of *S. aureus* (Cretenet et al., 2011).

4.8 Procedures to Improve Cheese Safety

Raw milk contains different systems with antimicrobial properties that inhibit the growth of microorganisms in raw milk and/or contribute to the immunity of the calf, among which are enzymes and proteins.

Lactoperoxidase (EC 1.11.1.7, LPS) exerts its antimicrobial activity in combination with hydrogen peroxide and thiocyanate. Although both are present in milk, addition is needed to achieve its antibacterial benefits (Reiter et al., 1976; Uceda et al., 1994). Lysozyme (EC 3.1.2.17) has bactericidal effects against Gram-positive bacteria (Fox & Kelly, 2006). Xanthine oxidase or oxidoreductase (EC 1.17.3.2) contributes to the activation of LPS by supplying hydrogen peroxide (Fox & Kelly, 2006).

Lactoferrin is an iron-binding protein with bacteriostatic effects due to the deprivation of some ions (Fe, Mg and Ca) needed for microbial growth and survival. Lactoferrin-derived peptides are 100–1000 times more potent than intact lactoferrin (Steijns & van Hooijdonk, 2000). Bovine immunoglobulins transfer immunity to calves and may provide some immunity in the gut (Korhonen, Marnila & Gill, 2000).

4.8.1 Biopreservatives of Microbial Origin

Biopreservation is defined as the extension of shelf life and enhanced safety of food by the use of natural or controlled microbiota and/or antimicrobial compounds. It has become increasingly popular because it should have a smaller impact on food nutritional and sensory properties than chemical treatments, can reduce processing costs, does not require advanced technological equipment or skills and could improve animal productivity by natural means or control emerging pathogens (Reis et al., 2012).

LAB have been widely used to improve the safety of fermented products due to the production of a broad variety of antimicrobial compounds.

In situ production of organic acids increases the shelf life and safety of the final product. Their antimicrobial effects on both Gram-positive and Gram-negative bacteria are the result of the undissociated molecules. Lactic acid is generated by LAB as the end product of glucose fermentation. Acetic acid is produced by heterofermentative LAB (Reis et al., 2012). Benzoic acid is formed by certain strains of *L. acidophilus*, *L. casei*, *S. thermophillus* and *L. helveticus* from some acids present in milk (Garmiene et al., 2010). Propionic acid, produced by propionibacteria from lactate, also has fungistatic properties (Reis et al., 2012).

In situ production of hydrogen peroxide (H_2O_2) inhibit the growth of psychrotrophic and pathogenic microorganisms at refrigeration temperatures. Most *Lactobacilli* species are able to form hydrogen peroxide by oxidising lactate (Reis et al., 2012).

Diacetyl (2,3-butanodione) has a broad antimicrobial activity, exerted through interaction with the microbial cell membrane. Gram-positive bacteria are more resistant, while Gram-negative and yeasts exhibit a higher sensitivity (Lanciotti et al., 2003).

Reuterin (β-hydroxypropionaldehyde), produced by some *L. reuteri* strains, shows antimicrobial activity towards a broad spectrum of foodborne pathogens and spoilage organisms. It has a higher antimicrobial activity on Gram-negative than on Gram-positive pathogens (Arqués et al., 2008a and b; El-Ziney & Debevere, 1998). In situ reuterin production by *L. reuteri* plus glycerol in dairy products has been proved (Langa et al., 2013) and was effective in preventing the late-blowing defect caused by *Clostridium tyrobutyricum* in hard cheese (Gómez-Torres et al., 2014).

Bacteriocins are antimicrobial peptides active against related (narrow spectrum) or nonrelated (broad spectrum) bacteria. They form transitory pores in the cytoplasmic membrane of sensitive cells that cause dissipation of the proton motive force. They are classified into class I (posttranslational modified) and class II (unmodified) bacteriocins (Cotter, Hill & Ross, 2005). The potential of bacteriocins to control the growth of pathogens in foodstuffs has been extensively reviewed (e.g. Cotter, Hill & Ross, 2005; Favaro, Barretto Penna & Todorov, 2015; Holzapfel, Geisen & Schillinger, 1995).

Nisin is effective against the genera *Clostridium*, *Bacillus* and *Listeria* but not against Gram-negative bacteria, yeasts or moulds (Medina & Nuñez, 2010). A nisin-producing starter culture effectively inhibited the growth of *L. monocytogenes* in Camembert during the first 24 h and until the end of the second week of ripening, although regrowth of the pathogen was observed later (Maisnier-Patin et al., 1992). In raw milk Manchego cheese, nisin-producing *Lc. lactis* ssp. *lactis* strains (ESI 515 and TAB 50) decreased *Listeria* levels throughout a 60-day ripening period (Rodríguez et al., 1998), and the TAB 50 strain lowered *S. aureus* levels during cheese ripening (Rodríguez et al., 2000). A nisin-producing *Lc. lactis* ssp. *cremoris* M104 strain inhibited the growth of a three-strain cocktail of enterotoxigenic *S. aureus* in Graviera cheese, although the pathogen survived during ripening and storage (Samelis et al., 2014).

Lacticin 3147 has a broad activity spectrum (Ryan et al., 1996). A Lacticin 3147–producing *Lc. lactis* strain used as a starter culture in the manufacture of Cottage cheese reduced numbers of *L. monocytogenes* to <10 cells/g within five days at 4°C (McAuliffe, Hill & Ross, 1999). This strain, applied to the surface of smear-ripened cheese, resulted in a 3 Log unit reduction in *L. monocytogenes* levels with respect to control cheese (O'Sullivan et al., 2006).

Pediocin PA1/AcH is a broad-spectrum bacteriocin produced by strains of *Pediococcus acidilacti*. A pediocin-producing *L. plantarum* strain (WHE 92) successfully inhibited *L. monocytogenes* growth for 21 days when sprayed on the surface of Munster cheese (Ennahar, Assobhei & Hasselmann, 1998). However, resistant *L. monocytogenes* strains developed when used in red-smear cheeses (Loessner et al., 2003). A pediocin-producing *Lc. lactis* strain (CL1) added as adjunct to the starter culture in cheese manufacture resulted in a 3 and 1 Log reduction of *L. monocytogenes* and *S. aureus* populations, respectively, after 30 days of ripening (Rodríguez et al., 2005b).

Many strains of the genus *Enterococcus* produce bacteriocins (named enterocins) with strong antilisterial activity (Giraffa, 2003). Enterocin AS-48, a cyclic bacteriocin produced by *E. faecalis* INIA 4 (Rodríguez et al., 1997), decreased *L. monocytogenes* counts by 3 Log units after 8 h and by 6 Log units after 7 days in raw milk Manchego cheese (Nuñez et al., 1997). Enterocin 1146, produced by *E. faecium* DPC 1146, exhibited a rapid bactericidal effect on *L. monocytogenes* in milk and was stable throughout the ripening of Cheddar cheese (Parente & Hill, 1992). The enterocin produced by *E. faecium* 7C5 during Talegio cheesemaking was stable until the end of the ripening and inactivated *L. monocytogenes* on the cheese surface (Giraffa & Carminati, 1997).

Although several bacteriocin-producing *Lactobacillus* strains have been characterised (Favaro, Barretto Penna & Todorov, 2015), works dealing with their application are scarce. Bacteriocinogenic *Lactobacillus sakei* 2a strain inhibited the growth of *L. monocytogenes* in cheese spreads by in situ bacteriocin production (Ruiz Martínez et al., 2015).

Bacteriocins produced by streptococci isolated from food-related environments have also been characterised. A thermophilin-producing *S. salivarius* ssp. *thermophilus* B used in the manufacture of yogurt exhibited higher activity against *L. monocytogenes* than against *S. aureus* (Benkerroum, Oubel & Ben Mimoun, 2002). *S. macedonicus* ACA-DC 198, used as starter or adjunct culture in Kasseri cheese, produced the lantibiotic macedocin that was active during the 90 days of ripening (van den Berghe et al., 2006).

Only nisin use as additive has received regulatory approval by the FDA and the EFSA. Several attempts to incorporate nisin in different packaging systems have been tested: nisin-coated films exhibited antimicrobial activity during milk storage (Mauriello et al., 2005). Cellulose-based bioactive inserts with nisin adsorbed to the surface reduced the levels of *Listeria innocua* and

S. aureus in sliced Cheddar cheese stored in modified atmosphere packaging (MAP) at refrigeration temperatures (Scannell et al., 2000). Polymer films releasing nisin and/or natamycin were unsuitable for the packaging of a surface-ripened cheese but prevented the growth of spoilage microorganisms on the surface of a packaged soft cheese (Hanusova et al., 2010).

Natamycin or pimaricin is a natural antifungal compound produced by *Streptomyces natalensis* that specifically binds to ergosterol, altering membrane functions. It is active against yeasts and moulds and has no effect on bacteria, viruses or protozoa. Natamycin has been approved as a food additive in over 40 countries (Ollé Resa, Jagus & Gerschenson, 2014). It can be applied on the cheese surface by spraying, dipping or brushing. Entrapment in edible films decreases diffusion and maintains high in situ concentrations (Kristo, Koutsoumanis & Biliaderis, 2008).

Plant extracts have antimicrobial properties against bacteria, moulds and yeast. In general, Gram-positive bacteria are more sensitive than Gram-negative bacteria. Out of 21 plant essential oils, those of bay, cinnamon, clove and thyme were the most inhibitory ones against foodborne pathogens (Smith-Palmer, Stewart & Fyfe, 1998). These four oils reduced *L. monocytogenes* and *Salmonella* Enteritidis levels to less than or equal to 1.0 Log CFU/mL in low-fat cheese. In contrast, in full-fat cheese, only clove oil was able to achieve such *L. monocytogenes* reduction, and all of them proved ineffective against *S.* Enteritidis (Smith-Palmer, Stewart & Fyfe, 2001).

Bacteriophages or phages are viruses that specifically infect and multiply in bacteria and, therefore, can be considered natural antibacterial agents. The narrow host specificity of phages makes them very attractive candidates as biopreservatives in fermented products to avoid interference with starter performance or the development of the secondary microbiota. A bacteriophage preparation effective against *L. monocytogenes*, Listex™ P100 (EBI Food Safety) received FDA approval for use on cheese in 2006, and on all food products in 2007 (García et al., 2010).

Several successful examples of phage use to combat pathogens in milk and cheese are available: Complete eradication of *L. monocytogenes* was achieved on surface-ripened smear cheese by surface application of the virulent phage P100 (Carlton et al., 2005). A cocktail of two lytic phages inactivated *S. aureus* in milk and curd (García et al., 2007) and was also successfully applied to fresh and hard cheese (Bueno et al., 2012).

4.8.2 Physical Treatments

Thermal treatments destroy foodborne pathogens and are one of the primary techniques used to ensure food safety. The heat resistance of any pathogen is influenced by many factors such as strain variation, previous growth conditions, exposure to heat, acid, and other stresses and composition of the matrix. In addition, the number of surviving cells detected depends on the recovery medium and incubation conditions used.

The heat resistance of *L. monocytogenes* was reviewed by Doyle et al. (2001). Studies on the survival of *L. monocytogenes* during the manufacture of non-fat dry milk, Cottage and Mozzarella cheeses revealed that it can survive these processes if the inoculum is high or processing temperatures are too low.

E. coli O157:H7 has been reported as being not particularly resistant to heat. The thermal inactivation of *E. coli* O157:H7, mainly in meat or meat products, was reviewed by Stringer, George and Peck (2000).

The heat resistance of *Salmonellae* was reviewed by Doyle & Mazzotta (2000). Certain serotypes of *Salmonella enterica*, like *Salmonella* Senftenberg 775W, are known for their resistance to thermal treatments. Studies in milk have shown that a greater concentration of total solids increased the heat resistance of *Salmonella*. More *Salmonellae* survive the drying process when the final moisture content is 6% rather than 3%. *Salmonellae* in dried milk powders are extremely heat resistant, with viable cells still detected after 10 h at 76.6°C.

Johnson, Nelson and Johnson (1990) referred to one study of milk heat treatment for cheesemaking in which pathogens were inoculated into raw milk at levels of 10^5 CFU/mL and milk was heat-treated in a commercial HTST pasteuriser. All strains of *Yersinia enterocolitica, Campylobacter* spp., *E. coli* O157:H7 and all but one *Salmonella* species were destroyed at 65°C. *Salmonella* Senftenberg (rarely isolated from cheese) was inactivated at 69°C. *L. monocytogenes,* at levels of 10^4 CFU/mL was inactivated at 66°C and at 10^5 CFU/mL required 69.0°C.

A large number of bacteria subjected to a heat treatment are not killed but injured and they pose a potential threat, since they may, under suitable conditions, repair themselves (McMahon et al., 2000). Several studies about factors affecting resuscitation of stressed *L. monocytogenes* have shown that repair can occur in a wide range of environmental conditions (pH 4.2–9.6, 0.5–10% NaCl contents and temperatures 4°C–43°C). The combination of several adverse conditions acts synergistically in preventing the repair of injured *L. monocytogenes* cells (Besse, 2002).

High hydrostatic pressure (HHP) processing is a non-thermal technology that produces microbiologically safe products, modifies the functional properties of proteins and polysaccharides and alters biochemical reactions with minimal impact on their nutritional and sensory characteristics (Farkas & Hoover, 2000). The resistance of microorganisms is highly variable, with Gram-positive bacteria generally being more resistant than Gram-negative ones (Russell, 2002). The bactericidal efficacy of HHP processing is affected by the pressure level, holding time length, treatment temperature and matrix composition, as well as the bacterial growth temperature and growth phase.

Milk shows a baroprotective effect on bacteria: Inactivation of *E. coli* MG1655 in phosphate buffer and milk at 600 MPa for 15 min at 20°C reached 8.3 and 1.6 Log units, respectively (García-Graells, Masschalck & Michiels, 1999). However, treatments of *L. monocytogenes* in skim and whole milk at 600 MPa for 1.5 min resulted in a higher survivor rate in skim milk than in whole milk (Bull et al., 2005). An HHP treatment at 400 MPa in combination with the activated LPS achieved complete *L. innocua* inactivation (>7 Log units), whereas a 2 to 5 Log unit reduction was achieved with the HHP treatment alone. However, the same combination did not increase *E. coli* inactivation with respect to the HHP treatment alone (García-Graells, Valckx & Michiels, 2000).

In cheese, HHP treatments, applied at different stages of ripening, achieved destruction or significant reduction in the levels of pathogens and spoilage bacteria. HHP treatments (450 MPa for 10 min or 500 MPa for 5 min) of 14-day-old Sainte Maure cheese resulted in *L. monocytogenes* reductions of more than 5.6 Log units (Gallot-Lavallée, 1998). HHP treatments, applied after 2 or 50 days of ripening, of cheeses made with raw milk inoculated with *L. monocytogenes* Scott A achieved pathogen reductions of 0.9 and 6.2 Log units at 300 MPa for 10 min and of 5.0 and above 6.3 Log units at 500 MPa for 5 min, respectively (Arqués et al., 2005). Gram-negative pathogens such as *E. coli* O157:H7 and *Salmonella* are less baroresistant than *Listeria*. Pressurisation of 2- or 50-day-old cheeses made from raw milk inoculated with *E. coli* O157:H7 achieved reductions of 1.3 and 3.8 Log units at 300 MPa for 10 min and of 3.7 and above 6 Log units at 500 MPa for 5 min, respectively (Rodríguez et al., 2005a). Reductions above 5.4 Log units were obtained for two *Salmonella* strains when model cheeses made from contaminated milk were HHP-treated at 400 MPa and 22°C for 10 min (De Lamo-Castellví et al., 2007).

Combinations of HHP treatments with antimicrobial-producing LAB have been investigated for synergistic effects (Medina & Nuñez, 2010).

High-voltage pulsed electric field (PEF) treatment is a non-thermal processing method consisting in treating liquid foods with high-voltage pulsed electric fields to inactivate microorganisms and enzymes. PEF treatment is applied at ambient temperature for a short (microseconds) time (Sepulveda-Ahumada, Ortega-Rivas & Barbosa-Canovas, 2000). The electric field causes a potential difference across the cell membrane and induces a sharp increase in membrane conductivity and permeability. Membrane destruction occurs when the induced membrane potential exceeds a critical value of 1 V for a short duration (2 ms), which corresponds to an external field of roughly 10 kV/cm for *E. coli*. The principal processing variables influencing the PEF treatment are electric field intensity, number of pulses, treatment time and microbial cell concentration (Bendicho, Barbosa-Canovas & Martín, 2002).

No significant differences in the inactivation of *L. monocytogenes* Scott A in whole fat, 2% fat or skim milk were observed. Reductions of 1–3 Log units were obtained with a PEF treatment of 400 pulses at 30 kV/cm and 25°C, and a higher field strength or longer treatment time resulted in a greater reduction. Increasing the temperature to 50°C achieved a 4 Log reduction (Reina et al., 1998). A 3 Log reduction was obtained in pasteurised milk inoculated with *E. coli* by a PEF treatment of 23 pulses at 42.8 kV/cm (Bendicho, Barbosa-Canovas & Martín, 2002). *Salmonella* Dublin, inoculated at 3.6 Log CFU/mL in raw milk, disappeared after a PEF treatment of 40 pulses at 36.7 kV/cm and was not detected in treated milk for 8 days at 7°C–9°C (Sensoy, Zhang & Sastry, 1997). Raw milk shelf life was greatly increased (up to two weeks) when subjected to several PEF treatments (Qin et al., 1995).

The inactivation rates achieved by PEF can be improved by combining it with other processes. Moderate heating applied after the PEF treatment increased the inactivation rate (Sobrino et al., 2001). Nisin (10 or 100 IU/mL) addition after PEF treatment had an additive or synergistic effect with low or high field strengths, respectively (Calderón-Miranda, Barbosa-Cánovas & Swanson, 1999).

Cheddar cheese was made with PEF-treated milk, HTST or LTLT pasteurised in an attempt to test PEF as an alternative to milk pasteurisation. HTST and LTLT pasteurisation reduced microbial levels in raw milk (2.4 Log CFU/mL) by 0.7 Log units, whereas only a 0.22 Log units reduction was achieved by the PEF treatment (Sepulveda-Ahumada, Ortega-Rivas & Barbosa-Canovas, 2000).

Irradiation or ionising radiation (γ-rays, X-rays, electron beam) is a decontamination technology that upgrades food safety and extends the shelf life of food (Farkas, 2006). About 100 irradiated food products (red meat and poultry, seafood, fruits and vegetables, egg products and spices) have been approved for consumption in countries worldwide (Mahapatra, Muthukumarappan & Julson, 2005). The irradiation dose applied should be balanced to provide the highest antimicrobial effectiveness with a reasonable safety margin along with retention of food wholesomeness (Konteles et al., 2009).

Ras cheese made from irradiated milk was free from pathogens and coliforms and had enhanced body and flavour intensity in comparison to the cheese made with non-irradiated milk. Although an oxidised flavour was detected on fresh and one-month-old Ras cheese, it disappeared during ripening (Abd El Baky et al., 1986). In a study on Feta pre- or post-process contamination, milk or the packaging brine was contaminated with *L. monocytogenes* at 10^3 CFU/mL, and contaminated Feta samples were vacuum-packaged, exposed to irradiation doses of 1.0, 2.5 and 4.7 kGy and stored at 4°C for one month. None of the irradiation doses eliminated *L. monocytogenes* in pre-process contaminated samples, but the highest dose reduced its population to a level compliant with EU regulations. In the post-process contamination, the 2.5 and 4.7 kGy doses reduced its counts below the detection limit. Irradiation at 4.7 kGy altered colour and affected the aroma profile, although it was restored after 30 days of cold storage (Konteles et al., 2009).

4.9 Conclusions and Future Trends

Although dairying started more than eight millennia ago, there is still place for innovation in the field as we expand our knowledge about cheese microbial communities and the beneficial effects of these microorganisms on cheese and on our own health.

The consumption of raw milk cheeses implies a higher microbiological risk than that of pasteurised milk cheeses due to the possible presence of pathogens. Prevention of milk contamination should start at the farm level.

A better understanding of the molecular mechanisms underlying pathogenesis and resistance will help us combat these pathogens in food. Combined treatments of biopreservatives and physical treatments will increase the possibilities of fighting these pathogens with minimal effects on nutritional and sensory characteristics, opening up new ways to manage microbial risks in dairy products.

References

Abd El Baky, A. A., Farahat, S. M., Rabie, A. M. & Mobasher, S. A. (1986). The manufacture of Ras cheese from gamma irradiated milk. *Food Chemistry*, 20, 201–212.

Addis, E., Fleet, G. H., Cox, J. M., Kolak, D. & Leung, T. (2001). The growth, properties and interactions of yeasts and bacteria associated with the maturation of Camembert and blue-veined cheeses. *International Journal of Food Microbiology*, 69, 25–36.

Arfi, K., Leclercq-Perlat, M. N., Spinnler, H. E. & Bonnarme P. (2005). Importance of curd-neutralising yeasts on the aromatic potential of *Brevibacterium linens* during cheese ripening. *International Dairy Journal*, 15, 883–891.

Arocha, M. M., McVey, M., Loder, S. D., Rupnow, J. H. & Bullerman, L. (1992). Behaviour of hemorrhagic *Escherichia coli* O157:H7 during the manufacture of Cottage cheese. *Journal of Food Protection*, 55, 379–381.

Arqués, J. L., Rodríguez, E., Gaya, P., Medina, M. & Nuñez, M. (2005). Effects of combinations of high-pressure treatment and bacteriocin-producing lactic acid bacteria on the survival of *Listeria monocytogenes* in raw milk cheese. *International Dairy Journal*, 15, 893–900.

Arqués, J. L., Rodríguez, E., Nuñez, M. & Medina, M. (2008a). Antimicrobial activity of nisin, reuterin, and the lactoperoxidase system on *Listeria monocytogenes* and *Staphylococcus aureus* in cuajada, a semisolid dairy product manufactured in Spain. *Journal of Dairy Science*, 91, 70–75.

Arqués, J. L., Rodríguez, E., Nuñez, M. & Medina, M. (2008b). Inactivation of Gram-negative pathogens in refrigerated milk by reuterin in combination with nisin or the lactoperoxidase system. *European Food Research and Technology*, 227, 77–82.

Banjara, N., Suhr, M. J. & Hallen-Adams, H. E. (2015). Diversity of yeast and mold species from a variety of cheese types. *Current Microbiology*, 70, 792–800.

Bendicho, S., Barbosa-Canovas, G. V. & Martín, O. (2002). Milk processing by high intensity pulsed electric fields. *Trends in Food Science and Technology*, 13, 195–204.

Benkerroum, N., Oubel, R. & Ben Mimoun, L. (2002). Behavior of *Listeria monocytogenes* and *Staphylococcus aureus* in yogurt fermented with a bacteriocin-producing thermophilic starter. *Journal of Food Protection*, 65, 799–805.

Beresford, T. P., Fitzsimons, N. A., Brennan, N. L. & Cogan, T. M. (2001). Recent advances in cheese microbiology. *International Dairy Journal*, 11, 259–274.

Beresford, T. P. (2004). The microbiology of cheese ripening. In Fox, P. F., McSweeney, P. L. H., Cogan, T. M. & Guinee, T. P. (eds.), *Cheese: Chemistry, Physics and Microbiology*. Elsevier Academic Press, San Diego, California, USA, pp. 287–317.

Bernardeau, M., Vernoux, J. P., Henri-Dubernet, S. & Gueguen, M. (2008). Safety assessment of dairy microorganisms: The *Lactobacillus* genus. *International Journal of Food Microbiology*, 126, 278–285.

Besse, N. G. (2002) Influence of various environmental parameters and of detection procedures on the recovery of stressed *Listeria monocytogenes*: A review. *Food Microbiology*, 19, 221–234.

Beuchat, L. R. & Golden, D. A. (1989). Antimicrobials occurring naturally in foods. *Food Technology*, 43, 134–142.

Bhowmik, T. & Marth, E. H. (1990). Role of *Micrococcus* and *Pediococcus* species in cheese ripening: a review. *Journal of Dairy Science*, 73, 859–866.

Bintsis, T., Litopoulou-Tzanetaki, E., Davies, R. & Robinson, R. K. (2000). Microbiology of brines used to mature feta cheese. *International Journal of Dairy Technology*, 53, 106–112.

Bolotin, A., Quinquis, B., Renault, P. Sorokin, A., Ehrlich, S. D., Kulakauskas, S., Lapidus, A., Goltsman, E., Mazur, M., Pusch, G. D., Fonstein, M., Overbeek, R., Kyprides, N., Purnelle, B., Prozzi, D., Ngui, K., Masuy, D., Hancy, F., Burteau, S., Boutry, M., Delcour, J., Goffeau, A. & Hols, P. (2004). Complete sequence and comparative genome analysis of the dairy bacterium *Streptococcus thermophilus*. *Nature Biotechnology*, 22, 1554–1558.

Buazzi, M. M., Johnson, M. E. & Marth, E. H. (1992). Fate of *Listeria monocytogenes* during the manufacture of Mozzarella cheese. *Journal of Food Protection*, 55, 80–83.

Bueno, E., García, P., Martínez, B. & Rodríguez, A. (2012). Phage inactivation of *Staphylococcus aureus* in fresh and hard-type cheeses. *International Journal of Food Microbiology*, 158, 23–27.

Bull, M. K., Hayman, M. M., Stewart, C. M., Szabo, E. A. & Knabel, S. J. (2005). Effect of prior growth temperature, type of enrichment medium, and temperature and time of storage on recovery of *Listeria monocytogenes* following high pressure processing of milk. *International Journal of Food Microbiology*, 101, 53–61.

Calderón-Miranda, M. L., Barbosa-Cánovas, G. V. & Swanson, B. G. (1999). Inactivation of *Listeria innocua* in skim milk by pulsed electric fields and nisin. *International Journal of Food Microbiology*, 51, 19–30.

Calzada, J., Del Olmo, A., Picon, A., Gaya, P. & Nuñez, M. (2013). Reducing biogenic amine-producing bacteria, decarboxylase activity and biogenic amines in raw milk cheese by high pressure treatments. *Applied and Environmental Microbiology*, 79, 1277–1283.

Capodifoglio, E., Centola Vidal, A. M., Lima, J. A. S., Bortoletto, F., D'Abreu, L. F., Siqueira Goncalves, A. C., de Carvalho Balieiro, J. C. & Netto, A. S. (2016). Lipolytic and proteolytic activity of *Pseudomonas* spp. isolated during milking and storage of refrigerated raw milk. *Journal of Dairy Science*, 99, 5214–5223.

Carlton, R. M., Noordman, W. H., Biswas, B., de Meester, E. D. & Loessner, M. J. (2005). Bacteriophage P100 for control of *Listeria monocytogenes* in foods: Genome sequence, bioinformatic analyses, oral toxicity study, and application. *Regulatory Toxicology and Pharmacology*, 43, 301–312.

Claeys, W. L., Cardoen, S., Daube, G., De Block, J., Dewettinck, K., Dierick, K., De Zutter, L., Huyghebaert, A., Imberechts, H., Thiange, P., Vandenplas, Y. & Herman, L. (2013). Raw or heated cow milk consumption: Review of risks and benefits. *Food Control*, 31, 251–262.

Corsetti, A., Rossi, J. & Gobbetti, M. (2001). Interactions between yeasts and bacteria in the smear surface-ripened cheeses. *International Journal of Food Microbiology*, 69, 1–10.

Cotter, P. D., Hill, C. & Ross, R. P. (2005). Bacteriocins: Developing innate immunity for food. *Nature Reviews Microbiology*, 3, 777–788.

Cretenet, M., Even, S. & Le Loir, Y. (2011). Unveiling *Staphylococcus aureus* enterotoxin production in dairy products: a review of recent advances to face new challenges. *Dairy Science & Technology*, 91, 127–150.

Cretenet, M., Nouaille, S., Thouin, J., Rault, L., Stenz, L., Francois, P., Hennekinne, J.-A., Piot, M., Maillard, M. B., Fauquant, J., Loubiere, P., Le Loir, Y. & Even, S. (2011). *Staphylococcus aureus* virulence and metabolism are dramatically affected by *Lactococcus lactis* in cheese matrix. *Environmental Microbiology Reports*, 3, 340–351.

Crow, V., Curry, B. & Hayes, M. (2001). The ecology of non-starter lactic acid bacteria (NSLAB) and their uses as adjuncts in New Zealand Cheddar. *International Dairy Journal*, 11, 275–283.

D'Amico, D. J., Druart, M. J. & Donnelly, C. W. (2010). Behavior of *Escherichia coli* O157:H7 during the manufacture and aging of Gouda and stirred-curd Cheddar cheeses manufactured from raw milk. *Journal of Food Protection*, 73, 2217–2224.

Dalmasso, M. & Jordan, K. (2014). Absence of growth of *Listeria monocytogenes* in naturally contaminated Cheddar cheese. *Journal of Dairy Research*, 81, 46–53.

De Buyser, M. L., Dufour, B., Maire, M. & Lafarge, V. (2001). Implication of milk and milk products in food-borne diseases in France and in different industrialised countries. *International Journal of Food Microbiology*, 67, 1–17.

De Filippis, F., Genovese, A., Ferranti, P., Gilbert, J. A. & Ercolini, D. (2016). Metatranscriptomics reveals temperature-driven functional changes in microbiome impacting cheese maturation rate. *Scientific Reports*, 6, 21871.

De Lamo-Castellví, S., Roig-Sagués, A. X., López-Pedemonte, T., Hernández-Herrero, M. M., Guamis, B. & Capellas, M. (2007). Response of two *Salmonella enterica* strains inoculated in model cheese treated with high hydrostatic pressure. *Journal of Dairy Science*, 90, 99–109.

Delavenne, E., Mounier, J., Asmani, K., Jany, J.-L., Barbier, G. & Le Blay, G. (2011). Fungal diversity in cow, goat and ewe milk. *International Journal of Food Microbiology*, 151, 247–251.

Delbes, C., Alomar, J., Chougui, N., Martin, J. F. & Montel, M. C. (2006). *Staphylococcus aureus* growth and enterotoxin production during the manufacture of uncooked, semihard cheese from cows' raw milk. *Journal of Food Protection*, 69, 2161–2167.

Doyle, M. E. (2015). Multidrug-resistant pathogens in the food supply. *Foodborne Pathogens and Disease*, 12, 261–279.

Doyle, M. E. & Mazzotta, A. S. (2000). Review of studies on the thermal resistance of salmonellae. *Journal of Food Protection*, 63, 779–795.

Doyle, M. E., Mazzotta, A. S., Wang, T., Wiseman, D. W. & Scott, V. N. (2001). Heat Resistance of *Listeria monocytogenes*. *Journal of Food Protection*, 64, 410–429.

Drevets, D. A. & Bronze, M. S. (2008). *Listeria monocytogenes*: epidemiology, human disease, and mechanisms of brain invasion. *FEMS Immunology Medical Microbiology*, 53, 151–165.

EC (2016) *RASFF – The Rapid Alert System for Food and Feed – 2015 annual report*. RASFF online: http://ec.europa.eu/food/safety/rasff/index_en.htm.

Eckner, K. F., Roberts, R. F., Strantz, A. A. & Zottola, E. A. (1990). Characterization and behavior of *Salmonella* javiana during manufacture of Mozzarella-type cheese. *Journal of Food Protection*, 53, 461–464.

EFSA (2013). Scientific opinion on the risk posed by pathogens in food of non-animal origin. Part 1 (outbreak data analysis and risk ranking of food/pathogen combinations). *EFSA Journal*, 11, 3025.

EFSA-ECDC (2015). The European Union summary report on trends and sources of zoonoses, zoonotic agents and food-borne outbreaks in 2013. *EFSA Journal*, 13, 3991.

El-Gazzar, F. E. & Marth, E. H. (1992). *Salmonellae*, Salmonellosis, and dairy foods: A review. *Journal of Dairy Science*, 75, 2327–2343.

El-Ziney, M. G. & Debevere, J. M. (1998). The effect of reuterin on *Listeria monocytogenes* and *Escherichia coli* O157:H7 in milk and cottage cheese. *Journal of Food Protection*, 61, 1275–1280.

Ennahar, S., Assobhei, O. & Hasselmann, C. (1998). Inhibition of *Listeria monocytogenes* in a smear-surface soft cheese by *Lactobacillus plantarum* WHE 92, a pediocin AcH producer. *Journal of Food Protection*, 61, 186–191.

Ercolini, D., Hill, P. J. & Dodd, C. E. R. (2003). Bacterial community structure and location in Stilton cheese. *Applied and Environmental Microbiology*, 69, 3540–3548.

Ercolini, D., Russo, F., Ferrocino, I. & Villani, F. (2009). Molecular identification of mesophilic and psychrotrophic bacteria from raw cow's milk. *Food Microbiology*, 26, 228–231.

Falentin, H., Deutsch, S. M., Jan, G., Loux, V., Thierry, A., Pararyre, S., Maillard, M. B., Dherbécourt, J., Cousin, F. J., Jardin, J., Siquier, P., Couloux, A., Barbe, V., Vacherie, B., Wincker, P., Gibrat, J. F., Gaillarch, C. & Lortal, S. (2010a). The complete genome of *Propionibacterium freudenreichii* CIRM-BIA1T, a hardy actinobacterium with food and probiotic applications. *PLoS One*, 5, e11748.

Falentin, H., Postollec, F., Parayre, S., Henaff, N., Le Bivic, P., Richoux, R., Thierry, A. & Sohier, D. (2010b). Specific metabolic activity of ripening bacteria quantified by real-time reverse transcription PCR throughout Emmental cheese manufacture. *International Journal of Food Microbiology*, 144, 10–19.

FAO/WHO Food and Agriculture Organization of the United Nations, World Health Organization (2001). *Health and Nutritional Properties of Probiotics in Food Including Powder Milk with Live Lactic Acid Bacteria*. Córdoba, Argentina, 34 pp.

FAO/WHO (2002). *Guidelines for the Evaluation of Probiotics in Food*. London, Ontario, Canada, 11 pp.

Farkas, J. (2006). Irradiation for better foods. *Trends in Food Science and Technology*, 17, 148–152.

Farkas, D. & Hoover, D. G. (2000). High pressure processing. *Journal of Food Science*, Supplement – Kinetics of Microbial Inactivation for Alternative Food Processing Technologies, 65, 47–64.

Favaro, L., Barretto Penna, A. L. & Todorov, S. D. (2015). Bacteriocinogenic LAB from cheeses – Application in biopreservation? *Trends in Food Science & Technology*, 41, 37–48.

Fernández, M., Hudson, J. A., Korpela, R. & de los Reyes-Gavilán, C. G. (2015). Impact on human health of microorganisms present in fermented dairy products: An overview. *BioMed Research International*, 2015, 412714.

Ferreira, A. D. & Viljoen, B. C. (2003). Yeasts as adjunct starters in matured Cheddar cheese. *International Journal of Food Microbiology*, 86, 131–140.

Foulquié Moreno, M. R., Sarantinopoulos, P., Tsakalidou, E. & de Vuyst, L. (2006). The role and application of enterococci in food and health. *International Journal of Food Microbiology*, 106, 1–24.

Fox, P. F. & Kelly, A. L. (2006). Review: Indigenous enzymes in milk: Overview and historical aspects – part 1 & 2. *International Dairy Journal*, 16, 500–532.

Fox, P. F. & McSweeney, P. L. H. (2004). Cheese: an overview. In Fox, P. F., McSweeney, P. L. H., Cogan, T. M. & Guinee, T. P. (eds.), *Cheese: Chemistry, physics and Microbiology*, Elsevier Academic Press, San Diego, California, USA, pp. 1–18.

Fox, P. F., Wallace, J. M., Morgan, S., Lynch, C. M., Niland, E. J. & Tobin, J. (1996). Acceleration of cheese ripening. *Antonie van Leeuwenhoek International Journal of General and Molecular Microbiology*, 70, 271–297.

Galli, B. D., Prado Martin, J. G., Moreira da Silva, P. P., Porto, E. & Fillet Spoto, M. H. (2016). Sensory quality of Camembert-type cheese: Relationship between starter cultures and ripening molds. *International Journal of Food Microbiology*, 234, 71–75.

Gallot-Lavallée, T. (1998). Efficiency of high pressure treatment for destruction of *Listeria monocytogenes* in goat cheese from raw milk. *Sciences des Aliments*, 18, 647–655.

García, P., Rodríguez, L., Rodríguez, A. & Martínez, B. (2010). Food biopreservation: promising strategies using bacteriocins, bacteriophages and endolysins. *Trends in Food Science & Technology*, 21, 373–382.

García, P., Madera, C., Martínez, B. & Rodríguez, A. (2007). Biocontrol of *Staphylococcus aureus* in curd manufacturing processes using bacteriophages. *International Dairy Journal*, 17, 1232–1239.

García-Graells, C., Masschalck, B. & Michiels, C. W. (1999). Inactivation of *Escherichia coli* in milk by high-hydrostatic-pressure treatment in combination with antimicrobial peptides. *Journal of Food Protection*, 62, 1248–1254.

García-Graells, C., Valckx, C. & Michiels, C. W. (2000). Inactivation of *Escherichia coli* and *Listeria innocua* in milk by combined treatment with high hydrostatic pressure and the lactoperoxidase system. *Applied and Environmental Microbiology*, 66, 4173–4179.

Garmiene, G., Salomskiene, J., Jaseitiene, I., Macioniere, I. & Miliauskiene, I. (2010). Production of benzoic acid by lactic acid bacteria from *Lactobacillus, Lactococcus* and *Streptococcus* genera in milk. *Milchwissenschaft*, 65, 295–298.

Gaya, P., Medina, M., Bautista, L. & Nuñez, M. (1988). Influence of lactic starter inoculation, curd heating and ripening temperature on *Staphylococcus aureus* behaviour in Manchego cheese. *International Journal of Food Microbiology*, 6, 249–257.

Giraffa, G. (2003). Functionality of enterococci in dairy products. *International Journal of Food Microbiology*, 88, 215–222.

Giraffa, G. & Carminati, D. (1997). Control of *Listeria monocytogenes* in the rind of Taleggio, a surface-smear cheese, by a bacteriocin from *Enterococcus faecium* 7C5. *Science des Aliments*, 17, 383–391.

Gkatzionis, K., Yunita, D., Linforth, R. S. T., Dickinson, M. & Dodd, C. E. R. (2014). Diversity and activities of yeasts from different parts of a Stilton cheese. *International Journal of Food Microbiology*, 177, 109–116.

Gobbetti, M., De Angelis, M., Di Cagno, R., Mancini, L. & Fox, P. F. (2015). Pros and cons for using non-starter lactic acid bacteria (NSLAB) as secondary/adjunct starters for cheese ripening. *Trends in Food Science & Technology*, 45, 167–178.

Gómez-Torres, N., Ávila, M., Gaya, P. & Garde, S. (2014). Prevention of late blowing defect by reuterin produced in cheese by a *Lactobacillus reuteri* adjunct. *Food Microbiology*, 42, 82–88.

Govaris, A., Koidis, P. & Papatheodorou, K. (2001). The fate of *Escherichia coli* O157 : H7 in Myzithra, Anthotyros, and Manouri whey cheeses during storage at 2 and 12 degrees C. *Food Microbiology*, 18, 565–570.

Granato, D., Branco, G. F., Gomes Cruz, A., Fonseca Faria, J. & Sha, N. P. (2010). Probiotic dairy products as functional foods. *Comprehensive Reviews in Food Science and Food Safety*, 9, 455–470.

Gravani, R. B. (1984). Salmonellosis-Salmonella food poisoning. *Dairy Food Sanitation*, 4, 227.

Halász, A., Baráthá, A., Simon-Sarkadi, L. & Holzapfel, W. (1994). Biogenic amines and their production by microorganisms in food. *Trends in Food Science and Technology*, 5, 42–49.

Hannon, J. A., Wilkinson, M. G., Delahunty, C. M., Wallace, J. M., Morrissey, P. A. & Beresford, T. P. (2003) Use of autolytic starter systems to accelerate the ripening of Cheddar cheese. *International Dairy Journal*, 13, 313–323.

Hanusova, K., Stastna, M., Votavova, L., Klaudisova, K., Dobias, J., Voldrich, M. & Marek, M. (2010). Polymer films releasing nisin and/or natamycin from polyvinyldichloride lacquer coating: Nisin and natamycin migration, efficiency in cheese packaging. *Journal of Food Engineering*, 99, 491–496.

Hargrove, R. E., McDonough, F. E. & Mattingly, W. A. (1969). Factors affecting survival of *Salmonella* in Cheddar and Colby cheese. *Journal of Milk and Food Technology*, 32, 480–484.

Hemme, D. & Foucaud-Scheunemann, C. (2004). *Leuconostoc*, characteristics, use in dairy technology and prospects in functional foods. *International Dairy Journal*, 14, 467–494.

Himery, N., Vasseur, V., Coton, M., Mounier, J., Jany, J.-L., Barbier, G. & Coton, E. (2014). Filamentous fungi and mycotoxins in cheese: A review. *Comprehensive Reviews in Food Science and Food Safety*, 13, 437–456.

Holzapfel, W. H., Geisen, R. & Schillinger, U. (1995). Biological preservation of foods with reference to protective cultures, bacteriocins and food-grade enzymes. *International Journal of Food Microbiology*, 24, 343–362.

Ibrahim, G. F., Baldock, A. K., Radford, D. R. & Ireland, L. B. (1981a). Inhibition of *Staphylococcus aureus* growth and enterotoxin A production in cheddar cheese produced with variable starter activity. *Journal of Food Protection*, 44, 263–267.

Ibrahim, G. F., Radford, D. R., Baldock, A. K. & Ireland, L. B. (1981b). Inhibition of growth of *Staphylococcus aureus* and enterotoxin A production in cheddar cheese produced with induced starter failure. *Journal of Food Protection*, 44, 189–193.

Irlinger, F. & Bergère, J.-L. (1999). Use of conventional biochemical tests and analyses of ribotype patterns for classification of micrococci isolated from dairy products. *Journal of Dairy Research*, 66, 91–103.

Irmler, S., Bavan, T., Oberli, A., Roetschi, A., Badertscher, R., Guggenbühl, B. & Berthoud, H. (2013). Catabolism of Serine by *Pediococcus acidilactici* and *Pediococcus pentosaceus*. *Applied and Environmental Microbiology*, 79, 1309–1315.

Jakobsen, R. A., Heggebo, R., Sunde, E. B. & Skjervheim, M. (2011). *Staphylococcus aureus* and *Listeria monocytogenes* in Norwegian raw milk cheese production. *Food Microbiology*, 28, 492–496.

Johnson, E. A., Nelson, J. H. & Johnson, M. (1990). Microbiological safety of cheese made from heat-treated milk I. *Journal of Food Protection*, 53, 441–452..

Karper, J. B., Nataro, J. P. & Mobley, H. L. (2004). Pathogenic *Escherichia coli*. *Nature Reviews Microbiology*, 2, 123–140.

Kasrazadeh, M. & Genigeorgis, C. (1995). Potential growth and control of *Escherichia coli* O157:H7 in soft Hispanic type cheese. *International Journal of Food Microbiology*, 25, 289–300.

Konteles, S., Sinanoglou, V. J., Batrinou, A. & Sflomos, K. (2009). Effects of γ-irradiation on *Listeria monocytogenes* population, colour, texture and sensory properties of Feta cheese during cold storage. *Food Microbiology*, 26, 157–165.

Korhonen, H., Marnila, P. & Gill, H. S. (2000). Milk immunoglobulins and complement factors. *British Journal of Nutrition*, 84, S75–S80.

Kristo, E., Koutsoumanis, K. & Biliaderis, C. (2008). Thermal, mechanical and water vapor barrier properties of sodium caseinate films containing antimicrobials and their inhibitory action on *Listeria monocytogenes*. *Food Hydrocolloids*, 22, 373–386.

Kwan, P. S. L., Birtless, A., Bolton, F. J., French, N. P., Robinson, S. E., Newbold, L. S., Upton, M. & Fox, A. J. (2008). Longitudinal study of the molecular epidemiology of *Campylobacter jejuni* in cattle on dairy farms. *Applied and Environmental Microbiology*, 374, 3626–3633.

Lanciotti, R., Patrignani, F., Bagnolini, F., Guerzoni, M. E. & Gardini F. (2003). Evaluation of diacetyl antimicrobial activity against *Escherichia coli, Listeria monocytogenes* and *Staphylococcus aureus*. *Food Microbiology*, 20, 537–543.

Langa, S., Landete, J. M., Martín-Cabrejas, I., Rodríguez, E., Arqués, J. L. & Medina, M. (2013). *In situ* reuterin production by *Lactobacillus reuteri* in dairy products. *Food Control*, 33, 200–206.

Le Loir, Y., Baron, F. & Gautier, M. (2003). *Staphylococcus aureus* and food poisoning. *Genetics and Molecular Research*, 2, 63–76.

Le Maréchal, C., Thiéry, R. Vautor, E. & Le Loir, Y. (2011). Mastitis impact on technological properties of milk and quality of milk products – a review. *Dairy Science & Technology*, 91, 247–282.

Leroy, F. & de Vuyst, L. (2004). Lactic acid bacteria as functional starter cultures for the food fermentation industry. *Trends in Food Science and Technology*, 15, 67–78.

Loessner, M., Guenther, S., Steffan, S. & Scherer, S. (2003). A pediocin-producing *Lactobacillus plantarum* strain inhibits *Listeria monocytogenes* in a multispecies cheese surface microbial ripening consortium. *Applied and Environmental Microbiology*, 69, 1854–1857.

López-Pedemonte, T., Brinez, W. J., Roig-Sagues, A. X. & Guamis, B. (2006). Fate of *Staphylococcus aureus* in cheese treated by ultrahigh pressure homogenization and high hydrostatic pressure. *Journal of Dairy Science*, 89, 4536–4544.

Mahapatra, A. K., Muthukumarappan, K. & Julson, J. L. (2005). Applications of ozone, bacteriocins and irradiation in food processing: a review. *Critical Reviews in Food Science and Nutrition*, 45, 477–461.

Maisnier-Patin, S., Deschamps, S. N., Tatini, S. R. & Richard, J. (1992). Inhibition of Listeria monocytogenes in Cammembert cheese made with a nisin-producing starter. *Lait*, 72, 249–263.

Marchand, S., de Block, J., de Jonghe, V., Coorevits, A., Heyndrickx, M. & Herman, L. (2012). Biofilm formation in milk production and processing environments; influence on milk quality and safety. *Comprehensive Reviews in Food Science and Food Safety*, 11, 133–147.

Masoud, W., Vogensen, F. K., Lillevang, S., Abu Al-Soud, W., Sorensen, S. J. & Jakobsen, M. (2012). The fate of indigenous microbiota, starter cultures, Escherichia coli, Listeria innocua and Staphylococcus aureus in Danish raw milk and cheeses determined by pyrosequencing and quantitative real time (qRT)-PCR. *International Journal of Food Microbiology*, 153, 192–202.

Mauriello, G., De Luca, E., La Storia, A., Villani, F. & Ercolini, D. (2005). Antimicrobial activity of a nisin plastic film for food packaging. *Letters in Applied Microbiology*, 41, 464–469.

McAuliffe, O., Hill, C. & Ross, R. P. (1999). Inhibition of Listeria monocytogenes in Cottage cheese manufactured with a lacticin 3147-producing starter culture. *Journal of Applied Microbiology*, 86, 251–256.

McDonough, F. E., Hargrove, R. E. & Tittsler, R. P. (1967). The fate of Salmonella in the manufacture of cottage cheese. *Journal of Milk and Food Technology*, 30, 354–357.

McMahon, C. M. M., Byrne, C. M., Sheridan, J. J., McDowell, D. A., Blair, I. S. & Hegarty, T. (2000). The effect of culture growth phase on induction of the heat shock response in Yersinia enterocolitica and Listeria monocytogenes. *Journal of Applied Microbiology*, 89, 198–206.

McSweeney, P. L. H. & Sousa, M. J. (2000). Biochemical pathways for the production of flavour compounds in cheeses during ripening: A review. *Lait*, 80, 293–324.

Medina, M. & Nuñez, M. (2010). Applications of protective cultures, bacteriocins and bacteriophages in foods and beverages. In Lacroix, C. (ed.), *Protective Cultures, Antimicrobial Metabolites and Bacteriophages for Food and Beverage Biopreservation*. Woodhead Publishing Limited, Cambridge, UK, pp. 267–296.

Meyrand, A., Boutrand-Loei, S., Ray-Gueniot, S., Mazuy, C., Gaspard, C. E., Jaubert, G., Perrin, G., Lapeyre, C. & Vernozy-Rozand, C. (1998). Growth and enterotoxin production of Staphylococcus aureus during the manufacture and ripening of Camembert-type cheeses from raw goats' milk. *Journal of Applied Microbiology*, 85, 537–544.

Mills, S., O'Sullivan, O., Hill, C., Fitzgerald G. & Ross, R. P. (2010). The changing face of dairy starter culture research: From genomics to economics. *International Journal of Dairy Technology*, 63, 149–170.

Montel, M.-C., Buchin, S., Mallet, A., Delbes-Paus, C., Vuitton, D. A., Desmasures, N. & Berthier, F. (2014). Traditional cheeses: Rich and diverse microbiota with associated benefits. *International Journal of Food Microbiology*, 177, 136–154.

Morales, P., Calzada, J., Juez, C. & Nuñez, M. (2010). Volatile compounds in cheeses made with Micrococcus sp. INIA 528 milk culture or high enzymatic activity curd. *International Journal of Dairy Technology*, 63, 538–543.

Morgan, M. (2008) Methicillin-resistant Staphylococcus aureus and animals: Zoonosis or humanosis? *Journal of Antimicrobial Chemotherapy*, 62, 1181–1187.

Naguib, M. M., Sabbour, M. M. & Nour, M. M. (1979). The effect of salting rate and initial count of Salmonella typhi in Domiati cheese-milk on its longevity during pickling. *Archiv fur Lebensmittelhygiene*, 30, 150–151.

Nataro, J. P. & Karper, J. B. (1998). Diarrheagenic Escherichia coli. *Clinical Microbiology Reviews*, 11, 142–201.

Novella-Rodríguez, S., Veciana-Nogués, M. T., Trujillo-Mesa, A. J. & Vidal-Carou, M. C. (2002). Profile of biogenic amines in goat cheese made from pasteurized and pressurized milks. *Journal of Food Science*, 67, 2940–2944.

Nuñez, M., Bautista, L., Medina, M. & Gaya, P. (1988). Staphylococcus aureus, thermostable nuclease and staphylococcal enterotoxins in raw ewes milk manchego cheese. *Journal of Applied Bacteriology*, 65, 29–34.

Nuñez, M., Rodríguez, J. L., García, E., Gaya, P. & Medina, M. (1997). Inhibition of Listeria monocytogenes by enterocin 4 during the manufacture and ripening of Manchego cheese. *Journal of Applied Microbiology*, 83, 671–677.

Oliveira, A. P., Nielsen, J. & Forster, J. (2005). Modeling Lactococcus lactis using a genome-scale flux model. *BMC Microbiology*, 5, 39.

Oliver, S. P., Boor, K. J., Murphy, S. C. & Murinda, S. E. (2009). Food safety hazards associated with consumption of raw milk. *Foodborne Pathogens and Disease*, 6, 793–806.

Oliver, S. P., Jayarao, B. M. & Almeida, R. A. (2005). Foodborne pathogens in milk and the dairy farm environment: food safety and public health implications. *Foodborne Pathogens and Disease*, 2, 115–129.

Ollé Resa, C. P., Jagus, R. J. & Gerschenson, L. N. (2014). Natamycin efficiency for controlling yeast growth in models systems and on cheese surfaces. *Food Control*, 35, 101–108.

O'Sullivan, L., O'Connor, E. B., Ross, R. P. & Hill, C. (2006). Evaluation of live-culture-producing lacticin 3147 as a treatment for the control of Listeria monocytogenes on the surface of smear-ripened cheese. *Journal of Applied Microbiology*, 100, 135–143.

Padilla, B., Manzanares, P. & Belloch, C. (2014). Yeast species and genetic heterogeneity within Debaryomyces hansenii along the ripening process of traditional ewes' and goats' cheeses. *Food Microbiology*, 38, 160–166.

Papageorgiou, D. K. & Marth, E. H. (1989a). Fate of Listeria monocytogenes during the manufacture, ripening and storage of Feta cheese. *Journal of Food Protection*, 52, 82–87.

Papageorgiou, D. K. & Marth, E. H. (1989b). Fate of Listeria monocytogenes during the manufacture and ripening of Blue cheese. *Journal of Food Protection*, 52, 459–465.

Parente, E. & Hill, C. (1992). Characterization of enterocin 1146, a bacteriocin from Enterococcus faecium inhibitory to Listeria monocytogenes. *Journal of Food Protection*, 55, 497–508.

Quigley, L., O'Sullivan, O., Stanton, C., Beresford, T. P., Ross, R. P., Fitzgerald, G. F. & Cotter, P. D. (2013). The complex microbiota of raw milk. *FEMS Microbiology Reviews*, 37, 664–698.

Qin, B. L., Pothakamury, U. R., Vega, H., Martín, O., Barbosa-Cánovas, G. V. & Swanson, B. G. (1995). Food pasteurization using high-intensity pulsed electric fields. *Food Technology*, 49, 55–60.

Ramsaran, H., Chen, J., Brunke, B., Hill, A. & Griffiths, M. W. (1998). Survival of bioluminescent Listeria monocytogenes and Escherichia coli O157:H7 in soft cheeses. *Journal of Dairy Science*, 81, 1810–1817.

Rasolofo, E. A., St-Gelais, D., LaPointe, G. & Roy, D. (2010). Molecular analysis of bacterial population structure and dynamics during cold storage of untreated and treated milk. *International Journal of Food Microbiology*, 138, 108–118.

Reina, L. D., Jin, Z. T., Zhang, Q. H. & Yousef, A. E. (1998). Inactivation of Listeria monocytogenes in milk by pulsed electric field. *Journal of Food Protection*, 61, 1203–1206.

Reis, J. A., Paula, A. T., Casarotti, S. N. & Penna, A. L. B. (2012). Lactic acid bacteria antimicrobial compounds: Characteristics and applications. *Food Engineering Reviews*, 4, 124–140.

Reiter, B., Marshall, V. M. E., Bjorck, L. & Rosen, C. G. (1976). Nonspecific bactericidal activity of lactoperoxidase-thiocyanate-hydrogen peroxide system of milk against *Escherichia coli* and some Gram-negative pathogens. *Infection and Immunity*, 13, 800–807.

Rodríguez, E., Arqués, J. L., Gaya, P., Tomillo, J., Nuñez, M. & Medina, M. (2000). Behaviour of *Staphylococcus aureus* in semi-hard cheese made from raw milk with nisin-producing starter cultures. *Milchwissenschaft*, 55, 633–635.

Rodríguez, E., Arqués, J. L., Nuñez, M., Gaya, P. & Medina, M. (2005a). Combined effect of high-pressure treatments and bacteriocin-producing lactic acid bacteria on inactivation of

Escherichia coli O157-H7 in raw-milk cheese. *Applied and Environmental Microbiology*, 71, 3399–3404.
Rodríguez, E., Calzada, J., Arqués, J. L., Rodríguez, J. M., Nuñez, M. & Medina, M. (2005b). Antimicrobial activity of pediocin-producing *Lactococcus lactis* on *Listeria monocytogenes, Staphylococcus aureus* and *Escherichia coli* O157:H7 in cheese. *International Dairy Journal*, 15, 51–57.
Rodríguez, E., Gaya, P., Nuñez, M. & Medina, M. (1998). Inhibitory activity of a nisin-producing starter culture on *Listeria innocua* in raw ewes milk Manchego cheese. *International Journal of Food Microbiology*, 39, 129–132.
Rodríguez, J. L., Gaya, P., Medina, M. & Nuñez, M. (1997). Bactericidal effect of enterocin 4 on *Listeria monocytogenes* in a model dairy system. *Journal of Food Protection*, 60, 28–32.
Rosengren, A., Fabricius, A., Guss, B., Sylven, S. & Lindqvist, R. (2010). Occurrence of foodborne pathogens and characterization of *Staphylococcus aureus* in cheese produced on farm-dairies. *International Journal of Food Microbiology*, 144, 263–269.
Ruegg, P. L. (2003). Practical food safety interventions for dairy production. *Journal of Dairy Science*, 86, E1–E9.
Ruiz Martínez, R. C., Staliano, C. D., Silva Vieira, A. D., Morales Villarreal, M. L., Todorov, S. D., Isay Saad, S. M. & Gombossy de Melo Franco, B. D. (2015). Bacteriocin production and inhibition of *Listeria monocytogenes* by *Lactobacillus sakei* subsp. *sakei* 2a in a potentially synbiotic cheese spread. *Food Microbiology*, 48, 143–152.
Russell, N. J. (2002) Bacterial membranes: the effects of chill storage and food processing. An overview. *International Journal of Food Microbiology*, 79, 27–34.
Ryan, M. P., Rea, M. C., Hill, C. & Ross, R. P. (1996). An application in Cheddar cheese manufacture for a strain of *Lactococcus lactis* producing a novel broad-spectrum bacteriocin, lacticin 3147. *Applied and Environmental Microbiology*, 62, 612–619.
Ryser, E. T. & Marth, E. H. (1987a). Behaviour of *Listeria monocytogenes* during the manufacture and ripening of Cheddar cheese. *Journal of Food Protection*, 50, 7–13.
Ryser, E. T. & Marth, E. H. (1987b). Fate of *Listeria monocytogenes* during the manfacture and ripening of Camembert cheese. *Journal of Food Protection*, 50, 372–378.
Saarela, M., Mogensen, G., Fonden, R., Matto, J. & Mattila-Sandholm, T. (2000). Probiotic bacteria: Safety, functional and technological properties. *Journal of Biotechnology*, 84,197–215.
Salque, M., Bogucki, P. I., Pyzel, J., Sobkowiak-Tabaka, I., Grygiel, R., Szmyt, M. & Evershed, R. P. (2013). Earliest evidence for cheese making in the sixth millennium BC in northern Europe. *Nature*, 493, 522–525.
Samelis, J., Lianou, A., Pappa, E. C., Bogovic-Matijasic, B., Parapouli, M., Kakouri, A. & Rogelj, I. (2014). Behavior of *Staphylococcus aureus* in culture broth, in raw and thermized milk, and during processing and storage of traditional Greek Graviera cheese in the presence or absence of *Lactococcus lactis* subsp *cremoris* M104, a wild, novel nisin A-producing raw milk isolate. *Journal of Food Protection*, 77, 1703–1714.
Scannell, A. G. M., Hill, C., Ross, R. P., Marx, S., Hartmeier, W. & Arendt, E. K. (2000). Development of bioactive food packaging materials using immobilised bacteriocins Lacticin 3147 and Nisaplin. *International Journal of Food Microbiology*, 60, 241–249.
Sensoy, I., Zhang, Q. & Sastry, S. K. (1997). Inactivation kinetics of *Salmonella* dublin by pulsed electric fields. *Journal of Food Processing and Engineering*, 20, 367–381.
Sepulveda-Ahumada, D. R., Ortega-Rivas, E. & Barbosa-Canovas, G. V. (2000). Quality aspects of cheddar cheese obtained with milk pasteurized by pulse electric fields. *Food and Bioproducts Processing*, 78, 65–71.
Settanni, L. & Moschetti, G. (2010). Non-starter lactic acid bacteria used to improve cheese quality and provide health benefits. *Food Microbiology*, 27, 691–697.

Shah, N. P. (2007). Functional cultures and health benefits. *International Dairy Journal*, 17, 1262–77.

Shrestha, S., Grieder, J. A., McMahon, D. J. & Nummer, B. A. (2011a). Survival of *Listeria monocytogenes* introduced as a post-aging contaminant during storage of low-salt Cheddar cheese at 4, 10, and 21 degrees C. *Journal of Dairy Science*, 94, 4329–4335.

Shrestha, S., Grieder, J. A., McMahon, D. J. & Nummer, B. A. (2011b). Survival of *Salmonella* serovars introduced as a post-aging contaminant during storage of low-salt Cheddar Cheese at 4, 10, and 21 degrees C. *Journal of Food Science*, 76, M616–M621.

Smith-Palmer, A., Stewart, J. & Fyfe, L. (1998) Antimicrobial properties of plant essential oils and essences against five important food-borne pathogens. *Letters in Applied Microbiology*, 26, 118–122.

Smith-Palmer, A., Stewart, J. & Fyfe, L. (2001) The potential application of plant essential oils as natural food preservatives in soft cheese. *Food Microbiology*,18, 463–470.

Sobrino, A., Rosell, J. R., Bendicho, S., Sanchis, V. & Martín, O. (2001). Inactivation of *Staphylococcus aureus* and *Bacillus cereus* in milk by pulsed electric fields and moderate heating. In *Nizo Dairy Conference*, Ede, The Netherlands (Cited by Bendicho et al., 2002).

Spano, G., Goffredo, E., Beneduce, L., Tarantino, D., Dupuy, A. & Massa, S. (2003). Fate of *Escherichia coli* O157:H7 during the manufacture of Mozzarella cheese. *Letters in Applied Microbiology*, 36, 73–76.

Steijns, J. M. & van Hooijdonk, A. C. M. (2000). Occurrence, structure, biochemical properties and technological characteristics of lactoferrin. *British Journal of Nutrition*, 84, S11–S17.

Stringer, S. C., George, S. M. & Peck, M. W. (2000). Thermal inactivation of *Escherichia coli* O157:H7. *Journal of Applied Microbiology*, Symposium Supplement 88, 79S–89S.

Sweeney, M. J. & Dobson, A. D. W. (1998). Mycotoxin production by *Aspergillus*, *Fusarium* and *Penicillium* species. *International Journal of Food Microbiology*, 43, 141–158.

Thierry, A., Maillard, M.-B., Hervé, C., Richoux, R. & Lortal, S. (2004). Varied volatile compounds are produced by *Propionibacterium freudenreichii* in Emmental cheese. *Food Chemistry*, 87, 439–446.

Tzanetakis, N., Litopoulou-Tzanetaki, E. & Vafopoulou-Mastrogiannaki, A. (1991). Effect of *Pediococcus pentosaceus* on microbiology and chemistry of Teleme cheese. *LWT-Food Science and Technology*, 24, 173–176.

Uceda, R., Guillen, A. M., Gaya, P., Medina, M. & Nuñez, M. (1994). The effect of ewe milk lactoperoxidase system on *Pseudomonas fluorescens* growth, casein breakdown, peptide formation and milk coagulation characteristics. *Milchwissenschaft*, 49, 139–143.

Van de Guchte, M., Penaud, S., Grimaldi, C. Barbe, V., Bryson, K., Nicolas, P., Robert, C., Oztas, S., Mangenot, S., Couloux, A., Loux, V., Dervyn, R., Bossy, R., Bolotin, A., Batto, J. M., Walunas, T., Gibrat, J. F., Bessières, P., Weissenbach, J., Ehrlich, S. D. & Maguin, E. (2006). The complete genome sequence of *Lactobacillus bulgaricus* reveals extensive and ongoing reductive evolution. *Proceedings of the National Academy of Sciences of the United States of America*, 103, 9274–9279.

Van den Berghe, E., Skourtas, G., Tsakalidou, E. & de Vuyst, L. (2006). *Streptococcus macedonicus* ACA-DC 198 produces the lantibiotic, macedocin, at temperature and pH conditions that prevail during cheese manufacture. *International Journal of Food Microbiology*, 107, 138–147.

Van Kranenburg, R., Kleerebezem, M., Van Hylckama Vlieg, J. E., Ursing, B. M., Boekhorst, J., Smit, B. A., Ayad, E. H. E., Smit, G. & Siezen, R. J. (2002). Flavour formation from amino acids by lactic acid bacteria: predictions from genome sequence analysis. *International Dairy Journal*, 12, 111–121.

Vedamuthu, E. R. (1994). The dairy *Leuconostoc*: Use in dairy products. *Journal of Dairy Science*, 77, 2725–2737.

Verdier-Metz, I., Gagne, G., Bornes, S., Monsallier, F., Veisseire, P., Delbès-Paus, C. & Montel, M.-C. (2012). Cow teat skin, a potential source of diverse microbial populations for cheese production. *Applied and Environmental Microbiology*, 78, 326–333.

Verraes, C., Vlaemynck, G., Van Weyenberg, S., De Zutter, L., Daube, G., Sindic, M., Uyttendaele, M. & Herman, L. (2015). A review of the microbiological hazards of dairy products made from raw milk. *International Dairy Journal*, 50, 32–44.

Warriner, K. & Namvar, A. (2009). What is the hysteria with *Listeria*? *Trends in Food Science and Technology*, 20, 245–254.

Yousef, A. E. & Marth, E. H. (1988). Behaviour of *Listeria monocytogenes* during the manufacture and storage of Colby cheese. *Journal of Food Protection*, 51, 12–15.

Yousef, A. E. & Marth, E. H. (1990). Fate of *Listeria monocytogenes* during the manufacture and ripening of Parmesan cheese. *Journal of Dairy Science*, 73, 3351–3356.

5

Cheeses with Protected Land- and Tradition-Related Labels: Traceability and Authentication

Luiz Javier R. Barron[1], Noelia Aldai[1], Mailo Virto[2] and Mertxe de Renobales[2]

[1] *Food Technology and Biochemistry and Molecular Biology, Faculty of Pharmacy – University of the Basque Country/EHU, Vitoria-Gasteiz, Spain*
[2] *Biochemistry and Molecular Biology, Faculty of Pharmacy – University of the Basque Country/EHU, Vitoria-Gasteiz, Spain*

5.1 Introduction: Protected Land- and Tradition-Related Labels

Cheese is one of the traditional foods which humanity has been preparing for the last several millennia, primarily in Europe and the areas surrounding the Mediterranean Sea. Cheesemaking technology was well established in Europe about 7,000 years ago (Curry, 2013). Currently, sheep and goat production predominate in the Mediterranean basin, influenced most likely by the dry climatic conditions which did not favour intensive agriculture. Cattle, in contrast, are more abundant in Central and Northern Europe (Flamant, 1992). It is, thus, not surprising that through time a truly impressive diversity of cheeses has been developed in Europe and the Mediterranean area. Several famous European cheeses have been linked to their production region for centuries, as described in the documents to register their Protected Land- and Tradition-related Labels (PLTLs). These labels provide a system to ascertain that the characteristics of a given product which originated in a specific place arise from having been traditionally produced in that particular location following time-honoured processes. According to Licitra (2010), a traditional food product 'expresses a symbiotic relationship between the culture of rural communities and physical features of the area.' The use of these PLTL labels originated in Europe, and European countries have traditionally maintained active policies to promote these food products.

Because PLTLs are considered a kind of intellectual property (World Intellectual Property Organization), similar, or equivalent, systems are appearing in other countries as well. Quite often these cheeses were made by European immigrants in the manner of the cheeses they knew, using local ingredients from their countries of adoption and were named after the place where they were first made, and are now being recognised. Monterey Jack (Anonymous, 2016a) is a good example in the United States, a country with 'Standards of Identity' for cheeses in Title 21 of its Code of Federal Regulation (J. Bauer, American Cheese Association, personal communication). Another example is Minas Cheese from Minas Gerais state in Brazil, recognised as 'immaterial heritage' (Arcuri et al., 2013). Queso Oaxaca is made in Mexico in a similar way to Italian Mozzarella. Today a very large number of cheeses are manufactured in many different countries (Anonymous, 2016b). In this chapter, we deal primarily with cheeses bearing PLTL labels.

Global Cheesemaking Technology: Cheese Quality and Characteristics, First Edition.
Edited by Photis Papademas and Thomas Bintsis.

Figure 5.1 European Union official labels for geographical indications and traditional specialties. From left to right: Protected Designation of Origin (PDO; red and yellow), Protected Geographical Indication (PGI; blue and yellow) and Traditional Specialty Guaranteed (TSG; blue and yellow).Taken from the EU Agriculture and Rural Development Web page 'Geographical Indications and Traditional Specialties'. After: EC (2016).

In Europe there are three different categories of PLTL labels described in Council Regulation (EEC) 2081/92 (EC, 1992): Protected Designation of Origin (PDO), Protected Geographical Indication (PGI), and Traditional Specialties Guaranteed (TSG). Figure 5.1 shows their officially approved labels. In addition, Regulation (EU) No. 1151/2012 (EC, 2012) establishes the general quality schemes applicable to each of these three categories of PLTL products, as well as general guidelines for verification and control.

PDO is the name of a clearly defined region, or specific place, used to describe an agricultural product or foodstuff originating in that region. In the case of cheeses, it means that the milk used is produced and processed and transformed in that area using traditional and recognised know-how. In addition, the characteristics of that cheese are 'essentially or exclusively due to a particular geographical environment with its inherent natural and human factors'. These characteristics are not reproducible outside this area.

PGI is the name of a given region, or specific place which describes a cheese that is traditionally produced or the milk used in its manufacture is obtained or transformed in that area, and enjoys a traditional good reputation. That is, at least one of the stages of the production of a given cheese takes place in the geographical area.

In the case of TSG, the agricultural or food product has specific features which distinguish it from other similar products of the same category. It must be a traditional product, produced using traditional raw materials and by a traditional method. In other words, the traditional character of the cheese could be either its composition or the method and means of production.

Currently, in Europe, there are 245 different cheeses with PLTL labels. Table 5.1 summarises this information.

The specific physico-chemical characteristics, production methods and geographical area of each PLTL cheese are described in varying degrees of detail in the corresponding documents, which can be accessed for each cheese through the European Commission Registry (EC, 2016). The geographic area for the production of each cheese is very clearly described in all cases. The species of milk origin (single or mixed) is stated in all cases, and sometimes even the breed(s) allowed is(are) specified. Feed materials are often described, specifically indicating those that are not allowed (for example, in several cases, maize silage or ingredients from genetically modified organisms are forbidden), the percentage of ingredients that should come from the geographical area and whether or not grazing is required for a minimum number of days.

Table 5.1 Cheeses bearing Protected Land- and Tradition-related Labels (PLTL) from Europe and some other countries.[a]

Country	No. of cheeses and type of label	Some examples
Austria	6 PDO	Tiroler, Vorarlberger, Gailtaler.
Belgium	1 PDO	Fromage de Herve.
Czech Republic	3 PGI	Olomoucké tvarůžky, Jihočeská Zlatá Niva.
Denmark	2 PGI	Danablu, Esrom.
France	45 PDO, 7 PGI	Roquefort, Comté, Ossau-Iraty, Camembert de Normandie, Rocamadour, Gruyère, Chevrotin.
Germany	6 PDO, 3 PGI	Altenburger Ziegenkäse, Allgäuer Emmentaler,
Greece	21 PDO	Feta, Galotyri, Kefalograviera, Anevato, Batzos.
Ireland	1 PDO	Imokilly Regato.
Italy	49 PDO, 1 PGI	Parmigiano Reggiano, Grana Padano, Pecorino Romano, Fiore Sardo, Mozzarella, Ragusano, Bitto.
Lituania	2 PGI	Liliputas.
Latvia	1 TSG (applied)	Janu siers.
Netherlands	4 PDO, 3 PGI	Edam, Gouda, Kanterkaas, Leiden farmers' cheese.
Poland	3 PDO, 2 PGI	Redykolka, Oscypek, Wielkopolski ser smazony.
Portugal	11 PDO, 1 PGI	Serra da Estrella, Azeitao, Beira Baixa, Terrincho.
Romania	1 PDO	Telemea da Ibanesti.
Slovenia	4 PDO	Nanoski sir, Tolminc, Mohant, Bovski sir.
Slovakia	8 PGI	Slovenská parenika, Slovensky ostiepok, Oravsky.
Spain	26 PDO, 2 PGI	Manchego, Idiazabal, Torta del Casar, Cabrales, Mahón, Flor de Guía, Palmero, Tetilla, Zamorano.
Sweden	1 PGI	Svecia.
Switzerland[b]	10 PDO	Raclette, Appenzeller, Tilsiter, Le Gruyère, Tomme.
United Kingdom	10 PDO, 6 PGI	Stilton, Staffordshire, Beacon Fell, Swaledale.
Georgia[c]	12	Sulguni, Kartuli Klevi, Chogi, Kuba.
Brazil		Minas.
United States[d]		Colby, Monterey Jack.

a) Data were obtained primarily from EC (2016) and other references listed in the text.
b) Switzerland: Anonymous (2016c).
c) Georgia: Anonymous (2016d).
d) United States: Anonymous (2016e).

Production methods are specified with varying degrees of detail: thermal milk treatment, type of rennet, a particular strain(s) of microorganism(s) (such as *Penicillium roqueforti,* specially for blue cheeses), an autocthonous starter culture that is specifically mentioned in certain cases and a minimum ripening period in other cases. Most documents describe the sensory and external characteristics of the cheese. In all cases, a regulatory entity keeps a register of cheesemakers and carries out the necessary administrative control measures to ascertain that the cheese is produced according to specifications.

5.2 Traceability

All European food systems must have a traceability system (EC, 2002). Traceability is usually defined as the ability to reconstruct the history of a given cheese in two opposing directions: (1) from producer to consumer ("farm-to-fork") and (2) from consumer to producer ("fork-to-farm") (Bechini et al., 2008). Thus, a good traceability system should comprise three elements: (1) unequivocal identification of each production lot, and of each of the cheeses in each lot; (2) date, time and location to which lots or individual cheeses have been transferred (for ripening, marketing, etc) and (3) a method to relate all data. Thus, a traceability system should provide the necessary information to know where, when, with the milk of which animals, and by whom a particular cheese was produced and sold, and also facilitate recall of the cheeses in a specific lot should a problem be detected. Regulatory bodies for PLTLs have a traceability system with varying degrees of complexity to keep track of the production history of each cheese. Consumers are thus assured that the milk and the cheese were produced according to the information provided on the label and following the specifications of the particular PLTL.

Several traceability systems have been reported in the literature. Casein labels, or casein disks, provide an economic and easy-to-use traceability system, very popular with many PDOs (Labelys, 2016). A casein disk is placed on each individual cheese at manufacturing time. Disks may carry a significant amount of information, such as the logo of the corresponding PDO, a bar code or a consecutive numbering system which identifies the individual cheese, the name of the dairy and the manufacturing date. Code numbers are recorded on the daily production report. This system is adequate for small- and medium-size productions of cheeses made and ripened on the premises, with the milk from the same farmhouse, or few milk producers. However, it breaks down at the retail stage if cheeses are cut into smaller portions and packaged. In some cases, like Montasio or Grana Padano cheeses, in addition to the casein label, the name and trademark of the cheese are printed on the side of the cheese wheel to ascertain the identity of portions, even though the portion cannot be traced back to a specific lot.

In contrast, the traceability system of Parmigiano Reggiano cheese is a good example of a modern system, applicable to a large and complex production–marketing system (Regattieri, Gamberi & Manzini, 2007). A radio frequency identification (RFID) system using wireless microchips stores all necessary information (milk origin, cheese manufacturing data and ripening place and distribution centre) in a very small tag which is ultimately applied to the large cheese wheel. At portioning time, the tag is read into a database. Each portion is individually wrapped and assigned an alphanumerical code which is printed on the envelope. This information is also read into the database so that each portion is clearly linked to the cheese wheel from which it was cut. Currently, various electronic traceability systems are under development that combine sensing devices, software packages and data sharing (Papetti et al., 2012). A specific sequence inserted in a uniquely located site of carefully selected lactic acid bacteria as a "biological" marker serves as a "proof-of-origin" culture to unequivocally authenticate cheeses, even if cut into pieces and grated. It has been proved effective in Swiss PDO cheeses, Emmental, Tete de Moine and Appenzeller (Ludin et al., 2016).

5.3 Authentication: What Should Be Authenticated?

Authentication refers to ascertaining that the specific food item is as described in its label and/or in its product specifications. The final product produced in a specific way is the subject to authentication. Thus, both the cheese and its production process must be authenticated. To ascertain that a given cheese conforms to the specifications described in its registration document, two main types of methodologies are needed: (1) assessment of the documentation kept by individual cheesemakers regarding all the stages and processes to manufacture the cheese in its commercial form and (2) analytical procedures to verify the aspects that each PLTL regulatory body considers essential for the identity of its product (Manning & Soon, 2014). As mentioned earlier, regulatory entities of PLTL cheeses assess the corresponding documentation.

The broad objective of food authentication is, therefore, to identify unique markers, or groups of markers, to establish the authenticity of the food or to identify the potential presence of forbidden materials to resolve authenticity issues (Mafra, Ferrerira & Oliveira, 2008). For authentication purposes, a marker should be a measurable indicator (such as a chemical compound), or group of measurable indicators, which can be quantitatively determined in the final product and which links that product with a specific production feature. Because different markers or groups of markers are often needed to authenticate a PLTL cheese or a specific production process, multivariate statistical analysis is necessary to obtain a fingerprint for the process or cheese (Karoui & De Baerdemaeker, 2007). Due to the complexity of the cheese matrix, many methodologies are initially developed and tested at various production stages prior to obtaining the cheese in its final commercial state, particularly in the milk used for cheesemaking.

In this chapter, recently published analytical methodologies to ascertain that specifications of PLTL cheeses are fulfilled are reviewed. For aspects not covered here, we refer the reader to the following excellent comprehensive review (De la Fuente & Juárez, 2003). The following aspects are considered: (1) raw materials, (2) geographical location, (3) animal management and feeding systems, (4) cheesemaking technologies and (5) sensory characteristics.

5.3.1 Raw Materials

Milk origin (species and/or breeds). Adding cow's milk, often more abundant and less expensive, to sheep's, goat's or buffalo's milk in cheeses which explicitly forbid it is the most frequent breach of product specifications. Although this is mostly done for economic reasons, the presence of undeclared types of milk in a cheese also has serious health implications because many consumers are allergic to cow's or sheep's milk. Consequently, a number of analytical procedures have been investigated to authenticate the species of milk origin and to identify the potential presence of other milks in PLTL cheeses. These procedures are generally grouped in two main categories depending on which molecule is being measured: (1) DNA-based and (2) protein-based techniques. The animal species from which the milk was obtained can be identified through an analysis of the DNA from the somatic cells that remain in the cheese. DNA methodologies are most reliable due to the specificity of the DNA markers used and the stability of the DNA under cheesemaking conditions. They have been successfully used for the past couple of decades to determine animal species in many different types of foods, including milk and cheese (Bottero & Dalmasso, 2011; Mafra, Ferrerira & Oliveira, 2008). The use of species-specific primers to detect different species in cheeses made with binary or ternary milk mixtures is summarised in Table 5.2. High-sensitivity methods to detect low amounts (<1%) of bovine milk are primarily interesting for health reasons. Polymorphisms of the melanocortin1 receptor gene proved to be adequate for identifying breeds of the same species in cheeses made with milks from one or several breeds (Table 5.2). Randomly Amplified Polymorphic DNA (RAPD) and Sequence Characterised Amplified Region (SCAR) markers have been recently developed for the identification of non-autochthonous sheep breeds in Serra da Estrela cheeses (Cunha et al., 2016).

Species-specific differences in caseins and in their proteolytic fragments have also been used to identify the milk of a non-allowed species in a cheese. The EU reference method to determine the presence of cow's milk in cheese made with sheep's, goat's or buffalo's milk (EC, 2008)

Table 5.2 Some currently developed DNA methodologies to authenticate the milk used in cheese manufacture.

Species/ breeds detected	DNA region analysed	Methodology	Cheese(s) analysed (references)
Bovine milk in mixtures of sheep and goat milk	Mitochondrial genome: D-loop region and tRNAlys; species-specific primers	High-resolution melting (HRM) analysis	PDO Feta, commercial and experimental with different percentages of sheep and goat milk (Ganopoulos et al., 2013)
Bovine (≥1%) in water buffalo milk	12S rRNA and D-loop: species-specific primers	HRM	PDO and commercial non-PDO Mozzarella (Sakaridis et al., 2013)
Cow, sheep, goat and buffalo milks: simultaneous detection; 1% (v/v) in binary mixtures;	Species-specific mitochondrial DNA sequences	Multiplex PCR,[a] capillary electrophoresis (CE)	PDO and non-PDO commercial cheeses (Gonçalves et al., 2012)
Cow, sheep, and goat milks	Species-specific primers	PCR electrophoretic patterns[a]	PDO Portuguese cheeses made with milk from single or mixed species (Santos Guerreiro, Fernandes & Bardsley, 2012)
Reggiana cow breed	Melanocortin 1 Receptor (MC1R) gene polymorphisms	PCR-RFLP, PCR-APLP[b]	Parmigiano Reggiano PDO cheese (Russo et al., 2007)
Massese sheep breed	MC1R single nucleotide polymorphisms (SNP)		Cheeses made with Massese sheep milk, or mixtures of other breeds (Fontanesi et al., 2011)
Girgentana, Maltese and Derivata di Siria goat breeds	Microsatellite markers (20)	Multiplex PCR, CE	Girgentana goat cheeses with Maltese and Derivata di Siria milks (Sardina et al., 2015)

a) PCR: Polymerase chain reaction;
b) PCR-RFLP: PCR restriction fragment length polymorphism; PCR-APLP: PCR amplified product length polymorphism.

is based on the detection of species-specific γ_2-casein proteolytic fragments obtained by in vitro plasmin hydrolysis of cheese samples. This method is reported to yield reliable results for evaluating the presence of cow's milk in sheep's and goat's cheeses, although in water buffalo cheeses results may sometimes be uncertain (Pizzano et al., 2011). Bovine-specific fragment 1–23 of α_{s1}-casein served as marker for the presence of cow's milk in two sheep's Pecorino cheeses, and fragment 16–25 of β-casein confirmed the inclusion of water buffalo's milk in cheeses made with a required mixture of cow's and water buffalo's milk (Sforza et al., 2011). The 149–162 peptide of β-lactoglobulin has been used to detect the presence of cow's milk in PDO buffalo Ricotta cheeses by matrix-assisted laser desorption time-of-flight mass spectrometry (MALDI-TOF MS) (Russo, Rega & Chambery, 2016). Signature tryptic fragments of bovine α_{s1}- and β-caseins, identified by MALDI-TOF MS, proved instrumental in guaranteeing the authenticity of PDO water buffalo Mozzarella cheese (Caira et al., 2016). An excellent review of immunological methods is that by Pizzano et al. (2011). To the best of our knowledge, protein-based methodologies to identify breeds within a given species have not been reported.

Non-milk fat. The official European Union method for detecting other fats in cow's milk based on the analysis of triacylglycerols (TAG) by gas chromatography was successfully adapted by Fontecha et al. (2006) to analyse non-milk fat in Manchego and Mahón PDO cheeses. The direct analysis of TAG in direct analysis in real time ionisation–high resolution mass spectrometry (DART-HRMS) can detect sunflower, rapeseed and/or soybean oil in soft cheeses at levels as low as 1% (w/w) with minimum sample preparation (Hrbek et al., 2014).

5.3.2 Geographical Location

The geographical origin of a PLTL cheese is obviously the most important aspect to authenticate. Vegetation and soil are a unique combination for each area, thus providing a wealth of marker compounds which can be transferred from the animal ingesting the plants to the milk and, ultimately, to the cheese. Plant compounds frequently used are terpenes and long-chain hydrocarbons, and fatty acids and their biohydrogenation metabolites produced in the rumen. The soil and the environment contribute trace elements, stable isotopes and occasionally radioactive elements. The agro-animal system, therefore, can identify the geographical origin of a cheese. Some of the recent research efforts to define various markers to authenticate the geographical origin of cheeses are summarised in Table 5.3.

Table 5.3 Some methodologies to authenticate the geographical location of cheese or milk used for cheese manufacture.

Geographical origin of cheese or milk	Molecular marker(s) or global approaches	Analytical methodology	Cheese(s) or milk(s) analysed (references)
Lowlands and highlands (mountains)	1-Phytene, 2-phytene, neophytadiene	GC-MS[a]	Italian cheeses: lowland or alpine mountain farms (Povolo et al., 2009)
Specific Alpine areas using different types of pastures	Vaccenic acid, t11c15 CLA, α-linolenic acid, eicosapentaenoic, pentadecanoic, heptadecanoic, heptadecenoic acids. 1-Phytene/2-phytene, nonacosano/heptacosano. δ-Carene, β-pinene, total terpenes	GC-FID[b] GC-MS	Cheeses (Bitto PDO, Nostrale d'Alpe; Asiago) produced at different altitude, or predominant botanical species (*Festuca-Agrostis)* (De Noni & Battelli, 2008; Falchero et al., 2010; Povolo et al., 2012; Povolo et al., 2013)
Milk from grazing cows	Microbial r16S-23S intergenic transcribed spacer: 17 markers	Automated ribosomal intergenic spacer analysis (ARISA)	Alpine pasture (700–2000 m), valleys (350–600 m), or lowland plains (<135 m) (Bonizzi et al., 2007)
Milk from grazing cows	Volatile compounds profile	Electronic nose	Pastures with *Trifolium alpinum* and *Festuca nigrensis* (Falchero et al., 2010)
PDO or commercial non-PDO cheese	Mass spectral data (ions) for volatile compounds	Proton-transfer mass spectrometry (PT-MS)	PDO Cumin cheeses or commercial non-PDO brands (Galle et al., 2011)
PDO cheeses from specific regions	Volatile composition; Sensory flavour and odour attributes	GC-MS; Sensory analysis	Manchego, Zamorano, Roncal and Idiazabal PDO (Barron et al., 2005) Canestrato Pugliese, Fiore Sardo and Pecorino Romano PDO (Di Cagno et al., 2003)
Different productions of a PDO cheese area;	Ba, Cu, Cr, Al, Fe, Zn, Hg, Pb, Mg, Mn, Ni, Se and V	Atomic emission spectroscopy	Bryndza cheeses from different Slovak regions (Suhaj & Korenovská, 2008)

Table 5.3 (Continued)

Geographical origin of cheese or milk	Molecular marker(s) or global approaches	Analytical methodology	Cheese(s) or milk(s) analysed (references)
	$\delta^{13}C$, $\delta^{15}N$	Inductively coupled plasma MS (ICP-MS) Isotopic ratio MS (IRMS)	Milks for Mozzarella PDO cheeses from different sites (Brescia et al., 2005)
	β-Galactose, β-lactose, acetic acid and glycerol	^1H NMR	Mozzarella PDO cheeses from different sites (Mazzei & Piccolo, 2012)
Different production areas making the same cheese type	Cheese infrared spectra; $\delta^{13}C$, $\delta^{15}N$, $\delta^{18}O$, $\delta^{87}Sr$	Near and mid-Infrared spectroscopy (NIR, MIR) IRMS	Emmental type cheeses made in different European countries (Karoui et al., 2005)
	Volatile profile	Electronic nose	Emmental type cheeses made in different European countries (Pillonel et al., 2003a; Pillonel et al., 2003b)
	V3 Region of the 16S rDNA	Denaturing gradient gel electrophoresis (DGGE)	Minas cheeses made in different regions of Brazil (Arcuri et al., 2013)
Cow's milk from farms of same region	16S-32S Intergenic transcribed spacer: 64 markers (279–756 bp)	ARISA	Farms on a wide plain: resolution with less than 5 km difference (Feligini et al., 2015)

a) GC-MS: Gas chromatography–mass spectrometry;
b) GC-FID: Gas chromatography with flame ionisation detection.

A handful of studies used volatile profiles to differentiate among cheeses belonging to separate PDOs (Barron et al., 2005), or between a PDO cheese and its non-PDO commercial counterpart (Galle et al., 2011).

Perhaps the most powerful analytical tools to differentiate cheese (or milk) produced in various countries, regions or particular sites are the spectroscopic (near and mid-infrared, ultraviolet–visible, atomic emission or nuclear magnetic resonance) and spectrometric ($\delta^{13}C$, $\delta^{15}N$, $\delta^{18}O$, and $\delta^{87}Sr$ isotopic ratios, inductively coupled plasma or proton-transfer mass spectrometry) techniques, because these spectral techniques provide a global approach to identifying the geographical origin of a cheese, giving its "fingerprint" without specifying a particular marker compound(s). In addition, they facilitate screening many samples in a short time. The following are examples of the usefulness of these techniques: the identification of Emmental type cheese (PDO and non-PDO) made in different European countries (Karoui et al., 2005), buffalo milk for PDO Mozzarella Campana cheese made in different sites (Brescia et al., 2005) or the identification of Bryndza cheese produced in different regions of Slovakia (Suhaj & Korenovská, 2008). Two stable isotope methods were recently validated by an international collaborative study to authenticate Parmigiano Reggiano and Grana Padano cheeses (Camin et al., 2015).

Analysis of the genetic diversity of microbial cheese populations by PCR methodologies can be used to authenticate the geographical origin of cheeses because cheese microbiota depend, in part, on the location of the farm. For example, sheep's cheeses produced in the four regions of the Minas Gerais state (Brazil) were identified through an analysis of the V3 region of the bacterial 16S ribosomal DNA (Arcuri et al., 2013) extracted from cheese. The microbial fingerprint obtained from the automated ribosomal intergenic spacer analysis of milk differentiated cow's milk from farms located less than 5 km apart (Feligini et al., 2015), or at different alpine altitudes (Bonizzi et al., 2007).

5.3.3 Animal Management and Feeding Systems

Specific animal management and/or feeding systems are sometimes required by different PLTL labels. In several cases, such as Roquefort or Pont-l'Evêque cheeses, 75%–80% of feedstuffs must be produced in the specific geographical area, whereas in other cases (Parmigiano Reggiano, Picodon or Laguiole) certain components such as maize silage are specifically prohibited. Other PLTL regulatory bodies (Pecorino di Picinisco, Chevrotin or Pont-l'Evêque) require several months of pasture grazing (weather permitting). Fatty acids, terpenoids, tocopherols and carotenoids, and phenolic compounds have been proposed as molecular markers to characterise and authenticate milk and cheese obtained from grazing animals (Prache et al., 2005) (Table 5.4). Some of those compounds can also be used to ascertain geographical location, as described in the previous section. Major research efforts have been made to characterise milk and cheese from grazing cows, and to a much lesser extent, from grazing ewes and goats. But there is still much to do to unequivocally differentiate cheeses made with milk from grazing animals from those manufactured with milk from animals managed under other feeding regimes. The main difficulty stems from the use of concentrates obtained from various plant materials which can provide several of the same compounds found in fresh grass, thus yielding compound profiles in milk or cheese similar to those derived from fresh grass (Shingfield, Bonnet & Scollan, 2013). Care must be exercised in the interpretation of this type of result.

Table 5.4 Some methodologies to authenticate animal management and feeding system in cheese, or in the milk used for cheese manufacture.

Animal management and feeding system	Molecular marker(s) or global approaches	Analytical methodology	Cheese(s) or milk(s) analysed (references)
Animal feeding with maize silage	Cyclopropyl-, and ω-cyclohexyl fatty acids	GC-FID[a]	Milk used for Parmigiano Reggiano PDO cheese (Caligiani et al., 2016)
Grazing management and/or other feeding systems	Vaccenic, rumenic, α-linolenic, pentadecanoic, heptadecanoic, eicosapentaenoic acids, total CLA[b], n-3/n-6 PUFA	GC-FID	Milk for Cantal PDO cheeses: different grazing (Coppa et al., 2011). Fatuli cheeses: mountain grazing or valley indoor feeding (Valnegri et al. 2011) Idiazabal PDO cheeses: indoor, valley part-time or mountain grazing (Abilleira et al., 2009). Grana Padano PDO cheeses: lowlands or mountain areas (Prandini et al., 2009)
	β-Caryophyllene, α-copaene, α-pinene, camphene, β-pinene, δ-3-carene	GC-MS[c]	Toma Piemontese PDO cheeses: indoor and extensive grazing (Revello-Chion et al., 2010). Milk for Cantal PDO cheeses: strip/paddock grazing (Tornambé et al., 2006) Milk for Idiazabal PDO cheese: indoor/valley part-time (Abilleira et al., 2011)

Table 5.4 (Continued)

Animal management and feeding system	Molecular marker(s) or global approaches	Analytical methodology	Cheese(s) or milk(s) analysed (references)
	γ-Curcumene, α-curcumene		La Serena PDO cheeses: indoor or Mediterranean pastures (Fernández-García et al., 2008)
	α-Tocopherol, α-tocopherol/ cholesterol	HPLC-DAD[d], Fluorescence	Caciotta cheeses: indoor feeding or extensive grazing (Pizzoferrato et al., 2007) Goat milk from mountain region of South of Spain (Delgado-Pertíñez et al., 2013) Idiazabal PDO cheeses: indoor, valley part-time, or mountain grazing (Valdivielso et al., 2015)
	All-trans-β-carotene, lutein		Milk used for Ragusano PDO cheese: grazing time *per* day (Marino et al., 2012)
			Milk from Central France: silages/grazing based feeding (Agabriel et al., 2007)
	Milk spectra	FTIR: 925–5008 cm^{-1}	Milk from The Netherlands: different feeding systems (Capuano et al., 2014)
		UV/Vis: 450–530 nm	Milk for Canestrato Pugliese PDO cheese: different feeding systems (Priolo et al., 2003)
	$\delta^{18}O$, relative proportions of PUFA, MUFA and SFA[e]	IRMS and NMR	Milk from Brittany and Massif Central (France): mixed rations of pasture, grass silage and hay (Renou et al., 2004)
	Hippuric, phenylacetic, 4-hydroxybenzoic acids.	HPLC-DAD	Milk for PDO Cantal cheeses: indoor forages or grazing (Besle et al., 2010)
	Fatty acid composition	GC-FID	Cow's milk from 10 European countries: different systems (Coppa et al., 2014)
Indoor feeding: different concentrate formulations	Milk spectra	Front-Face Fluorescence: 250–290 nm excitation/ 380–410 nm emission	Goat milk from northern Tunisia (Hammami et al., 2010)

a) Gas chromatography with flame ionisation detection;
b) Conjugated linoleic acid;
c) Gas chromatography–mass spectrometry;
d) High-performance liquid chromatography with diode array detection;
e) Saturated fatty acids.

Ruminal biohydrogenation reactions of linoleic (*c*9,*c*12-C18:2) and α-linolenic (*c*9,*c*12,*c*15-C18:3) acids prevalent in fresh grass yield different positional and geometrical isomers of oleic (*c*9-C18:1) and linoleic acids, vaccenic (*t*11-C18:1) and rumenic (*c*9,*t*11-C18:2) acids being the major compounds (Mohammed et al., 2009). Other metabolites such as *t*11,*c*13-C18:2 and *t*11,*c*15-C18:2 have also been described as indicative of grass feeding (Alves and Bessa, 2014). However, when the diet of the animals includes concentrates with highly digestible carbohydrates, the content of isomers *t*10-C18:1, *t*10,*c*15-C18:2 and *t*7,*c*9-C18:2 increases with a concomitant decrease in that of vaccenic and rumenic acids, as extensively reviewed in Aldai et al. (2013) and Bessa, Alves and Santos-Silva (2015). To the best of our knowledge, the *t*11-C18:1/*t*10-C18:1 ratio has not yet been used to investigate the possible supplementation of pasture diet with concentrate in authentication studies.

Terpenoids are mostly transferred directly from the ingested plant to the milk and cheese, although, according to some authors, terpenoids can also be degraded or modified by rumen microbiota and lactic acid bacteria used for cheesemaking (Belviso et al., 2011). Some specific compounds like β-caryophyllene, α-coapene and α- and β-pinene or camphene are often found in cheese from animals grazing in the humid grasslands of Europe (Revello-Chion et al., 2010; Tornambé et al., 2006). In contrast, γ- and α-curcumene were the predominant terpenes in cheeses made with milk from animals grazing in southwestern Mediterranean pastures (Fernández-García et al., 2008). The application of accurate quantitative methods is still a challenge (Abilleira et al., 2009) because most analytical procedures for terpenoids use solid-phase microextraction (SPME) semi-quantitative techniques. As an advantage, terpenoid profiles in cheese from grazing animals are more difficult to reproduce by adding other feedstuffs to concentrates than fatty acid profiles, due to the very low terpenoid content found in milk from ruminants fed conserved forages and/or concentrates (Prache et al., 2005).

The methodologies used to authenticate cheeses from different animal managements or feeding systems are based on specific chemical fingerprints, or ratios between compounds, rather than on the presence or absence of one or various isolated molecular markers. For example, the α-tocopherol/cholesterol ratio was used to differentiate Caciotta cheeses produced with milk from goats under grazing or indoor feeding systems (Pizzoferrato et al., 2007). The terpenoid profile found in Toma Piamontese PDO mountain cheeses successfully differentiated winter cheeses (hay-based indoor-fed cows) from summer cheeses (grazing cows) (Revello-Chion et al., 2010). Fatty acid profiles classified PDO Idiazabal cheeses according to the changes in feeding systems during lactation, from indoors in winter to extensive mountain grazing in late spring (Valdivielso et al., 2015). In general, most authors indicate that high variability in the potential marker composition occurs due to one or more of the following reasons: feedstuffs and pasture diversity, geographical grassland altitude and/or location, seasonal feeding variations linked to lactation stage and pasture availability, and type of management systems (indoor feeding based on concentrates and forages, fresh grass indoors, part-time grazing supplemented with forages and/or mixed rations, extensive grazing on mountain or valley, paddock or strip grazing, etc.).

Perhaps the most clear, unequivocal and highly discriminant markers for a type of feed are cyclopropyl- and ω-cyclohexyl-fatty acids found in maize silage (Caligiani et al., 2016). The presence of these very unusual fatty acids in milk samples definitely establishes the use of maize silage in animal feeding. These could be the only markers defined so far that could be generally used, but to date no investigations have been reported in cheese.

As mentioned above for cheese geographical location, other effective methodologies to authenticate animal management and feeding are based on spectral information from spectroscopic techniques. However, to date, the application of spectroscopic techniques to authenticate animal feeding systems has been studied only in milk.

In the last few years, cheeses are appearing in the market bearing two different labels from two very different organisms: a PLTL and an organic production label which represents extra requirements on top of those of the corresponding PLTL regulatory entity. Differentiating organic cheese (or milk) from its conventional counterpart under the same PLTL label by analytical methods does not appear to be feasible. Even though attempts have been made to differentiate them using fatty acid profiles or even isotopic ratios, those potential markers are affected by a combination of factors other than pasture grazing, like animal breed, stage of lactation, health of animals and the overall farming systems linked to specific regions. There is a lack of valid comparative studies (see Schwendel et al. (2015) for a review), and therefore care must be exercised in the interpretation of these type of results.

5.3.4 Cheesemaking Technologies

There are very few studies on various cheesemaking technologies aimed at differentiating PLTL from non-PLTL cheeses (Table 5.5). Although commercially available bovine rennet is most frequently used, certain cheeses are characterised by the use of a specific type of rennet produced by the cheesemaker. A characteristic peptide present in Serpa cheeses made with cardoon flower rennet, identified by capillary electrophoresis, allowed detection of Serpa cheeses made with non-permitted animal rennet (Roserio et al., 2003). Lamb or kid rennet pastes are used in a few PDO cheeses from the Mediterranean area, characterised by a marked increase in the proportion of short-chain free fatty acids as well as by a distinct piquant taste as a result of the presence of pregastric lipase in these rennets (Addis et al., 2005; Virto et al., 2003). A high lipolysis and a quantity of short-chain free fatty acids (up to 10 carbon atoms) over 50% of the total are very good analytical indications of the use of pregastric lipase-containing rennet.

The use of an autochthonous starter instead of a commercial one was investigated in Fiore Sardo cheese during 90 days of ripening. Metabolomics using Proton Nuclear Magnetic Resonance (^{1}H NMR) techniques with multivariate analysis differentiated both ripening time

Table 5.5 Some methodologies to authenticate cheesemaking technologies in cheese.

Technological parameters	Molecular marker(s) or global approaches	Methodology	Cheese or milk analysed (references)
Type of rennet	Butanoic, hexanoic, octanoic, decanoic acids, total SFA[a]	GC-FID[b]	Idiazabal PDO cheeses: lamb rennet paste or commercial rennets (Virto et al., 2003) Pecorino Romano PDO cheeses: lamb rennet paste or commercial rennets (Addis et al., 2005). Feta PDO cheeses: liquid lamb rennet or commercial calf rennet (Georgala et al., 2005) Goat cheeses: kid rennet paste or commercial animal rennet (Fontecha et al., 2006)
	Nitrogen fractions	Physico-chemical analysis	Sheep cheeses: cardoon flower or commercial bovine rennet (Galán et al., 2008)
	Peptide composition	CE[c]	Serpa PDO cheeses: cardoon flower or commercial bovine rennet (Roserio et al., 2003)
Starter culture	Carbohydrates, organic acids, amino acids	^{1}H NMR[d]	Fiore Sardo PDO cheeses: commercial or autochthonous starter (Piras et al., 2013)
Milk heat treatment	Cheese spectra	Tryptophan Fluorescence: 290 nm ex.	Emmental type cheeses: raw/thermised milk (Karoui et al., 2005)
	V3 Region of the 16S rDNA	DGGE[e]	Minas cheeses: raw/pasteurised milk (Arcuri et al., 2013)

a) SFA: Saturated fatty acids
b) GC-FID: Gas chromatography with flame ionisation detection
c) CE: Capillary electrophoresis
d) ^{1}H NMR: Proton nuclear magnetic resonance
e) DGGE: Denaturing gradient gel electrophoresis.

and the type of culture used (Piras et al., 2013). The geographic origin of PDO Mozzarella and Caciocavallo Silano cheeses could also be related to the microbial population of natural whey cultures used in their manufacture (Ercolini et al., 2008).

Tryptophan fluorescence spectra has been proposed as a good tool to differentiate Emmental cheese made with raw milk from those made with thermised cow's milk (Karoui et al., 2005). Similarly, analysis of band profiles of the bacterial V3 region of the 16S rDNA by denaturing gradient gel electrophoresis differentiated Minas Gerais cheese made with raw cow's milk (Arcuri et al., 2013). These two methodologies could be useful to authenticate cheeses made with raw milk.

5.3.5 Sensory Characteristics

The sensory characteristics of all PLTL cheeses are described in their corresponding registration documents. To the best of our knowledge, only two regulatory bodies conduct mandatory sensory analysis by an accredited external laboratory to ensure the expected sensory quality of the product: (1) the Regulatory Council of Idiazabal (Perez-Elortondo et al., 2007) and (2) the Regulatory Council of Roncal (Mendia et al., 2003), both of which are PDO cheeses from Spain. In both cases, a panel of trained experts periodically analyses random cheeses from each producer throughout the production season. Only cheeses receiving a score above a set value are allowed to use the PDO label. A PLTL cheese produced according to specifications will not always meet consumer expectations from a sensory point of view.

Sensory analysis could also differentiate sheep cheeses of three separate Spanish PDOs produced in the same season (Barron et al., 2005). Therefore, sensory analysis can be an important factor in the establishment of the authenticity of a cheese.

5.4 Innovation, Modern Technologies and Traditional Cheeses

Cheeses protected under geographical labels are traditional products strongly linked to the area where they have been manufactured for the last many years or even several centuries. Each has its own distinct personality. The unwritten assumption is that the characteristics of today's cheeses resemble the original ones developed, in most cases, long time ago. But it is not possible to confirm it, except for certain physical or technological features which may have been recorded in old documents, such as shape, size, colour, use of traditional rennets or smoking. Throughout time, cheesemakers have adapted their manufacturing practices to the technological developments of their period in an attempt to improve their products and continue making a living. Here are a few examples of past technological innovations which today have become widely accepted.

Until 1874 when Christian Hansen commercialised the first standardised liquid rennet, cheesemakers prepared their own rennet using very different recipes (Bozzetti, 2001). Nowadays the majority of artisan cheesemakers use commercially available rennets in their PLTL-labelled products. Similarly, automatic vats and temperature- and humidity-controlled ripening chambers are standard equipment. Before milk pasteurisation was introduced, endogenous microbiota in raw milk acidified it. In some cases, the cheesemaker prepared a starter culture in whey (unaware of what was going on) by mixing part of the acidified milk with whey and using it the following day, a practice which is still followed in a few cases. Today, sensor, information and DNA-based technologies are widely used in many phases of traditional cheesemaking. For example, a variety of mixtures of different microorganisms and starter cultures that have been improved to enhance the development of certain aromatic compounds during ripening are used by most artisan and industrial PDO cheesemakers alike. Sensoring equipment is used in automated vats, ripening chambers and traceability systems. Yet, PLTL cheeses maintain their own distinct personalities.

New formulations of animal coagulants (kosher or halal), even if not widely used as yet in PLTL cheeses, can bring traditional cheeses to certain ethnic and/or religious communities. The use of genetically modified feed components (some of which can be grown in Europe) could help cheesemakers to keep production costs of PLTL cheeses within the reach of many consumers.

Thus, regulatory bodies of PLTL-labelled cheeses, while protecting the characteristics of their traditional products, can favour adaptation to new technological developments and situations, ultimately contributing to maintain rural populations, preserving the environment and the diversity of traditional products.

5.5 Conclusions

Taking into consideration the analysis of the methodologies that have been reported in the literature to authenticate different aspects of a PLTL cheese, it is clear to us that there is no general marker (chemical compound or group of compounds) or no unique methodology that could be used in every case, with the possible exception of maize silage markers. In other words, authentication has to be carried out at the local or, at the most, the regional level. The cheese in question must be well characterised, and a database with the most important discriminating parameters should be constructed. It should also include distinctive aspects of its manufacturing technology and animal feeding requirements. Therefore, local pastures and other feed materials need to be analysed to identify plant compounds that may be transferred to the milk and on to the cheese. Most of the methodologies described in this article to authenticate PLTL cheeses are still at the research level. To the best of our knowledge, few, if any, of them are currently being implemented by PLTL regulatory bodies.

Regulatory bodies for each PLTL cheese need to define which features are essential for the personality of their particular cheese and which ones can, realistically, be authenticated in a consistent manner by analytical techniques.

Acknowledgements

Financial support was provided in part by the University of the Basque Country (Unesco Cathedra 09/07), the Basque (IT-766-13) and Spanish Governments (AGL2013-48361-C1-1-R).

References

Abilleira, E., Collomb, M., Schlichtherle-Cerny, H., Virto, M., de Renobales, M. & Barron, L. J. R. (2009). Winter/spring changes in fatty acid composition of farmhouse Idiazabal cheese due to different flock management systems. *Journal of Agricultural Food Chemistry*, 57, 4746–4753.

Abilleira, E., Virto, M., Nájera, A. I., Albisu, M., Pérez-Elortondo, F. J., Ruiz de Gordoa, J. C., de Renobales, M. & Barron, L. J. R. (2011). Effects of seasonal changes in feeding management under part-time grazing on terpene concentrations of ewes' milk. *Journal Dairy Research*, 78, 129–135.

Addis, M., Piredda, G., Pes, M., Di Salvo, R., Scintu, M. F. & Pirisi, A. (2005). Effect of the use of three different lamb rennet pastes on lipolysis of the Pecorino Romano cheese. *International Dairy Journal*, 15, 563–569.

Agabriel, C., Cornu, A., Journal, C., Sibra, C., Grolier, P. & Martin, B. (2007). Tanker milk variability according to farm feeding practices: Vitamins A and E, carotenoids, color, and terpenoids. *Journal of Dairy Science*, 90, 4884–4896.

Aldai, N., de Renobales, M,, Barron, L. J. R. & Kramer, J. G. (2013). What are the trans-fatty acid issues in foods after discontinuation of industrially produced trans fats? Ruminant products, vegetable oils, and synthetic supplements. *European Journal of Lipid Science and Technology*, 115, 1378–1401.

Alves, S. P. & Bessa, R. J. B. (2014). The trans-10,cis15 18:2: A missing intermediate of trans-10 shifted rumen biohydrogenation pathway? *Lipids*, 49, 527–541.

Anonymous (2016a). http://www.cheese.com/monterey-jack/. Accessed 1/9/2016.

Anonymous (2016b). http://www.cheese.com//. Accessed 1/9/2016.

Anonymous (2016c). http://www.cheesesfromswitzerland.com/cheese-assortment/swiss-alp-cheese-and-specialities.html. Accessed 8/7/2016.

Anonymous (2016d). https://en.wikipedia.org/wiki/Protected_Geographical_Indications_for_Georgian_products. Accessed 10/7/2016.

Anonymous (2016e). https://en.wikipedia.org/wiki/List_of_foods_and_drinks_named_after_places#Cheese_names_from_the_United_States. Accessed 11/7/2016.

Arcuri, E. F., El Sheikha, A. F., Rychlik, T., Piro-Métayer, I. & Montet, D. (2013). Determination of cheese origin by using 16S rDNA fingerprinting of bacteria communities by PCR-DGGE: Preliminary application to traditional *Minas Cheese. Food Control*, 30, 1–6.

Barron, L. J. R., Redondo, R., Flanagan, C., Pérez-Elortondo, F. J., Albisu, M., Nájera, A. I., de Renobales, M. & Fernández-García, E. (2005). Comparison of the volatile composition and sensory characteristics of Spanish PDO cheeses manufactured from ewes' raw milk and animal rennet. *International Dairy Journal*, 15, 371–382.

Bechini, A., Cimino, M. G. C. A., Marcelloni, F. & Tomasi, A. (2008). Patterns and technologies for enabling supply chain traceability through collaborative e-business. *Information and Software Technology*, 50, 342–359.

Belviso, S., Giordano, M., Dolci, P. & Zeppa, G. (2011). Degradation and biosynthesis of terpenoids by lactic acid bacteria: First evidence. *Dairy Science & Technology*, 91, 227–236.

Besle, J. M., Viala, D., Martin, B., Pradel, P., Meunier, B., Berdagué, J. L., Fraisse, D., Lamaison, J. L. & Coulon, J. B. (2010). Ultraviolet-absorbing compounds in milk are related to forage polyphenols. *Journal of Dairy Science*, 93, 2846–2856.

Bessa, R. J. B., Alves, S. P. & Santos-Silva, J. (2015). Constraints and potentials for the nutritional modulation of the fatty acid composition of ruminant meat. *European Journal of Lipid Science and Technology*, 117, 1325–1344.

Bonizzi, I., Feligini, M., Aleandri, R. & Enne, G. (2007). Genetic traceability of the geographical origin of typical Italian water buffalo mozzarella cheese: A preliminary approach. *Journal of Applied Microbiology*, 102, 667–673.

Bottero, M. T. & Dalmasso, A. (2011). Animal species identification in food products. Evolution of biomolecular methods. *The Veterinary Journal*, 190, 34–38.

Bozzetti, V. (2001). L'evoluzione storica dei cagli vegetali e animali. *Scienza e tecnica lattiero-casearia*, 52, 227–245.

Brescia, M. A., Monfreda, M., Buccolieri, A. & Carrino, C. (2005). Characterisation of the geographical origin of buffalo milk and Mozzarella cheese by means of analytical and spectroscopic determinations. *Food Chemistry*, 89, 139–147.

Caira, S., Pinto, G., Nicolai, M. A., Chianese, L. & Addeo, F. (2016). Simultaneously tracing the geographical origin and presence of bovine milk in Italian water buffalo Mozzarella cheese using MALDI-TOF data of casein signature peptides. *Analytical Bioanalytical Chemistry*, 408, 5609–56021.

Caligiani, A., Nocetti, M., Lolli, V., Marseglia, A. & Palla, G. (2016). Development of a quantitative GC-MS method for the detection of cylcopropane fatty acids in cheese as new molecular markers for Parmigiano Reggiano authentication. *Journal of Agricultural Food Chemistry*, 64, 4158–4164.

Camin, F., Bertoldi, D., Santato, A., Bontempo, L., Perini, M., Ziller, L., Stroppa, A. & Larcher, R. (2015). Validation of methods for H, C, N and S stable isotopes and elemental analysis of cheese: Results of an international collaborative study. *Rapid Communications in Mass Spectrometry*, 29, 415–423.

Capuano, E., Rademaker, J., van den Bijgaart, H. & van Ruth, S. (2014). Verification of fresh grass feeding, pasture grazing and organic farming by FTIR spectroscopy analysis of bovine milk. *Food Research International*, 60, 59–65.

Coppa, M., Chassaing, C., Ferlay, A., Agabriel, C., Laurent, C., Borreani, G., Barcarolo, R., Baars, T., Kusche, D., Harstad, O. M., Verbic, J., Golecky, J., Delavaud, C., Chilliard, Y. & Martin, B. (2014). Potential of milk fatty acid composition to predict diet composition and authenticate feeding systems and altitude origin of European bulk milk. *Journal of Dairy Science*, 98, 1539–1551.

Coppa, M., Verdier-Metz, I., Ferlay, A., Pradel, P., Didienne, R., Farruggia, A., Montel, M. C. & Martin, B. (2011). Effect of different grazing systems on upland pastures compared with hay diet on cheese sensory properties evaluated at different ripening times. *International Dairy Journal*, 21, 815–822.

Cunha, J. T., Ribeiro, T. I., Rocha, J. B., Nunes, J., Teixeira, J. A. & Domingues, L. (2016). RAPD and SCAR markers as potential tools for detection of milk origin in dairy products: Adulterant sheep breed in Serra da Strela cheese production. *Food Chemistry*, 211, 631–636.

Curry, A. (2013). The milk revolution. *Nature*, 500, 20–22.

De la Fuente, M. A. & Juárez, M. (2003). Authenticity assessment of dairy products. *Critical Reviews in Food Science and Nutrition*, 45, 563–585.

De Noni, I. & Battelli, G. (2008). Terpenes and fatty acid profiles of milk fat and "Bitto" cheese as affected by transhumance of cows on different mountain pastures. *Food Chemistry*, 109, 299–309.

Delgado-Pertíñez, M., Gutiérrez-Peña, R., Mena, Y., Fernández-Cabanás, V. M. & Laberye, D. (2013). Milk Production, fatty acid composition and vitamin E content of Payoya goats according to grazing level in summer on Mediterranean shrublands. *Small Ruminant Research*, 114, 167–175.

Di Cagno, R., Banks, J., Sheenan, L., Fox, P. F., Brechany, E. Y., Corsetti, A. & Gobbetti, M. (2003). Comparison of the microbiological, compositional, biochemical, volatile profile and sensory characteristics of three Italian PDO ewes' milk cheeses. *International Dairy Journal*, 13, 961–972.

EC (1992). Council Regulation (EC) no. 2081/92 of 14 July 1992 on the protection of geographical indications and designations of origin for agricultural products and foodstuffs. *Official Journal of the European Union, L* 208, 1–8.

EC (2002). Regulation (EC) 178/2002 of the European Parliament and of the Council of 28 January 2002 laying down the general principles and requirements of food law, establishing the European Food Safety Authority and laying down procedures in matters of food safety (art. 18). *Official Journal of the European Union*, L 31, 1–24.

EC (2008). Commission Regulation (EC) 273/2008 of 5 March 2008. Reference method for the detection of cow's milk and caseinate in cheeses from ewe's milk, goat's milk and buffalo's milk or mixtures. *Official Journal of the European Union, L* 88, 1–115.

EC (2012). Regulation (EC) No 1151/2012 of the European Parliament and of the Council of 21 November 2012 on quality schemes for agricultural products and foodstuffs. *Official Journal of the European Union, L* 343, 1–29.

EC (2016). *Geographical indications and traditional specialities*. http://ec.europa.eu/agriculture/quality/schemes/index_en.htm. Accessed 6/7/2016.

Ercolini, D., Frisso, G., Mauriello, G., Salvatore, F. & Coppola, S. (2008). Microbial diversity in natural whey cultures used for the production of Caciocavallo Silano PDO cheese. *International Journal of Food Microbiology*, 124, 164–170.

Falchero, L., Lombardi, G., Gorlier, A., Lonati, M., Odoardi, M. & Cavallero, A. (2010). Variation in fatty acid composition of milk and cheese from cows grazed on two alpine pastures. *Dairy Science & Technology*, 90, 657–672.

Feligini, M., Panelli, S., Sacchi, R., Ghitti, M. & Capelli, E. (2015). Tracing the origin of raw milk from farm by using automated ribosomal intergenic spacer analysis (ARISA) fingerprinting of microbiota. *Food Control*, 50, 51–56.

Fernández-García, E., Imhof, M., Schlichtherle-Cerny, H., Bosset, J. O. & Núñez, M. (2008). Terpenoids and benzenoids in La Serena cheese made at different seasons of the year with a Cynara cardunculus extract as coagulant. *International Dairy Journal*, 18, 147–157.

Flamant, J. C. (1992). Les systèmes d'élevage méditerranéennes dans leurs rapports aux systems céréaliers: Diversité et evolution (the Mediterranean breeding systems in relation to the cereal systems: Diversity and evolution). In Guessous, F., Kabbali, A., & Narjisse, H. (eds.), *Livestock in the Mediterranean Cereal Production Systems*, vol. 49. EAAP Publication. Wageningen, pp 30–49.

Fontanesi, L., Beretti, F., Dall'Olio, S., Portolano, B., Matassino, D. & Russo, V. (2011). A melanocortin 1 receptor (MC1R) gene polymorphism is useful for authentication of Massese sheep dairy products. *Journal of Dairy Research*, 78, 122–128.

Fontecha, J., Mayo, I., Toledano, G. & Juárez, M. (2006). Triacylglycerol composition of protected designation of origin cheeses during ripening. Authenticity of milk fat. *Journal of Dairy Science*, 89, 882–887.

Galán, E., Prados, F., Pino, A., Tejada, L. & Fernández-Salguero, J. (2008). Influence of different amounts of vegetable coagulant from cardoon Cynara cardunculus and calf rennet on the proteolysis and sensory characteristics of cheeses made with sheep milk. *International Dairy Journal*, 18, 93–98.

Galle, S. A., Koot, A., Soukoulis, C., Cappellin, L., Biasioli, F., Alewjin, M. & van Ruth, S. M. (2011). Typicality and geographical origin markers of protected origin cheese from The Netherlands revealed by PTR-MS. *Journal of Agricultural Food Chemistry*, 59, 2554–2563.

Ganopoulos, I., Sakaridis, I., Argiriou, A., Madesis, P. & Tsaftaris, A. (2013). A novel closed-tube method based on high resolution melting (HRM) analysis for authenticity testing and quantitative detection in Greek PDO Feta cheese. *Food Chemistry*, 141, 835–840.

Georgala, A., Moschopoulou, E., Aktypis, A., Massouras T., Zoidou, E., Kandarakis, I. & Anifantakis, E. (2005). Evolution of lipolysis during the ripening of traditional Feta cheese. *Food Chemistry*, 93, 73–80.

Gonçalves, J., Pereira, F., Amorim, A. & van Asch, B. (2012). New method for the simultaneous identification of cow, sheep, goat, and water buffalo in dairy products by analysis of short species-specific mitochondrial DNA targets. *Journal of Agricultural Food Chemistry*, 60, 10480–10485.

Hammami, M., Rouissi, H., Slah, N., Selmi, H., Al-Otaibi, M., Blecker, C. & Karoui, R. (2010). Fluorescence spectroscopy coupled with factorial discriminant analysis technique to identify sheep milk from different feeding systems. *Food Chemistry*, 122, 1344–1350.

Hrbek, V., Vaclavik, L., Elich, O. & Hajslova, J. (2014). Authentication of milk and milk-based foods by direct analysis in real time ionization-high resolution mass spectrometry (DART-HRMS) technique: A critical assessment. *Food Control*, 36, 138–145.

Karoui, R. & De Baerdemaeker, J. (2007). A review of the analytical methods coupled with chemometrics tools for the determination of the quality and identity of dairy products. *Food Chemistry*, 102, 621–640.

Karoui, R., Dufour, E., Pillonel, L., Schaller, E., Picque, D., Cattenoz, T. & Bosset, J. O. (2005). The potential of combined infrared and fluorescence spectroscopies as a method of determination of the geographic origin of Emmental cheeses. *International Dairy Journal*, 15, 287–298.

Labelys (2016). http://www.labelys.com/index.php?cat=28&lang=en. Accessed 30/7/2016.

Licitra, G. (2010). World wide traditional cheeses. Banned for business? *Dairy Science & Technology*, 90, 357–374.

Ludin, P., von Ah, U., Rollier, D., Roetschi, A. & Eugster, E. (2016). Lactic acid bacteria as markers for the authentication of Swiss Cheeses. *Chimia*, 70, 349–353.

Mafra, I., Ferrerira, I. M. P. L. V. O. & Oliveira, M. B. P. P. (2008). Food authentication by PCR-based methods. *European Food Research Technology*, 227, 649–665.

Manning, L. & Soon, J. M. (2014). Developing systems to control food adulteration. *Food Policy*, 49, 23–32.

Marino, V. M., Schadt, I., La Terra, S., Manenti, M., Caccamo, M., Licitra, G. & Carpino, S. (2012). Influence of season and pasture feeding on the content of α-tocopherol and β-carotene in milk from Holstein, Brown Swiss and Modicana cows in Sicily. *Dairy Science & Technology*, 92, 501–513.

Mazzei, P. & Piccolo, A. (2012). 1H HRMAS-NMR metabolomic to assess quality and traceability of mozzarella cheese from Campania buffalo milk. *Food Chemistry*, 132, 1620–1627.

Mendia, C., Larráyoz, P., Ordóñez, A. I., Ibáñez, F. C. & Torre, P. (2003). Monitoring taste panel efficiency during evaluation of the sensory quality of Roncal cheese. *Journal of Sensory Studies*, 18, 91–102.

Mohammed, R., Stanton, C. S., Kennelly, J. J., Kramer, J. K. G., Mee, J. F., Glimm, D. R., O'Donovan, M. & Murphy, J. J. (2009). Grazing cows are more efficient than zero-grazed and grass silage-fed cows in milk rumenic acid production. *Journal of Dairy Science*, 92, 3874–3893.

Papetti, P., Corrado, C., Antonucci, F., Figorilli, S., Solani, S. & Menesatti, P. (2012). A RFID web-based infotracing system for the artisanal Italian cheese quality traceability. *Food Control*, 27, 234–241.

Perez-Elortondo, F. J., Ojeda, M., Albisu, M., Salmerón, J., Etaio, I. & Molina, M. (2007). Food quality certification. An approach for the development of accredited sensory evaluation methods. *Food Quality & Preferences*, 18, 425–439.

Pillonel, L., Ampuero, S., Tabacchi, R. & Bosset, J. O. (2003a). Analytical methods for the determination of the geographic origin of Emmental cheese. Volatile compounds by GC/MS-FID and electronic nose. *European Food Research Technology*, 216, 179–183.

Pillonel, L., Badertscher, R., Froidevaux, P., Haberhauer, G., Holzl, S., Horn, P., Jakob, A., Pfammatter, E., Piantini, U., Rossmann, A., Tabacchi, R. & Bosset, J. O. (2003b). Stable isotope ratios, major, trace and radioactive elements in Emmental cheeses of different origins. *Lebensmittel Wissenschaft & Technologie*, 36, 615–623.

Piras, C., Marincola, F. C., Savorani, F., Engelsen, S. B., Consentino, S., Viale, S. & Pisano, M. B. (2013). A NMR metabolomics study of the ripening process of the Fiore Sardo cheese produced with autochthonous adjunct cultures. *Food Chemistry*, 141, 2137–2147.

Pizzano, R., Nicolai, M. A., Manzo, C. & Addeo, F. (2011). Authentication of dairy products by immunochemical methods: A review. *Dairy Science & Technology*, 91, 77–95.

Pizzoferrato, L., Manzi, P., Marconi, S., Fedele, V., Claps, S. & Rubino, R. (2007). Degree of antioxidant protection: A parameter to trace the origin and quality of goat's milk and cheese. *Journal of Dairy Science*, 90, 4569–4574.

Povolo, M., Pelizzola, V., Lombardi, G., Tava, A. & Contarini, G. (2012). Hydrocarbon and fatty acid composition of cheese as affected by pasture vegetation type. *Journal of Agricultural Food Chemistry*, 60, 299–308.

Povolo, M., Pelizzola, V., Passolungo, L., Biazzi, E., Tava, A. & Contarini, G. (2013). Characterization of two Agrostis-Festuca alpine pastures and their influence on cheese composition. *Journal of Agricultural Food Chemistry*, 61, 447–455.

Povolo, M., Pelizzola, V., Ravera, D. & Contarini, G. (2009). Significance of the nonvolatile minor compounds of the neutral lipid fraction as markers of the origin of dairy products. *Journal of Agricultural Food Chemistry*, 57, 7837–7394.

Prache, S., Cornu, A., Berdagué, J. L. & Priolo, A. (2005). Traceability of animal feeding diet in the meat and milk of small ruminants. *Small Ruminant Research*, 59, 157–168.

Prandini, A., Sigolo, S., Cerioli, C. & Piva, G. (2009). Survey on conjugated linoleic acid (CLA) content and fatty acid composition of Grana Padano cheese produced in different seasons and areas. *Italian Journal of Animal Science*, 8, 531–540.

Priolo, A., Lanza, M., Barbagallo, D., Finochiario, L. & Biondi, L. (2003). Can the reflectance spectrum be used to trace grass feeding in ewe milk? *Small Ruminant Research*, 48, 103–107.

Regattieri, A., Gamberi, M. & Manzini, R. (2007). Traceability of food products: General framework and experimental evidence. *Journal of Food Engineering*, 81, 347–356.

Renou, J. P., Deponge, C., Gachon, P., Bonnefoy, J. C., Coulon, J. B., Garel, J. P., Veritgé, R. & Ritz, P. (2004). Characterization of animal products according to geographical origin and feeding diet using nuclear magnetic resonance and isotope ratio mass spectrometry: Cow milk. *Food Chemistry*, 85, 63–66.

Revello-Chion, A., Tabacco, E., Giaccone, D., Peiretti, P. G., Battelli, G. & Borreani, G. (2010). Variation of fatty acid and terpene profiles in mountain milk and "Toma Piamontese" cheese as affected by diet composition in different seasons. *Food Chemistry*, 121, 393–399.

Roserio, B. L., Gómez-Ruiz, J. A., García-Risco, M. & Molina, E. (2003). Vegetable coagulant (Cynara cardunculus) use evidenced by capillary electrophoresis permits PDO Serpa cheese authentication. *Lait*, 83, 343–350.

Russo, R., Rega, C. & Chambery, A. (2016). Rapid detection of water buffalo ricotta adulteration or contamination by matrix-assisted laser desorption/ionization time-of-flight mass spectrometry. *Rapid Communications in Mass Spectrometry*, 30, 497–503.

Russo, V., Fontanesi, L., Scotti, E., Tzaaoli, M., Dall'Olio, S. & Davoli, R. (2007). Analysis of melanocortin 1 receptor (MC2R) gene polymorphisms in some cattle breeds: Their usefulness and application for breed traceability and authentication of Parmigiano Reggiano cheese. *Italian Journal of Animal Science*, 6, 257–272.

Sakaridis, I., Ganopoulos, I., Argiriou, A. & Tsaftaris, A. (2013). High resolution melting analysis for quantitative detection of bovine milk in pure water buffalo mozzarella and other buffalo dairy products. *International Dairy Journal*, 28, 32–35.

Santos Guerreiro, J., Fernandes, P. & Bardsley, R. G. (2012). Identification of the species of origin of milk in cheeses by multivariate statistical analysis of polymerase chain reaction electrophoretic patterns. *International Dairy Journal*, 25, 42–45.

Sardina, M. T., Tortorici, L., Mastrangelo, S., Di Gerlando, R., Tolone, M. & Portolano, B. (2015). Application of microsatellite markers as potential tools for traceability of Girgentana goat breed dairy products. *Food Research International*, 74, 115–122.

Schwendel, B. H., Wester, T. J., Morel, P. C. H., Tavendale, M. H., Deadman, C., Shadbolt, N. M. & Otter, D. E. (2015). Invited review: Organic and conventionally produced milk – an evaluation of factors influencing milk composition. *Journal of Dairy Science*, 98, 721–746.

Sforza, S., Lambertini, F., Manzini, L., Galaverna, G., Dossena, A. & Marchelli, R. (2011). Proteolytic peptides as molecular markers of species authenticity in cheeses. In Ebeler, S. E., Takeoka, G. R. & Winterhalter, P. (eds.), *Progress in Authentication of Food and Wine*. ACS Symposium Series, pp. 215–226.

Shingfield, K. J., Bonnet, M. & Scollan, N. D. (2013). Recent developments in altering the fatty acid composition of ruminant-derived foods. *Animal*, 7, 132–162.

Suhaj, M. & Korenovská, M. (2008). Correlation and distribution of elemental markers of origin in the production of Bryndza sheep cheese. *Food Chemistry*, 107, 551–557.

Tornambé, G., Cornu, A., Pradel, P., Kondjoyan, N., Carnat, A. P., Petit, M. & Martin, B. (2006). Changes in terpene content in milk from pasture-fed cows. *Journal of Dairy Science*, 89, 2309–2319.

Valdivielso, I., Bustamante, M. A., Buccioni, A., Franci, O., Ruiz de Gordoa, J. C., de Renobales, M. & Barron, L. J. R. (2015). Commercial sheep flocks – fatty acid and fat-soluble antioxidant composition of milk and cheese related to changes in feeding management throughout lactation. *Journal of Dairy Research*, 82, 334–343.

Valnegri, L., Panseri, S., Franzoni, M., Antoniazzi, V., Soncin, S., Soncini, G. & Chiesa, L. M. (2011). Chemical composition, fatty acid profile and food safety criteria to characterize an artisanal Italian mountain cheese from goats grazing on pastures. *Milk Science International*, 66, 286–289.

Virto, M., Chávarri, F., Bustamante, M. A., Barron, L. J. R., Aramburu, M., Vicente, M. S., Perez-Elortondo, F. J., Albisu, M. & de Renobales, M. (2003). Lamb rennet paste in ovine cheese manufacture. Lipolysis and flavor. *International Dairy Journal*, 13, 391–399.

6

An Overview of the Cheesemaking Process

Thomas Bintsis[1] *and Photis Papademas*[2]

[1] *11 Parmenionos, 50200 Ptolemaida, Greece*
[2] *Department of Agricultural Sciences, Biotechnology and Food Science, Cyprus University of Technology, Limassol, Cyprus*

6.1 Introduction

It is well known that the primary objective of cheesemaking was to convert milk into a stable product via fermentation, and the evolution of the process has resulted in a great variety of cheeses, more than 1000; however, many of them have common characteristics and must be considered as variants. In addition to the use of different milks, the development of distinctly different cheese varieties is due to some modifications of the basic steps of the cheesemaking process. These modifications were the result of the specific geographical, climatological and environmental characteristics of the region in which each cheese appeared. The total world-wide milk production in 2015 amounts to 818 mil tonnes, and the projection for 2016 is 826 mil tonnes. As expected, cow's milk represents 82% (674 mil tonnes) of the total, and the rest is buffalo's milk (13%, 110 mil tonnes), goat's milk (2.4%, 19.6 mil tonnes) and sheep's milk (1.3%, 10.4 mil tonnes). Global cheese production in 2015 amounted to 19.9 mil tonnes, while the EU-28 produced 9 mil tonnes. Germany, France, Italy and the Netherlands are the major cheese producers, accounting for 65% of the total EU production (IDF, 2016).

The science of cheese manufacture has been reviewed extensively (Fox & McSweeney, 2004; Fox & Guinee, 2013; Freitas et al., 2013; Johnson & Law, 2010; Kindstedt, 2013a; Kindstedt, 2013b; Nielsen, 2004; Robinson & Wilbey, 1998), and an overview of the cheesemaking process is presented in this chapter. A detailed description of the cheesemaking process, together with their characteristics, for each of 100 (98 cheese varieties plus 2 variants) cheeses is given in Part II of the book.

Cheesemaking is, in fact, a dehydration process in which fat and casein in milk are concentrated between 6- and 12-fold, depending on the variety (Fox & McSweeney, 2004). The first and most important factor for the manufacture of cheese is the chemical composition of the milk, as discussed in Section 6.2. The components of milk are concentrated, and the degree of this concentration is regulated by the extent of the five basic steps of cheesemaking: (1) acidification, (2) coagulation, (3) dehydration, which includes cutting, cooking, stirring, pressing and other processes that promote syneresis (i.e. release of the whey), (4) moulding and (5) salting (Figure 6.1). The composition of the curd is determined during the first one or two days of cheesemaking, and the extent of syneresis is the most important phenomenon that the cheesemaker is challenged to adjust, by controlling the conditions for each basic step. The organoleptic characteristics of each cheese will be determined through the whole process of cheesemaking, and

Global Cheesemaking Technology: Cheese Quality and Characteristics, First Edition.
Edited by Photis Papademas and Thomas Bintsis.

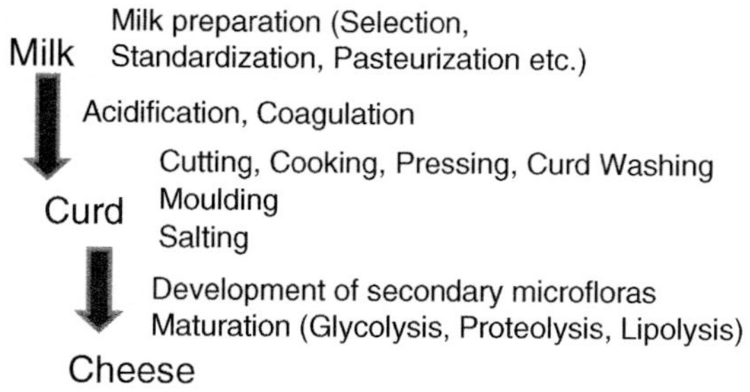

Figure 6.1 A simplified diagram of the cheesemaking and the processes involved.

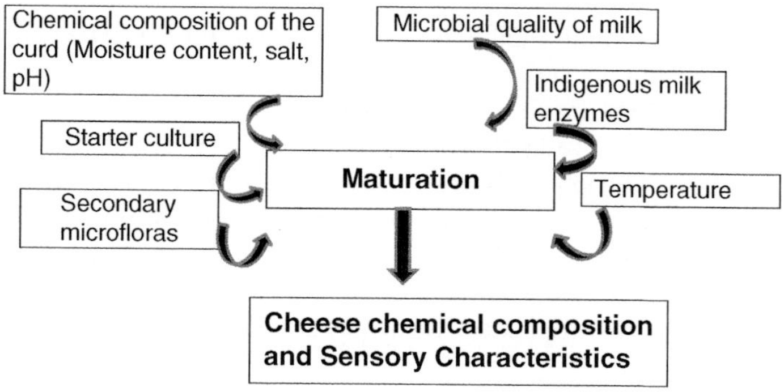

Figure 6.2 Main factors contributing to the final cheese characteristics.

ultimately, through the complex process of maturation (Figure 6.2), which may last up to three years. Therefore, this chapter aims at describing the most important parameters that one should take into consideration when embarking on the journey of transforming milk into cheese.

6.2 Milk Types and Composition

Cow's milk, the major ingredient used in global cheesemaking, is produced by several breeds. A high-milk-producing cow breed is the Holstein-Friesian, representing 90% of total cow's milk production. Other important breeds, especially for producing milk for cheesemaking, are Swiss Brown, Montbéliarde (cheeses produced: Comté, Epoisses), Ayrshire (cheese produced: Cheddar) and Salers (cheeses produced: Cantal, Salers). Sheep breeds whose milk is well known for their use in cheese production are Lacaune (cheese produced: Roquefort), Awassi, Assaf, Manchega (cheese produced: Manchego) and Chios (cheese produced: Feta). Important goat breeds for milk production are Alpine (French lactic cheeses), Damascus, Saanen and Vlahiki. Finally, Mozzarella PDO is made from buffalo's milk that comes from the Mediterranean Italian breed. The basic milk components and how they affect cheesemaking are now discussed.

6.2.1 Casein

Casein is the major nitrogenous fraction of all the milk types described in this chapter (i.e. cow, sheep, goat and buffalo) and is typically present at a level of 2.5%. There are four types of casein: α_{s1}-, α_{s2}-, β- and κ-casein. Casein exists in milk in the form of micelles, which in turn are made up of a number of sub-micelles bound together by colloidal calcium phosphate (CCP). As extensively described by Guinee and O'Brien (2010), micelles have κ-casein on the surface with the protruding chain of the hydrophilic part of the protein stabilising the system. The hydrophobic phosphorylated and calcium-sensitive caseins, that is, α_{s1}-, α_{s2}- and β-casein, form the inner part of the micelle. The total casein of goat's and sheep's milks is 2.3%–3.5% and 4.4%, respectively. There is a well-reported polymorphism of caseins for cow, sheep and goat; for example, five variants of α_{s1}-casein have been identified for cow's and sheep's milk and 10 for goat's milk.

The destabilisation of the casein micelle (by acidification or renneting) will cause the gelation of milk and the formation of curds as described later in the chapter.

6.2.2 Whey Protein

The whey fraction of cow's milk protein is typically ~0.7 g/100 g (i.e. ~80% of the total protein). The main types of whey protein are β-lactoglobulin (54 g/100 g), α-lactalbumin (21 g/100 g), immunoglobulin(s) (14 g/100 g) (Ig) and bovine serum albumin (BSA) (6 g/100 g), and they exist as globular soluble proteins. The whey fraction of goat's milk ranges from 0.37 to 0.7 g/100 g, of which 39.2–72.1 g/100 g constitutes β-lactoglobulin and 17.8–33.3 g/100 g α-lactalbumin (Amigo & Fontecha, 2011). Sheep's milk whey proteins account for 17%–22% (i.e. ~0.98 g/100 g) of the total protein. As with other milk types, α-lactalbumin and β-lactoglobulin are the major whey proteins (Ramos & Juarez, 2011). Whey proteins are denatured by heat and acidity, a function that is technologically exploited in the production of whey cheeses (i.e. Ricotta and Anari) and smooth-textured cheeses such as Quark.

6.2.3 Lipids

Cow's milk has an average fat content of 3.7 g/100 g (Holstein breed), but this level varies significantly from 3.0 to 5.0 g/100 g with breed, diet, health, stage of lactation and animal husbandry. For example, Swiss Brown, Ayrshire Simmental, Montbéliarde and Salers breeds have a milk fat of 4.0 g/100 g, while Guersney and Jersey cows produce less milk yearly, but the fat content is very high: 5.0 g/100 g. The fatty acid composition of cow's milk is characterised by a saturated fat content of 66 g/100 g, of which the short chain fatty acids C4 to C10 contribute 11.1 g/100 g. Palmitic acid represents 28 g/100 g of the saturated fat in cow's milk. From the unsaturated fatty acids, the most abundant is oleic acid at 21.1 g/100 g (Guinee & O'Brien, 2010). Goat's milk has a range of milk fat depending on the breed, that is, Vlahiki of Greece 5.6 g/100 mL, Damascus of Cyprus 4.3 g/100 mL and Alpine of France at 3.3 g/100 mL. Sheep's milk fat also ranges from 5.1 to 8.7 g/100 mL with an average of 7 g/100 mL. Finally, buffalo's milk has a fat content of 6.7 g/100 mL (Sindhu & Arora, 2011). It is well known that the different fatty acid composition (especially the short chain fatty acids) of goat's and sheep's milk compared to cow's milk also differentiates the flavour/odour of these milks and consequently the organoleptic characteristics of the cheeses produced. For example, the caprylic C8 and capric C10 percentage in sheep's milk is almost double than that of cow's milk, while for goat's milk the capric acid C10 is three times higher than in cow's milk. The short chain fatty acid (SCFA) fraction of goat's milk represents 15%–18% of the total fatty acids compared to 5%–8% in cow's milk. In addition, goat's milk owes its characteristic odour to the branched chained free fatty acids (fewer than 11 atoms of carbon), which are completely absent from cow's milk (Amigo & Fontecha, 2011).

Milk fat is readily hydrolysed by lipases either naturally present in the milk or derived from bacteria. Damage to the milk fat globule membrane (MFGM) by excessive handling and turbulence will render the membrane accessible to lipases, and hence undesired lipolysis might occur. This is highly unacceptable, especially for milk that is to be used for cheesemaking.

6.2.4 Minerals

Cow's milk typically contains 120 mg/100 mL (~30 mM) of calcium, which exists as colloidal inorganic calcium (12.5 mM), caseinate calcium (8.5 mM), soluble unionised calcium (6.5 mM) and serum ionic calcium (2.5 mM). Goat's milk has a higher reported content of calcium (134 mg/100 mL) than cow's milk, but their distribution between colloidal and ionic forms is similar (Amigo & Fontecha, 2011). On the other hand, sheep's milk contains 193 mg/100 mL, having a higher concentration in the colloidal phase, which positively influences cheese yield (Ramos & Juarez). The influence of calcium on the physico-chemical properties of milk and the impact on cheesemaking is extensively discussed in Chapter 2.

6.2.5 Lactose

The lactose content in cow's (Holsten-Friesian), sheep's, goat's and buffalo's milk is very similar and ranges from 4.5% to 4.7%. Other cow breeds such as the Jersey and Ayrshire have slightly higher lactose contents at 5.0% and 4.95%, respectively. In the manufacture of cheese, most (96%–98%) of the lactose is removed in the whey, and the residual lactose is readily converted during cheesemaking by starter cultures to lactic acid, and its lactose content in cheeses varies according to the cheesemaking procedure. In the manufacture of some cheese varieties, for example, Dutch cheeses, the curds are washed to reduce their lactose content and thereby regulate the pH of the pressed curd to ~5.3. In most other varieties, for example, Cheddar and Emmental, the level of lactose, and hence of lactic acid, in the curd is not controlled by washing. Excessive lactic acid in cheese curd leads to a low pH, a strong, acidic, harsh taste and a brittle texture. On the other hand, Halloumi cheese retains its lactose content as no starter cultures are used during its manufacture. The relation between high lactose content in cheese, CCP and ultimately cheese texture is discussed in Chapter 2. Acid-coagulated cheeses and long-matured hard cheeses contain a very low concentration of lactose; therefore, they could be consumed by lactose-intolerant people with no side effects (Guinee & O'Brien, 2010).

6.3 Raw Milk Quality for Cheesemaking

6.3.1 Animal Nutrition and the Effect on Milk Composition

Animal nutrition systems employed in the rearing of cows, sheep, goats or even buffaloes vary according to environmental and other factors (i.e. availability of pastures for free grazing, climatic conditions prevailing, animal breed). It is also well understood that whichever system is used, that is, intensive, semi-extensive or extensive, the aim is to cater to the well-being of the animal and to optimise milk production. Milk quality is certainly affected by animal nutrition, and taints in milk occur when are animals are fed unsuitable types of herbage. In the case of sheep and goats, the consumption of seasonally available herbs is a common cause of taints. Examples include garlic, garlic mustard, onions, cabbage turnip leaves, kale mint and others. If the above are to be consumed by the dairy animals, then the milk will be tainted and most probably cheese too made with the same milk (Robinson & Wilbey, 1998). It is also well documented that low-quality silage may be the source for anaerobes (i.e. *Clostridium tyrobutyricum*) in the milk and ultimately in the cheese, causing late blowing in extra mature cheeses (Aureli, Scalfaro & Franciosa, 2011). For this reason, silage of any kind is strictly regulated for the production of Parmigiano Reggiano (Consorzio del Formaggio 'Parmigiano Reggiano'-feeding regulations for dairy cows), the Swiss Sbrinz and Asiago.

Animal feeding strategies could also lead to enrichment of milk with important fatty acids such as conjugated linoleic acid (CLA). Amigo and Fontecha (2011) have reported the effects of feeding fresh forages and the seasonal changes of Mediterranean natural pastures on fatty acids. Apart from the increase of ω3 fatty acids, a feeding regime involving free grazing of the animals had a positive effect on the aromatic profile of milk which is carried over to cheeses. Papademas and Robinson (2002) studied the volatile fraction of Halloumi cheese made with a sheep's/goat's milk mixture versus an 'industrial' Halloumi where the major milk in the mixture was cow's milk. The authors observed differences in volatile compounds originating from plants (i.e. forage), and terpenes and sesquiterpenes were isolated and quantified in considerable concentrations from the sheep/goat milk Halloumi cheese, illustrating the effect that free grazing has on the milk's aromatic profile. Cheeses of Alpine origin, that is, Asiago, Berner Alpkäse and Le Gruyère d'Alpage, are manufactured with milk originating from a mountainous region and are described in Part II of this book, while feeding systems are discussed from the standpoint of traceability and authenticity in Sections 5.2 and 5.3.

6.3.2 Microbial Activity of Milk

6.3.2.1 Hygienic Raw Milk Production

Raw milk must come from animals that do not show any symptoms of infectious diseases communicable to humans through milk and that are in a good general state of health (no inflammation or wound of the udder, no disease such as enteritis with diarrhoea and fever). Regarding brucellosis and tuberculosis, raw milk must come from herds that are officially free from the mentioned zoonosis. The criteria for raw cow's milk production are a total plate count (TPC) at 30°C ≤100,000 cfu/mL and a somatic cell count (SCC) at ≤400,000 (mL). For raw milk of other species (i.e. sheep, goat and buffalo), the only criterion set is a TPC at 30°C of ≤1,500,000 cfu/mL.

6.3.2.2 Milk Storage and Transport Conditions

According to the requirements of the Regulation 853/2004 (EU, 2004), immediately after milking, milk must be cooled immediately to not more than 8°C (in the case of daily collection) or not more than 6°C if collection is not daily. During transport, the cold chain must be maintained and, on arrival at the establishment of destination, the temperature of the milk must not be more than 10°C. The aforementioned temperature requirements do not apply if the milk is processed within 2 h of milking (i.e. in cases where the farm is adjacent to or very near the dairy) or a higher temperature is necessary for technological reasons related to the manufacture of certain dairy products.

6.3.2.3 Microbial Contamination

It is generally accepted that a healthy animal produces a healthy, pathogen-free milk Microbial contamination of milk may arise through the premises (i.e. the farm), the staff involved in milking and handling of the milk and the equipment that milk comes into contact before, during and after milking (i.e. milk storage tanks) (Angelides, 2014).

6.3.2.4 Raw Milk Cheeses

If raw milk is intended to be used for the production of raw milk cheeses and their technology does not involve any heat treatment (i.e. French lactic acid cheeses), then the TPC at 30°C must be ≤500,000 cfu/mL. In addition, in case the milk comes from herds of animals that are not officially free of brucellosis but are tested and shown to be free of the disease, the raw milk can be used for cheesemaking provided the cheese will mature for at least 60 days (EC 853/2004). Important raw milk cheeses from different categories (i.e. extra-hard, fresh lactic acid, propionic acid and semi-hard) are described in this book, that is, Parmigiano Reggiano, Sbrinz, Crottin de Chavignol, Robiola di Riccaverano, Allgäu, Emmentaler, Appenzeller, Herrgård and others.

6.3.3 Other Factors Affecting Milk Composition

6.3.3.1 Stage of Lactation

The effect that late lactation has on the quality of milk and consequently on cheese quality has been reviewed. The main effects on milk are summarised as follows: (1) late lactation milk has been found to have lower casein as a percentage of true protein and a higher level of FFA than milk from cows in early lactation and (2) high SCC. The detrimental effects on cheese are reported as high moisture content and pH, impact on degradation of α_{s1}-casein in cheese during ageing, reduced yield (gel formation not ideal) in high-SCC milk and quality of Cheddar cheese. A solution to the problem described here could be the elimination of late lactation milk from cheesemaking by not milking animals when their milk yield reduces dramatically (Guinee & O'Brien, 2010).

6.3.3.2 Genetic Variants of Milk Proteins

Animal genetics affects the milk's performance during cheesemaking as it is reported that the BB genotypes (compared to the AA variants), of both κ-casein and β-Lg are generally associated with a higher concentration of casein and superior rennet coagulation properties, as reflected by higher curd firming rates and gel firmness after a given renneting time. The BB variants of κ-casein and β-Lg have also been associated with superior cheesemaking properties, as reflected by the higher recovery of fat, a lower level of curd fines in cheese whey, and higher actual and moisture-adjusted cheese yields for a range of varieties, including Cheddar, Svecia, Parmigiano Reggiano, Edam and Gouda, low-moisture Mozzarella and Camembert (Guinee & O'Brien, 2010). The well-known polymorphism of caseins in goat's milk could also affect milk composition and ultimately cheesemaking; Vacca et al. (2014) reported that the casein genotypes from goat's milk of the Sarda breed were associated with highest percentages in protein and fat. Another study reported that the LALBA gene isolated from Sarda goat had affected its renneting properties (Dettori et al., 2015). The genetic polymorphism of β-casein, β-lactoglobulin and κ-casein and milk production traits in Merino breed sheep were studied by Corral, Padilla and Izquierdo (2010). They concluded that certain genes had an effect on the protein and fat percentage of milk as well as on the milk's yield.

6.3.4 Enzymatic Activity of Milk

6.3.4.1 Proteinases

The milk of several species (cow, goat, sheep, horse, donkey, bison, pig and human) was subsequently shown to contain proteolytic activity. There are two main proteinases present in milk, plasmin and cathepsin D. In milk, plasmin is associated with casein micelles and is concentrated in rennet-coagulated cheese, curds and casein. Its activity increases in situations where there is an increased influx of blood constituents into milk; that is, during mastitic infection and in late lactation (O'Mahony, Fox & Kelly, 2013), it may contribute to proteolysis in cheese during ripening, especially in varieties that are scalded at a high temperature and in which the coagulant is extensively or completely denatured, for example, Emmental and Parmigiano Reggiano (Fox, 2003). O' Mahony et al. (2013) reported that partially purified cathepsin D from milk hydrolysed α_{s1}-casein to a peptide with the same molecular mass or electrophoretic mobility as α_{s1}-CN (f24-199), which is one of the primary peptides produced from α_{s1}-casein by chymosin. The proteolytic specificity of cathepsin D on β-casein is also similar to that of chymosin, and it can also cleave κ-casein (very poor milk clotting properties) while β–lactoglobulin is resistant to cleaving. Cathepsin D (which increases in mastitic milk) is resistant to thermal treatments and could survive high temperature short time (HTST) pasteurisation and scalding temperatures of Swiss cheeses. In conclusion, it should be taken into consideration that as plasmin proteolysis is an uncontrolled process, the quality of raw milk (i.e. low SCC) is imperative for producing quality cheeses.

6.3.4.2 Lipases

O'Mahony, Fox and Kelly (2013) gives a detailed description of the activity of lipoprotein lipase (LPL), stating that it is considered the most important indigenous milk enzyme from a technological point of view. Fox (2003) also points out that LPL may cause hydrolytic rancidity in milk and butter but at the same time contributes positively to the ripening of raw milk cheese. Interestingly, the activity and the distribution pattern of the enzyme differs with milk type; that is, goat's milk contains 4% of the lipolytic activity of cow's milk but most of the enzyme (45%) is distributed on the cream and serum phase when compared to cow's milk, where 80% of LPL is associated with casein micelles instead. This difference in LPL distribution may explain the greater susceptibility of goat's milk to spontaneous lipolysis and the characteristic goat's milk flavour.

6.3.5 Milk Residues

6.3.5.1 Antibiotics

Antimicrobial drugs are administered to dairy animals in order to treat several bacterial diseases. Residues are secreted in the milk, and each drug is given a withdrawal period. Since the presence of antibiotic residues in milk could affect its technological properties (i.e. inhibit the activity of starter cultures) and contribute to the development of antibiotic-resistant pathogens, the dairy industry is rigorously testing the milk upon reception at the factory. Milk that tests positive is separated and discarded by the industry, and usually the implicated producer will be fined. A report by FDA (2015) stated that 99.22% of the milk samples tested negative against 31 known animal drugs including the most common β-lactams, sulphonamides and tetracyclines.

6.3.5.2 Mycotoxins

Aflatoxin M1 in milk is of major concern as the particular mycotoxin is considered to be toxic to humans. Aflatoxin M1 appears in milk and milk products as a metabolite of aflatoxin B1, which contaminates animal feeds. Aflatoxin B1 can be produced by *Aspergillus flavus* and *Aspergillus parasiticus*. Rapid testing is performed by the dairy industry at milk reception in the factory, and milk that exceeds legal limits is discarded. The maximum European Commission (EC) limit of residue for aflatoxin M1 is 50 ng/kg in dairy milk and 25 ng/kg for milk-based baby foods. MRL levels in other countries such as the US, China and Brazil are higher: 500 ng/L (Jawaid et al., 2015).

6.4 Additives in Cheese Milk

6.4.1 Calcium Chloride

Calcium chloride is usually added (20–35 g/100 L) in cow's or goat's milk in order to provide the necessary ionic calcium for fine milk coagulation during the second phase of the enzymic coagulation (Horne, 2011). Calcium chloride has been used as an additive in milk for many cheeses; that is, semi-hard cheeses (e.g. Havarti-Dutch, Präst-Sweden, Dutch goat cheese, Edam, Gouda), Cheddar-type cheeses and ultrafiltrated cheeses.

6.4.2 Preservatives

Some preservatives are permissible in cheeses; the use of sodium and potassium nitrate (E251, E252, respectively) is allowed in cheeses up to 50 mg/kg. Natamycin (E235-Pimaricin) and nisin (E234) can be used on cheese surfaces to prevent mould growth (Ritvanen, 2013). Exogenous lysozyme (E1105) may be added to milk in the manufacture of semi-hard cheese varieties, for example, Gouda, Edam and Emmental, as an alternative to nitrate to prevent the growth of *Cl. tyrobutyricum*, which can cause late gas blowing and off-flavour defects during ripening (O'Mahony, Fox & Kelly, 2013).

6.4.3 Colourings

Natural colours are added in some cheeses (i.e. Cheshire). The reason for adding colour to cheese was primarily to alleviate colour differences in cow's milk owing to different feeding regimes according to the season (i.e. green fodder intake will increase the uptake of β-carotene in milk and hence yellow colour intensity). There are commercially available natural colours; carmine is one of the most stable natural colours available. The colour shade is pH dependent, orange in acid solutions and purple in alkaline. Carminic acid (labelled E120 in Europe) is the natural, active colour pigment extracted with water from the female cochineal insect (*Dactylopius coccus costa*), which lives mainly in Latin America and a few other countries. The cochineal lives on the *Opuntia fiscus Indica* cactus. Another well-known natural colour is the annatto colour pigments bixin and norbixin (E 160b), which are extracted from the bright crimson seeds of a tropical shrub (*Bixa orellana L.*) growing in South and Central America, India and Africa. Annatto seeds have long been valued as a spice for flavouring and colouring savoury dishes. The major proportion of the world production of annatto comes from the collection of seeds from wild trees or trees planted on family farms (Chr. Hansen, 2016b).

6.5 Milk Standardisation

Although the cheese moisture content – and thus the amount of protein and fat – is determined by the cheesemaking process, the fat/casein ratio of the cheese is determined by the fat/casein ration of the milk used. Since the milk components vary in composition, the cheesemaker needs to standardise the milk. Standardisation may be achieved in small-scale plants by addition of skimmed milk to whole milk to reduce the fat content or by partially skimming the cheesemilk (e.g. Parmigiano Reggiano); less frequently, and only in high-fat cheese varieties (e.g. Manouri), by adding cream to increase the fat content. However, in most large cheese factories, standardisation is achieved using a centrifugal separator, which separates whole milk into skim milk and cream, which may then be mixed, perhaps in-line in an automated process, in proportions calculated to achieve the desired ratio of constituents (Kelly, 2007a). Standardisation ensures that manufacturers deliver levels of fat in dry matter (FDM) required by legal specifications or standards of identity for specific varieties.

6.6 Treatments of Raw Milk for Cheesemaking

6.6.1 Thermisation

If milk is to be stored before processing, then it is treated to lower than pasteurisation temperatures (50°C–70°C for 5–30 s) in order to reduce (1) the viable bacterial load in the milk and

(2) control the development of bacterial-associated enzymatic activities. In general, thermisation improves the quality of cheeses prepared from milks that have been cold stored.

6.6.2 Pasteurisation

Cheesemilk is usually pasteurised; HTST 72°C–75°C/15–20 s in a continuous flow plate heat exchanger or low temperature long time (LTLT) 63°C–65°C for 30 min in a cheese vat. Pasteurisation will inactivate the all known pathogens (i.e. *Listeria monocytogenes*, *Escherichia coli* O157 H7, *Campylobacter*, *Staphylococcus* and *Salmonella* spp.) including the most pathogenic bacteria (i.e. *Mycobacterium tuberculosis* and *Coxiella burnetii*). Pasteurisation of milk affects the coagulation and cheesemaking characteristics of milk as it will increase the yield by 0.1%–0.4% (partial denaturation of whey proteins, maximum 5%, will complex with κ-casein retained in the cheese curd and hence increase the yield). Heat treatment to denature whey proteins is widely used in the commercial manufacture of fresh acid curd cheeses, for example, Quark and Fromage Frais.

6.6.3 Microfiltration

Microfiltration is employed by the cheese industry to produce a cheesemilk of the highest possible standard from the standpoint of hygiene. The milk passes under pressure through membranes of pore size 0.01 to 10 mm, retaining fat particles, endospores and vegetative bacterial cells (retentate). The permeate will be a fat-free milk of very low to zero bacterial load. The retentate, after UHT treatment, could be added back to the permeate and make up the cheesemilk (Robinson & Wilbey, 1998). Examples of cheeses that are made with microfiltrated milk are the Swedish Greve, Herrgård and the Swiss Raclette.

6.6.4 Ultrafiltration

Ultrafiltration has the potential to increase the concentration of protein and fat in milk (i.e. total solids in milk), which can be useful in the manufacture of cheese. Whole milk can be concentrated by a factor of up to five times, while skimmed milk can be concentrated up to seven times (Goulas & Grandison, 2008). By concentrating the milk prior to cheesemaking, and hence reducing or eliminating curd syneresis (whey drainage), some or all of the whey protein is retained in the curd. These proteins have high water-binding capacities, and so more water and water-soluble components of the milk may also be retained in the curd. Intermediate concentrated retentates (milk concentrated in the range two- to fivefold) have been applied to a wide variety of cheeses. Reported yield increases of 6%–8% with Cheddar and 14% with Feta cheese have been achieved while maintaining good quality (Goulas & Grandison, 2008). The calcium content of cheeses produced with the ultrafiltration process is higher than in their conventional counterparts. It is reported that the calcium content of dry and brine-salted Mozzarella cheeses produced with ultrafiltration increased by 41% and 32%, respectively, in comparison to conventional cheeses. An about fourfold higher calcium content was reported in the ultrafiltrated variety of Cottage cheese than in the conventional one (Gursoy et al., 2013).

6.6.5 Bactofugation

One of the most important defects of hard matured cheeses is the 'late blowing' phenomenon. This involves the presence of bacterial spores such as *Cl. tyrobutyricum* which ferment lactic acid to butyric acid, acetic acid, carbon dioxide (CO_2) and hydrogen. Excessive gas will produce numerous large eyes, splits and cracks, literally destroying the cheese. The industry has employed bactofugation to remove the endospores from cheese milk by centrifugation of milk to 8,000–10,000g since 90%–95% of the endospores have a higher density than milk. The sludge (fat-rich) that contains the endospores amounts to 2%–3% of initial feed and is treated at 130°C–140°C for a few seconds, then added back to the milk. Cheeses like Gouda and Maasdammer described in this book are made by bactofugated milk.

6.6.6 Homogenisation

Milk is homogenised when is passed under high pressure (15–25 MPa) through small valves. As a result, the fat globule size is reduced, increasing at the same time its surface area by a factor of 5–6. The native fat globule membrane is altered through the shearing effect, and a new membrane including casein micelles, sub-micelles and whey proteins is formed. Although homogenisation is vital for the production of stable (i.e. no flocculation) pasteurised bottled milk, it is hardly used in cheesemaking. It is performed when fresh, acid curd cheese is to be manufactured because it contributes to the formation of a homogeneous, thick, creamy texture in the end product. Homogenisation is generally avoided in rennet curd cheeses because it will produce curds that are easy to break/crack and ultimately lead to less elastic, more brittle, more fracturable cheeses. In addition, homogenisation affects the cooking properties of the cheese where a lower degree of flow/spread and stringiness is recorded. On the other hand, in some applications, for example, blue cheeses (see Danablu), where flavour formation depends on lipolysis, milk is homogenised in order to allow lipases to attack the modified fat globule membrane more readily (Robinson & Wilbey, 1998). Also, homogenisation is necessary when cheeses are made from recombined milk (Guinee & O'Callaghan, 2010).

6.6.7 High-Pressure Processing (HPP)

Nunez and Medina (2013) describe how HPP of raw milk at low temperature eliminates non-spore-forming pathogens while maintaining most of the activity of native milk enzymes. The cheesemaking properties of milk are influenced by HPP processing, inducing changes in milk proteins and mineral balance. More specifically, pressurisation causes the denaturation of β-lactoglobulin and alters the size of casein micelles. These changes modify the technological properties of milk and may reduce the coagulation time. HPP treatment of milk increases cheese yield, since the denaturation of whey proteins and their incorporation into the curd results in a higher water-holding capacity. The effect of HPP on cheese characteristics is summarised as follows: intracellular enzymes are released because of cell lysis, there is an increased water retention due to microstructural changes, proteolysis is enhanced by improved substrate accessibility, lipolysis is reduced by inactivation of lipases and esterases and some volatile compound are not formed; therefore, either the flavour is not affected or the cheeses are slightly milder. Finally, the texture is affected as the cheeses are more elastic with a less crumbly texture, and in some cheese varieties the colour is more yellowish.

6.7 Acidification

Milk for cheesemaking may be acidified by its indigenous LAB or by using a 'backslop' culture (i.e. natural whey culture), that is, a volume of whey retained from a previous day's cheesemaking. In such cases, the rate of acidification is unpredictable, and the growth of undesirable bacteria very often leads to defective cheeses. While these traditional techniques continue to be used for certain artisanal varieties (e.g. Berner Alpkäse, Parmigiano Reggiano, Reggianito and other cheeses), the addition of selected starter cultures is now globally used in industrial cheesemaking. Starter cultures are now produced and supplied to the cheese industry by a number of companies and may be mixed-strain starters (containing unknown combinations of unknown strains of LAB) or defined-strain cultures (containing known combinations of known strains of LAB) (Bintsis & Athanasoulas, 2015). Starter cultures used by the cheese industry can be divided into three broad groups: mesophilic, thermophilic and mixed, and the types of starter cultures added in some cheeses are shown in Table 6.1. More details for the ecology of starter cultures in cheese can be found in Section 4.4.1.

Acidification starts with the addition of the starter, and the rate of acidification depends on the amount and type of starter added, as well as environmental conditions. Acidification, at the appropriate rate and time, is probably the most critical step in cheese manufacture and

Table 6.1 Types of starter cultures and species of lactic acid bacteria used in some cheeses.[a]

Type of culture	Species	Cheese
Mesophiles		
Type O	*Lc. lactis* ssp. *lactis* and *Lc. lactis* ssp. *cremoris*	Anevato, Cheddar, Danbo, Feta, San Simon da Costa, Svecia.
Type DL	*Lc. lactis* ssp. *lactis*, *Lc. lactis* ssp. *cremoris*, *Lc. lactis* ssp. *lactis* biovar. *diacetylactis*, and *Leuc. mesenteroides* ssp. *cremoris*	Dutch goat cheese, Edam, Esrom, Gouda, Havarti, Hohenheim Trappisten, Herrgård, Präst, Västerbottensost.
Thermophiles		
	S. thermophilus and *Lb. delbrueckii* ssp. *lactis*	Appenzeller, Emmentaler, Sbritz.
	S. thermophilus and *Lb. delbrueckii* ssp. *bulgaricus*	Acid curd, Feta, Mont-d'Or, Quark, Vacherin.
	S. thermophilus, *Lb. delbrueckii* ssp. *bulgaricus* and *Lb. delbrueckii* ssp. *lactis*	Tête de Moine
	S. thermophilus *Lb. delbrueckii* ssp. *bulgaricus* and *Lb. helveticus*	Le Gruyère, Graviera Kritis, Kefalograviera, Kefalotyri.
Mixed		
	S. thermophilus and *Leuconostoc* sp.	Pategrás
	Lb. delbrueckii ssp. *lactis*, *S. thermophilus* and *Lc. lactis* ssp. *lactis*	Berner Alpkäse
	Lc. lactis ssp. *lactis*, *Lc. lactis* ssp. *cremoris*, *S. thermophilus* and *Lb. delbrueckii* ssp. *bulgaricus*	Kefalograviera, Kefalotyri.
Natural whey cultures		Parmigiano Reggiano, Reggianito, Sbrinz, Berner Alpkäse, Fiore Sardo, Le Gruyère, Appenzeller®, Raclette, Caciocavallo, Mozzarella di Bufala

a) Data obtained from Part II of the current book.

plays a number of important roles, as it (1) controls or prevents the growth of spoilage or pathogenic microorganisms; (2) influences the activity of enzymes during maturation and hence affects cheese flavour and quality; (3) solubilises CCP and thus helps determine the level of calcium in the cheese curd and the ratio of soluble to colloidal calcium, which greatly influence cheese texture; (4) promotes syneresis and hence helps determine cheese composition (especially the moisture content of the cheese) and thus the yield and growth of the microfloras and (5) affects the activity of the coagulant during manufacture and the retention of coagulant activity in the cheese curd, which influences the rate of proteolysis throughout maturation (McSweeney, 2007a).

Acidification using acid (e.g. lactic acid and glucono-δ-lactone) can be used as an alternative to acidification by starter culture or indigenous LAB. It is common practice in the manufacture of soft cheeses (e.g. Quark and other acid curd cheeses). Direct acidification is more controllable than biological acidification and, unlike starter cultures, is not susceptible to bacteriophage infection (Fox & Guinee, 2013). The limited application of direct acidification is mainly due to the lack of the additional effects of the starter cultures on the biochemistry of the maturation and safety aspects (see Part I, Section 4.4.1).

6.8 Coagulation

The coagulation of milk is brought about by destabilising the casein micelles, either by acidification, that is, by neutralising the negative charge on the micelle surface, or enzymically, that is, by removing the hydrophilic glyco-macro-peptide through enzymatic action (i.e. rennet). As discussed later in this chapter (see the categorisation in Section 6.14), the two different ways to coagulate the milk were used for the basic categorisation of cheeses: (1) acid-coagulated and (2) rennet-coagulated, with the latter accounting for more than 75% of the total cheeses produced worldwide (Fox & Guinee, 2013).

The manufacture of acid-coagulated cheeses (e.g. Cottage, Quark and Galotyri) is based on the property that caseins are insoluble at their isoelectric point (pH 4.6) and thus form a gel in which fat is entrapped. The acidification may be caused by the adventitious microbiota, starter culture or acidogens (e.g. lactic acid and glucono-δ-lactone) at temperatures of 20°C–30°C (Fox & McSweeney, 2004; Guinee & O'Brien, 2010). Acidification results in a number of physico-chemical changes promoting hydration/dispersion or dehydration/aggregation effects on the casein micelle, with the ratio of these effects changing as the pH declines during the acidification process. The onset of coagulation typically occurs at pH 5.1, and further reduction in pH towards 4.6 coincides with the eventual formation of a continuous gel structure with sufficient rigidity to enable separation of whey from the curd by physical means (e.g. cutting, cooking and pressing). The majority of cheeses are manufactured by enzymic coagulation, that is, the addition of rennet. Traditionally, rennet from the stomachs of young animals (calves, kids, lamps and buffalo) were used (rennet paste), and these contain mainly chymosin (95%) and a little pepsin. Rennet paste is a traditional rennet preparation used to coagulate milk for certain Spanish, Italian and Greek cheeses (e.g. Idiazabal, Majorero, Cabrales and Caciocavallo, Feta). In addition to chymosin and pepsin, rennet paste contains a lipase, pregastric esterase, which is very important for lipolysis in these cheeses. Rennet pastes (particularly those made locally) have been the subject of public health concerns; in addition, lack of standardisation of their enzymatic activities is responsible for a large variability in the quality of the produced cheeses. The majority of cheeses are nowadays produced using commercially prepared liquid rennet. In addition, there has been active research interest in producing alternatives to rennet paste by blending commercial lipases into normal rennet extracts (McSweeney, 2007b). The great demand for cheesemaking has led to the use of several substitutes. Microbial acid proteinases (e.g. those produced naturally by the organisms *Rhizomucor meihei, Rhizomucor pusillus* and *Cryphonectria parasitica*) are available on the market and are used widely, as is pure calf chymosin produced by fermentation of organisms modified genetically to produce this enzyme (McSweeney, 2007b). Plant rennets are used to coagulate milk for some cheese (e.g. Serra da Estrella, Torta del casar). The most successful plant rennets are from the flowers of the Cardoon thistle (*Cynara cardunculus*), which grows wild in Spain and Portugal.

The manufacture of rennet-coagulated cheeses is based on the fact that κ-casein is hydrolysed with chymosin (i.e. rennet) at the peptide bond Phe_{105}-Met_{106}, causing it to coagulate; the optimum pH is 5.1–5.5. κ-Casein is hydrolysed to produce para-κ-casein (κ-casein fragment 1-105, κ-CN f1-105) and macropeptides (also called glycomacropeptides or caseinomacropeptides; κ-CN f106-169). Macropeptides (~30% κ-casein or 4%–5% total casein) diffuses into the aqueous phase, and para-κ-casein remains attached to the micelle core (McSweeney, 2007a).

The enzymatic coagulation of milk has been extensively reviewed (Dalgleish, 1993; Green & Grandison, 1993; Hyslop, 2003). Rennet coagulation is a two-phase process, a primary enzymatic phase and a secondary non-enzymatic phase (Fox & Guinee, 2013). The first phase begins with the enzymes hydrolysing casein, especially along the micelle surface. This activity triggers a non-enzymatic phase, during which the spherical micelles progressively lose their ability to interact with water molecules, due to the reduction in steric and charge repulsion (between particles) caused by the loss of the macropeptide from κ-casein forcing them to interact instead with one another to form aggregates and chains (Lucey, Johnson & Horne, 2003). As coagulation progresses, the micellar chains increase in length and thickness, interlocking to form a three-dimensional, netlike matrix that entraps water and other major components of milk, including lactose, fat, whey proteins and minerals (Kindstedt, 2013a).

The coagulation is greatly influenced by pH, calcium concentration, protein content and temperature (Lomholt & Qvist, 1999; Lucey, 2002). In fact, the extent of acidification will determine the final pH of the coagulant and subsequently the pH of the final cheese. As the pH decreases, there is a concomitant loss of CCP from the casein sub-micelles and, below about pH 5.5, the sub-micelles dissociate into smaller aggregates (Lawrence et al., 2004). CCP is the main buffer in milk (and in cheese; see Part I, Section 2.2.3). When the pH decreases before the whey is drained, the calcium phosphate will be lost in the whey. However, if the whey is separated from the curd before acidification, then the CCP will remain in the curd and will act as a buffer throughout the next steps of the cheesemaking process, thus preventing a very low pH in the final cheese. The addition of $CaCl_2$ to milk can reduce the rennet coagulation time and improve gel properties, but the effect appears to be less than might be expected and may be negative at certain pH values and at high calcium concentrations, especially if the gel is held for a long period before cutting (Fox et al., 2000).

6.9 Post-Coagulation Processes

The resultant coagulum, that is the curd, is subjected to a number of operations that promote syneresis, and an approximate tenfold concentration of the casein, fat and micellar calcium phosphate components is achieved. These operations include (1) cutting the gel into pieces (referred to as curd particles, cubes of 0.5–5 cm), (2) cooking (scalding), that is, heating the particles in the whey, (3) cheddaring, that is, blocks of curds are piled on top of each other and then put through a mill and ground into small pieces, (4) curd washing, that is, removing some whey and adding some water, (5) stretching, (6) moulding the pieces to the desired shape and weight of finished cheese, (7) pressing to enhance the whey drainage and (8) salting. Not all cheeses undergo these steps, and some cheeses are manufactured using additional steps (see Part II).

6.9.1 Cutting

When the gel has attained sufficient firmness, it is cut with special knives. Traditionally, the appropriate firmness is determined subjectively by the cheesemaker, but more recently it is determined objectively using rheological measurements or light scattering techniques. The size of the curd particles is determined mainly by the cutting time and rate, and varies with the cheese type from the size of the rice or wheat (for extra-hard varieties, e.g. Parmigiano Reggiano and Sbrinz) to corn (for hard, e.g. Kefalograviera and semi-hard, e.g. Raclette du Valais) to 8 cm^3 cubes (for soft, e.g. Quark). In practice, if the curd is cut when it is very soft, the moisture content of the resulting cheese is lower (Johnson, Chen & Jaeggi, 2001; Johnson & Law, 1999). If the gel is left for a longer time before cutting, the moisture content of the cheese is higher.

Presumably, this change in the moisture content is a reflection of the extent of bonding between and within casein particles, which increases with time (Lucey, Johnson & Horne, 2003). Additional events, such as fusion of particles, rearrangement processes and further incorporation of micelles into the network all contribute to the growth in gel strength.

Cutting results in an expanding of the curd surface area and thus promotes syneresis. The greater the surface area to volume ratio of the cut particles, that is, the smaller the curd particles, the greater the release of whey and the lower the moisture content of the cheese. The appropriate particle size will be vital for the next step, that is, cooking, which includes heating and stirring the curds and whey.

6.9.2 Cooking (Scalding)

Most cheeses are cooked, that is, heated to temperatures higher than those used for coagulation, mainly as a means of increasing the syneresis of the curd particles, which begins after the coagulum is cut. The temperature and the rate are characteristic of the variety; for cheeses in which a low moisture content is required, the coagulum is cut into small curd particles, and a high cook temperature is used (e.g. Parmigiano Reggiano at 55°C for 30 min, Sbrinz 54°C–57°C for 10–20 min). Characteristic scalding temperatures are Gouda at 33°C–38°C, Cheddar about 40°C, Präst at 42°C, Kefalotyri at about 44°C, Kefalograviera at 48°C, Reggianito at 50°C, Graviera Kritis at 50°C–52°C, Emmentaler and Parmigiano Reggiano at 55°C, Sbrinz at 57°C and Le Gruyère at 54°C–59°C. Other cheeses such as Västerbottensost (hard cheese) are cooked at 40°C but for several hours. The starter must be appropriate for the cook temperature: a mesophilic *Lactococcus* spp. for cheeses cooked at a temperature below 40°C and a thermophilic *Lactobacillus* spp. with or without *S. thermophilus* for a cheese cooked at a temperature above 40°C.

After cutting, the curd particles are continuously stirred during the cooking and holding stage in the vat, and the collisions between particles also increase syneresis. After whey drainage, the curd particles start to fuse together (depending on the pH), unless fusion is prevented by stirring in the vat.

Cooking to higher temperatures, longer times and with more stirring allows the curd to contract and expel more of the whey. Moreover, cooking influences how rapidly LAB strains produce lactic acid, and that in turn affects syneresis because curd particles contract and expel whey more readily as the pH decreases. The effects of cooking temperature can be complex, affecting not only lactic acid production rates but also curd demineralisation and buffering capacity. Thus, cooking times and temperatures need to be delicately balanced to shrink, dehydrate, acidify and demineralise the curd particles as they are prepared for the maturation.

6.9.3 Cheddaring

Cheddaring is a cheesemaking step, used for some cheese varieties (e.g. Cheddar and Cheshire), where the curd is cut into pieces (~5 kg) and cheddared until the pH reaches about 5.4. The pieces are flowed and developed a fibrous texture like chicken-breast meat, which was considered to be critical for development of the characteristic texture of Cheddar cheese. However, the principal physic-chemical reaction during cheddaring is the decrease in pH, which dissolves the CCP and makes the curd stretchable.

6.9.4 Curd Washing

The curds for several varieties of cheese are 'washed' with water either before whey drainage (in the vat) or after whey drainage (on drainage tables, finishing vats or belts). Curd washing after whey drainage in traditional or farmhouse cheese manufacture may be done in the cheese vat once the whey is drained to the level of the curd bed. Washing before whey drainage involves removing a portion of the whey and diluting the remaining whey with warm (35°C–55°C) water, while washing after whey drainage entails adding cold water (less than ~18°C) to the curd particles or curd chips and stirring the mixture. Cheeses that are washed in the vat before whey drainage mostly belong to the category of Dutch-type cheeses (e.g. Gouda and Edam). In addition, some semi-hard (e.g. Murcia al Vino), bacteria-surface-ripened (e.g. Danbo and Esrom), and blue-veined (e.g. Danablu) cheeses also undergo curd washing.

Traditionally, the objective of curd washing was to cook the curds in unjacketed vats, but it also reduces the lactose content of the curds and thus avoids over-acidification and yields a milder, sweeter cheese. Washing has little effect on the calcium content of curd because much of the calcium is colloidal at the pH (~6.15–6.35) and temperature (36°C–39°C) at whey drainage. Also, washing does not influence the level of proteolysis, as measured by levels of pH 4.6- soluble nitrogen and free amino acids (Fox & Guinee, 2013). Hence, altering the level of curd washing provides a useful means of altering curd pH, which influences ripening, appearance, flavour and physical properties (e.g. sliceability, shreddability and elasticity).

6.9.5 Stretching

In the manufacture of pasta-filata cheeses, stretching (i.e. heating, kneading and mixing) of curd with added hot whey or water (~70°C–80°C) was used originally to 'remake' a defective Pecorino cheese (see Part II, Section 8.9). In fact, it is an additional heat treatment to improve the microbiological quality of the cheese. The stretching of the curd in hot water imparts the desired textural and melting characteristics to the cheese, and thus a number of pasta-filata cheeses were developed (e.g. Mozzarella, Kasseri and Kachkaval). Hot water treatment of the curd at the desired pH (~5.2–5.4) raises the temperature to about 58°C–60°C and promotes a number of structural changes that impart the characteristic fibrous, stringy, 'oily', glistening texture of the cooked cheese by aggregation of the para-casein matrix into fibres of high tensile strength; coalescence of fat globules into pools of free fat, which are entrapped between the casein fibres; and inactivation of residual coagulant in the curd, which reduces the level of proteolysis in the cheese during storage, and thereby minimises deterioration of the stringiness (Fox & Guinee, 2013).

6.9.6 Moulding/Drainage

When the desired degree of syneresis has occurred, as judged subjectively by the cheesemaker, the curds are separated from the whey, usually on some form of perforated metal screen. For further syneresis, the curd is then subjected to some extra manufacturing steps (e.g. cheddaring, stretching, moulding, pressing and salting) which are more or less variety specific. For small, high-moisture cheeses, the curd is scooped from the vat and transferred to perforated moulds, of variable size, where syneresis occurs; the cheese may be inverted in the moulds but is subjected to no other treatment.

Interestingly, the size and shape of the mould corresponds to the characteristic shape and size of the cheese Presumably, the traditional shapes reflected the moulds available when the particular cheese evolved or match the shape of the pot that the cheeses were packed in (e.g. Feta in wooden barrels). Many of the traditional shapes persist, but some have been replaced (e.g. hard and semi-hard cheeses in rectangular blocks), mainly for convenience and for marketing reasons. The size of a cheese is not only cosmetic: surface-ripened cheeses must be small because ripening is dominated by the surface microflora and the enzymes and products secreted by them, which diffuse to only small distances into the cheese. At the other extreme are cheeses with propionic acid fermentation, which must have a close texture and be large enough to retain sufficient CO_2 for eye development (Fox & Guinee, 2013).

Cheese cloths are used, as an additional tool for moulding and drainage, for a number of cheeses such as Castelmagno, Mahón-Menorca, San Simon da Costa, Anevato and Brynza.

6.9.7 Pressing

During moulding, and after the whey has drained, the curd particles are fused, or knitted, to form a larger entity. Knitting continues the dehydration and demineralisation processes that began during cooking. Knitting is often accompanied or followed by pressing, that is, applying pressure to the curd to release additional whey (i.e. syneresis). In addition to the effect on syneresis, pressing promotes a more complete fusion of the curd particles, resulting in a more closed texture and a surface with fewer openings.

Pressing is applied to extra-hard, hard and semi-hard cheeses, and the pressure varies with the cheese variety; usually ~0.5 bar (1200 kg/wheel) for most hard cheeses and up to 6 bar (2000 kg/wheel) for Emmentaler. Interestingly, some hard and semi-hard varieties such as Mahón-Menorca, Majorero, Manchego and Tulum are pressed but not cooked.

6.9.8 Salting

Salt (i.e. sodium chloride) is added to all cheese varieties at some point of manufacture to a content varying from 0.3 to 0.4 (Emmentaler) and 0.7 (Mozzarella di Bufala) to 8%–10% (Mihaliç), whereas, most varieties have a salt content of 1.5%–2.5% salt. In terms of methods of salting, the cheese is either immersed in brine solution 18°–22° Bé (i.e. 19%–24% NaCl) – in most cases in saturated brine, that is 21.3° Bé (i.e. 23% NaCl), or salt is rubbed on the surface of the cheese or salt is added on the milled curd (Cheddar cheese). The amount of salt taken up by the cheese depends on the concentration of brine (or the amount of salt added in the case of dry salting), the moisture content of the curd, the temperature of the environment and the ratio of surface area to volume of the cheese (Bintsis, 2006).

The salting in cheese ripening performs a variety of functions (Fox, 1987; Fox & McSweeney, 1998; Guinee & Fox, 1993): (1) it affects the growth and survival of bacteria; (2) it affects the activity of enzymes (indigenous milk enzymes, rennet, enzymes from starter and secondary microflora); (3) salt promotes syneresis of the curd, resulting in whey expulsion and thus in a reduction in the moisture content of cheese – mainly for cheeses that are not cooked (e.g. Feta); (4) it modifies the hydration of proteins; (5) it has a positive effect on flavour and taste, either directly to give the characteristic flavour, or indirectly by masking off-flavours or inhibiting those undesirable bacteria that cause off-flavours and (6) it contributes to the dietary intake of sodium (Na). However, salting has not been given the necessary attention from the cheesemaker, and it is very common for defective cheese to be produced due to an improper salting step.

Different concentrations of salt occur in cheeses, mainly due to the different methods (i.e. dry salting or brining), and the stage at which the curd is salted. These differences in salt concentration consequently result in differences in microfloras, water activity (a_w) and pH values, together with differences in the degree of fermentation, proteolysis and flavour development.

When dry salt is rubbed onto the cheese, a rind is formed (e.g. Graviera Kritis) and this is practised by artisanal cheesemakers. The rind is low in moisture, high in salt and the caseins form a very dense network. Fat may be squeezed out and the surface becomes 'oily', and this happens when stored in low RH. However, the rind actually slows moisture loss from the rest of the cheese, and the interior of the cheese does not become too dry. The rind protects the interior as a natural packaging material, whereas metabolites of the surface microfloras contribute to the development of the sensory characteristics.

6.10 Control of Cheesemaking Steps

The appropriate control of the above steps is of great importance for the cheesemaker, in order to manage the cheese to expel the correct amount of whey from the curd and to achieve a moisture content ranging from 30% for extra-hard (e.g. Parmigiano Reggiano) to 70%–80% for acid curd cheeses (e.g. Quark), depending on the cheese variety being made. The moisture content (which determines the a_w) is the main parameter influencing the activity of the starter and secondary microfloras and their enzymes in the cheese, and thus the biochemistry and microbiology of the maturation process and finally the quality and safety of the cheese. The higher the moisture content of a cheese, the faster it will mature, but the less stable it will be (Fox et al., 2000). In addition, the producer needs to control the rates of acidification and curd demineralisation so that the buffering capacity of the newly made cheese is compatible with a pH ranging from 5.4 to 4.6 (Kindstedt, 2013a). The pH, moisture and salt-in-moisture (S/M) content of the unripened curd collectively shape the chemical environment of the cheese. The chemical environment and physical parameters such as temperature, relative humidity, exposure to oxygen and air movement and physical handling such as smearing, washing with brine and turning, together with the addition of adjunct culture, determine which microbiota (NSLAB, yeasts and moulds) within the cheese and on its surface will be developed. The kinetics of the milk, curd and cheese microbiota will determine the progress of the maturation process and thus the development of cheese flavour, aroma, texture and appearance. In addition, the chemical environment will determine which enzymes within the cheese are switched on or suppressed, and sometimes their activities.

6.11 Cheese Maturation

The texture, flavour and aroma of cheese are directly dependent on the curd pH, curd composition (moisture, protein, fat and mineral content), and maturation conditions (temperature and RH). During the maturation of cheese, the curd, which is rubbery and tasteless, is transformed, by changes in the composition and structure, to a desirable cheese with a characteristic appearance, taste and aroma. Since the different curds have different compositions and may be modified by a variety of agents under various manufacturing conditions, a very wide range of final products is manufactured. The changes are very complex, but it is generally recognised that three primary events occur during maturation: glycolysis, proteolysis and lipolysis. In addition, various secondary changes occur concomitantly, and it is these secondary transformations that are mainly responsible for the finer aspects of cheese flavour (Fox et al., 1993). The volatile flavour compounds in cheese originate from degradation of the major milk constituents, namely, lactose, citrate, milk lipids and milk proteins (collectively called caseins) during ripening, which, depending on the variety, can take a few weeks to more than 30 months long (e.g. Parmigiano Reggiano – gold label; see Part II, Section 1.1).

The specific characteristics of typical products arise mainly from the specific raw materials employed, the area of production, the environmental conditions, the traditional tools (see Part I, Section 7.2) and the cheesemaking process. In general, soft, high-moisture cheeses are matured for relatively short periods, whereas hard, strongly flavoured cheeses are matured for up to three years. Surface-ripened cheeses require to be matured in a high-RH environment, but most hard cheeses must be kept in dry conditions to inhibit surface microbial growth. The main biochemical reactions (i.e. glycolysis, proteolysis and lipolysis) which lead to the development of typical textures and flavours are discussed in Part I, Chapter 3.

6.12 Adjunct Cultures and Acceleration of the Maturation Process

Since the maturation process is slow and consequently expensive, much research has been devoted to accelerating the process. The methods used to accelerate the ripening of cheeses are shown in Table 6.2, and all the methods have their advantages and disadvantages. Unlike other methods, the addition of adjuncts is cheap, since only a low initial number of cells is required, no additional equipment is involved and no legal difficulties are involved (Shakeel-Ur-Rehman, Fox & McSweeney, 2000).

Secondary cultures, or adjunct cultures or adjuncts, are defined as any cultures that are deliberately added at some point of the manufacture of cheese, but whose primary role is not acid production (Beresford & Cogan, 1997). The application of adjuncts could be seen as an attempt by the cheesemaker to give back to the cheese, in a scientific manner, some of the biodiversity removed by pasteurisation, improved hygiene and the addition of defined-strain starter culture. It is generally accepted that cheese made from raw milk matures faster and develops a more intense flavour than that made from pasteurised milk. Adjunct cultures are commonly used as shown in Table 6.3.

Bio-protective cultures may be incorporated in cheesemaking, either as starters or as adjunct cultures (Chanos, Morgenstern & Dalgas, 2013). Bio-protective starter cultures must possess a range of physical and biochemical characteristics, and most importantly, the ability to achieve growth and sufficient production of antimicrobial metabolites, which must be demonstrated in the specific cheese environment. In addition, adjunct bio-protective cultures may or may not contribute to the maturation of cheese. Propionic acid bacteria are the most common adjunct cultures in hard cheese ripening, and recently they have also been used as protective bio-preservatives and probiotics (Grattepanche et al., 2008; Meile, Le Blay& Thierry, 2008). However, the proliferation of and the ability to produce antimicrobial metabolites throughout the maturation needs to be demonstrated, as well as, any effect on the starter culture. Nowadays, starter culture companies have launched a variety of bio-protective cultures for many types of cheeses, for example, BioSafe cultures (Chr. Hansen, 2016a).

Table 6.2 Methods for accelerating the maturation of cheese.

Method	Advantages	Limitations
1. Elevated temperatures	Effective; no legal barriers; simple; low cost	Non-specific; risk of microbial spoilage
2. Exogenous enzymes		
Rennet	Natural additive; cheap	Not effective
Plasmin	Indigenous milk enzyme; effective	Expensive
Proteinases/peptidases	Low cost; specific action; flavour options	Difficult to incorporate uniformly; risk of over-ripening; limited commercial use to date
Lipases	Traditional for certain cheeses	Risk of rancidity; limited use
3. Selected/attenuated starters		
Selected starters	Normal additives	None
Attenuated starters	Easily incorporated; safe	Expensive; limited flavour range
4. Adjuncts	Natural microflora; relatively inexpensive; flavour options	Possible atypical flavour; careful selection required
5. Genetically modified starters	Desirable enzyme profiles; absolute control of ripening	Cost of regulatory compliance; legal barriers
6. Cheese slurries	Rapid flavour development; commercially used	Risk of microbial spoilage; limited for processed cheeses
7. Addition of free amino acids	Choice of flavour	Expensive; limited work to date
8. High-pressure technology	Potential to give rapid ripening	Limited work and knowledge to date

After: Fox et al. (1996), Law (2001) and El-Soda (1993).

Table 6.3 Cheese varieties in which adjunct cultures are commonly used.

Cheese category	Adjunct culture used	
Swiss-type cheese varieties	Propionic acid bacteria (PAB)	*Propionibacterium freudenreichii, Propionibacterium acidipropionici, Propionibacterium jensenii* and *Propionibacterium thoenii*
Mould surface-ripened cheeses	Yeasts and moulds	*Penicillium camemberti, Geotrichum candidum*
Blue-veined	Moulds	*Penicillium roqueforti*
Bacterial surface-ripened cheeses	Yeasts and coryneforms	*Debaryomyces hansenii, Kluyveromyces lactis, Candida utilis, Saccharomyces cerevisiae* and *Rhodosporium infirmominiatum, Brevibacterium linens* and *Brevibacterium casei*

6.13 Packaging

Packaging of cheese is the most important step in attracting consumers when cheeses are displayed on the shelves of cheese shops or other retail outlets. Any material used for packaging of a particular cheese should afford general protection, prevent moisture loss, improve appearance, protect against microorganisms and prevent oxygen transmission (Goyal & Goyal, 2012).

Thus, factors that must be considered in selecting a cheese packaging material include permeability to water vapour, gases and light, potential for migration of compounds from food to packaging or vice versa and practical considerations including suitability for labelling and compatibility with conditions during distribution and sale.

For soft cheeses which are eaten fresh and have short shelf lives, the packaging should protect the cheese from microbial contamination, while for hard cheeses that are matured, the packaging needs to have specific moisture and gas permeability characteristics. Protection against dehydration can be achieved by using packaging films with low water vapour transmission: semi-barrier (polypropylene, low-density polyethylene) or barrier films (aluminium, polyvinylidene chloride, polyvinyl chloride, orientated polypropylene, high-density polyethylene) (Jalilzabeh, Tunçtürk & Hesari, 2015). Hard cheeses are typically packaged in polyethylene/polyamide vacuum-pack bags, which retard the growth of aerobic spoilage bacteria and contamination of the cheese from the outside. There is a difference for hard cheeses that do not develop rind, and these varieties are packed and ripened within a plastic film wrap so that moisture loss and the inedible fraction of the cheese are significantly reduced (Schneider et al., 2010). For certain varieties (e.g. Gouda), paraffin wax was traditionally used as a packaging material, while today a latex emulsion (plastic coat) may be used. Care must be taken in packaging very soft cheeses, as mechanical stresses may result in collapse of the cheese structure within the package. High-moisture fresh cheese varieties are sensitive to dehydration, and must be packaged in suitable barrier materials, which also provide light and oxygen barriers (Kelly, 2007b).

Soft cheeses which respire, such as surface mould-ripened varieties, may be manipulated by the use of modified atmosphere packaging (MAP) combined with appropriate permeability characteristics of a plastic overwrap. Nowadays, many cheeses are sold in small portions of variable size and shape, or cut as slices, dices, stripes and so on. The procedure and mechanisation of cutting have been reviewed by Schneider et al. (2010). As with packaging material, the composition of the gas mixture is dependent on the cheese variety. Thus, for fresh cheese, the best mixture was 0% O_2–75% CO_2–25% N_2, as it inhibited mould growth and protected the cheese from losing its softness, while for matured cheese the best mixture was 10% O_2–0% CO_2–90% N_2, as it decreased lipolysis (Kirkin, Gunes & Kilic-Akyilmaz, 2013).

MAP technologies increase the commercial life of cheeses because they combine protection against oxidation and dehydration with the inhibition of undesirable microorganisms (Juric et al., 2003; Olivares et al., 2012). MAP technologies have been studied for extra-hard cheeses such as Parmigiano Reggiano (Romani et al., 2002), hard cheeses such as Graviera (Arvanitoyannis, Kargaki & Hadjichristodoulou, 2011a; Trobetas, Badeka & Kontominas, 2008) soft and pasta-filata cheeses such as Mozzarella (Olivares et al., 2012), surface mould-ripened cheeses (Rodrigues-Aguilera et al., 2011a; Rodrigues-Aguilera et al., 2011b) and whey cheeses such as Anthotyros (Arvanitoyannis, Kargaki & Hadjichristodoulou, 2011b; Papaioannou et al., 2007).

Packaging for hard cheese in new biopolymers which give it an extended shelf life have been studied (Goyal & Goyal, 2012). The use of renewable biopolymers may benefit the environment and at the same time improve the utilisation of agricultural by-products based on proteins like casein; on carbohydrates like starch and cellulose; on lipids, and also on polymers from surplus monomers produced in agriculture such as polylactate; and finally, on bacterial produced polymers from microorganisms grown on waste, like poly 3-hydroxy-butyrate. Edible coatings and edible films have been successfully applied to several cheeses (Cerqueira & Vicente, 2013; Conte et al., 2007; Conte et al., 2009; Duan et al., 2007), and the incorporation of nisin, natamycin and other antimicrobial agents has resulted in the improvement of the safety of certain cheeses (Cao-Hoang et al., 2010; Fajardo et al., 2010; Martins et al., 2010; Pintado, Ferreira & Sousa, 2010; Ture et al., 2011).

Nowadays, improvements in cheese packaging materials have enabled consumers to enjoy the finest cheeses of global origin all over the world. Protection during transportation and improvement in the appearance of the cheese are a general rule, and cheeses are packed differently according to the special needs of each variety. In addition, packaging material is now recyclable and biodegradable and incorporates special agents that can increase cheese shelf life.

6.14 Main Cheese Categories

The number of cheese varieties produced worldwide is probably not known yet. The great range of cheese varieties, even excluding the local variants, makes classification of cheese a very difficult and complicated task. Classification schemes have described, but they are not entirely successful.

Sanders (1953) described more than 400 cheeses and classified them in four groups according to the moisture content and eight sub-groups according to the ripening microorganisms (Table 6.4).

Davis (1965) suggested a classification scheme based primarily on rheological properties (moisture content), while a second scheme classified cheese into hard, semi-hard and soft. Varieties were then listed within each category according to milk type, method of coagulation, cutting of coagulum, scalding of the curds, drainage of whey and method of salting and moulding. Scott (1986) classified cheeses primarily based on moisture content, for example, hard, semi-hard and soft, and subdivided these groups on the basis of cooking (scalding) temperature and/or secondary microflora (Table 6.5).

The National Committees of the International Dairy Federation (IDF) listed 395 cheese varieties in the *Bulletin of the IDF* (IDF, 1971) in order to promote the fair use of cheese designations in international trade and improve protection of consumers against fraudulent practices. The cheeses were categorised according to different characteristics, such as (Robinson & Wilbey, 1998): (1) country of origin, (2) type of milk (i.e. cow's, sheep's, goat's and buffalo's), (3) moisture of cheese (i.e. hard, semi-hard and soft), (4) internal characteristics (i.e. close or open texture, large, medium or small eye-holes, slit openings in the curd, ripened with blue or white moulds, colour of curd and presence of herbs or spices), (5) external

Table 6.4 Classification of cheeses according to the moisture content and the ripening microorganisms.

1) **Very hard (grating):**
   a) Ripened by bacteria: Asiago old, Parmesan, Romano, Sapsago and Spalen.
2) **Hard:**
   a) Ripened by bacteria, without eyes: Cheddar, Granular or Stirred-curd and Caciocavallo.
   b) Ripened by bacteria, with eyes: Swiss, Emmentaler and Gruyère.
3) **Semi-soft:**
   a) Ripened principally by bacteria: Brick and Münster.
   b) Ripened by bacteria and surface microorganisms: Limburger, Port du Salut and Trappist.
   c) Ripened principally by blue mould in the interior: Roquefort, Gorgonzola, Blue, Stilton and Wensleydale.
4) **Soft:**
   a) Ripened: Bel Paese, Brie, Camembert, Cooked, Hand and Neufchâtel (as made in France).
   b) Unripened: Cottage, Pot, Baker's, Cream, Neufchâtel (as made in the United States), Mysost, Primost and fresh Ricotta.

After: Sanders (1953).

Table 6.5 Classification of cheese according to moisture content, scalding temperature and ripening microfloras.

Hard cheese (moisture 20%–42%)				
Low scald, lactic starter	Medium scald, lactic starter	High scald, propionic culture	Plastic curd, lactic starter, propionic culture	
Edam (NL) Gouda (NL) Cantal (F) Fontina (I) Cheshire (UK)	Cheddar (UK) Gloucester (UK) Derby (UK) Leicester (UK) Svecia (S)	Parmesan (I) Emmental (CH) Gruyère (CH) Beaufort (F)	Scarmoza (I) Provolone (I) Caciocavallo (I) Mozzarella (I) Cecil (USSR)	
Semi-hard cheese (moisture 45%–55%)				
Lactic starter	Smear coat	Blue-veined mould		
St. Paulin (F) Caerphilly (UK) Lancashire (UK) Trappist (H) Providence (F)	Herve (B) Limburg (B) Romadur (G) Münster (F) Tilsit (G)	Stilton (UK) Roquefort (F) Gorgonzola (I) Danablue (D) Wensleydale (UK)		
Soft (moisture > 55%)				
Acid coagulated	Smear coat or surface mould	Surface mould	Normal lactic starter	Unripened fresh
Cottage (USA) Mozzarella (I) Pizza cheese (I) Queso-Blanco (P)	Brie (F) Bel Paese (I) Marolles (F)	Camembert (F) Carre d'est (F) Neufchatei (F) Chaource (F)	Colwich (UK) Lactic (UK) Quarg (G) Petit Swisse (F)	Cottage (UK) York (UK) Cambridge (UK)

After: Scott (1986).

characteristics (i.e. rind that is hard or soft, smooth or rough, smear or mould-coated, dusted with spices, herbs or ash and type of final coating – plastic, wax or leaves), (6) weight of cheese (i.e. shapes and sizes), (7) FDM, (8) moisture content and (9) moisture in fat-free substance. The 395 cheese varieties listed originated from 27 countries, but it is noteworthy that the same variety was entered by several countries. Cheddar was listed by 20 countries, Camembert by 15, Gouda by 15, Edam by 13, Emmental by 12, Tilsiter by 9, Gruyère by 8, Parmesan by 5 and Brie by 6. The list was revised by the IDF in 1981 (IDF, 1981) and remains the most exhaustive list available. In the revised list, 510 cheese varieties from 29 countries were included, since 157 new cheeses were added and 39 cheeses had to be eliminated, and corrections were made in respect of 130 cheeses. Again, the same variety was entered by several countries. Cheddar was listed by 19 countries, Camembert by 17, Gouda by 16, Edam by 11, Emmental by 11, Tilsiter by 8, Gruyère by 6, Parmesan by 5, Brie by 8, Mozzarella by 5 and Feta by 4.

Fox et al. (2000) categorised cheeses into three super-families on the basis of the method of milk coagulation: (1) rennet cheeses (most major international cheese varieties, approximately 75% of total cheese produced), (2) acid cheeses (approximately 25% of the total cheese production is generally consumed fresh, e.g. Cottage and Quarg) and (3) heat/acid cheeses (e.g. Ricotta). In addition, Fox et al. (2000) subdivided the rennet-coagulated cheeses into further groups on the basis of characteristic ripening agents or manufacturing technology, namely, internal bacterial ripened cheese, mould-ripened and surface-ripened cheeses. The same authors subdivided the internal bacterial ripened group on the basis of certain criteria: moisture (extra hard, hard and semi-hard), the presence of eyes, or a characteristic technology such as cooking/stretching or maturation in brine. Internal bacterial ripened cheese with eyes was further subdivided into hard (Swiss type, e.g. Emmentaler) or semi-hard (Dutch type, e.g. Gouda and Edam) varieties. Soft cheeses were not included in the family of internal bacterial ripened cheeses, as they have characteristic secondary microfloras which have a major effect on the characteristics of the cheese (McSweeney, Ottogalli & Fox, 2004). Mould-ripened cheeses were subdivided into surface mould, for example, Brie and Camembert, and internal mould, for example, Roquefort and Stilton. Smear-ripened cheeses are characterised by the development of a complex microflora consisting of yeasts and, later, bacteria (e.g. coryneforms) on the cheese surface during ripening (McSweeney, Ottogalli & Fox, 2004).

McSweeney (2007c) classified cheeses on the basis of method of coagulation of the milk and various technological parameters into about 12 major families, within the three super-families suggested by Fox et al. (2000).

On the basis of the aforementioned information, the classification scheme used in the current book, together with the cheese varieties described in Part II, are shown in Table 6.6. Although it is fairly comprehensive and helpful for a better understanding and study of the cheeses, it suffers from a number of limitations, the most important being the inclusion of

Table 6.6 Classification of cheeses which is used in the current book.

Category	Cheeses
1. Extra-hard	Parmigiano Reggiano PDO, Reggianito, Sbrinz PDO.
2. Hard	Allgäu mountain, Asiago PDO, Berner Alpkäse PDO, Cantal PDO, Cheddar, Cheshire, Fiore Sardo PDO, Graviera Kritis PDO, Idiazabal PDO, Kefalograviera PDO, Kefalotyri, Le Gruyère PDO, Ossau Iraty PDO, Tulum, Västerbottensost, Würchwitzer mite cheese.
3. Semi-hard	Appenzeller®, Arzua-Ulloa PDO, Castelmagno PDO, Comté PDO, Flaouna, Formaggio di Fossa PDO, Havarti, Herrgård, Mahón-Menorca PDO, Majorero PDO, Manchego PDO, Murcia al Vino PDO, Präst, Raclette du Valais PDO, Raclette Suisse®, San Simon da Costa PDO, Svecia PGI, Serpa, Sombor, Tête de Moine PDO, Tuma Persa.
4. Soft	Afuega l'Pitu PDO, Anevato PDO, Bryndza, Cremoso, Galotyri PDO, Kopanisti PDO, Maltese Ġbejna, Serra da Estrela PDO, Torta del casar PDO.
5. Dutch-type	Edam, Edam Holland PGI, Noordhollandse Edam PDO, Gouda, Gouda Holland PGI, Noordhollandse Gouda PDO, Hollandse Geitenkaas PGI (Dutch goat cheese).
6. Swiss-type (Propionic acid)	AllgäuEmmental, Emmentaler PDO, Grevé, Maasdammer, Pategrás.
7. White-brined	Batzos PDO, Beyaz Peynir, Feta PDO, Halitzia, Halloumi, Mihaliç, Sjenica, Urfa.
8. Pasta-filata	Caciocavallo Podolico PDO, Kachkaval, Kashar, Kasseri PDO, Mozzarella di Bufala Campana PDO, Parenica, Provolone Valpadana PDO, Ragusano PDO, Vasteddadella Valle del Belìce PDO.
9. Mould surface-ripened	Altenburger PDO, Camembert de Normandie PDO.
10. Bacterial surface-ripened (smear)	Danbo, Epoisses PDO, Esrom PGI, Hohenheim Trappisten, Maroilles or Marolles PDO, Reblochon (de Savoie) PDO, Vacherin Mont-d'Or PDO.
11. Blue-veined	Cabrales PDO, Danablu PGI, Fourme d' Ambent PDO, Fourme de Montbrison PDO, Gamonedo PDO, Roquefort PDO, Stilton PDO.
12. Acid coagulated	Acid curd, Crottin de Chavignol PDO, Quark, Robiola di Roccaverano PDO.
13. Heat coagulated (whey cheeses)	Anari, Anthotyros, Manouri PDO, Mesost & Messmor.

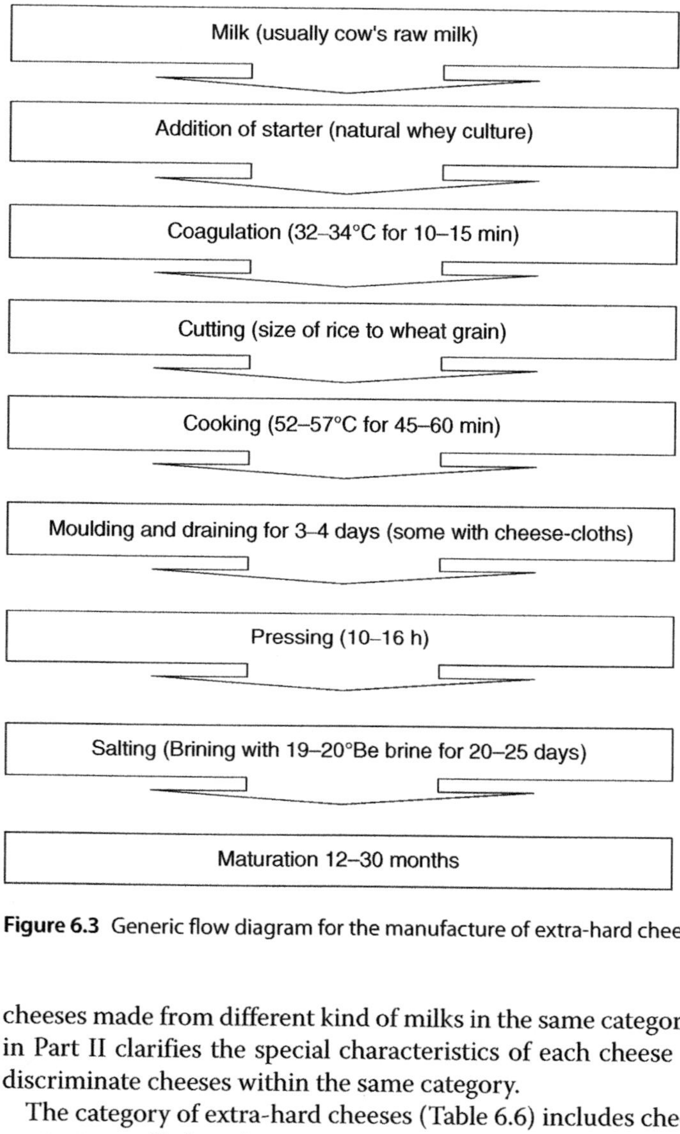

Figure 6.3 Generic flow diagram for the manufacture of extra-hard cheeses.

cheeses made from different kind of milks in the same category. The description of the cheeses in Part II clarifies the special characteristics of each cheese variety, and thus the reader can discriminate cheeses within the same category.

The category of extra-hard cheeses (Table 6.6) includes cheese varieties with a low moisture content, usually less than 30%. A generic flow diagram for the cheeses of this category is presented in Figure 6.3. Extra-hard cheeses are matured for a long period (up to 30 months) and are characterised by a granular texture and a strong flavour. They are produced as large cheese wheels, weighing up to 48 kg. Extra-hard cheeses are made from cow's raw milk which is either used as is (e.g. Sbrinz with a FDM 45%–55%) or partially skimmed (e.g. Parmigiano Reggiano with an FDM of 40%). The acidification is carried out with natural whey cultures, and the curds are cut at the size of rice to wheat grains. They are cooked at high temperatures (e.g. 57°C for Sbrinz) and are subjected to various pressures.

Hard cheeses (Table 6.6) usually have a moisture content between 30% and 40%. A generic flow diagram for the cheeses of this category is presented in Figure 6.4. In general, they are manufactured with a process that is very similar to the extra-hard process. The acidification is carried out with natural whey cultures, thermophilic or mesophilic starters, and the coagulum is cooked at various temperatures (i.e. 38°C–45°C for 25–40 min), whereas Le Gruyère is cooked at 57°C and Västerbottensost at 40°C but for several hours. Hard cheeses are pressed, usually brine-salted and matured for several months. A special cheesemaking process is

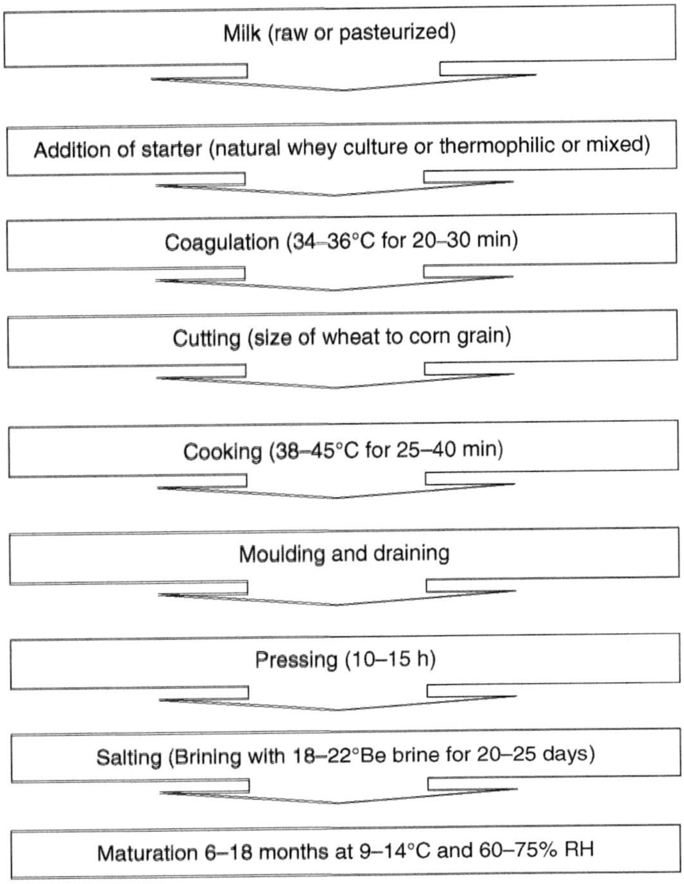

Figure 6.4 Generic flow diagram for the manufacture of hard cheeses.

followed for Cheddar, with cheddaring, milling, dry salting within the curd and pressing (see Part II, Section 2.5). There are some smear-ripened cheeses included in this category, such as Le Gruyère and Ossau Iraty, and these are matured in high RH, for the smear microflora to develop.

The category of semi-hard cheeses (Table 6.6) is quite a heterogeneous category, and many semi-hard cheeses, with a moisture content of 44%–55%, could belong to more than one category, while they present some important differences in the cheesemaking process. Some of them may be consumed as semi-hard, or at a longer maturation period, as hard cheese (e.g. Manchego and Majorero). Sombor cheese, which has also been classified as a semi-hard cheese, has a moisture content of 44%–50% in the upper part but 52%–58% in the inner part. A generic flow diagram for the cheeses of this category is presented in Figure 6.5. Semi-hard cheeses are acidified by the adventitious milk microflora when raw milk is used, or by the starter culture (mesophilic or mixed) when pasteurised milk is used. The coagulum is cut at a size of corn to hazelnut, and is cooked at 36°C–45°C, or higher temperatures is cooked at 44°C–53°C). Some semi-hard cheeses, such as Mahón-Menorca, Murcia al Vino and Sombor, are not cooked. Most of them (e.g. Appenzeller, Havarti and Sombor) are curd washed, with 10%–20% of warm water added to dilute the lactose content in the curd. They are pressed, salted, usually in brine of 18°–22° Bé and matured for 2–8 months. A number of semi-hard cheeses (e.g. Raclette, are smear-ripened cheeses.

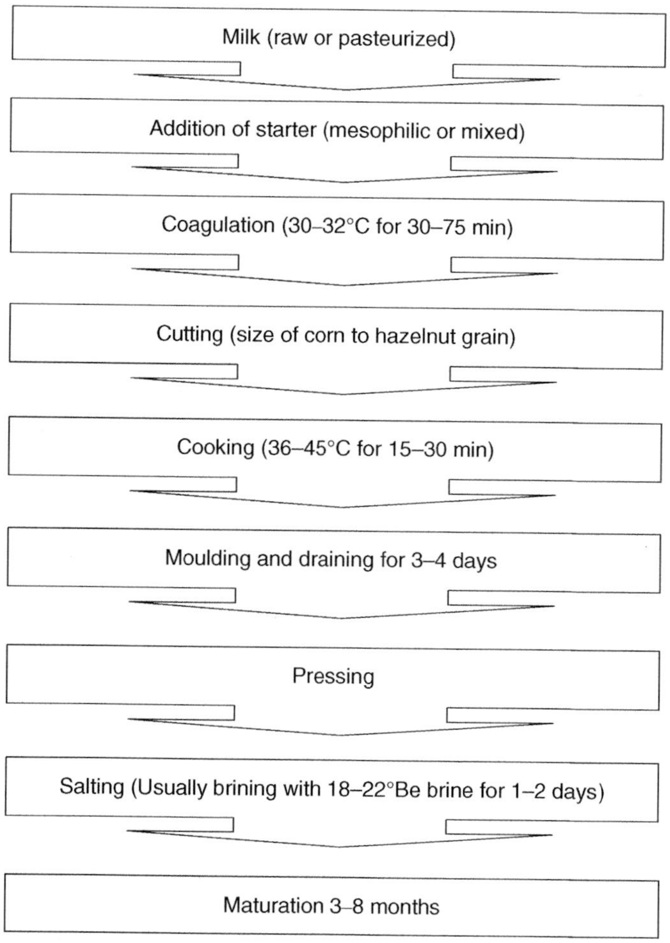

Figure 6.5 Generic flow diagram for the manufacture of semi-hard cheeses.

Soft cheeses coagulated with rennet are another cheese category (Table 6.6), with moisture content higher than 55%. They have a soft texture, and some of them are spreadable cheeses. The cheesemaking process is quite simple, and a generic flow diagram for the cheeses of this category is presented in Figure 6.6. They are acidified with mesophilic or thermophilic starter cultures and coagulated at low temperatures for long periods. The curd is cut at various sizes. Some soft cheeses are consumed fresh; some are ripened for 5–60 days. There are cases where soft cheeses are, after maturation, categorised as semi-soft, as their moisture decreases significantly; for example, Serra da Estrella is a soft cheese (moisture 43%–48%) when ripenened for 30 days, but, when matured for more than 120 days, it is classified as semi-hard (Serra da Estrella Velho with moisture 35%–39%).

Dutch-type cheeses (Table 6.6) are cheese varieties in which a few small eye-holes are formed by the CO_2 produced from fermentation of citrate by the starter culture. A generic flow diagram for the cheeses of this category is presented in Figure 6.7. Mesophilic starter culture, containing citrate-positive bacteria, is used and for some cheeses thermophilis plus adjuncts. After coagulation, the curd is cut at the size of corn to hazelnut, and the curd is washed with warm water. Dutch-type cheeses are cooked at 37°C–45°C and pressed. They are salted in brines with 18°–20° Bé and matured for 3–4 weeks to 1 year.

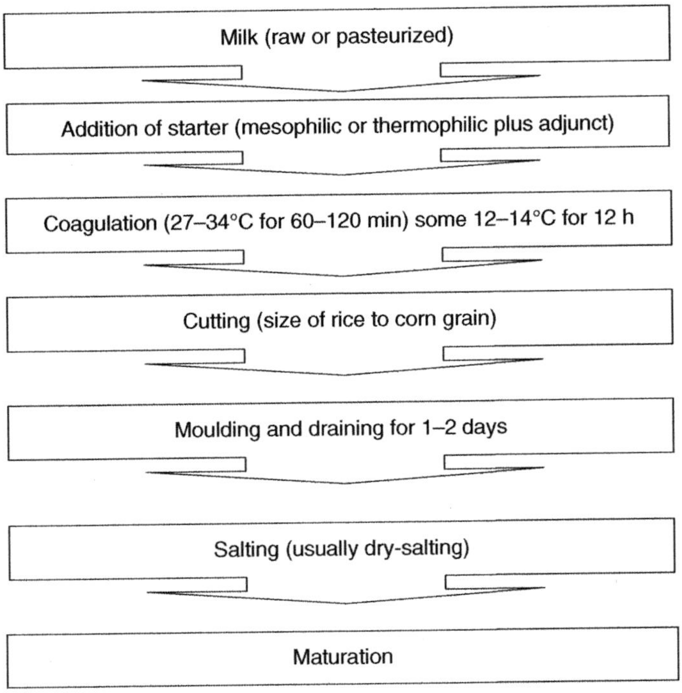

Figure 6.6 Generic flow diagram for the manufacture of soft cheeses.

Swiss-type cheeses are cheeses which traditionally were made as large cheese wheels weighing 60–120 kg. The characteristic of these cheeses is the numerous large eye-holes, which are formed by CO_2 produced from the fermentation of lactate by propionic acid bacteria (PAB) (e.g. *Propionibacterium freudenreichii* ssp. *shermanii*). A generic flow diagram for the cheeses of this category is presented in Figure 6.8. The cheesemaking process shares many similarities with that of hard cheeses; adjunct cultures containing PAB are added together with the starter culture.

White-brined cheeses (Table 6.6) constitute a separate family of cheeses, the characteristics of which is that they are ripened and preserved in brine until delivered to the customer. They are soft to semi-hard cheeses, traditionally produced under various names in the Balkans and East Mediterranean and neighbouring countries. Traditionally, brined cheeses are made mainly from sheep's, goat's and buffalo's milks; therefore, they retain the white colour of these milks. A generic flow diagram for the cheeses of this category is presented in Figure 6.9. Mesophilic or thermophilic starter cultures are used, and some varieties are cooked, while most are not. The characteristic step in the cheesemaking process of these cheeses is that maturation takes place with the cheese submerged in brine.

Pasta-filata cheeses (Table 6.6) are manufactured from curd cooked at 60°C–70°C, kneaded and stretched to form a smooth, plasticised, fibrous texture. This technique was 'discovered' in Italy, when a cheesemaker, trying to reuse a defective Pecorino cheese with cracks on the surface, cut it into slices and dipped it in hot whey (85°C–90°C); he tried to remake the same type of cheese, but, while waiting for the whey to cool, he dipped his hands into the container to control the cheese, and to his surprise he saw that the cheese mass began to stretch, and various pieces of cheese began to melt. Thus, the Vastedda della Valle del Belìce was made (see Part II, Section 8.9). A generic flow diagram for the cheeses of this category is presented in Figure 6.10. The acidification is carried out by the autochthonous microflora when raw milk is used, or by thermophilic starter culture when pasteurised milk is used. The curd (coagulated with rennet)

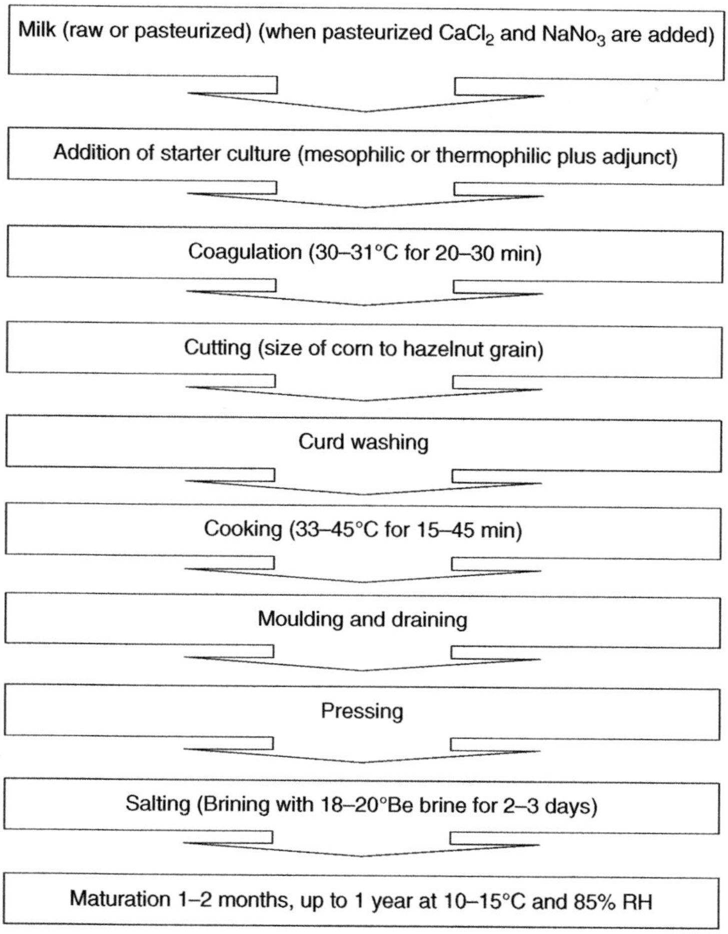

Figure 6.7 Generic flow diagram for the manufacture of Dutch-type cheeses.

is cut at the size of rice to hazelnut and is cooked at 38°C–42°C for 20–30 min, or at 50°C–65°C for 5–6 min. Then the curd is stretched at 60°C–70°C, moulded and salted with brine. Some past-filata cheeses, such as Mozzarella di Bufala, are consumed fresh, and most of them are matured for 2–4 months, and even up to 2 years (e.g. Caciocavallo Podolico).

Mould surface-ripened cheese varieties (Table 6.6) have a special characteristic: the growth of *Penicillium camemberti* (spores of *P. camemberti* are added as an adjunct culture) on the surface of the cheese, causing the characteristic softening of the cheese. A generic flow diagram for the cheeses of this category is presented in Figure 6.11. The acidification is carried out by a mesophilic starter, and the milk is coagulated using rennet extract. Then the coagulum is transferred, with or without cutting, into moulds for drainage. The cheeses are brine-salted and ripened at 9°C–16°C for 10–30 days. During ripening, the mould is being developed on the surface of the cheese and lactate is catabolised, increasing the pH. At the same time, lactate migrates from the core, calcium phosphate precipitates at the elevated pH of the surface and soluble calcium phosphate migrates from the core towards the surface. The foregoing phenomena, together with the extensive proteolysis, cause the characteristic softening of the cheese.

Bacterial surface (smear)-ripened cheeses (Table 6.6) are characterised by the development of special bacterial microflora on the surface of the cheese. A generic flow diagram for the cheeses of this category is presented in Figure 6.12. The milk is acidified using mesophilic,

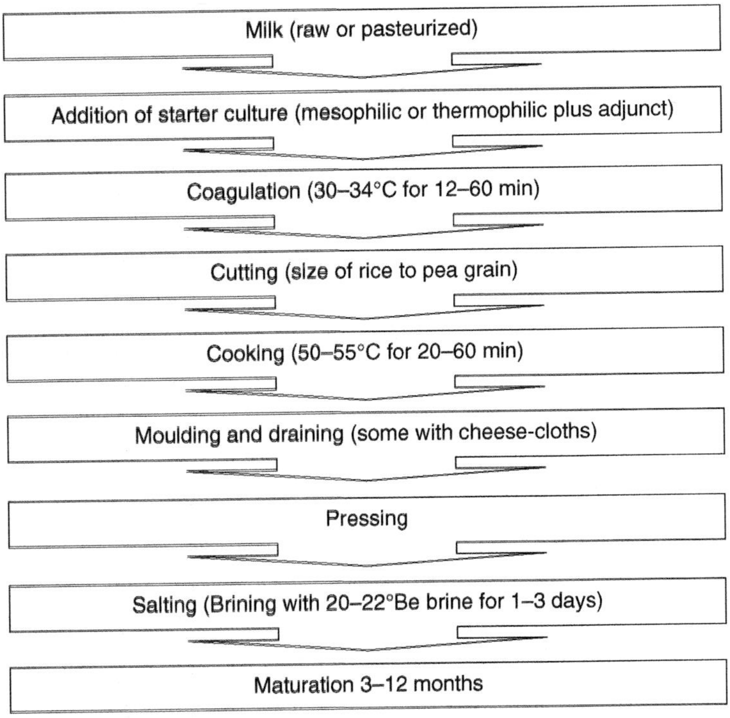

Figure 6.8 Generic flow diagram for the manufacture of Swiss-type cheeses.

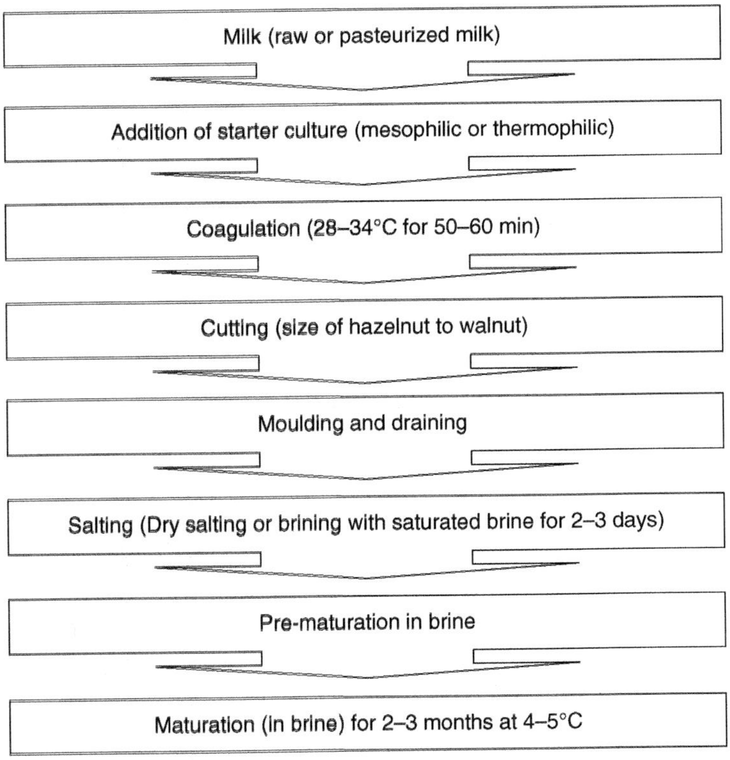

Figure 6.9 Generic flow diagram for the manufacture of white-brined cheeses.

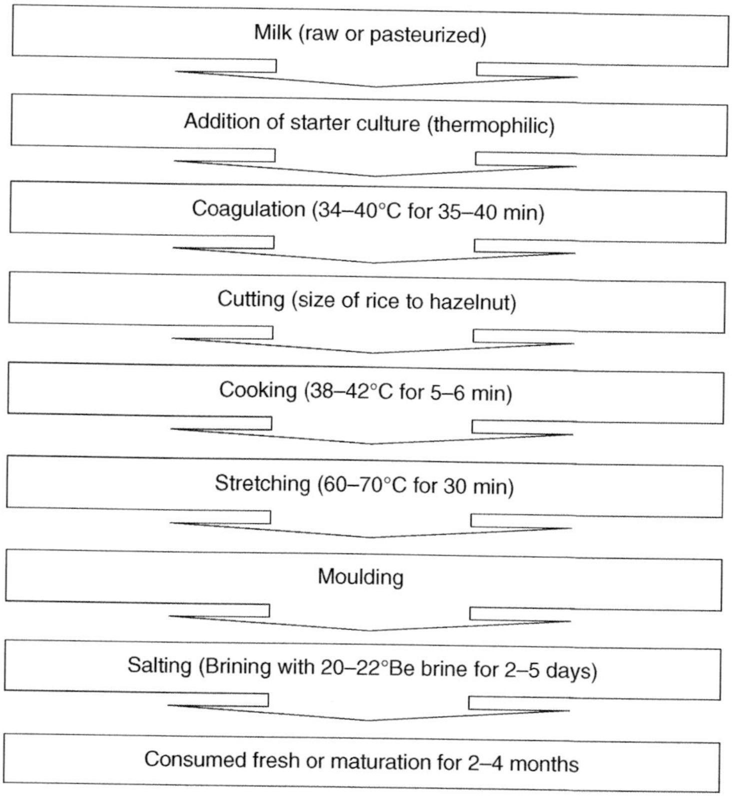

Figure 6.10 Generic flow diagram for the manufacture of pasta-filata cheeses.

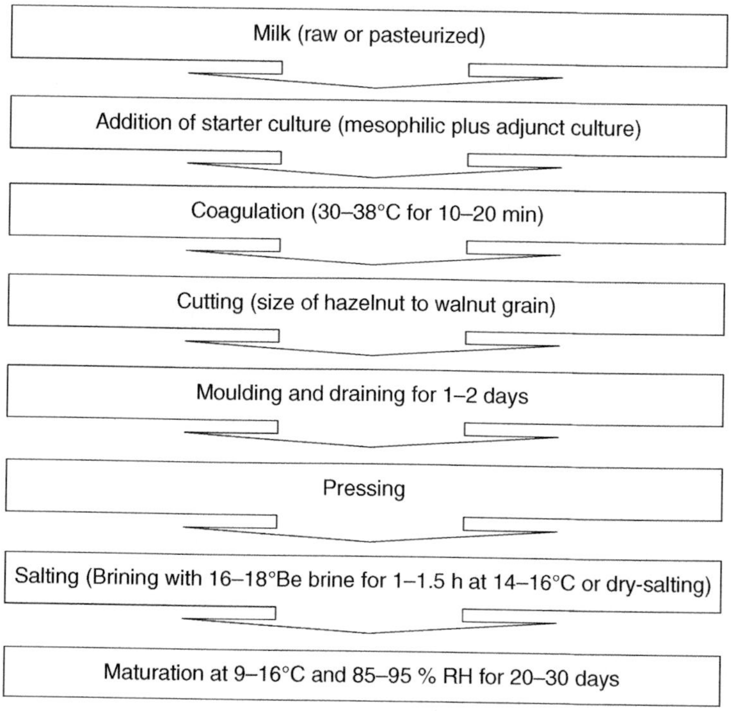

Figure 6.11 Generic flow diagram for the manufacture of mould surface-ripened cheeses.

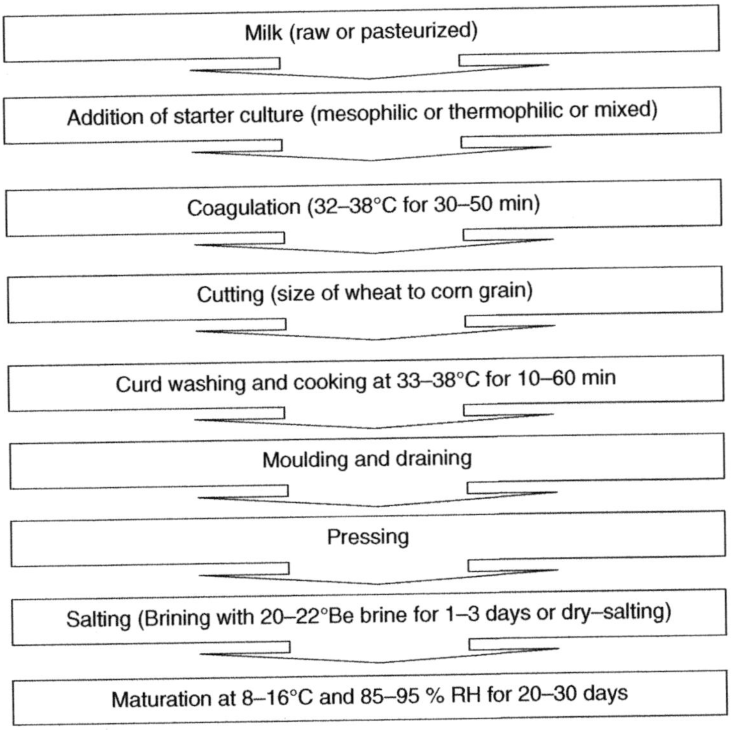

Figure 6.12 Generic flow diagram for the manufacture of bacterial surface-ripened cheeses.

thermophilic or mixed starter culture. It is coagulated with rennet, and cut at the size of wheat to corn grain. Some are curd washed and cooked 34°C–41°C, depending on the fat content, and moulded. Some are pressed and salted in brine. Throughout ripening (at 18°C–20°C and 90%–95% RH), the surface of the cheese is washed with a brine solution containing special bacterial microflora; that is called smearing. After the manufacturing process, the surface microflora is dominated by yeasts (e.g. *D. hansenii* and *G. candidum*). Theses yeasts metabolise lactate and thus deacidify the cheese surface and encourage the growth of coryneform bacteria (e.g. *Corynebacterium* spp., *Arthrobacter* spp. and *Brevibacterium* spp.). Coryneforms are high proteolytic and lipolytic bacteria. The cheeses are then stored at 8°C–12°C and 85% RH to complete the maturation process.

Blue-veined cheeses (Table 6.6) are characterised by blue veins caused by the growth of *Penicillium roqueforti* in the interior of the cheese. A generic flow diagram for the cheeses of this category is presented in Figure 6.13. The acidification is carried out by mesophilic starter cultures, and the milk is coagulated by rennet extracts. The curds is cut at the size of hazelnut to walnut and, for some varieties, is washed with warm water before being transferred in the moulds and after that it is dry-salted. The curd is not pressed, since the growth of the mould needs oxygen, and thus an open texture. The curd is pierced with needles containing mould spores. Blue-veined cheeses have a strong flavour, caused by the extensive lipolysis and the presence of n-methyl-ketones, which are produced from fatty acids.

Acid-coagulated cheeses (Table 6.6) are characterised by the fact that the milk is coagulated from the acid at a pH of 4.6. A generic flow diagram for the cheeses of this category is presented in Figure 6.14. The acidification is caused by either the adventitious LAB or by the addition of an acid (e.g. lactic acid or acidic acid), and the coagulation

Whey cheeses (Table 6.6) are characterised by the fact that the coagulation of the milk (or whey) in caused by heating at 85°C–90°C. A generic flow diagram for the cheeses of this category is presented in Figure 6.15. Whey from the manufacture of hard or other cheeses is mixed with milk and/or cream and heated at 88°C–90°C for 40–45 min; heating causes aggregation of the whey proteins and formation of the curd. Salt is added, and the curd is moulded, drained and cooled. It may be packed and consumed immediately, or left for drying and used for grating.

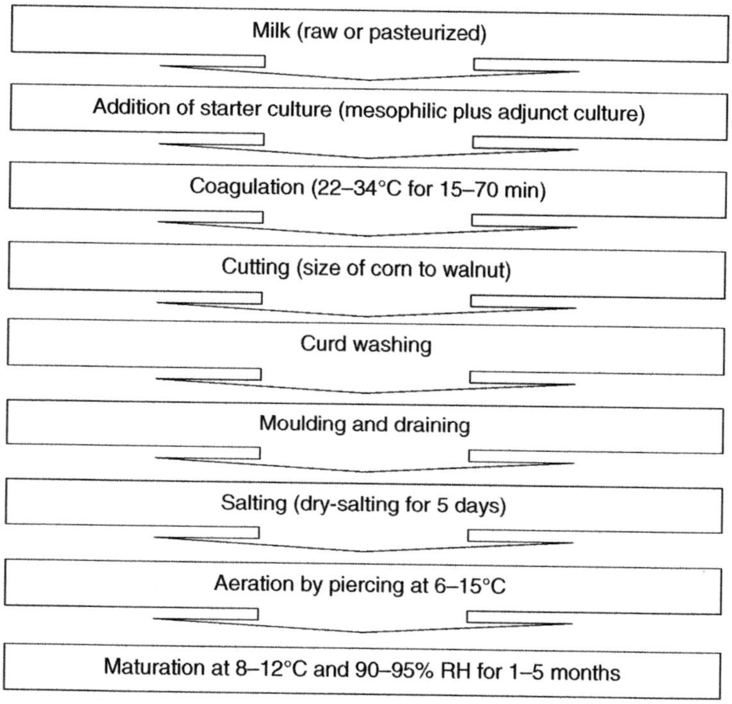

Figure 6.13 Generic flow diagram for the manufacture of blue-veined cheeses.

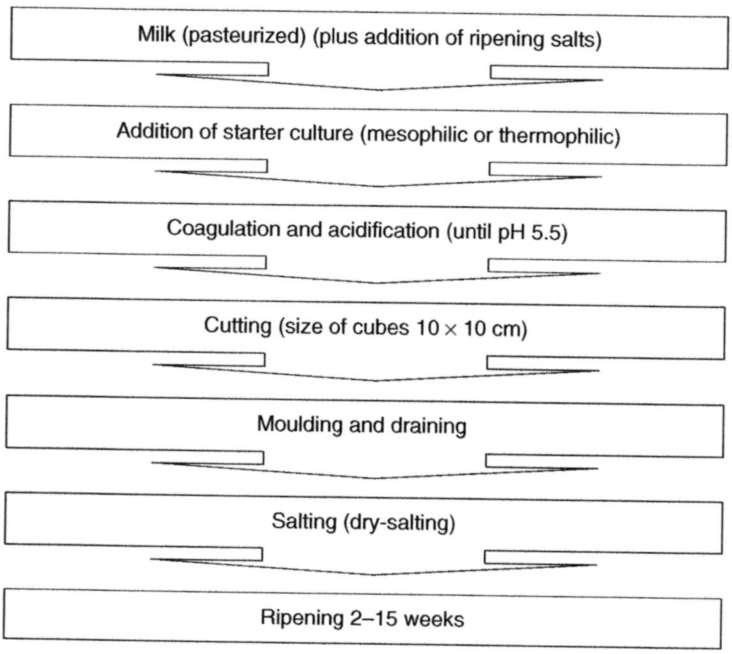

Figure 6.14 Generic flow diagram for the manufacture of acid-coagulated cheeses.

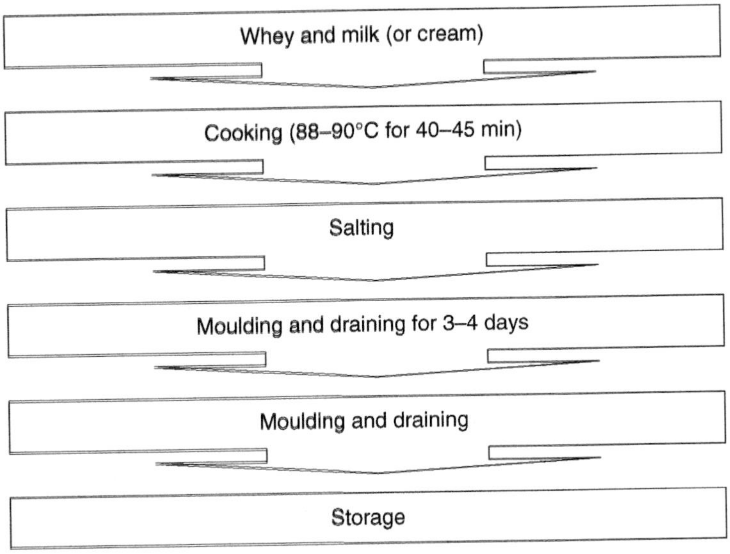

Figure 6.15 Generic flow diagram for the manufacture of whey cheeses.

References

Amigo, L., Fontecha, J. (2011). Goat milk. In Fuquay, J. W., Fox, P. F., & McSweeney, P. L. H. (Eds.), *Encyclopedia of Dairy Sciences*, 2nd edition. Academic Press, Waltham, MA, pp. 484–493.

Angelides, A. S. (2014) The microbiology of raw milk. In Papademas, P. (ed.), *Dairy Microbiology: A Practical Approach*. CRC Press, pp. 22–68.

Arvanitoyannis, I. S., Kargaki, G. K. & Hadjichristodoulou, C. (2011a). Effect of several MAP compositions on the microbiological and sensory properties of Graviera cheese. *Anaerobe*, 17, 310–314.

Arvanitoyannis, I. S., Kargaki, G. K. & Hadjichristodoulou, C. (2011b). Effect of three MAP compositions on the physical and microbiological properties of a low fat Greek cheese known as Anthotyros. *Anaerobe*, 17, 295–297.

Aureli, P., Scalfaro, C. & Franciosa, G. (2011). Clostridium spp. In Fuquay J. W., Fox P. F. & McSweeney, P. L. H. (eds.), *Encyclopedia of Dairy Sciences*, 2nd edition. Elsevier, UK, pp. 47–53.

Beresford, T. & Cogan, T. (1997). Improving Cheddar cheese flavour. In *Proceedings of the 5th Cheese Symposium*. Teagasc/University College Cork, Cork, Ireland, pp. 53–61.

Bintsis, T. & Athanasoulas, A. (2015). Dairy starter cultures. In Papademas P. (ed.), *Dairy Microbiology, A Practical Approach*. CRC Press, Boca Raton, FL, pp. 114–154.

Bintsis, T. (2006). Quality of brine. In Tamime, A. Y. (ed.), *Brined Cheeses*. Blackwell Publishing, Oxford, UK, pp. 264–301.

Cao-Hoang, L., Chaine, A., Gregoire, L. & Wache, Y. (2010). Potential of nisin-incorporated sodium caseinate films to control *Listeria* in artificially contaminated cheese. *Food Microbiology*, 27, 940–944.

Cerqueira, M. A. & Vicente, A. A. (2013). Shelf-life extension of cheese using edible packaging materials. In Preedy, V. R., Watson, R. R. & Patel, V. B. (eds.), *Handbook of Cheese in Health: Production, Nutrition and Medical Sciences*. Wageningen Academic Publishers, Wageningen, the Netherlands, pp. 123–135.

Chanos, P., Morgenstern, H. U. & Dalgas, A. C. (2013). In Park, Y. W. & Haenlein, G. F. W. (eds.), *Milk and Dairy Products in Human Nutrition: Production, Composition and Health*. John Wiley & Sons Ltd., Oxford, pp. 167–180.

Chr. Hansen (2016a). *BioSafe cultures*. http://www.chr-hansen.com/food-cultures-and-enzymes/cheese/cards/collection-cards/protection/biosafe. Accessed: 20-06-2016.

Chr. Hansen (2016b). *Annatto*. http://www.chr-hansen.com/en/natural-colors/a-rainbow-of-colors/yellow-orange/annatto. Accessed 01-11-2016.

Conte, A., Gammariello, D., Di Giulio, S., Attanasio, M. & Del Nobile, M. A. (2009). Active coating and modified atmosphere packaging to extend the shelf-life of Fior di Latte cheese. *Journal of Dairy Science*, 92, 887–894.

Conte, A., Scrocco, C., Sinigaglia, M. & Del Nobile, M. A. (2007). Innovative active packaging systems to prolong the shelf-life of Mozzarella cheese. *Journal of Dairy Science*, 90, 2126–2131.

Corral, J. M., Padilla, J. A. & Izquierdo, M. (2010). Associations between milk protein genetic polymorphisms and milk production traits in Merino sheep breed. *Livestock Science*, 129, 73–79.

Dalgleish, D. G. (1993). The Enzymatic Coagulation of Milk. In Fox, P. F. (ed.), *Cheese: Chemistry, Physics and Microbiology*, 2nd edition, Vol. 1. Chapman and Hall, London, pp. 69–100.

Davis, J. G. (1965). *Cheese, Vol. 1, Basic Technology: Vol. 2: Bibliography*. Churchill Livingstone, London.

Dettori, M. L., Pazzola, M., Paschino, P., Pira, M. G. & Vacca, G. M. (2015). Variability of the caprine whey protein genes and their association with milk yield, composition and renneting properties in the Sarda breed. 1. The LALBA gene. *Journal of Dairy Research*, 82 (4), 434–441.

Duan, J., Park, S.-I., Daeschel, M. A. & Zhao, Y. (2007). Antimicrobial chitosan-lysozyme (CL) films and coatings for enhancing microbial safety of Mozzarella cheese. *Journal of Food Science*, 72, M355–M362.

El-Soda, M. (1993). The role of lactic acid bacteria in accelerated cheese ripening. *FEMS Microbiology Reviews*, 12, 239–252.

EC (2004). EC Regulation No 853/2004 of the *European Parliament laying down specific hygiene rules for on the hygiene of foodstuffs. Official Journal of the European Union*, L139, 55–205.

Fajardo, P., Martins, J. T., Fucinos, C., Pastrana, L., Teixeira, J. A. & Vicente, A. A. (2010). Evaluation of a chitosan based edible film as carrier of natamycin to improve the storability of Saloio cheese. *Journal of Food Engineering*, 101, 349–356.

Food and Drug Administration (2015). *Milk Drug Residue Sampling Survey*. Department of Health and Human Services, USA.

Fox, P. F. (1987). Significance of salt in cheese ripening. *Dairy Industries International*, 52 (9), 20–22.

Fox, P. F. (2003). The major constituents of milk. In Smit, G. (ed.), *Dairy Processing-Improving Quality*. Woodhead Publishing Ltd, pp. 5–42.

Fox, P. F. & Guinee, T. P. (2013). Cheese Science and Technology. In Park, Y. W. & Haenlein, G. F. W. (eds.), *Milk and Dairy Products in Human Nutrition: Production, Composition and Health*. John Wiley & Sons Ltd., Oxford, pp. 384–400.

Fox, P. F. & McSweeney, P. L. H. (1998). *Dairy Chemistry and Biochemistry*. Blackie Academic & Professional, London.

Fox, P. F. & McSweeney, P. L. H. (2004). Cheese: An overview. In Fox, P. F., McSweeney, P. L. H., Cogan, T. M. & Guinee, T. P. (eds.), *Cheese: Chemistry, Physics and Microbiology: General Aspects*, 3rd edition, Vol. 1. Elsevier Academic Press, London, pp. 1–18.

Fox, P. F., Guinee, T. P., Cogan, T. M. & McSweeney, P. L. H. (2000). Principal families of cheese. In *Fundamentals of Cheese Science*. Aspen Publishers, pp. 388–428.

Fox, P. F., Law, J., McSweeney, P. L. H. & Wallace, J. (1993). Biochemistry of cheese ripening. In Fox, P. F. (ed.), *Cheese: Chemistry, Physics and Microbiology*, 2nd edition, Vol. 2. Chapman & Hall, pp. 389–438.

Fox, P. F., Wallace, J. M., Morgan, S., Lynch, S., Niland, C. M. & Tobin, J. (1996). Acceleration of cheese ripening. *Antonie van Leeuwenhoek*, 70, 271–297.

Freitas, A. C., Rodrigues, D., Duarte, A. C. & Gomes, A. M. (2013). The principals of cheese making: an overview. In Preedy, V. R., Watson, R. R. & Patel, V. B. (eds.), *Handbook of Cheese in Health: Production, Nutrition and Medical Sciences.* Wageningen, Academic Publishers, pp. 53–72.

Goulas, A. & Grandison, A. S. (2008). Applications of membrane separation. In Britz, T. J. & Robinson, R. K. (eds.), *Advanced Dairy Science and Technology.* Blackwell Publishing Ltd, Oxford, UK. doi: 10.1002/9780470697634.ch2.

Goyal, S. & Goyal, G. K. (2012). Recent technologies for dairy products packaging. *Scientific Journal of Agricultural,* 1 (6), 164–167.

Grattepanche, F., Miescher-Schwenninger, S., Meile, L. & Lacroix, C. (2008). Recent developments in cheese cultures with protective and probiotic functionalities. *Dairy Science & Technology,* 88 (4–5), 421–444.

Green, M. L. & Grandison, A. S. (1993). Secondary (non-enzymatic) phase of rennet coagulation and post-coagulation phenomena. In Fox, P. F. (ed.), *Cheese: Chemistry, Physics and Microbiology,* 2nd edition, Vol. 1. Chapman and Hall, London, pp. 101–140.

Guinee, T. P. & Fox, P. F. (1993). Salt in cheese: Physical, chemical and biological aspects. In Fox, P. F. (ed.), *Cheese: Chemistry, Physics and Microbiology,* 2nd edition, Vol. 1. Chapman and Hall, London, pp. 257–302.

Guinee, T. P. & O'Brien, B. (2010). The quality of milk for cheese manufacture. In Law, B. A. & Tamime, A. Y. (eds.), *Technology of Cheesemaking,* 2nd edition. Blackwell Publishing Ltd., Chichester, UK, pp. 1–67.

Gursoy, O., Kesenkas, H. & Yilmaz, Y. (2013). White Cheese. In: Preedy, V. R., Watson, R. R. & Patel, V. B. (eds.), *Handbook of Cheese in Health: Production, Nutrition and Medical Sciences.* Wageningen Academic Publishers, pp. 181–192.

Horne, D. S. (2011). Casein Micellar structure. In Fuquay, J. W., Fox, P. F., McSweeney, P. L. H. (eds.), *Encyclopedia of Dairy Sciences,* 2nd edition. Elsevier, UK, pp. 772–779.

Hyslop, D. (2003). Rennet coagulation of milk. In Fox P. F. & McSweeney, P. L. H. (eds.), *Advanced Dairy Chemistry, Vol. 1. Proteins,* 3rd edition. Aspen, Gaithersburg, pp. 839–878.

IDF (1971). *Cheese names. Annual Bulletin.* International Dairy Federation, Brussels.

IDF (1981). *Catalogue of cheeses. IDF Bulletin No 141.* International Dairy Federation, Brussels.

IDF (2016). *The world dairy situation. Bulletin of the IDF 485.* International Dairy Federation, Brussels.

Jalilzabeh, A., Tunçtürk, Y. & Hesari, J. (2015). Extension shelf life of cheese: A review. *International Journal of Dairy Science,* 10 (2), 44–60.

Jawaid, S., Talpur, F. N., Nizamani, S. M. & Afridi, H. I (2015). Contamination profile of Aflatoxin M1 residues in milk supply chain of Sindh, Pakistan. *Toxicology Reports,* 2, 1418–1422.

Johnson, M. & Law, B. A. (2010). The origins, development and basic operations of cheesemaking technology. In Law, B. A. & Tamime, A. Y. (eds.), *Technology of Cheesemaking,* 2nd edition. Blackwell Publishing Ltd., Chichester, UK, pp. 68–97.

Johnson, M. & Law, B. A. (1999). The origins, development and basic operations of cheesemaking technology. In Law, B. A. (ed.), *Technology of Cheesemaking.* Sheffield Academic Press, Sheffield, UK, pp. 1–32.

Johnson, M. E., Chen, C. M. & Jaeggi, J. J. (2001). Effect of rennet coagulation time on composition, yield, and quality of reduced-fat Cheddar cheese. *Journal of Dairy Science,* 84, 1027–1033.

Juric, M., Bertelsen, G., Mortensen, G. & Petersen, M. A. (2003). Light-induced colour and aroma changes in sliced, modified atmosphere packaged semi-hard cheeses. *International Dairy Journal,* 13, 239–249.

Kelly, A. L. (2007a). Why must milk be standardised for cheesemaking? In McSweeney, P. L. H. (ed.), *Cheese Problems Solved.* Woodhead Publishing Ltd, Cambridge, UK, pp. 18–19.

Kelly, A. L. (2007b). Introduction: How may cheese be packaged? In McSweeney, P. L. H. (ed.), *Cheese Problems Solved.* Woodhead Publishing Ltd, Cambridge, UK, pp. 159–160.

Kindstedt, P. S. (2013a). The making of great cheeses. *Microbe*, 8 (9), 361–367.

Kindstedt, P. S. (2013b). The making of great cheeses, Part 2. *Microbe*, 8 (10), 401–407.

Kirkin, C., Gunes, G. & Kilic-Akyilmaz, M. (2013). Preservation of precut white cheese by modified atmosphere packaging. *International Journal of Dairy Technology*, 66, 576–586.

Law, B. A. (2001). Controlled and accelerated cheese ripening: The research base for new technologies. *International Dairy Journal*, 11, 383–398.

Lawrence, R. C., Gilles, J., Crow, V. L., Heap., H. A., Honoré, C. G., Johnston, K. A. & Samal, P. K. (2004). Cheddar cheese and related dry-salted cheese varieties. In Fox, P. F., McSweeney, P. L. H., Cogan, T. M. & Guinee, T. P. (eds.), *Cheese: Chemistry, Physics and Microbiology: General Aspects*, 3rd edition, Vol. 2. Elsevier Academic Press, London, pp. 71–102.

Lomholt, S. B. & Qvist, K. B. (1999). The formation of cheese curd. In Law, B. A. (ed.), *Technology of Cheesemaking*. Sheffield Academic Press, Sheffield, UK, pp. 66–98.

Lucey, J. A. (2002). Formation and physical properties of milk protein gels. *Journal of Dairy Science*, 85, 281–294.

Lucey, J. A., Johnson, M. E. & Horne, D. S. (2003). Invited review: Perspectives on the basis of the rheology and texture properties of cheese. *Journal of Dairy Science*, 86 (9), 2725–2743.

Martins, J. T., Cerqueira, M. A., Souza, B. W. S., Avides, M. C. & Vicente, A. A. (2010). Shelf-life extension of Ricotta cheese using coatings of galactomannans from non conventional sources incorporating nisin against *Listeria monocytogenes*. *Journal of Agricultural and Food Chemistry*, 58, 1884–1891.

McSweeney, P. L. H. (2007a). Introduction: How does rennet coagulate milk? In McSweeney, P. L. H. (ed.), *Cheese Problems Solved*. Woodhead Publishing Limited, Cambridge, UK, pp. 50–51.

McSweeney, P. L. H. (2007b). What enzymes are in rennet? In McSweeney, P. L. H. (ed.), *Cheese Problems Solved*. Woodhead Publishing Limited, Cambridge, UK, pp. 55–56.

McSweeney, P. L. H. (2007c). Principal families of cheese. In McSweeney, P. L. H. (ed.), *Cheese Problems Solved*. Woodhead Publishing Limited, Cambridge, UK, pp. 176–177.

McSweeney, P. L. H., Ottogalli, G. & Fox, P. F. (2004). Diversity of cheese varieties: An overview. In Fox, P. F., McSweeney, P. L. H., Cogan, T. M. & Guinee, T. P. (eds.), *Cheese: Chemistry, Physics and Microbiology: General Aspects*, 3rd edition, Vol. 2. Elsevier Academic Press, London, pp. 2–22.

Meile, L., Le Blay, G. & Thierry, A. (2008). Safety assessment of dairy microorganisms: *Propionibacterium* and *Bifidobacterium*. *International Journal of Food Microbiology*, 126, 316–320.

Nielsen, E. W. (2004). Principles of cheese production. In Hui, Y. H., Meunier-Goddik, L., Hansen, A. S., Josephsen, J., Nip, W.-K., Stanfield, P. S. & Toldrá, F. (eds.), *Handbook of Food and Beverage Fermentation Technology*. Markel Dekker Inc, New York, pp. 219–239.

Nunez, M. & Medina, M. (2013). Elimination of pathogenic bacteria in cheese by high pressure processing. In Preedy, V. R., Watson, R. R. & Patel, V. B. (eds.), *Handbook of Cheese in Health: Production, Nutrition and Medical Sciences*. Wageningen Academic Publishers, pp. 329–343.

Olivares, M. L., Sihufe, G. A., Capra, M. L., Rubiolo, A. C. & Zorrilla, S. E. (2012). Effect of protective atmospheres on physicochemical, microbiological and rheological characteristics of sliced Mozzarella cheese. *LWT-Food Science and Technology*, 47, 465–470.

O'Mahony, J. A., Fox, P. F. & Kelly, A. L. (2013). Indigenous enzymes of milk. In McSweeney, P. L. H. & Fox, P. F. (eds.), *Advanced Dairy Chemistry Volume 1A: Proteins: Basic Aspects*, 4th edition, pp. 337–387.

Papademas, P. & Robinson, R. K. (2002). Some volatile plant compounds in Halloumi cheeses made from ovine or bovine milk. *Lebensmittel-Wissenschaft und-Technologie/FST*, 35 (6), 512–516.

Papaioannou, G., Chouliara, I., Karatapanis, E. A., Kontominas, G. M. & Savvaidis, N. I. (2007). Shelf-life of a Greek whey cheese under modified atmosphere packaging. *International Dairy Journal*, 17, 358–364.

Pintado, C. M. B. S., Ferreira, M. A. S. S. & Sousa, I. (2010). Control of pathogenic and spoilage microorganisms from cheese surface by whey protein films containing malic acid, nisin and natamycin. *Food Control*, 21, 240–246.

Ramos, M. & Juárez, M. (2011). Sheep milk. In Fuquay, J. W., Fox, P. F., & McSweeney, P. L. H. (eds.), *Encyclopedia of Dairy Sciences*, 2nd edition. Academic Press, Waltham, MA, pp. 494–502.

Rehman, S. U., Fox, P. F. & McSweeney, P. L. H. (2000). Methods used to study non-starter microorganisms in cheese: A review. *International Journal of Dairy Technology*, 53 (3), 113–119.

Robinson, R. K. & Wilbey, R. A. (1998). *Cheesemaking Practice – R. Scott*, 3rd edition. Aspen Publishers, Maryland, USA.

Rodrigues-Aguilera, R., Oliveira, J. C., Montanez, J. C. & Mahajan, P. V. (2011b). Effect of modified atmosphere packaging on quality factors and shelf-life of surface mould ripened cheese: Part I constant temperature. *LWT-Food Science and Technology*, 44, 330–336.

Romani, S., Sacchetti, G., Pittia, P., Pinnavaia, G. G. & Rosa, M. D. (2002). Physical, chemical, textural and sensorial changes of portioned Parmigiano Reggiano cheese packed under different conditions. *Food Science and Technology International*, 8, 203–211

Ritvanen, T. (2013). Reduced-fat cheeses: Composition and health potential. In Preedy, V. R., Watson, R. R., & Patel, V. B. (eds.), *Handbook of Cheese in Health: Production,* Nutrition and Medical Sciences. Wageningen Academic Publishers, pp. 637–650.

Sanders, G. P. (1953). *Cheese Varieties and Descriptions*. Agriculture Handbook, US Department of Agriculture, Washington, DC.

Schneider, Y., Kluge, C., Weiβ, U. & Rohm, H. (2010). Packaging materials and equipment. In Law, B. A. & Tamime, A. Y. (eds.), *Technology of Cheesemaking*, 2nd edition. Blackwell Publishing Ltd., Chichester, UK, pp. 413–439.

Scott, R. (1986). *Cheesemaking Practice*. Applied Science Publishers, London.

Sindhu, J. S. & Arora, S. (2011). Buffalo milk. In Fuquay, J. W., Fox, P. F., & McSweeney, P. L. H. (eds.), *Encyclopedia of Dairy Sciences*, 2nd edition. Academic Press, Waltham, MA, pp. 503–511.

Trobetas, A., Badeka, A. & Kontominas, M. G. (2008). Light induced changes in grated Graviera cheese packaged under modified atmospheres. *International Dairy Journal*, 18, 1133–1139.

Ture, H., Eroglu, E., Ozen, B. & Soyer, F. (2011). Effect of biopolymers containing natamycin against *Aspergillus niger* and *Penicillium roquefortii* on fresh kashar cheese. *International Journal of Food Science and Technology*, 46 (1), 154–160.

Vacca, G. M., Dettori, M. L., Piras, G., Manca, F., Paschino, P. & Pazzola, M. (2014). Goat casein genotypes are associated with milk production traits in the Sarda breed. *Animal Genetics*, 45 (5), 723–731.

7

Traditional Wooden Equipment Used for Cheesemaking and Their Effect on Quality

Giuseppe Licitra[1], *Margherita Caccamo*[2], *Florence Valence*[3] *and Sylvie Lortal*[3]

[1] *Department of Agriculture, Nutrition and Environment, University of Catania, Catania, Italy*
[2] *CoRFiLaC, Ragusa, Italy*
[3] *INRA, Agrocampus Ouest, Science et Technologie du lait et de l'oeuf, Rennes, France*

7.1 Introduction to Traditional Cheeses

Experts on pre-history speculate that cheesemaking dates back more than 10,000 to 18,000 years in Central Asia in the steppe, in Mesopotamia, Anatolia and the Middle East. In these places, society founded its activities on agriculture and breeding. Archaeological surveys have established that in the Neolithic age, the ancient Sumerian and Mesopotamian cultures of the Tigris-Euphrates basin raised animals and engaged in dairy production as early as 7000–8000 BC. Cheesemaking was definitely practised in ancient Egypt and other parts of the Middle East and Europe at least 5,000 and possibly as far back as 8,000 years ago (Salque et al., 2013).

Cheese culture gradually evolved and spread in the Mediterranean basin, Europe and all around the world . The geographical atlas of the *Treccani Encyclopaedia* reports that in our days almost 50% of the population lives in rural areas (about 3.5 billion), most of them in developing countries. The World Bank (2007) underscores the fact that 3 billion people live in rural areas, with limited access to land and other capital and technological resources required to make farming a viable endeavour. The rest (0.5 billion) live in less favoured areas of the developed countries.

Hundreds of traditional cheeses are produced worldwide, in direct relationship with nature, and they are examples of sustainable agriculture – even if it is not a conscious goal; these products have fed billions of people for centuries.

The Family Farming World Conference of 2011 remarked that 'family farming represents a sector of strategic value because of its economic, social, cultural, environmental, and territorial functions'. Women and men engaged in family farming produce 70% of the world's food. Family farming is the basis of sustainable food production aimed at food security and food sovereignty, environmental management of land and its biodiversity, and preservation of the important socio-cultural heritage of rural communities and nations.

Traditional products have a strong linkage to the territory of origin (i.e. orography, landscape, rural environment and human resources) and therefore testify to the history, culture and lifestyle of those communities that produce them, handed down over the generations, usually orally. It follows that traditional cheeses are a unique expression of the symbiotic interaction between human resources, the culture of rural communities and nature (Licitra, 2010). Worldwide, every traditional cheese originates from complex systems which draw on the specific bio-organoleptic characteristics tied to several 'biodiversity factors', such as the environment; the macro- and

Global Cheesemaking Technology: Cheese Quality and Characteristics, First Edition.
Edited by Photis Papademas and Thomas Bintsis.

micro-climate; the natural pasture; the breed of the animals (often autochthonous breeds); the use of raw milk and its natural microflora; the use of natural coagulants; the use of natural ingredients (e.g. saffron, pepper, herbs, sugar, flour and spice); the use of traditional equipment and the natural aging conditions, including the ancestral practice of sun drying and the overall expertise of the cheesemakers and ripeners (Licitra, 2010).

Every traditional production system is characterised by a sequence of countless biological and natural processes, each one marked by its own natural rhythms. The cheesemaker has to understand, support and coordinate the delicate harmony of the sequence of actions and timing of cheesemaking and of the aging process in order to produce the most exciting form of milk: cheese. Each biodiversity factor will synergistically influence the quality of the final products, with intense and diversified flavours even within a cheese variety, depending on the specific origin of the product (Licitra, 2010).

Even in the twenty-first century, in Europe (France, Italy, Spain and Portugal), most traditional cheeses are produced in less favoured areas (including most of the PDO cheeses), in an environment that would otherwise be abandoned, where livestock and farming are often the only way to use the land. If 'the overall world governance' in the era of globalisation will not be able to stop the rural exodus, and is unable to offer to the rural population alternatives to new employment opportunities and lodging in the cities, the conservation of natural resources of the planet will be compromised.

Farmers of disadvantaged rural areas (mountains, high hills and developing countries) also play the role of guardian of the environment by protecting natural resources, reducing soil erosion, reducing deforestation and desertification, and maintaining animal and vegetal bio-diversity.

Most of the time, small family farmers living in the aforementioned areas produce with limited resources, and so cheesemakers also historically have had to adapt to the equipment and places that nature offered them. Traditional equipment represents one of the most important sources of biodiversity to produce traditional cheeses.

It is reasonable to assume that daily practice led cheesemakers to select the equipment and the places that would maximise the production of a good product. Therefore, the equipment and practices that have been passed down for thousands of years over the generations to the present day need more attention and respect, and should not be banned by law owing to a hypothetical food safety risk which has never been rigorously scientifically demonstrated (Licitra, 2010).

7.2 Traditional Equipment

Farmers and cheesemakers since prehistoric times have used tools from natural materials to collect, transform milk and to age the cheeses. Wood has been safely allowed to directly contact food for centuries. Fruit and vegetables as well as fresh or smoked fish have been stored in wooden crates. In cheese- and winemaking, wooden boards and barrels have been indispensable in traditional production. There are many other examples where wood has been used as a lightweight and still rough or porous packaging material from natural sources. Nevertheless, there are objections to the use of wood to directly contact food, as it is usually considered less hygienic than other smooth or synthetic materials. To date, there is no evidence that any foodborne disease can be attributed to wood given the proper use of wood and hygienic standards for production, storage and applications (Aviat et al., 2016).

Some examples of equipment still used in traditional cheese production, grouped according to the major transformation processes, are now listed:

Milk collection: Calebasse wooden bucket (pumpkin container for wagashi cheese in Benin used by small family farmers in rural less favoured areas).

Milk coagulation: Vats called wooden *tina* for Sicilian cheeses (Ragusano PDO, Vastedda della Valle del Belice PDO, Palermitano and Provola dei Nebrodi), *gerle* for French Cantal PDO; *tinaccio* for Caciocavallo podolico and bell-shaped copper cauldrons for Parmigiano Reggiano PDO and Grana Padano PDO.

Breaking the curd: Wooden stick called *ruotula* for Ragusano PDO, *rotula* for Caciocavallo Palermitano and Vastedda della valle del Belice PDO and wooden *spino* for Maiorchino cheese and Parmigiano Reggiano (nowadays also made with stainless steel).

Acidification of the curd and resting phases: A wooden box called *mastredda* for Ragusano PDO, *tavoliere* for Tuma persa cheese, wood containers for Castelmagno PDO, *cisca* for Provola dei Nebrodi and Pecorino Siciliano PDO and trellis of canes *cannara* for Caciocavallo Palermitano. Wooden tables are also used for holding (drying and to form the rind) the wheels of Parmigiano Reggiano and Grana Padano.

Stretching the curd: Round wooden containers are used for the stretching phase of Caciocavallo podolico, Ragusano PDO (called *staccio*) and Caciocavallo Palermitano (called *piddiaturi*). To stretch the curd, different wooden sticks, given different names according to the local culture, are used.

Moulding/shaping: Wrapped in a vegetable canvas tissue, the wheels of Castelmagno PDO are moulded on wooden *fascelle*; wrapped in linen cloth, the wheels of Parmigiano Reggiano and Grana Padano, Asiago, Swiss cheeses are moulded in wood hoops *fascere*, called *garbua* for the Sicilian Maiorchino cheese. Rush baskets are used for the Pecorino Siciliano PDO. Baskets made from a plant that grows in marshes called *sklinitzi* are used to drain Halloumi or Anari cheeses in Cyprus.

Maturation: Wooden shelves are used worldwide to ripen the cheeses, and we give a few examples: Italian Asiago PDO, Parmigiano Reggiano PDO, Grana Padano PDO Castelmagno PDO, Tuma persa, Maiorchino, Pecorino toscano PDO, Fiore sardo PDO and Pecorino siciliano PDO; French Cantal PDO, Beaufort PDO, Sales PDO and Reblochon de Savoie PDO; Swiss Emmentaler PDO and Le Gruyere PDO; Brazilian Minas and Turkey Kulek cheese.

Other natural materials are also used to age cheeses, such as a wooden beam (horseback) on which cheeses are hung in pairs with ropes called *liama* or strings of *Cannu*, of *Zammarra* or cotton for Ragusano PDO, or ropes for Provolone valdostana, Caciocavallo podolico, provole and so on.

Bags of cloth filled with cheeses of Formaggio di Fossa PDO were laid in an ancient pit (flask-shaped and filled with straw), and carved into the sandstone rock for a depth of about 3 m.

Direct packaging: In France, the sale of cheese in wooden containers such as Vacherin Mont-d'Or, Brie, Camembert and many others is widespread.

During the last decades, scientists started to study the complexity and the scientific value of using traditional equipment for cheesemaking and ripening of products. The traditional equipment (from wood or other natural material) cannot be considered simple containers, or insignificant tools with mechanical functions, and therefore easy to replace. It is equally true that not all traditional equipments have the same importance in cheesemaking and ripening. The introduction of small technologies, compatible with their limited economic resources, of most producers of less favoured areas, it is not only possible but also necessary, but only after learning what needs to be known before the change is implemented. Wooden equipment is an example of a great need for scientific knowledge. The complexity of biofilms formed on the wood surface, with hundreds of species, clones and strains, is an excellent example of biodiversity (Licitra et al., 2007; Lortal et al., 2009; Mariani et al., 2007). Wood therefore cannot be replaced by metals or plastics materials, without providing any demonstration of obvious risk or imposing a law, by simply stating that appropriate use of starter cultures (with a few microorganisms) on pasteurised milk solves the problems. Fortunately, even in the most advanced societies, consumers are fully aware that traditional production imparts a more complex flavour profile, more pronounced sensory and health properties, in summary a better quality compared to mass products. In agreement with other studies, there is evidence that European consumers may trade off some degree of inconvenience in the purchase, expensiveness and preparation of traditional food products in order to enjoy the specific taste, quality, appearance, nutritional value, healthfulness and safety (Kupiec & Revell, 1998).

Of particular interest is the way in which the wood, or any other natural material, in contact with the milk or the cheese interacts with it and influences the quality of the final products. The determining factors are the type of contact (solid vs. liquid, or solid vs. solid or semi-solid), the contact time, the pH, the ambient humidity and nutrient availability. From wood vat biofilm to milk (solid vs. liquid), the transfer of microorganisms takes place very quickly, in just a few minutes, for example, and can include heating steps (Licitra et al., 2007). The interaction between wooden shelves and cheese is between a solid versus a semi-solid structure, takes a longer time (weeks or months) and occurs in cold and humid conditions in ripening cellars (Lortal, Licitra & Valence, 2014). Besides the wooden casks that rapidly transfer microorganisms to milk, much of the equipment mentioned earlier have a solid versus solid contact, and in general the contact time is short, a few hours or at most a few days (acidification in a wooden *tavoliere*, or in the phase of moulding in a wooden *fascere*), and therefore it is reasonable to hypothesise that they are important compared to vats and shelves. Mainly in the ripening stages, where cheeses are placed on wooden shelves, the conditions for effective interaction between wood and cheese are created, where the contact is extended for weeks and months, with stabilised pH and ambient humidity. Wood spruce is most frequently used for dairy products, in particular, for shelves, and Douglas fir and chestnut for vats. It is therefore reasonable to think that the most important traditional equipment is the wooden tinas and shelves, and their role in cheesemaking has been scientifically studied. It would however be desirable to further explore other traditional materials of minor equipment, for example, bamboo, bulrush and textiles.

7.2.1 Wood Characteristics

Wood properties vary according to the species, growth conditions and moisture content. The fundamental characteristics that distinguish wooden material from other materials are mainly heterogeneity, hygroscopicity, anisotropy, elasticity, impregnability and acidity. *Anisotropic* means that its physical properties vary according to the orientation of the fibres. *Hygroscopic* means that water may be bound in the wood cells by either molecular or capillary forces. Electron microscopy reveals a very specific structure with tubular fibres that can be easily seen in a tangential cut (Aviat et al., 2016).

The physical properties of wood (for example, mechanical strength, elasticity and thermal conductivity) are strongly related to the water content below the fibre saturation point (FSP). The FSP varies from species to species, but is around 28% to 30% water content (weight of water/weight of dry wood), but hardly above the FSP (Monteiro, 2014). Dry wood, as used in most technical applications, has a moisture content below 19%, while green wood contains from 60% to 200% water (Greer & Pamberton, 2008). Thus, water in dry wood must be considered to be bound water and is adsorbed in amorphous regions of the cell walls consisting of cellulose (40% to 50%), hemicellulose (10% to 30%) and lignin (15% to 30%). A considerable swelling of these amorphous regions can be noticed during uptake of water (Engelund et al., 2013). The relationship between the water content the equilibrium relative humidity is given by the sorption isotherm. A typical sorption isotherm shows a water content of 7% at 30% RH, 10% at 60% RH, and 14% at 80% RH (TIS, 2015). Thus, wood can take up a considerable amount of water and release it again, depending on the RH of the environment. The uptake of water causes a swelling of the wood structure due to water adsorption, while the removal of water leads to shrinking. This macroscopic shrinkage has to be taken into account whenever wood is exposed to variable humidity.

Wood contains in its porous structure a number of free low-molecular-weight compounds, called extractables. They include both volatile and non-volatile organic compounds. Although volatile compounds represent only a small percentage of wood extractables, they influence wood acidity (Stevanovic & Perrin, 2009), hygroscopicity, colour (Amusant, Morett & Richard, 2007; Gierlinger, Jacques & Grabner, 2004), odour, mechanical properties and also the natural durability of wooden material (Aloui et al., 2004; Schultz & Nicholas, 2000).

Because of its irregular surface (crevices, cracks, etc.) and its high porosity, wood is not theoretically easy to clean, at least according to the hygiene standards imposed by regulations adopted by various states in different continents. In the following parts of this chapter, we will review scientific research showing that wood offers greater – or at least not lower than – cleaning guarantees, plastic and stainless steel. Even though wood has never been documented to be involved in any foodborne disease outbreak, the question of its safety must still be raised (Lortal, Licitra & Valence, 2014).

7.3 Biofilms of Wooden Vats

Wood in contact with raw milk and cheeses is rapidly covered by a microbial biofilm. Wooden vats have been studied for Italian Ragusano PDO (Licitra et al., 2007; Lortal et al., 2009), Vastedda della valle del Belice PDO (Scatassa et al., 2015), Caciocavallo Palermitano (Di Grigoli et al., 2015; Gaglio et al., 2015; Scatassa et al., 2015; Settanni et al., 2012) and for the French Cantal and Salers cheeses (Didienne et al., 2012; Richard, 1997). In all these studies, the researchers showed a high biodiversity of microorganisms present in biofilms not only between the different types of cheeses studied, but also within the same cheese variety, between different wooden vats (tina or gerle) used in different cheese plants. The biodiversity confirms and demonstrates the strong characterisation of the territorial origin of the products.

On the wooden vat called the *tina* of Ragusano PDO, the microscopic observations of the biofilm from 15 tinas (Douglas fir) revealed a continuous layer of microorganisms entrapped in a thick polysaccharidic matrix. Molecular characterisation performed on five different tinas demonstrated the predominance of LAB, in particular *S. thermophilus,* and the presence of thermophilic lactobacilli, lactococci and a few high-GC-content bacteria, such as coryneform bacteria (Licitra et al., 2007). Variable but usually very low levels of moulds and yeasts were detected. The microbial profiles of the tina had common and specific features depending on the farm; they exhibited 2 to 10 co-dominant species in all cases, thus representing a valuable source of biodiversity. More than 200 clones of *S. thermophilus* were isolated from four tinas, and after identification by 16S sequencing, they were analysed by pulsed-field gel electrophoresis (PFGE) and multilocus sequence typing (MLST) in order to assess the number of strains. Several strains were shown to cohabit each tina, and seven new sequence types were found. By comparing these new strains isolated from tinas to the MLST profile of a collection of 160 other strains of *S. thermophilus,* it was found that the Sicilian strains formed a completely separate cluster (Valence et al., 2016) and were thus unique. This might suggest a selection pressure exerted by the biology within the wooden biofilm. Regarding the contribution of the strains to the ripening, the PFGE profiles of several clones of *S. thermophilus* isolated from Ragusano cheese during ripening were identical to those of the corresponding tina. The tina's biofilm was also shown to be a very efficient delivery system for LAB, releasing in a few minutes 10^5 to 10^6 CFU of LAB per ml of milk poured into the vat (Lortal et al., 2009). Lactic acid is thus produced by a combination of the natural raw milk ecosystem and its inoculation by the lactic flora from the tina biofilm.

Settanni et al. (2012) studied the influence of the wooden equipment used for traditional cheese manufacturing from raw milk on the variations of the physico-chemical characteristics and microbial populations during the ripening of Caciocavallo Palermitano, that is, raw cow milk pasta filata cheese produced in Sicily. The wooden vat (tina) was found to be a reservoir of LAB, while undesirable (spoilage and/or pathogenic) microorganisms were not hosted or were present at very low levels. LAB, especially thermophilic cocci, dominated the whole cheesemaking process of all productions. Undesired microorganisms decreased in number or disappeared during the transformation, particularly after curd stretching. LAB were isolated from the wooden vat surface and subjected to phenotypic and genetic characterisation and identification. *S. thermophilus* dominated the microbial community of the wooden vat. Fourteen other LAB species belonging to six genera (*Enterococcus, Lactobacillus, Lactococcus, Leuconostoc, Streptococcus* and *Weissella*) were also detected. All *S. thermophilus* isolates were genetically differentiated, and a consortium of four strains persisted during the whole traditional production process. As confirmed by the pH and the total acidity after the acidification step, indigenous *S. thermophilus* strains acted as a mixed starter culture. Moreover, Scatassa et al. (2015), studying the Caciocavallo Palermitano wooden vats, isolated LAB with the predominance of *Lb. casei, E. faecium, Lb. rhamnosus, S. thermophilus* and *P. acidilactici*, whereas in Ragusano cheeses was detected a predominance of *S. thermophilus* and no *E. faecium*. In another study on wooden vats used to produce Caciocavallo Palermitano cheese (Di Grigoli et al., 2015), the persistence of LAB from the wooden vat during ripening was evaluated by a direct comparison of the polymorphic profiles. Three strains belonging to the species *Enterococcus faecalis, Enterococcus casseliflavus* and *Enterococcus gallinarum* were found to be present during cheese maturation only in the traditional samples. In particular, *E. faecalis* FMA288 was found to dominate the enterococcal population at the end of the ripening period, evidencing the defining role of the wooden vat in the modification of LAB composition during Caciocavallo Palermitano cheese ripening.

The LAB predominant species isolated from the wooden vats utilised for the production of the Vastedda della valle del Belice PDO (fresh pasta filata cheese from raw sheep's milk produced in Sicily, Italy) were *E. faecium* and *Lb. casei* (Scatassa et al., 2015).

Cantal and Salers are PDO cheeses produced in France. The raw milk is placed directly into a traditional wooden vat called the gerle. This cheese is made without lactic starters, and the use of the wooden gerle is mandatory in its production regulation. It is a cylindrical or conical wooden vat made of chestnut wood with a capacity of 100 to 1000 L (Lortal, Licitra & Valence, 2014). The presence of yeasts and bacteria within the biofilm of wooden gerles was first observed by Richard (1997), using scanning electron microscopy. The biofilm microbial composition was then extensively explored in different gerles used for "Salers" production from 10 different farms, using several samples per gerle at four different periods of the year (Didienne et al., 2012). In contrast to that of the tina, the gerle biofilm is dominated by lactobacilli (4 to 6 log cm^{-2}), leuconostocs (1.4 to 5.2 log cm^{-2}), gram-negative bacteria (data not published), yeasts (3 to 5.5 log cm^{-2}) and moulds (1.7 to 4.5 log cm^{-2}), which is due to a different cheese technology. Once more, a large biodiversity in the biofilm composition was observed among gerles and was correlated with various environmental and production processes, including the sanitation practices of the equipment after use, of the different producers. As with tina wooden vats, the gerles were shown to very efficiently inoculate desirable lactic acid and ripening bacteria into milk, contributing to acidification and the final cheese character, and thus are safe for cheese ripening (Lortal, Licitra & Valence, 2014).

7.4 Wooden Shelves

The amount of cheeses ripened on wood shelves is estimated to be greater than 350,000 tonnes per year in France, about 300,000 tonnes per year in Italy, 100,000 tonnes per year in Switzerland and considerable amounts in Spain, Greece and Portugal, especially in PDO and PGI productions.

Wood, owing to its intrinsic hygroscopic properties, can lose or retain humidity depending on the temperature and ambient humidity. Cheesemakers are very careful in choosing the wooden shelves; wood must be correctly dried (15% to 18% humidity), which might require 3 to 5 months when this is done outside. Drying under vacuum is also possible and is much faster (a few days). No chemical treatments should have been applied to wood. A shelf which is too humid favours mould defects on the surface of cheese, and sometimes *Pseudomonas fluorescens.* If it is too dry, it favours the development of thick, strong rinds and red defects (Serratia). By using dry shelves with various levels of hygrometry (10% to 19% humidity), cheesemakers modulate with an ancestral empirical knowledge the kinetics of cheese drying, which has a strong influence on the correct setup of the rind and its microbial ecology. As mentioned before, wooden shelves have two different roles: one related to the biofilm's microbial ecology of the cheese surface and the other related to hydric exchange between cheese surfaces and cellar air humidity. Furthermore, most likely, the water flux from the cheese to the shelves also carries with it nutrients, microorganisms and probably antimicrobial components (Lortal, Licitra & Valence, 2014).

Several studies have been carried out on the antimicrobial properties of wood compounds. The compounds most frequently studied belong to a small number of classes: phenols, lignans, tannins, stilbenes, flavonoids and terpenoids (Pearce, 1996). Their effects are described as antimicrobial against bacteria, but it is not clear whether these are bacteriostatic or bactericidal, usually depending on the concentration of the antimicrobial component and the microorganism strain (Mourey & Canillac, 2002). Dumont et al. (1974) and Bosset et al. (1997), demonstrated the presence, in cheeses, of specific volatile compounds, in particular terpenic molecules, after contact with wood. Mariani et al. (2007) characterised the wooden shelves used for the ripening of Reblochon de Savoie PDO. Reblochon is a cow's milk smear-ripened cheese, mostly made from raw milk. After examining the biofilm composition of 50 shelves of three ages during two seasons, which were used to ripen cheeses from eight farms, the authors concluded that the biofilm was mainly composed of micrococci, corynebacteria, yeasts and moulds but also contained leuconostocs, lactobacilli, enterococci, coagulase-negative staphylococci and pseudomonas. This composition was constant over time and was similar to that of the cheese surface, supporting the hypothesis that wooden shelves represent an essential source of microbial flora for the rind of cheeses, as claimed by many cheesemakers. Schuler (1994) demonstrated the contribution of different wooden shelves to the development of the cheese rind microflora and to the final quality of a semi-hard Swiss cheese. The colonisation of the wood was estimated to occur in the first 2 mm of depth. The wooden shelves had a neutral pH (7 to 8.3), a high a_w and a low salt content (14 mg/cm^2). Regarding the wood vats, Mallia et al. (2005) presented results on the effect of the vat material (wooden, copper and stainless steel) the aroma profile of Piacentinu Ennese PDO cheese made from raw sheep milk. Solid-phase microextraction (SPME) coupled to gas chromatography/mass spectrometry/olfactometry (GC/MS/O) was used to determine volatile profile of Piacentinu cheese, produced by traditional (wood/copper) and stainless steel tools. Cheeses made by wooden tools showed a richer aroma profile, containing 2,3-butanedione, 3-hydroxy-2-butanone, 1,2-dimethyl hydrazine, limonene, copaene, a-caryophyllene, 3-methyl butanal, 2-furanmethanol, 2-hexene, dimethyl sulfone, and 2,6 dime-3-ethyl pyrazine, that were not found in copper-cheese and steel-cheese. The authors concluded that the natural microflora of the wooden vat clearly influenced and enriched the volatile profile of these cheeses.

Dervisoglu and Yazici (2001) studied the production process of Kulek cheese. Kulek, a ripened acid-curd cheese with added rennet, is one of the most important cheeses consumed in Turkey. For three months, they analysed the effect of packaging materials and examined microbiological changes during ripening. They compared the ripening process in wooden containers constructed from dry poplar boards, 1.5 cm thick, and plastic containers, 3 mm thick, purchased from a local store. Cheese samples on wood had more proteolytic microorganisms and psychotropic bacteria than cheese samples on plastic ($P < 0.05$). In general, the microbial results indicated that wooden material had a better permeability to air and moisture that enhanced microbial growth. Thus, the authors recommended using wooden packaging for ripening Kulek cheese to obtain better results.

7.5 Legislation Concerning Wood in Contact with Milk or Cheeses

In Europe, the first effort to harmonise the member rules about materials in contact with food was Council Directive 76/893/EEC of 23 November 1976 (EC, 1976), but wood was not cited. It was then replaced by directive 89/109/CEE of 21 December 1988 (EC, 1988), which cites the case of wood for the first time. Likewise, decision 96/536/EC of 29 July 1996 (EC, 1996), regarding milk-based products with traditional characteristics, mentioned that the instruments and equipment, whatever their nature, could be used if constantly maintained in a satisfactory state of cleanliness and regularly cleaned and disinfected. Finally, it was replaced by regulation 1935/2004 (EC, 2004a), in which wood is listed with other materials for which specific measures can be adopted, but member states can maintain or adapt national decisions.

In reality, it is not a generic derogation: the member states should identify which requirements of this directive are likely to affect the manufacture of milk-based products with traditional characteristics, by levelling off typical flavours, aromas and smells conferred by natural dairy microflora. The most important cases are related to traditional tools in cheesemaking and aging, and aging rooms. Lastly, the recent and very complete regulations 852/2004 (EC, 2004b) and 853/2004 (EC, 2004c), called the 'Hygiene Package' in Europe, defined specific hygienic rules for the production of food from animals, in particular, from milk (section IX: raw milk and dairy products). It is clearly stated that surfaces in contact with milk (tools, vats, etc.) should be easy to clean and well maintained.

The use of wood is mandatory in several PDO cheeses, such as the French Comté and Beaufort, and the Italian Ragusano and Vastedda valle del Belice. Wood is widely used in Italy and Spain. In non-European countries, the situation is variable: wood is allowed, for example, in Central Asia. Similar to Europe, the US regulations specify that equipment and utensils used to process milk and manufacture dairy products must be constructed in a way that allows for easy removal for cleaning and sanitising (Code of Federal Regulations, 2009). Use of wood in cheesemaking is nevertheless allowed, but this varies from state to state (Paxson, 2010). The FDA's current regulations state that utensils and other surfaces that contact food must be 'adequately cleanable' and properly maintained. In very recent years, in the US, some controversies between the US artisan cheese community and the FDA occurred regarding the information declaring that the FDA has issued an executive decree to end the long-standing practices in the cheesemaking industries of using wooden boards to age cheeses. After a strong reaction from the US artisan cheese community (Mc Neal, 2014) warning of a potentially catastrophic disruption in the market for artisanal, non-processed cheese, and also considering the effect of such a declaration on the importation of European and Canadian cheeses, which are mostly aged on wooden boards (Parmigiano Reggiano, Grana Padano, Pecorino Romano, Asiago, Comté, Beaufort, Reblochon, Munster, Cantal, to name a few), the FDA declared: 'We have not and are not prohibiting or banning the long-standing practice of using wood shelving in artisanal cheeses. Nor does the FDA and the Food Safety Modernization Act (FSMA) require any such action. Reports of the contrary are not accurate' (www.welch.house.gov). US artisanal cheeses producers and their cheese lovers are safe. This action will remain as a worldwide reference in the case of governmental or EU decisions that do not respect the history and centuries-old culture of the rural world.

7.6 Cleaning Systems

Aviat et al. (2016) in his recent review stated that, to date, wood in contact with food has not been found responsible for any foodborne outbreak, and yet wood tends to be considered less hygienic than other materials in contact with food, such as plastic, stainless steel and glass, all used in the food industry. This concept stems from the fact that wood is known to be a porous material, which is theoretically difficult to clean and decontaminate as required by the current legislation (EC, 2004a).

Since the 1990s, research on wood in contact with food has led to a partial reversal of this image. Since then, a number of scientific studies on wood and its microbiological status have been performed to investigate the impact of cleaning, disinfection, moisture content and wood properties on the survival and transfer of microorganisms. The procedures of cleaning after use usually involve a brushing step with water, or whey, and an adequate period of drying. A French survey in 2000 (ACTIA, 2000) revealed that brushing the shelves with water (cold or <35°C) and then subjecting the wood to water at high pressure at 85°C was the most frequently used method. By these two successive steps, a reduction of more than 5 logs of the total microbial flora can be reached (Lortal, Licitra & Valence, 2014). In addition, in Sicily, in order to better assess the method of cleaning and maintaining active tinas, a questionnaire was prepared for 15 farmers. The findings illustrated that (1) the farmers clean the tina daily after the cheesemaking, (2) a majority of farmers (11/15) brush the tina and clean it with hot water (55°C), (c) a minority (4/11) with hot water only without brushing and (d) all let the tina dry upside down until the next cheesemaking. The surface pH just after washing the tina was acid, between 4.5 and 5.0; the lowest values were in the bottom. This low surface pH is likely related to the local lactic acid production of LAB. No pathogens were detected within the biofilms of the 15 tinas (Lortal et al. 2009). The study of Scatassa et al. (2015) on 20 wooden vats (13 made of chestnut and 7 of Douglas fir) used for the production of Caciocavallo Palermitano cheese and the Vastedda della valle del Belice cheeses report that Sicilian cheese makers have traditionally managed wooden vat hygiene through washing of this container with the hot deproteinised whey resulting from the production of Ricotta cheese and/or hot water, sometimes with a careful brushing, leaving the wooden vat full with the whey for 12 h. These operations tend to empirically regulate, also in accordance with the environmental temperatures, the so-called *tina acidity*, influencing the microbial composition on the vat inner surface, which is in direct contact with the milk. The efficacy of the sanitation procedures applied during cheese production was demonstrated, as no indicator microorganisms (coliforms and *Escherichia coli*) or pathogens, such as *L. monocytogenes*, could be detected.

Besides the use of wood in the various production systems (cheese and wine making) it is important to report the results of the major scientific works related to the ability to clean and decontaminate cutting boars worldwide used in any family kitchen, restaurants, counters retailers of foods. In the 1990s, wooden cutting boards were suspected of being harder to clean because of the porosity of wooden material. Ak et al. (1994a) showed that wooden cutting boards covered with a multilayer of food residue did not absorb bacteria as quickly as new wooden cutting boards. Ak et al. (1994b) compared the cleaning and decontamination of plastic and wooden cutting boards. The objective was to prevent cross-contamination at home, and also in restaurants, retail butcher shops and the meat industry. The authors used new and used plastic and wooden boards cut into 5-cm-square blocks. Wooden cutting boards were made of ash, basswood, beech, birch, butternut, cherry, hard maple, oak and American black walnut. Plastic cutting boards were made of polyacrylic, polyethylene, foamed polypropylene, polystyrene and hard rubber. Ak et al. (1994b) concluded that with reasonable cleaning effort, new or used wooden cutting boards can be used safely in home kitchens. The authors concluded that the nature of the wood was responsible for the lethality of the bacteria tested in this study, whereas the nature of the plastic allowed bacteria to survive or even grow. A second hypothesis was that the wood had antimicrobial properties, whereas the plastic cutting board did not. Ak et al. (1994b) finally stated that wooden cutting boards are not associated with a high risk of cross-contamination of food.

An explanation of the nature of cutting boards and its effects on the survival of microorganisms comes from the study carried out by Chiu (2006) on the survival of *Vibrio parahaemolyticus* on different surfaces; rough and porous materials (bamboo and wood) were compared with smooth materials (plastic, stainless steel and glazed ceramic tile). *V. para haemolyticus* was shown to survive better on stainless steel, plastic and ceramic tile, representing smooth surfaces, and not so well on rough and porous surfaces such as bamboo and wood. The authors hypothesised that this was probably because smooth surfaces could maintain high moisture content and favour the survival of *V. parahaemolyticus.* In fact, porous surfaces may trap liquid and make it unavailable, so the moisture content on the wooden surface decreases and becomes an unfavourable environment for the survival of microorganisms.

The colonization of the wood was estimated to occur in the first 2 mm of depth. The wooden shelves had a neutral pH (7 to 8.3), a high a_w, and a low salt content (14 mg/cm^2). Examination by infrared spectroscopy of the zones of wood in contact with cheeses versus zones without contact (blank) revealed obvious spectral differences (Oulahal et al., 2009). In contrast, the spectra of the blanks and cheese zones were similar after cleaning, demonstrating the efficiency of the cleaning procedure.

With reference to what was the best way to clean and decontaminate the cutting boards, another study by Miller, Brown and Call (1996) compared the recoveries of beef bacterial microflora from plastic (polyethylene) and wooden (maple and/or beech laminated along the longitudinal direction) cutting boards. The major result was that no statistical difference ($P > 0.05$) was found between the cleaning step with water or chemical cleaners on wooden and plastic boards.

The reported results confirmed the traditional worldwide practice of cheese makers to use only hot water and brushing to clean wooden equipment. The very interesting and updated review of Aviat et al. (2016), based on 86 international references, demonstrates that the porous nature of wood, especially when compared with smooth surfaces, is not responsible for the limited hygiene of the material used in the food industry and that it may even be an advantage for its microbiological status. The rough or porous surface of wood is also an advantage for controlling the level of surface moisture. In addition, wood has the particular characteristic of producing antimicrobial components that are able to inhibit or limit the growth of pathogenic microorganisms. Indeed, there is a great deal of evidence that porosity is an advantage for the microbiological status of wood in contact with food, even when processing food. The study concluded that wood represents ecological ideas that are attractive to consumers (FEDEMCO & Partner España S.A., 2002; Gigon & Martin, 2006), and these have resulted in a new interest in wood for use in food packaging. It is clear that wooden packaging and wooden tool surfaces contribute beneficially to the final quality, safety and character of many food products.

7.7 Safety Assessment

For centuries, wood has been considered a natural package for the ripening of various food products, especially cheese (Aviat et al., 2016). New food safety regulations are contributing to the substitution of wood by other materials, like polypropylene, high-density polyethylene or stainless steel (Galinari et al., 2014; Scatassa et al., 2015). However, the replacement of wooden utensils by other materials changes the characteristics of cheese, affecting the traditional flavour and texture (Galinari et al., 2014). The presence of pathogens (Listeria, Salmonella, *E. coli O157* and *Staphylococcus aureus*) was analysed in more than 15 tinas (wooden vats) which came from different farms in the Ragusa region of Sicily. Except for very rare and very low levels of contamination with *S. aureus* (only seen after enrichment by the Bax system), none of these pathogens was detected within the biofilm (Lortal et al., 2009). Many hypotheses can explain this resistance to the establishment of pathogens on the wood. First, this observation is in agreement with findings for many other positive biofilms: the local pH of the wooden vat is below 5.0; the temperature cycle for Ragusano cheesemaking includes a heating step (never exceeding 45°C); nutritional competition with the positive microflora can occur; and the predominant species, *S. thermophilus*, can also produce bacteriocins (Fontaine & Hols, 2007). Lastly, brushing and washing can also limit the potential adhesion of milk pathogens on the surface of the biofilm. All these factors likely combine to form an efficient barrier to pathogens. The wood itself can release antimicrobial compounds towards pathogens, or it can have a bactericidal effect due to some of its physical properties (Miller, Brown & Call, 1996; Schulz, 1995). The complete absence of pathogens was also found in 10 gerles examined in Cantal cheesemaking. When raw milk before contact with the wooden vats was artificially contaminated with high levels of Listeria organisms and Staphylococcus (Didienne et al., 2012), pathogens were still unable to establish themselves within the biofilm. In a study conducted in the technical centre Actalia (ACTIA, 2000), samples (brushings of 25 cm^2) obtained from 90 different shelves were examined for the presence of pathogens. Neither Listeria nor Salmonella spp. were isolated, while *S. aureus* and *E. coli* were recovered <10 CFU/cm^2. On the other hand, in a study examining the distribution of *L. monocytogenes* within a cheese plant (Menendez et al., 1997), one sample from wooden shelves was positive (of five shelf samples examined). In contrast, in a Brazilian dairy plant manufacturing fresh cheeses (Silva et al., 2003) none of the five shelves explored was contaminated by *L. monocytogenes*. In another study concerning the survival of *L. monocytogenes* following cleaning and sanitation of wooden shelves, Zangerl et al. (2009) concluded that 'there is no reason to replace wood employed in cheese ripening processes with other materials' as long as cleaning procedures are appropriately followed. Mariani et al. (2011) carried out a study to characterise the development of *L. monocytogenes* on wooden shelves used for cheese ripening. The wooden shelves for the production process of the PDO 'Reblochon de savoie' were cut lengthwise in spruce wood (*Picea abies*). In this study, the traditional ripener collected cheeses from different farmhouse origins. The authors compared inoculations on native or autoclaved wooden samples after cleaning–drying, and also on native wooden samples before cleaning–drying after two incubations. Two strains of *L. monocytogenes* were selected according to their behaviour after inoculation on wooden shelves: the most and the least resistant. In the presence of a native microbial flora on the shelves, deposited populations of *L. monocytogenes* remained constant or even decreased by up to 2 log (CFU/cm^2) after 12 days of incubation at 15°C in all tested conditions. By contrast, *L. monocytogenes* populations increased by up to 4 log (CFU/cm^2) when the resident biofilm was thermally inactivated (autoclaved), suggesting a microbial origin of the observed inhibitory effect. Mariani et al. (2011) concluded that the resident microbial biofilm living on wooden ripening shelves displayed a stable anti-Listeria effect depending on the experimental ripening conditions. All together, results from that study suggest that biocontrol of pathogen multiplication on wooden shelves by resident biofilms should be considered for the microbiological safety of traditional ripened cheeses.

The correct maintenance of wooden vats, as part of good manufacturing practices of the Caciocavallo Palermitano (CP) and the Vastedda della valle del Belìce PDO (VB) cheese production, promotes the selection of a microbial flora able to play an active role in the achievement of the food safety objectives through the biocompetitive activity of LAB and the inhibitory activity against pathogenic bacteria, particularly *L. monocytogenes.* The percentage of lactic acid strains able to inhibit the growth of *L. monocytogenes* ATCC 7644 was 66.7% for the LAB isolated from VB vats and 55.7% for those isolated from CP vats. The presence of more than 50% of the LAB isolates with inhibitory activity versus *L. monocytogenes* ATCC 7644 constitutes an additional positive feature (Scatassa et al., 2015).

7.8 Conclusions

As wood is obviously a natural material, very few questions about its safety were addressed by scientists until 1970, when increased regulatory activity placed general hygienic pressure on food production. Even though no foodborne disease outbreak has ever been attributed to the use of wood, wood was suspect because of its porosity, and without any demonstration of obvious risk, it was replaced as often as possible by other materials (which ironically have been documented to carry risks). However, from the literature surveyed here, it seems clear that wooden vats and shelves act as a reservoir of microbial biodiversity contributing to the final quality, safety and character of dairy products. Moreover, the natural biofilms which form on wooden surfaces are safe and able to inhibit and limit pathogen implantation with mechanisms that still need to be further explored. Wood, as a tool to regulate cheese and cellar humidity, has also been proved to be difficult to replace by any other synthetic materials. Its role is crucial in the hydric balance and drying of cheese, which are subsequently crucial for the development of the expected microbial ecosystem on the rind; this ability has never been equalled by any other kind of shelving material. All these results would support the use of wood without restriction, provided that appropriate cleaning and drying procedures are utilised. General guidelines for wood management in the traditional dairy sector are missing and would be of great value to the future of artisanal cheese production. Many scientific questions remain and deserve exploration in order to understand the sustainability and microbial dynamic of this natural inoculating system as well as new tools for physic-chemical microexploration within the depths of the wood interior. In conclusion, this fascinating natural inoculating system, providing obvious microbial diversity in dairy traditional products, deserves much more investigation before being abandoned because of excessive and scientifically unjustified hygienic considerations. Globalisation represents a real risk of eliminating traditional systems (Licitra, 2010), and wooden tools are part of these systems all over the world. It is the responsibility of scientists to explore these systems and provide data, thus preventing inappropriate regulatory decisions. Wood should be considered in its natural technological environment as an essential component of traditional practices which provide the cheesemaking ecosystem with its original complexity and balance.

References

ACTIA. (2000). *Evaluation et maîtrise du risque microbiologique dans l'utilisation du bois pour l'affinage des fromages*. ACTIA, Paris, France.

Ak, N., Cliver, D. & Kaspar, C. (1994a). Cutting boards of plastic and wood contaminated experimentally with bacteria. *Journal of Food Protection*, 57, 16–22.

Ak, N., Cliver, D. & Kaspar, C. (1994b). Decontamination of plastic and wooden cutting boards for kitchen use. *Journal of Food Protection*, 57, 23–30.

Aloui, F., Ayadi, N., Charrier, F. & Charrier, B. (2004). Durability of European oak (*Quercus petraea* and *Quercus robur*) against white rot fungi (*Coriolus versicolor*): Relations with phenol extractives. *Holz als Roh- und Werkstoff*, 62 (4), 286–290.

Amusant, N., Morett, C. & Richard, B. (2007). Chemical compounds from *Eperua falcata* and *Eperua grandiflora* heartwood and their biological activities against wood destroying fungus (*Coriolus versicolor*). *Holz als Roh- und Werkstoff*, 65, 23–28.

Aviat, F., Gerhards, C., Rodriguez-Jerez, J. J., Michel, V., Le Bayon, I., Ismail, R. & Michel Federighi, M. (2016). Microbial safety of wood in contact with food: A review. Institute of Food Technologists. *Comprehensive Reviews in Food Science and Food Safety*, Vol. 00, 2016.

Bosset, J. O., Butikofer, U., Berger, T. & Gauch, R. (1997). Etude des composes volatils du Vacherin fribourgeois et du Vacherin Mont-d'Or. *Mitteilungen aus dem Gebiete der Lebensmittel-untersuchung un Hygiene*, 88, 233–258.

Chiu, T. H. (2006). Efficacy of electrolysed oxidizing water in inactivating *Vibrio parahaemolyticus* on kitchen cutting boards and food contact surfaces. *Letters in Applied Microbiology*, 43 (6), 666–672.

Code of Federal Regulations. (2009). *7 CFR 58 Part 58—Grading and inspection, general specifications for approved plants and standards for grades of dairy products, 58*.128. Equipment and utensils. 7 U.S.C. 1621–1627.

Dervisoglu, M. & Yazici, F. (2001). Ripening changes of Kulek cheese in wooden and plastic containers. *Journal of Food Engineering*, 48 (3), 243–249.

Didienne, R., Defargues, C., Callon, C., Meylheuc, T., Hulin, S. & Montel, M. C. (2012). Characteristics of microbial biofilm on wooden vats ('gerles') in PDO Salers cheese. *International Journal of Food Microbiology*, 156 (2), 91–101.

Di Grigoli, A., Francesca, N., Gaglio, R., Guarrasi, V., Moschetti, M., Scatassa, M. L. & Settanni, L. (2015). The influence of the wooden equipment employed for cheese manufacture on the characteristics of a traditional stretched cheese during ripening. *Food Microbiology*, 46, 81–91.

Dumont, J. P., Roger, S., Cerf, P. & Adda, J. (1974). Etude de composes volatils neutres presents dans le Vacherin. *Lait*, 54, 243–251.

Engelund, E. T., Thygesen, L. G., Svensson, S. & Hill, C. A. S. (2013). A critical discussion of the physics of wood-water interactions. *Wood Science and Technology*, 47, 141–161.

EC. (1976). Council Directive 76/893/EEC of 23 November 1976 on the approximation of the laws of the Member States relating to materials and articles intended to come into contact with foodstuffs *Official Journal of the European Union*, L 340, 19–24.

EC. (1989). Council Directive 89/109/EEC of 21 December 1988 on the approximation of the laws of the Member States relating to materials and articles intended to come into contact with foodstuffs. *Official Journal of the European Union*, L 040, 38–44.

EC. (1996). Commission Decision 96/536/EC of 29 July 1996 establishing the list of milk-based products in respect of which Member States are authorized to grant individual or general derogations pursuant to Article 8 (2) of Directive 92/46/EEC and the nature of the derogations applicable to the manufacture of such products (Text with EEA relevance). *Official Journal of the European Union*, L 230, 12–15.

EC (2004a). Regulation (EC) No. 1935/2004 of the European Parliament and of the Council of 27October 2004 on materials and articles intended to come into contact with food and repealing Directives 80/590/EEC and 89/109/EEC. *Official Journal of the European Union, L* 338, 4–14.

EC (2004b). Regulation (EC) No 852/2004 of the European Parliament and of the Council of 29 April 2004 on the hygiene of foodstuffs. *Official Journal of the European Union.*

EC (2004c). Regulation (EC) No. 853/2004 of the European Parliament and of the Council of 29 April 2004 laying down specific hygiene rules for the hygiene of foodstuffs. *Official Journal of the European Union*, L 139, 55–205.

FEDEMCO & Partner España, S. A. (2002). Computed assisted telephone consumer interviewing omnibus on Wood Packing, La limpieza de los envases hortofrutícolas preocupa a más del 80% de los consumidores. *Agroenvase*, 26.

Fontaine, L. & Hols, P. (2007). The inhibitory spectrum of thermophilin 9 from Streptococcus thermophilus LMD-9 depends on the production of multiple peptides and the activity of BlpGSt, a thiol-disulfide oxidase. *Applied and Environmental Microbiology*, 74, 1102–1110.

Gaglio, R., Cruciata, M., Di Gerlando, R., Scatassa, M. L., Cardamone, C., Mancuso, I., Sardina, M. T., Moschetti, G., Portolano, B. & Settannia, L. (2015). Microbial activation of wooden vats used for traditional cheese production and evolution of the neo-formed biofilms. *Applied and Environmental Microbiology*. Accepted manuscript posted online 6 November 2015, doi: 10.1128/AEM.02868-15.

Galinari, E., Escarião da Nóbrega, J., de Andrade, N. J. & de Luces Fortes Ferreira, C. L. (2014). Microbiological aspects of the biofilm on wooden utensils used to make a Brazilian artisanal cheese. *Brazilian Journal of Microbiology*, 45 (2), 713–720.

Gierlinger, N., Jacques, D. & Grabner, M. (2004). Colour of larch heartwood and relationships to extractives and brown-rot decay resistance. *Trees*, 18, 102–108.

Gigon, J. & Martin, B. (2006). *Le bois au contact alimentaire: peut-on s'en servir comme outil de communication* ? Report Univ. of Polytech'Lille, France.

Greer, L. & Pamberton, S. (2008). *The Structure and Mechanical Behaviour of Wood*. Univ. of Cambridge DoITPoMS. 2008. Available from: http://www.doitpoms.ac.uk/tlplib/wood/index.php

https://welch.house.gov/search/node/wooden%20boards%in%20cheese%20aging

http://www.treccani.it/enciclopedia/sviluppo-urbano-e-aumento-della-popolazione_(Atlante-Geopolitico)/

IYFF Final Declaration: Feeding the Word, Caring for the Earth http:/www.asiadhrra.org/wordpress/2011/11/20/iyff-final-declaration-feeding...; http://www.agriculturesnetwork.org/resources/extra/news/family-farming-world-conference

Kupiec, B. & Revell, B. (1998). Specialty and artisanal cheese today: The product and the consumer. *British Food Journal*, 100 (5), 236–243.

Licitra, G., Ogier, J. C., Parayre, S., Pediliggieri, C., Carnemolla, T. M., Falentin, H., Madec, M. N., Carpino, S. & Lortal, S. (2007). Variability of bacterial biofilms of the "tina" wood vats used in the Ragusano cheese-making process. *Applied and Environmental Microbiology*, 73, 6980–6987.

Licitra, G. (2010). World wide traditional cheeses: banned for business? *Dairy Science and Technology*, 90, 357–374.

Lortal, S., Di Blasi, A., Madec, M.-N., Pediliggieri, C., Tuminello, L., Tanguy, G., Fauquant, J., Lecuona, Y., Campo, P., Carpino, S. & Licitra, G. (2009). Tina wooden vat biofilm: A safe and highly efficient lactic acid bacteria delivering system in PDO Ragusano cheese making. *International Journal of Food Microbiology*, 132 (1), 1–8.

Lortal, S., Licitra, G. & Valence, F. (2014). Wooden tools: reservoirs of microbial biodiversity in traditional cheesemaking. *Microbiol Spectrum*, p. 420.

Mallia, S., Carpino, S., Corralo, L., Tuminello, L., Gelsonimo, R. & Licitra, G. (2005). Effects on aroma profile of Piacentinu and Ricotta cheese using different tools materials during cheese making. In Spanier, A. M., Shahidi, F., Parliment, T. H., Mussinan, C., Ho, C.-T. & Tratras Contis, E. (eds.), *Food Flavor and Chemistry: Explorations into the 21st Century*. Royal Society of Chemistry, Cambridge, UK, pp. 23–34.

Mariani, C., Briandet, R., Chamba, J.-F., Notz, E., Carnet-Pantiez, A., Eyoug, R. N. & Oulahal, N. (2007). Biofilm ecology of wooden shelves used in ripening the French raw milk smear cheese Reblochon de Savoie. *Journal of Dairy Science*, 90, 1653–1661.

Mariani, C., Oulahal, N., Chamba, J.-F., Dubois-Brissonnet, F., Notz, E. & Briandet, R. (2011). Inhibition of Listeria monocytogenes by resident biofilms present on wooden shelves used for cheese ripening. *Food Control*, 22, 1357–1362.

Mc Neal, G. S. (2014). http://www.forbes.com/sites/gregorymcneal/2014/06/09/fda-may-destroy-american-artisan-cheese-industry/#64d3614c549f

Menendez, S., Godinez, M. R., Rodriguez-Otero, J. L. & Centeno, J. A. (1997). Removal of Listeria spp. in a cheese factory. *Journal of Food Safety*, 17, 133–139.

Miller, A. Brown, T. & Call, J. (1996). Comparison of wooden and poly-ethylene cutting boards: Potential for the attachment and removal of bacteria from ground beef. *Journal of Food Protection*, 59, 854–858.

Monteiro, P. J. M. (2014). *The Structure and Properties of Civil Engineering Materials*. Univ. of California Berkeley, Lecture Notes CE 60. Available from: http://www.ce.berkeley.edu/#paulmont/CE60New/wood.pdf. Accessed 2012 August 17.

Mourey, A. & Canillac, N. (2002). Anti-*Listeria monocytogenes* activity of essential oils components of conifers. *Food Control*, 13 (4), 289–292.

Oulahal, N., Adt, I., Mariani, C., Carnet-Pantiez, A., Notz, E. & Degraeve, P. (2009). Examination of wooden shelves used in the ripening of a raw milk smear cheese by FTIR spectroscopy. *Food Control*, 20, 658–663.

Paxson, H. (2010). Locating value in artisan cheese: Reverse engineering terroir for New-World landscapes. *American Anthropologist*, 112, 444–457. doi:10.1111/j.1548-1433.2010.01251.x.

Pearce, R. (1996). Antimicrobial defences in the wood of living trees. *New Phytologist*, 132 (2), 203–233.

Richard, J. (1997). Utilisation du bois comme matériau au contact des produits laitiers. *Comptes rendus de l'Académie d'agriculture de France*, 83 (5), 27–34.

Salque, M., Bogucki, P. I., Pyzel, J., Sobkowiak-Tabaka, I., Grygiel, R., Szmyt, M. & Evershed, R. P. (2013). Earliest evidence for cheese making in the sixth millennium bc in northern Europe. *Nature*, 493, 522–525.

Settanni, L., Di Grigoli, A., Tornambè, G., Bellina, V., Francesca, N., Moschetti, G. & Bonanno, A. (2012). Persistence of wild Streptococcus thermophilus strains on wooden vat and during the manufacture of a traditional Caciocavallo type cheese. *International Journal of Food Microbiology*, 155, 73–81.

Scatassa, M. L., Cardamone, C., Miraglia, V., Lazzara, F., Fiorenza, G., Macaluso, G., Arcuri, L., Settanni, L. & Mancuso, I. (2015). Characterisation of the microflora contaminating the wooden vats used for traditional Sicilian cheese production. *Italian Journal of Food Safety*, 4 (4509), 36–39.

Schuler, S. (1994). Einfluss der Käseunterlage auf die Schmierebildung und die Qualität von Halbhartkäse. *Schweiz Milchwirtschaftl Forsch*, 23, 73–77.

Schulz, H. (1995). Holz im Kontakt mit Lebensmitteln. Hat Holz anti-bakterielle Eigenschaften? *Holz Zentralbl*, 84, 1395.

Schultz, T. P. & Nicholas, J. J. (2000). Naturally durable heartwood: Evidence for a proposed dual defensive function extractives. *Phytochemistry*, 54, 47–52.

Silva, I. M., Almeida, R. C., Alves, M. A. & Almeida, P. (2003). Occurrence of Listeria spp. in critical control points and the environment of Minas Frescal cheese processing. *International Journal of Food Microbiology*, 81, 241–248.

Stevanovic, T. & Perrin, D. (2009). *Chimie du bois.* In Presses Polytechniques et Universitaire Romandes.

Transport Information Service (TIS). (2015). *Lumber Properties.* Available from: http://www.tis-gdv.de/tis_e/verpack/holz/eigensch/eigensch.htm.

Valence, F., Delorme, C., Chuat, V., Pediliggieri, C., Madec, M.-N., Parayre, S., Carpino, S., Renault, P., Licitra, G. & Lortal, S. (2016). Specific cluster of *S. thermophilus* strains in Tina wooden vats used in Sicilian Ragusano PDO cheese. *Applied and Environmental Microbiology* (submitted).

Zangerl, P., Matlschweiger, C., Dillinger, K. & Eliskases-Lechner, F. (2009). Survival of Listeria monocytogenes after cleaning and sanitation of wooden shelves used for cheese ripening. *European Journal of Wood and Wood Products*, 68, 415–419.

Part II

Introduction

Cheeses from Argentina

Erica R. Hynes[1,2], *Maria Cristina Perotti*[1,2] *and Carina V. Bergamini*[1,2]

[1] *Facultad de Ingeniería Química (Universidad Nacional del Litoral), Santa Fe, Argentina*
[2] *Instituto de Lactología Industrial (Universidad Nacional del Litoral – Consejo Nacional de Investigaciones Científicas y Técnicas), Santa Fe, Argentina*

The dairy industry in Argentina is highly developed and is one of the most modern in Latin America. It is concentrated in the central and east-central parts of the country, in Córdoba, Santa Fe and Buenos Aires provinces.

Argentina is among the top 10 cheese-producing countries in the world and is in the second position in Latin America after Brazil. Cheese is by far the main dairy product as it accounts for approximately 45% of national milk production. Although most of the milk is processed in large- (1.5 million litres per day on average) and medium-scale dairy facilities, numerous small- and micro-scale companies make cheese almost exclusively, which process about 20% of the total milk production. Cheese production has grown significantly in the last decade (from 430,955 tonnes in 2001 to 508,000 tonnes in 2009), mainly due to a sustained increase in consumption. In fact, cheese consumption increased in the last decade, from 8.3 kg per capita in 2003 to 12.4 kg per capita in 2012. Argentina is the country with the highest consumption of cheeses of Latin America. Soft cheeses – Cremoso cheese and others – represent more than half of the total cheese production, followed by the semi-hard cheese varieties (30%) – Pategrás cheese being the more representative; hard cheeses (Reggianito among others) account for 15% of the total production (Capellini, 2011; Ministerio de Agroindustria, 2015).

In recent years, the domestic market has been more attractive than the external market. Cheeses are the third product of export (13% of total exports); only 2% of the total exports is of hard cheeses, while 5% and 6% are of semi-hard and soft cheeses, respectively. The major export markets are Brazil, accounting for 43% of the total exports in 2011, followed by Venezuela, Russia, Mexico, Chile and South Korea. The imports of cheeses remain at a very low level, between 2,000 and 8,000 million tonnes, mainly from Brazil and Uruguay, due to the wide range of high-quality cheese produced in Argentina (Palombo, 2012).

Acknowledgements

The authors thank Sucesores de Alfredo Williner S.A. for providing cheeses for the photographs.

Global Cheesemaking Technology: Cheese Quality and Characteristics, First Edition.
Edited by Photis Papademas and Thomas Bintsis.

No.	Cheese name	Milk used	Category	Section in Part II
1	Cremoso	Cow's	Soft	4.4
2	Reggianito	Cow's	Extra-hard	1.2
3	Pategrás	Cow's	Swiss-type	6.5

References

Capellini, O. R. (2011) Dairy development in Argentina. Food and Agricultural Organization of the United Nations (FAO), Rome. http://www.fao.org/docrep/013/al744e/al744e00.pdf. Accessed: June 2015.

Ministerio de Agroindustria. Presidencia de la Nación (2015). http://www.minagri.gob.ar/site/_subsecretaria_de_lecheria/lecheria/07_Estad%C3%ADsticas/index.php. Accessed: December 2015.

Palombo, A. (2012) Exportaciones de quesos. Salvavidas carioca. http://www.infortambo.com/admin/upload/arch/actualidad%20quesos.pdf. Accessed: September 2015.

Cheeses from Cyprus

Photis Papademas

Department of Agricultural Sciences, Biotechnology and Food Science, Cyprus University of Technology, Limassol, Cyprus

Dairying practices appear to have spread rapidly beyond their initial areas of origin such that by the 8th millennium BC Neolithic migrants from the northern Levantine mainland had transported domestic sheep and goats to Cyprus, where the animals were raised partly for milk production, as inferred from the early culling profiles observed.

The dairy sector (and especially the cheese industry) in Cyprus retains a significant share of the food sector contributing to the island's economy. In 2014, the value of cow's, sheep's and goat's milk production was approximately 127 million euros. The annual cow's milk production of 158,850 tonnes was almost 80% of the total milk production in Cyprus. Sheep's milk production amounted to 23,515 tonnes per year, followed closely by goat's milk production at 22,191 tonnes. Most of the milk is processed in medium- and large-sized dairies, but there are numerous small family-owned dairies that process a considerable amount of sheep's and goat's milk.

The main dairy product of Cyprus is Halloumi (white-brined cheese), with exports rising to 15,000 tonnes in 2015 generating a revenue of 102 million euros. Halloumi cheese exports are expected to rise by 15% in 2016. The cheese is exported to more than 12 countries worldwide, with the UK market absorbing approximately 40% of the total exports. The average Cypriot consumes annually 21.5 kg of cheese (Anonymous, 2014), of which Halloumi consumption is 8.3 kg. A local cheesemakers association has filed an application for Protected Designation of Origin (PDO) certification of the cheese, and at the moment a final decision from the European Commission is yet to be taken.

While Halloumi dominates the local cheese market, Anari (whey cheese), Flaouna cheese (seasonal – Greek Easter period) and Kefalotyri-type cheese (hard mature cheese – various types according to the cheesemaker) are also produced in much lesser quantities.

No.	Cheese name	Milk used	Category	Section in Part II
1	Anari	Whey of sheep's, goat's and cow's	Whey	**13.1**
2	Flaouna cheese	Mixtures of sheep's, goat's and cow's	Semi-hard	**3.5**
3	Halitzia	Mixtures of sheep's and goat's	White-brined	**7.4**
4	Halloumi	Mixtures of sheep's, goat's and cow's	White-brined	**7.5**

Reference

Anonymous (2014). Agricultural Statistics, Statistical Service of Cyprus, Nicosia, Cyprus.

Cheeses from Denmark

Ylva Ardö

Department of Food Science, University of Copenhagen, Frederiksberg, Denmark

During the last few years, the amount of cheese produced in Denmark increased from 300,000 to 370,000 tonnes, of which 313,000 tonnes was exported. The main part is soft to semi-hard varieties such as the internationally well-known Danablu, Danbo, Esrom and Havarti; however, the production of white salad cheese and pasta-filata is also important.

Rennet-coagulated cheese was made at least 1000 years ago at large Viking farms managed by women, and new cheesemaking procedures was introduced at monasteries. Both acid cheese varieties and cheeses coagulated by rennet are described in old tax documents. The larger their size (up to about 20 kg), the more they were prized. High-quality cheeses were described to be neither too salty nor too old, but with a few drops of whey in the eyes and with a ripened odour. Cheese was traditionally made on dairy farms in Denmark until the introduction of cooperatives during late 1800s. Semi-hard cheeses with an open texture were produced by mixing cheese curd with salt, and the curd was kneaded to press out the whey. Dutchmen, who settled in Denmark, introduced pre-pressing of the cheeses under the whey to produce a closed texture with small, round eyes as seen in Danbo. Cream was used for butter making, and the skimmed milk was made into half-fat cheeses. Improved separation technology resulted in more butter but poor quality of the traditional cheese varieties, and new cheesemaking recipes from Europe were adapted by Danish cheesemakers. In 1952, the names of the Danish cheese varieties were established by governmental legislation following an agreement between eight European countries (the Stresa Convention). Today the cheeses are made at large cooperatives and small private dairy companies, and one main actor accounts for more than 90% of the production, mainly using highly industrialised technology (Ardö, 2004; Jensen, 1974).

No.	Cheese name	Milk used	Category	Section in Part II
1	Danablu PGI	Cow's	Blue	**11.2**
2	Danbo	Cow's	Bacterial surface-ripened	**10.1**
3	Esrom PGI	Cow's	Bacterial surface-ripened	**10.3**
4	Havarti	Cow's	Semi-hard	**3.7**

References

Ardö, Y. (2004). Semi-hard Scandinavian cheese made with mesophilic DL-starter. In *Handbook of Fermented Food and Beverages*. Marcel Dekker Inc., New York, pp. 277–290.

Jensen, H. M. (1974). *Bidrag til den danske osteproduktions og de danske ostesoerters historie.* [*Contribution to the History of the Danish Cheese Production and the Danish Cheese Varieties*] Århus, Denmark

Cheeses from France

Françoise Berthier

Unité de Recherches en Technologie et Analyses Laitières Rue de Versailles, France

It is challenging to select a few cheeses for representing the numerous French cheeses, so high is the technological diversity among them. All main technologies are used, except the pasta-filata technology. Most French regions have their specific cheeses manufactured according to specific practices, which are more or less related to traditional ones; in addition, French cheeses with new names are regularly proposed on the market. Nevertheless, 10 cheeses have been selected that illustrate nine main technologies in use for manufacturing 786 different French mature cheeses while best covering the 11 different French geographical regions traditionally manufacturing cheeses. The selected cheeses are all PDO cheeses, which guarantees their territorial anchoring and allows for a reliable and detailed description. Since none of them have been recognised as a PDO cheese, fresh cheeses, as well as processed cheeses, made from milk have not been considered, although they represent a large volume of the total French production (in 2014, 140,000 tonnes and 666,000 tonnes, respectively; (*L'Economie Laitière en Chiffres*) and although some of them are traditional cheeses. Cheeses made from whey have also not been considered, although one of them is a PDO cheese, namely, PDO Brocciu. Eight of the 45 French PDO cheeses are made from cow's milk, which is the major milk for the volume of French mature cheeses produced from it and the diversity in cheese technology. In 2014, the volume of the mature French cheeses manufactured from cow's milk represented 87% (1,064, 000 tonnes) of the total production, while the volume manufactured from goat's milk represented 8% of it, and the volume manufactured from sheep's milk represented 5% of it (*L'Economie Laitière en Chiffres*). Similar proportions were recorded among the French PDO cheeses ('*Produits Laitiers AOP, les Chiffres clés 2014*'). Although goat's milk cheeses volume is low compared to that of cow's milk cheeses, France is, in the world, the first manufacturer and consumer of these cheeses ('Les fromages de chèvres'), which are the major outlet for French goat's milk ('La filière lait de chèvre').

No.	Cheese name	Milk used	Category	Pages
1	Camembert de Normandie PDO	Cow's	Mould surface-ripened	**9.2**
2	Cantal PDO	Cow's	Hard	**2.4**
3	Comté PDO	Cow's	Semi-hard	**3.4**
4	Crottin de Chavignol PDO	Goat's	Acid-coagulated	**12.2**
5	Epoisses PDO	Cow's	Bacterial surface-ripened	**10.2**
6	Fourme d'Ambert PDO	Cow's	Blue	**11.3**
7	Fourme de Montbrison PDO	Cow's	Blue	**11.4**
8	Maroilles PDO	Cow's	Bacterial surface-ripened	**10.5**
9	Ossau Iraty PDO	Sheep's	Hard	**2.13**
10	Reblochon de Savoie PDO	Sheep's	Bacterial surface-ripened	**10.6**

Note: Jean-Louis Maubois has contributed in this book with the description of Camembert de Normandie PDO and Cantal PDO

Cheeses from Germany

Katja Hartmann

Anton Paar GmbH, Germany

Cheese production in Germany has been increasing every year. In 2016, 2,474,460 tonnes of cheese was produced in Germany. Most of it was fresh cheese, which is used, for example, as bread spread or for cheesecake. In Germany, 75% of the produced cheeses are consumed in private households, and the rest is equally distributed between industrial consumers and bulk consumers, for example, canteens. The large amount of cheese being further processed is due to the fact that the demand for convenience products is increasing, and hard and semi-hard cheeses in particular offer a broad range of application. However, in Germany, cheese is still traditionally consumed to a large extent on a slice of bread, for example, resulting in a per capita cheese consumption of 23.5 kg per year. Emmental cheese, for example, is very popular because of its slight sweet flavour and is consumed as a sandwich or cut into cheese cubes and enjoyed with wine or crackers. In contrast, other German cheeses are mainly popular in their producing region, such as acid curd cheeses. Cheeses belonging to this group have a long tradition in certain regions in Germany. But even though these cheeses are worth trying, they are not appreciated by everyone due to the somewhat distinctive and sour taste.

Cheese name	Milk used	Category	Section in Part II
Acid curd	Cow's	Acid-coagulated	**12.1**
Allgau Emmental	Cow's	Hard	**6.1**
Allgau mountain	Cow's	Hard	**2.1**
Altenburger PDO	Cow's (<15% goat's)	Mould surface-ripened	**9.1**

Cheese name	Milk used	Category	Section in Part II
Hohenheim Trappisten	Cow's	Bacterial surface-ripened	10.4
Quark	Cow's	Acid coagulated	12.3
Würchwitzer mite	Cow's	Hard	2.17

Cheeses from Greece

Thomas Bintsis[1] and Efstathios Alichanidis[2]

[1] 11 Parmenionos, 50200 Ptolemaida, Greece
[2] Department of Food Science and Technology, School of Agriculture, Aristotle University of Thessaloniki, Thessaloniki, Greece

Cheesemaking in Greece has a long history and tradition, and the milk of small ruminants (sheep and goats) has always been the most prized for cheese production, from ancient years to the present day. According to Greek myths, Aristeus, the sun of Apollo, was sent by the Greek gods from Mount Olympus to ancient Greeks to teach them the art of cheesemaking. In 'return', cheese was included in bloodless offerings of the ancient Greeks to their gods. Cheesemaking was a well-established craft at the time of Homer's writings. Homer (1184 BC) referred to cheese made from the milk of sheep and goats in caves by the 'Cyclop' Polyphemus – probably because of the 'unsuitability' of the Greek terrain for cattle, and this may have been the ancestor of Feta cheese. Aristotle (384–322 BC) was familiar with the fact that milk consisted of fat, tyrine (that was the term Aristotle used for casein, tyri=cheese) and water, and that the curdling was carried out by fig juice and/or rennet; the latter was formed in the stomachs of young animals. Cheese has been a stable food of the Greek diet for centuries. The agora marketplace in ancient Athens had a section dedicated to the sale of fresh cheese, and cheese was also an item of the Spartan soldier's meal. It is not a coincidence then that the current per capita consumption of cheese is among the highest in the world. In antiquity, Greeks established a kind of appellation of origin for cheeses, and the best-known cheeses at that time were 'Cythneos tyros' from the Greek island of Cythnos.

Feta cheese is the most famous and the most widely consumed Greek cheese, and thus, the most extensively studied. However, there are many important cheeses produced in Greece. Twenty-one cheeses have been recognised as PDO cheeses. In addition, there is a number of other locally produced traditional cheeses (e.g. Arseniko, Mitato Kithiron, Niotiko, Vlachotyri, Chloro, Kithniako, Mastelo Chiou and Melichloro Limnou). It is worth noting that in the Cyclades Islands, at the Aegean sea, alone, more than 40 different cheeses are produced (two of them are PDO cheeses), nearly one cheese per island.

The main characteristic of cheese manufacture in Greece is the use of sheep's and goat's milks. Cheese manufacture together with livestock breeding are very important sectors for the Greek economy. More than 250,000 families are involved in cheese manufacture. Although the last year's industrial production has increased, there are many family cheesemaking enterprises producing artisanal cheeses for local markets all over the country.

More than 85% of the sheep's and goat's milk is transformed into approximately 190,000 tonnes of cheese, producing more than 100 different varieties. Production of soft cheeses, mainly Feta, has reached 125,500 tones, hard and semi-hard 42,800 tones and whey cheeses 21,700 tones (ELGO Dimitra, 2015). In this book, 11 cheeses have been selected to be described, the selection being based on specific sensory characteristics, unique cheesemaking procedures and economic importance.

No.	Cheese name	Milk used	Category	Section in Part II
1	Anevato PDO	Sheep's milk or mixtures	Soft	**4.2**
2	Anthotyros	Whey with the addition of milk or cream from sheep or goat	Whey cheese	**13.2**
3	Batzos PDO	Goat's or sheep's milk or mixtures	White-brined	**7.1**
4	Feta PDO	Sheep's or mixture with goat's up to 30%	White-brined	**7.3**
5	Galotyri PDO	Sheep's or goat's milk or mixtures of both	Soft	**4.5**
6	Graviera Kritis PDO	Sheep's, or mixture with goat's up to 20%	Hard	**2.8**
7	Kasseri PDO	Sheep's milk, or mixture of sheep's with goat's (up to 20%)	Pasta-filata	**8.4**
8	Kefalograviera PDO	Sheep's or mixture with goat's up to 10%	Hard	**2.10**
9	Kefalotyri	Cow's, sheep's, goat's and/or mixtures	Hard	**2.11**
10	Kopanisti PDO	Cow's, sheep's or goat's milk or a mixture	Soft	**4.6**
11	Manouri PDO	Whey with the addition of milk or cream from sheep or goat	Whey cheese	**13.3**

Reference

ELGO Dimitra (2015). Statistical data from the Hellenic Organization for Milk and Meat Hellenic Agricultural Organization Dimitra, Athens.

Cheeses from Italy

Giuseppe Licitra

University of Catania, Department of Agriculture and Food Science (DISPA), Italy

The Italian dairy section is the country's largest food sector and alone represents more than 12% of the total turnover of the national food. The production value exceeds 14.5 billion euros annually.

Italy's livestock is made up of 1,850 million dairy cows and 5 million sheep and goats. The Italian production of cow's milk is about 11 million tonnes and of sheep's and goat's milk is about 650,000 tonnes per year. This production is not sufficient to cover the internal demand and the exports of the country, which is approximately 20 million tonnes of milk per year, and therefore the country is forced to import about 40% of its needs.

The Italian dairy sector has a dualistic nature, one industrial and the other traditional.

Most of the industrial products (i.e. milk, butter, liquid whey, and other dairy products with a strong technological innovation) are undifferentiated. By contrast, Italy is the largest European country producer of traditional PDO cheeses (49 as of 2015), assuming a prominent position within the worldwide dairy sector. PDO and PGI cheeses absorb 70% of the national milk production.

Italy produces about 1225 million tonnes of cheeses, of which 497,000 tonnes is recognised as PDO, representing 41% of the total production. The first ten PDO cheeses for the quantity produced, Grana Padano (184,964 tonnes), Parmigiano Reggiano (132,684 tonnes), Gorgonzola (53,322 tonnes), Mozzarella di Bufala (37,308 tonnes), the Pecorino Romano (24,117 tonnes), Asiago (21,458 tonnes), Taleggio (8,956 tonnes), Montasio (6,896 tonnes), Provolone Valpadana (5,286 tonnes) and Fontina (4,400 tonnes), cover 93.6% of the total PDO production and are economically important for the Italian dairy sector.

Consequently, the other 39 PDO cheeses as a whole account for only 5.4% (26,665 tonnes) of the national PDO production, with an average production of about 680 tonnes per cheese.

Despite the limited quantity produced, these cheeses play a social and cultural role that goes far beyond the commercial value of the products.

Most of the PDO cheeses, excluding part of the production of Parmigiano Reggiano and Grana Padano (produced in the Po valley), are produced in less favoured areas (mountains and high hills), with extensive farming systems. All PDO cheeses have a strong linkage to the territory of origin (i.e. orography, landscape, rural environment and human resources) and therefore are testimony to the history, culture and lifestyle of those communities that produce them, handed down over the generations. It follows that traditional cheeses are a unique expression of the symbiotic interaction between human resources, the culture of rural communities and nature. Furthermore, they also play the role of 'guardian' of the environment by protecting natural resources, reducing soil erosion, reducing deforestation and desertification and preserving territorial biodiversity, including indigenous breeds and feeding native pasture.

No.	Cheese name	Milk used	Category	Section in Part II
1	Asiago PDO	Cow's	a) Semi-hard (Pressato), b) Hard (D'Allevo)	2.2
2	Caciocavallo Podolico PDO	Cow's	Pasta-filata	8.1
3	Castelmagno PDO	Cow's with the addition of sheep's and/or goat's 5%–20%	Semi-hard	**3.3**
4	Fiore Sardo PDO	Sheep's	Hard	**2.7**
5	Fossa PDO	Sheep's or cow's or mixtures with sheep's minimum 20%	Semi-hard	**3.6**
6	Mozzarella di Bufala Campana PDO	Buffalo's	Pasta-filata	**8.5**
7	Provolone Valpadana PDO	Cow's	Pasta-filata	**8.6**
8	Parmigiano Reggiano PDO	Cow's	Extra-hard	**1.1**
9	Ragusano PDO	Cow's	Pasta-filata	**8.8**
10	Robiola di Roccaverano PDO	Goat's, or mixtures with sheep's or cow's, with goat's >50%	Acid-coagulated	**12.4**
11	Tuma Persa PDO	Cow's	Semi-hard	**3.20**
12	Vasteddadella Valle del Belice PDO	Sheep's	Pasta-filata	**8.9**

Cheeses from Malta

Everaldo Attard[1], Anthony Grupetta[2] and Stefania Carpino[3]

[1] *Division of Rural Sciences and Food Systems, Institute of Earth Systems, University of Malta, Malta*
[2] *Veterinary Regulations Directorate, Marsa, Malta*
[3] *CoRFiLaC – Consorzio Ricerca Filiera Lattiero Casearia, Ragusa, Italy*

The three main dairy species reared in Malta are cows, sheep and goats. Dairy farms are distributed all around the Maltese islands. According to the Veterinary Regulation Directorate, up to November 2014 there were approximately 85 dairy bovine farms and approximately 1,339 ovine and caprine farms on mainland Malta. On the sister island, Gozo, there were approximately 35 dairy bovine farms and approximately 561 ovine and caprine farms. Many farms dedicated to dairy bovines; the rest are mixed dairy farms. In such instances, the producer may have ovines and caprines on the same farm. Most of the dairy bovine producers are affiliated with a dairy cooperative that processes the milk in a centralised dairy factory. Bovine milk and milk products are then distributed to retail outlets, where these are usually sold as fresh products. The sheep's and goat's milk is not consumed as such by the local population. Traditionally, sheep's milk was processed into cheese, while goat's milk was sold to customers at the door. The physic-chemical properties of sheep's milk facilitate the transformation and preservation of milk into cheese, traditionally known as Maltese Ġbejna. With the rise in incidence of brucellosis, the consumption of raw goat's milk was banned, and infected caprines were removed from herds. With improved disease and hygiene control, today goat's milk is processed into cheese, though this is not the typical traditional Maltese Ġbejna. Within this sector, the annual milk production is approximately 43 million litres, while cheese production is approximately 208 tonnes per year. Today, local consumers and tourists can enjoy these high-quality products manufactured in accordance with tradition but with good hygienic practices.

No.	Cheese name	Type of milk	Category	Section in Part II
1	Maltese Ġbejna	Sheep's	Soft	**4.7**

Cheeses from the Netherlands

Eva-Maria Düsterhöft, Wim Engels and Thom Huppertz

NIZO Food Research, The Netherlands

Of the milk delivered by Dutch dairy farmers to industry (12.7 billion kg), about 52% goes into the production of cheese. This results in a total annual production of 772,000 tonnes of cheese, most of which is made from pasteurised cow's milk, in large, modern, highly automated factories (2014). Nowadays, there are 18 industrial cheese factories in the Netherlands, whose annual production capacity varies from 30,000–150,000 tonnes of cheese. Only about 7,500 tonnes of cheese is still made from raw milk on dairy farms (Boerenkaas). Dutch goat's cheese production amounted to about 20,000 tonnes (2012), and represents ~2.5% of the total Dutch cheese production.

Approximately 75% of the Dutch cheese production is exported, mainly to European countries (Germany, France and Belgium being the largest destinations) and abroad. This represents an export value of about 3.5 billion EUR (50% and the total export value of Dutch dairy produce).

Royal Friesland Campina is, with seven production sites, the largest Dutch cheese producer, while other companies (Bel Leerdammer, DOC Kaas, CONO Kaasmakers, Delta Milk (de Graafstroom), Rouveen Kaasspecialiteiten and A-ware) operate from one or two factories. Most factories are located in the agriculture-dominated provinces (North Holland, Friesland, Groningen and Drenthe), which minimises transport time and costs. The majority of the dairy companies are cooperative organisations of dairy farmers, with member numbers varying from, for example, 19,000 (Royal Friesland Campina) to 150 (De Graafstroom). The scale of operation has increased steadily (e.g. in 1960, there were about 483 dairy producers in the Netherlands; in recent years, this has reduced to 51), and large improvements in efficiency have been achieved. Sustainability aspects govern more recent developments and when new production sites are built. The primary cheese producers usually take account of the very first phase of cheese ripening and storage only (e.g. two weeks). Further ripening and treatment of the cheese (which requires a large effort, notably for naturally ripened cheeses) is undertaken by companies specialised in cheese maturation and trading. Processing of the whey at most cheese factories is restricted to purification and concentration to a maximum of 30% dry matter. The concentrated whey is then transported to a few specialised factories, where it is processed to high-value dried dairy ingredients. Only a few companies have their own whey processing plants.

As the fifth largest cheese producer in the world, the Dutch are also in the top ten of cheese consumers, with about 20 kg per capita per annum. The prevailing use of cheese is as slices on a sandwich (for breakfast or lunch), or as blocks, accompanying a late afternoon or evening drink.

No.	Cheese name	Milk used	Category	Section in Part II
1	Edam	Cow's	Dutch-type	5.1
2	Gouda	Cow's	Dutch-type	5.2
3	Maasdammer	Cow's	Swiss-type	6.4
4	Hollandse geitenkaas' (Dutch goat cheese)	Goat's	Dutch-type	5.3

Cheeses from Portugal

Tânia G. Tavares[1,2] and F. Xavier Malcata[1,3]

[1] Laboratory of Engineering of Processes, Environment, Biotechnology and Energy (LEPABE), Portugal
[2] REQUIMTE/Department of Chemical Sciences, Faculty of Pharmacy, University of Porto, Portugal
[3] Department of Chemical Engineering, University of Porto, Portugal

There is a long tradition of artisanal cheesemaking in Portugal, via manufacturing techniques that have been optimised from generation to generation. Traditional Portuguese cheeses are known for their organoleptic uniqueness. Their production is essentially seasonal (usually in spring) and is still performed chiefly at the farmhouse level. Portuguese artisanal cheeses encompass 11 distinct types, produced from raw ewe's or goat's milks (or a mixture of both). They can be divided into four groups, based on the milk source and rennet type used:

1) *Azeitão, Castelo Branco, Évora, Nisa, Serpa* and *Serra da Estrela,* which are considered the most outstanding cheeses, are manufactured with raw sheep's milk and coagulated with a plant coagulant (*Cynara cardunculus*), without deliberate addition of any starter or non-starter culture. Such a combination leads to a soft paste due to extensive proteolysis, found in all cheeses manufactured with that specific plant coagulant, and a notable lipolysed character provided by the sheep's milk source.
2) *Terrincho* is manufactured with raw sheep's milk and coagulated with calf rennet.
3) *Cabra Transmontano* is manufactured with raw goat's milk and coagulated with calf rennet.
4) *Amarelo da Beira Baixa, Picante da Beira Baixa* and *Rabaçal* are manufactured with mixtures of raw sheep's and goat's milks and coagulated with calf rennet. Although all of them possess *Appélation d'Origine Protegée* (AOP) status, traditional cheeses represent about 22% of all cheese sales in Portugal – of which only 6% are formally sold with that label.

Cheesemaking protocols encompass milking of animals by early morning, brief storage of milk at room temperature, clotting with plant rennet (without addition of microbial cultures), cutting and manual working of the curd, spontaneous whey drainage and ripening for a minimum of 30 d. Although industrialisation of those traditional cheeses would be important in order to standardise cheese production, and more efficiently address safety issues (e.g. decrease contamination and survival of undesirable microorganisms such as coliforms), it should not be forgotten that the complex (and often unpredictable) microflora dynamics is responsible for several important differences among cheeses, as well as for some outstanding and unparalleled characteristics of Portuguese traditional cheeses.

No.	Cheese name	Milk used	Category	Section in Part II
1	Serra da Estrela PDO	Sheep's	Soft	**4.8**
2	Serpa PDO	Sheep's	Semi-hard	**3.18**

Cheeses from Serbia

Zorica Radulovic and Jelena Miocinovic

Department of Food Microbiology, Faculty of Agriculture, University of Belgrade, Serbia

The total cow's milk production in Serbia is around 1,500 million litres (ML) per annum (Yearbook, 2014), out of which approximately 700 ML is further delivered and processed in 200 dairy plants of different capacities (Analysis, 2012). The remaining milk is partly used and processed in households and small craft dairy plants and sold directly to consumers through various channels of direct sales. Sheep and goat milk production are 20 and 38 ML, respectively, usually processed in households and small-capacity dairy plants.

The five big dairy plants (>20 ML of milk/day) account for approximately 60% of the total milk intake, while middle- and small-sizes dairy plants process the remaining 40% of milk delivered to industry (Popovic, 2009). The assortment of dairy products manufactured is quite different depending on the plant size. The big dairy plants are oriented towards the production of pasteurised and UHT milks and fermented dairy products such as yoghurts (82%), while cheese and butter production accounts for about 18% (Popovic, 2009). On the other hand, middle- and small-capacity dairies process half the milk into cheeses and butter and the other half into other dairy products.

The most common cheeses in Serbia are white-brined cheeses and hard and semi-hard cheeses (Kachkaval type), but also in industry a significant part of production is based on acid-coagulated cheeses as well as Feta-type cheese from ultrafiltered milk.

No.	Cheese name	Milk used	Category	Section in Part II
1	Kachkaval	Cow's, sheep's or mixture	Pasta-filata	**8.2**
2	Sjenica	Cow's	White-brined	**7.7**
3	Sombor	Cow's	Semi-hard	**3.19**

References

Analysis (2012). *Sectoral Analysis of Raw Milk, Production and Processing of Milk and Dairy Products*. Commission for protection of competition of the Republic of Serbia, Belgrade, Republic of Serbia.

Popovic, R. (2009). Structural changes on Serbian milk market. *Prehrambena industrija – Mleko i mlečni proizvodi*, 20, 7–12.

Yearbook (2014). *Statistical Yearbook of the Republic of Serbia*. Statistical Office of the Republic of Serbia, Belgrade, Serbia.

Cheeses from Slovakia

Karol Herian[1] and Paul Jelen[2]

[1] Slovak Dairy Research Institute, Slovakia
[2] Department of Agricultural, Food and Nutritional Science, University of Alberta, Canada

Slovakia has about 5.4 million inhabitants. In the years when the country was a part of Czechoslovakia, under the communist regime, the annual milk production amounted to about 2 billion litres of milk. The annual cheese production was more than 50,000 tonnes of both cow's milk as well as sheep's milk cheeses. The annual cheese consumption was about 7.5 kg per capita; the rest of the cheese production was exported. After the 'Velvet Revolution' of 1989 and the subsequent division of Czechoslovakia, with the creation of the independent country of Slovakia in 2003, the production outputs changed drastically. At the present time, the annual milk production is scarcely 1 billion litres, while the total cheese production amounts to about 35,000 tons. This includes about 2,000 tonnes of sheep's milk cheeses. In contrast, the cheese consumption shows a continuously increasing tendency, with the present figure being about 9.5 kg per capita. The shortfall is being covered by increasing imports of cheese from abroad. Approximately 29% of all cheeses made in Slovakia are hard and semi-hard varieties, ripened with rind or under plastic seal. About 6% are mould-ripened cheeses (both white and green mould), while about 25% are cheeses of the pasta-filata type, both stretched and moulded. Soft cheeses – especially Quark (both soft and pressed) amount to over 26%; a further 6% is contributed by lump cheese (a raw material for Bryndza). Finally, about 8% is made up of sheep's milk cheeses, especially Bryndza, and fresh as well as ripened cheese, including predominantly the pasta-filata types. The last few years saw a significant increase in breeding of sheep and goats, leading to increases in the production of cheese specialties.

Slovakia, despite its small size, has a rich history of cheesemaking. Some of the Slovak cheese specialties are well known internationally, especially in the neighbouring countries, and especially the sheep's milk products. Some of these have been recently included in the EU list of PDO cheeses. As of 2016, there are presently seven Slovak specialties denoted as PDO: Slovak Bryndza, Slovak Ostiepok, Slovak Parenica, Zázrivecky korbáčik, Tekovský 'sausage', Salašnícky lump cheese and, most recently, the Klenovecky syrec.

No.	Cheese name	Milk used	Category	Section in Part II
1	Bryndza	Sheep's, or mixture with cow's	Soft	**4.3**
2	Parenica	Sheep's	Pasta-filata	**8.7**

Cheeses from Spain

Maria Belén López Morales

Food Science and Technology Department, International Excellence Campus for Higher Education and Research 'Campus Mare Nostrum', Veterinary Faculty, University of Murcia, Spain

Spain has more than 150 varieties of cheese. Compared with other countries such as France, this may seem few, but the variety of cheeses produced is huge due to the orography, weather conditions and especially the many local breeds of cattle, sheep and goats. These autochthonous breeds give the final sensory and physico-chemical properties to the cheeses produced, which differentiate them from cheeses of other countries. Cheese production in Spain has been increasing, reaching 400,000 tons in 2014, which represents a turnover of 2,700 million euros. The tradition of cheesemaking dates back to ancient times – the Tartessos, Phoenicians and Carthaginians elaborated cheeses, but the Celts had a particularly acute knowledge of cheesemaking technology. In 'De Re Rustica', Columella describes the technology of cheesemaking and also the conditions in which the Cantabros should send cheese to Rome. However, the Arabs gave more importance to goat's milk, and this tradition still remains in Andalucia and Extremadura. Pilgrimages along the Camino de Santiago or Saint James's Way allowed the exchange of cheeses between different communities.

Spain produces 32 types of cheese recognised with the Protected Designation of Origin (PDO) or Geographical Indication and Designations of Origin (PGI) status. PDO cheeses can be differentiated into 26 types, made from cow's, sheep's or goat's milk. Cheeses made from cow's milk are Afuega'lPitu, Cebreiro, Picón-Bejes-Tresviso, Casin, l'AltUrgell and La Cerdanya,

Nata de Cantabria and Tetilla produced in Galicia, Cantabria, Asturias and Cataluña. Cheeses made from sheep's milk are Idiazabal, Serena, Manchego, Roncal, Zamorano and Torta del Casar produced in Extremadura, León and Navarra. Cheeses made with goat's milk are Murcia al Vino, Ibores, Majorero, Palmero and Camerano produced in Murcia, Extremadura, Canarias and La Rioja. Cheeses made with mixtures of two or three milk types are Flor de Guía, Queso de Media Flor de Guía and Queso de Guía, Cabrales, Gamonedo, Mahón-Menorca and Liébana produced in the Canary and Balearic Islands, Asturias and Cantabria.

In this book, 11 types of cheeses are described, corresponding to those that are internationally well known such as the Manchego and Arzúa Ulloa cheeses; those with greater visibility in US markets (Murcia al Vino); or those which are considered different based on the cheesemaking technology (Torta del Casar and Gamonedo), shape (San Simon da Costa and Mahón-menorca cheese), sensory profile (Afuega´lPitu, Cabrales, Majorero) or for being considered part of the culinary heritage, such as Idiazabal cheese.

Acknowledgements

Professor Conchita Chamorro (Technical University of Madrid) is specially acknowledged for her deserving contribution to this work.

No.	Cheese name	Milk used	Category	Section in Part II
1	Afuega l'Pitu PDO	Cow's	Soft	**4.1**
2	Arzúa-Ulloa PDO	Cow's	Semi-hard	**3.2**
3	Cabrales PDO	Cow's or mixture of cow's, sheep's and/or goat's	Blue	**11.1**
4	Gamonedo PDO	Cow's or mixture of cow's, sheep's and/or goat's	Blue	**11.5**
5	Idiazabal PDO	Sheep's	Hard	**2.9**
6	Mahón-Menorca PDO	Cow's, or mixture with up to 5% sheep's	Semi-hard	**3.9**
7	Majorero PDO	Goat's	Semi-hard	**3.10**
8	Manchego PDO	Sheep's	Semi-hard	**3.11**
9	Murcia al Vino PDO	Goat's	Semi-hard	**3.12**
10	San Simon da costa PDO	Cow's	Semi-hard	**3.16**
11	Torta del Casar PDO	Sheep's	Soft	**4.9**

Cheeses from Sweden

Ylva Ardö

Department of Food Science, University of Copenhagen, Denmark

About 90,000 tonnes of cheese per year is produced in Sweden at primarily less than 10 medium-sized plants. The main type is semi-hard to hard cheese made from pasteurised cow's milk using calf rennet and mesophilic starter. The most popular cheese varieties are made with a closed texture and round eyes (Herrgård and Grevé) or with an open texture containing small, irregular eyes (Svecia, Präst and Västerbottensost). Only about 20,000 tonnes per year are exported.

In older times, cheese was mainly made from goat's milk, while cow's milk was used primarily for butter production. In Northern Sweden, the dairy animals were kept in the mountains during summer, and people moved with them into mountain chalets where the cheese and butter were made. Moisture was evaporated from whey by boiling to produce whey cheeses called Mesost if sliceable and Messmör if spreadable, depending on the moisture content.

Semi-hard cheese was made from cow's milk in the plain areas of southern Sweden, at least from the 1500s. Mainly small, half-fat, open-texture cheeses of about 1–2 kg but also up to 10–15 kg were made at farms and monasteries (Svecia). Large, full-fat, semi-hard cheeses that were famous for their delicious, aromatic and strong flavour were made especially for the vicar at special cheesemaking parties. An open-texture hard cheese variety with unique properties was developed during the 1800s by using the unique combination of a mesophilic starter and very long cooking times (Västerbottensost).

A round-eyed cheese variety was developed from Swiss protocols during the 1700s (Herrgård). Propionic acid bacteria contributed occasionally to flavour, and later a new round-eyed cheese variety named Grevé was developed to which they are added during production. Several local cheese varieties are made at about 100 farm dairies, which are especially popular at restaurants (Ardö, 1993; Magnus, 1555; Nilsson & Orre, 2015; Ränk, 1987).

No.	Cheese name	Milk used	Category	Section in Part II
1	Grevé	Cow's	Swiss-type	**6.3**
2	Herrgård	Cow's	Semi-hard	**3.8**
3	Mesost & Messmör	Whey of cow's milk	Whey	**13.4**
4	Präst	Cow's	Semi-hard	**3.13**
5	Svecia PGI	Cow's	Semi-hard	**3.17**
6	Västerbottensost	Cow's	Hard	**2.16**

References

Ardö, Y. (1993). Swedish cheese varieties. In Fox, P. F. (ed.), *Cheese: Chemistry, Physics and Microbiology, Vol 2, Major Cheese Groups*, 2nd edition. London, Chapman & Hall, pp. 254–257.

Magnus, O. (1555). [About the large and well-tasting cheeses]. *Historiadegentibusseptentrionalibus [History of the People of the Nordic Countries]*, 13th book, 46th chapter. Roma, Italy.

Nilsson, I. & Orre, I. (2015). *Svenska ostar & ostmakare. [Swedish Cheeses and Cheeseproducers]*. Bokförlaget Arena, Lund, sweden. ISBN 978-91-7843-476-3

Ränk, G. (1987) *[From Milk to Cheese]*. Nordiska Museets Handlingar, 66, 2nd edition. Berlings, Arlöv, Sweden. ISBN 91-7108-269-7.

Cheeses from Switzerland

Elisabeth Eugster-Meier[1], *Marie-Therese Fröhlich-Wyder*[2], *Ernst Jakob*[2] *and Daniel Wechsler*[2]

[1] *Bern University of Applied Sciences, School of Agricultural, Forest and Food Sciences HAFL, Zollikofen, Switzerland*
[2] *Agroscope, Research Division Food Microbial Systems, Federal Department of Economic Affairs, Education and Research EAER, Bern, Switzerland*

Approximately 80% of the cultivated land in Switzerland is unsuitable for crop cultivation. Therefore, livestock breeding and cheese manufacture have a long tradition in the country. The production of cheese was mentioned for the first time in the first century by Roman historian Pliny the Elder, who called the cheese *Caseus Helveticus*, the 'cheese of the Helvetians', one of the tribes living in Switzerland at the time. For centuries, the standard type of cheese was 'cottage cheese', produced by the spontaneous acidification of milk. The technique of adding rennet to milk for cheesemaking first appeared in Switzerland around the fifteenth century. Since such cheese could be stored for longer periods, it soon became used in travel provisions and food stocks for the winter. With time, cheesemaking, and also the ripening of cheeses, started to vary considerably across geographical areas, resulting in the regional development of characteristic cheese varieties named mostly after their area of production (e.g. Emmentaler and Gruyère). With the advent of the cheese trade in the eighteenth century, some of these cheese varieties were sold all over Europe and became very popular (FDFA, 2015).

Nowadays, 40% of the country's milk, which is predominantly cow's milk, is transformed into over 450 different cheese varieties. However, most of them are regional specialties that are produced only in small quantities and are consumed only by the local population. In this section of the book, a rather small selection of 10 characteristic Swiss cheese varieties that are of particular interest due to their traditional manufacturing techniques, outstanding sensory characteristics, unique methods of preparation for consumption, national and international popularity or economic importance are presented. Out of the 10 cheese varieties presented, Berner Alpkäse, Berner Hobelkäse, Emmentaler Le, Gruyère, Raclette du Valais, Sbrinz and Tête de Moine are still made from raw milk, underlining both the importance of raw milk cheeses for Switzerland's culinary heritage and also the high quality of Swiss cheeses with regard to food safety. However, Vacherin Mont-d'Or, a traditional soft cheese, is now made from thermised milk in order to combine optimal sensory quality and maximum food safety. With the exception of the two branded cheeses Appenzeller® and Raclette Suisse®, all the cheese varieties presented in this chapter are labelled as PDO. Moreover, all have one thing in common: they respect the growing consumer demand for natural food and do not contain any kind of food additives.

Acknowledgements

The authors acknowledge Rudolf Amrein, Nicolas Fehér, John Haldemann and Hans Winkler for their professional support in the preparation of this publication. Special thanks are due to the inter-professional organisations of these cheese varieties and all other partners for providing us with information, data, suitable pictures and for all other support of this work.

No.	Cheese name	Milk used	Category	Section in Part II
1	Appenzeller®	Cow's	Semi-hard	**3.1**
2	Berner Alpkäse PDO	Cow's	Hard	**2.3**
3	Berner Hobelkäse PDO	Cow's	Extra-hard	**2.3**
4	Emmentaler PDO	Cow's	Swiss-type	**6.2**
5	Gruyère PDO	Cow's	Hard	**2.12**
6	Raclette du Valais PDO	Cow's	Semi-hard	**3.14**
7	Raclette Suisse®	Cow's	Semi-hard	**3.15**
8	Sbrinz PDO	Cow's	Extra-hard	**1.3**
9	Tête de Moine PDO	Cow's	Hard	**2.14**
10	Vacherin Mont-d'Or PDO	Cow's	Bacterial surface-ripened	**10.7**

Cheeses from Turkey

İrem Uzunsoy[1] and Barbaros Özer[2]

[1] Bülent Ecevit University Caycuma Vocational High School, Department of Food Technology, Zonguldak, Turkey
[2] Ankara University, Faculty of Agriculture Department of Dairy Technology, Ankara, Turkey

In Turkey, the manufacture of cheese dates back to thousands of years ago (Ünsal, 1997). At present, more than 190 local cheese varieties are known, and the processing methods of these varieties vary from one region to another. Since methods of manufacturing show large diversity with locality, it is rather difficult to classify local cheeses according to their production practices. Originally, the production of traditional Turkish cheeses was limited to small-scale units, which made standardisation of the gross chemical composition and other properties of these products very difficult. Mechanisation and automation in cheesemaking have enabled much progress in the marketing of local cheese varieties. Modern technologies have replaced old-fashioned techniques in milk handling and the manufacture of cheese, including ripening the milk with starter cultures, coagulation, cutting the coagulum and drainage of the whey (Özer, 2014). In 2013, 600.266 tonnes of cheese was produced in Turkey (Anonymous, 2014). Cheese varieties of the most industrial importance are the classical Turkish Beyaz Peynir (white-brined cheese), Urfa (scalded white cheese matured in brine), Tulum (dry-salted white cheese matured in animal skin bags or in plastic containers), Kashar (similar to Kashkaval or Kasseri cheese) and Mihallıç cheeses. Depending on the ripening conditions and manufacturing practices, Turkish cheeses have different physical, chemical and microbiological characteristics. Examples of the local cheese varieties ripened in brine are Ezine, İzmir Salamura Tulum, Antep, Diyarbakır Örgü, Malatya, Van Otlu, Dil and Salamura Civil cheeses. Of these varieties, Dil, Diyarbakır Örgü, Malatya, Antep and Salamura Civil cheeses are scalded before being put into brine solution. These varieties are traditionally produced from sheep's or goat's milk or an appropriate mixture of both.

No.	Cheese name	Milk used	Category	Section in Part II
1	Beyaz Peynir	Sheep's, cow's or mixture	White-brined	7.2
2	Kashar	Cow's, sheep's or mixture	Pasta-filata	8.3
3	Mihalıç	Sheep's, goat's, cow's or mixture	Hard	7.6
4	Tulum	Sheep's, goat's, cow's or mixture	Hard	2.15
5	Urfa	Sheep's milk, cow's milk or mixture	White-brined	7.8

References

Anonymous (2014). National Milk Council of Turkey. USK report on dairy sector in the world and in Turkey.

Ünsal, A. (1997). [*Sut uyuyunca: Turkiye peynirleri.* Yapi Kredi Yayinlari, Istanbul, Turkey.]

Cheeses from the United Kingdom

Kimon-Andreas G. Karatzas[1] and Thomas Bintsis[2]

[1] *Department of Food and Nutrition Sciences, The University of Reading, United Kingdom*
[2] *11 Parmenionos, 50200 Ptolemaida, Greece*

Cheese production in Britain is known to date back to the times of the Romans, who are known to have introduced cheesemaking to the British. For centuries, cheesemaking was localised in farms and produced by peasants, until the Middle Ages, when production mainly moved to abbeys. The dissolution of monasteries in late 1530s by Henry VIII had an impact on cheese production until the seventeenth century, when the development of new techniques and larger markets resulted in the expansion of the cheese production.

Today, approximately 700,000 tonnes of cheese are consumed annually in the UK (including Cottage cheese and Fromage Frais). The average consumption of cheese in the UK is close to 10 Kg per capita annually, which is almost half of what an average European consumes. Cheddar is the UK's favourite cheese, accounting for 55% of household purchases, while the second most popular is Mozzarella, most of which is produced in the UK. Annual cheese production in the UK is close to 400,000 tonnes, while 70% of it is Cheddar. Cheese in the is mostly made from cow's milk, and there are approximately 700 types of cheese produced, of which 12 are PDOs.

The most consumed cheese in the UK, and probably in the world, is Cheddar, which originated from Somerset in the late twelfth century and is named after the gorges or caves in the town of Cheddar where the cheese was stored. There are records from the King of England's accounts (the 'Great Roll of the Pipe') back in 1170 showing that the King (Henry II) purchased 4.6 tonnes of Cheddar and also declared that the latter cheese was the best in Britain. Today, production of Cheddar takes place around the world; however, 14 Cheddar producers in the western part of the country (in the counties of Devon, Cornwall, Dorset and Somerset) produce a PDO named 'West Country Farmhouse Cheddar'. This cheese is produced using traditional techniques and it has to be matured for nine months.

The oldest cheese produced in Britain is thought to be Cheshire cheese, where all cheese produced in Cheshire County would carry this name. The first Cheshire cheese is believed to be produced by the Romans, in Chester. Cheshire cheese is even mentioned in the Domesday Book (end of the eleventh century), while it was the only cheese officially bought by the British Navy (since 1739). Later, in the nineteenth century, the popularity of Cheshire cheese increased, particularly in the industrial areas of England, where it was called the 'poor man's meat'. Today its annual production reaches close to 6,000 tonnes.

Blue Stilton, another well-known cheese that is also called the 'King of English Cheeses', is named after a village south of Peterborough which was a coaching stop on the Great North Road. Recent research has revealed that a cheese called 'Stilton' was made in the village in the early part of the eighteenth century. Originally, Stilton was a somehow different cheese that Daniel Defoe referred to as the 'English Parmesan'. Changes in the process were introduced due to increased demand, resulting in the Blue Stilton. Production of cheese ceased in the village during the course of the eighteenth century, and most of the cheese was then subsequently made in Leicestershire, Nottinghamshire and then Derbyshire. Today, Blue Stilton is a PDO cheese produced only in the aforementioned counties by locally produced milk that is pasteurised.

No.	Cheese name	Milk used	Category	Section in Part II
1	Cheddar	Cow's	Hard	**2.5**
2	Cheshire	Cow's	Hard	**2.6**
3	Stilton PDO	Cow's	Blue	**11.7**

1

Extra-Hard Cheeses

Giuseppe Licitra[1], *Erica R. Hynes*[2,3], *Maria Cristina Perotti*[2,3], *Carina V. Bergamini*[2,3], *Elisabeth Eugster-Meier*[4], *Marie-Therese Fröhlich-Wyder*[5], *Ernst Jakob*[5] *and Daniel Wechsler*[5]

[1] *Department of Agriculture, Nutrition and Environment, University of Catania, Catania, Italy*
[2] *Facultad de Ingeniería Química (Universidad Nacional del Litoral), Santa Fe, Argentina*
[3] *Instituto de Lactología Industrial (Universidad Nacional del Litoral – Consejo Nacional de Investigaciones Científicas y Técnicas), Santa Fe, Argentina*
[4] *Bern University of Applied Sciences, School of Agricultural, Forest and Food Sciences HAFL, Zollikofen, Switzerland*
[5] *Agroscope, Research Division Food Microbial Systems, Federal Department of Economic Affairs, Education and Research EAER, Bern, Switzerland*

1.1 Parmigiano Reggiano PDO – Italy

Name: Parmigiano Reggiano PDO	
Production area: Provinces of Parma, Reggio Emilia, Modena and parts of the provinces of Mantua and Bologna	
Milk: Cow's, raw	

1.1.1 Introduction

Parmigiano Reggiano is produced exclusively in the provinces of Parma, Reggio Emilia, Modena and parts of the provinces of Mantua and Bologna, on the plains, hills and mountains enclosed between the rivers Po and Reno. Since medieval times, when the Benedictine monks started producing these great cheeses specifically for long maturation, humankind has joined with nature, leaving it untouched and improving only the areas of man's intervention. Parmigiano Reggiano cheese has been a great cheese for at least nine centuries, which not only testifies to its

Global Cheesemaking Technology: Cheese Quality and Characteristics, First Edition.
Edited by Photis Papademas and Thomas Bintsis.

ancient origin but also underscores the fact that this cheese today is still identical to how it was eight centuries ago, having the same appearance and the same extraordinary fragrance, is made in the same way, in the same places and with the same expert ritual gestures. Historical evidence shows that already in 1200–1300 AD, Parmigiano Reggiano cheese had reached its perfect typical form that has remained unchanged up to the present day. Today, as in the past, cheese masters continue their tradition and willingly take a risk by proudly persisting in making their cheese solely with milk, rennet, fire and art, following the rigorous centuries-old methods and applications of the technique that are the result of devotion to their craft and long experience. The cheesemakers are the custodians and interpreters of the secrets of the true craft of milk processing, and although in hundreds of artisanal cheese dairies (about 350) they all work with their hands in the same way, the result of their work is inextricably linked to their personal experience and sensitivity, giving an appreciable diversity of taste and aromas. The milk produced in the area of origin undergoes an artisanal process that is entrusted to the skills and passion of the cheese masters. The result of a thousand years of expertise and culture, the choices made by the cheese masters are fundamental in the production of Parmigiano Reggiano, since craftsmanship is one of its distinctive features. In fact, every cheese master must "interpret" milk every day and turn it into cheese by favouring and enhancing the distinctiveness of its indigenous microflora.

1.1.2 Type

Parmigiano Reggiano PDO (Protected Designation of Origin) cheese is an extra-hard, semi-fat cheese. The content of fat in dry matter (FDM) is at least 32%. On average, a 24-month cheese has a moisture content of 30%, FDM of 40% and protein in dry matter (PDM) 47%.

1.1.3 Milk

Raw cow's milk is used to produce Parmigiano Reggiano PDO. The diet of dairy cows is based on the use of land for fodder production of Parmigiano Reggiano. In the daily feed, at least 50% of the dry matter of the fodder must be made from hay. The basic ration, consisting of forage, must be appropriately integrated with concentrates capable of balancing the intake of various nutrients in the diet. The dry matter of the feed as a whole must not exceed that supplied by the forage (i.e. a ratio of fodder to feed of not less than 1). In general it is not possible to use feeds that can transmit aromas and flavours from abnormal milk and alter the technological characteristics, feeds that are sources of contamination and feed in poor condition. Producers must adhere to the regulation specification for the PDO, where all the feeds that are allowed and not allowed to feed the dairy cows and other cattle in the herd are listed in detail.

1.1.4 Description and Sensory Characteristics

Parmigiano Reggiano PDO cheese has a cylindrical shape with a slightly convex or almost straight sides, with flat faces with slightly raised edge. The diameter of the flat faces is 35 to 45 cm, whereas the heel height is 20 to 26 cm. One form weighs at least 30 kg. The crust presents a natural straw colour and is approximately 6 mm thick. The colour of the cheese is between light straw to straw. The aroma and flavour of the paste is fragrant, delicate, tasty but not spicy. The structure of the dough is fine grained and flaky.

1.1.5 Method of Manufacture

Milk preparation: Every day, the milk from the evening milking is left in large vats until the morning, when the fatty part spontaneously rises to the surface. This is used for the production of butter.

Starter culture and rennet: As soon as the whole milk form the morning milking arrives from the farm, the skimmed milk from the night before is poured into the typical bell-shaped copper cauldrons where, after warming (33°C), natural calf rennet and fermented whey (natural whey starter), rich in natural lactic ferments, obtained from the processing of the day before are added. The use of any additives is not allowed, and the cheese obtained is completely natural.

Coagulation: The milk coagulates in around 10 min, and the curd is then broken down into minuscule granules until the size of a rice grain is reached, using a traditional tool called '*spino*'.

Cooking: This is where fire comes into the picture, in a cooking process which reaches 55°C, after which the cheesy granules sink to the bottom of the cauldron, forming a single mass. Raising the temperature will help expel whey from the granules and select mesophilic microflora. After around 30 min, the cheese mass is removed, with deft movements, by the cheesemaker.

Moulding: The curd is cut into two parts and wrapped in its typical cloth, and the curd is then placed in a mould that will give it its final shape. During the first day in the mould, called '*fascera*', the curd acidifies and more whey is removed by pressing the curd. Each cheese is given a unique, progressive number using a casein plate, and this number remains with it just like an identity card. A band is used overnight to mark the wheels with the month and year of production onto the cheese, as well as its cheese dairy registration number and the unmistakable dotted inscriptions around the complete circumference of the cheese wheel. The following day, the wheel will be transferred to a stainless steel mould with a convex side for 2–3 days.

Salting: The wheels (about 40 kg, for each wheel has used 600 litres of milk) are salted in saturated brine for 20–25 days.

Drainage: After salting, the wheels will rest in a wooden table for 24 hr to dry and to form the rind. The process of salting and drainage brings the production cycle to a close and inaugurates the no less fascinating cycle of maturation.

Maturation: The cheese wheels are laid out in long rows in the silent maturation rooms. The cheese is allowed to rest on wooden tables, where the outside of the cheese dries, forming a natural crust without being treated in any way and therefore remaining perfectly edible. The story of Parmigiano Reggiano is a long one, and also slow, following the natural rhythm of the seasons. In fact, the minimum maturation time is 12 months, and only at this point can it can be decided if each individual cheese is worthy of the name it was given at its birth.

1.1.6 Relevant Research

The Parmigiano Reggiano is one of the most studied cheeses in the world. Hundreds of papers are available, and some have been selected below. The following scientific papers present the chemical and microbial composition and evolution in raw milk and long ripening cheeses (Gatti et al., 2014; Neviani et al., 2013; Santarelli et al., 2013), such as the importance of natural whey starter for Parmigiano Reggiano (Bottari et al., 2010). Proteolysis and fatty acid composition have been studied extensively (Malacarne et al., 2009; Melia et al., 2009; Sforza et al., 2009, 2012), as also sensory evaluation and evolution (Zannoni & Hunter, 2013; Zannoni, 2010). Different studies have been carried out on the health benefits of Parmigiano Reggiano cheese, and in particular Pampaloni et al. (2011) reported a finding related to bone health.

Special emphasis is given to the results of Pellegrino and Resmini (2001), which are related to the issue of the safety of raw milk cheeses. The authors claim that production of cheese from raw milk is a controversial issue, taking into account the need to supply consumers with food having well-established safety. The debate is about two opposite approaches: (1) adopting mandatory pasteurisation of cheesemilk in every type of cheesemaking and (2) adopting tools other than milk pasteurisation that provide the same level of safety in raw milk cheese. Grana Padano cheese and Parmigiano Reggiano cheese, both known worldwide as Italian Grana Cheese (or Grana-type), are extra-hard cheeses with a recognised PDO, and their product specification provides raw milk as the raw material. However, conditions occurring during in vat-cheesemaking, early ripening and late ripening promote the compositional characteristic of mature IGC, which are expected to ensure safety for consumers.

1.2 Reggianito Cheese – Argentina

Name: Reggianito	
Production area: Pampeana region, Argentina	
Milk: Cow's, raw or pasteurised	

1.2.1 Introduction

Reggianito cheese is a popular Argentinean Grana-type cheese made exclusively from cow's milk. The cheesemaking technology is an adaptation from that used for Italian hard cheeses Grana Padano and Parmiggiano-Regiano, which were introduced to Argentina in the late nineteenth and early twentieth centuries by Italian immigrants. Unlike Italian Grana-type cheeses, Reggianito cheese has higher moisture and fat content, the ripening time is shorter and the size is much smaller.

The production area is extensive, covering mainly the Pampeana region (the provinces of La Pampa, Buenos Aires, Santa Fe, Córdoba and San Luis), in which numerous dairies are located.

1.2.2 Type

Argentinean legislation (Código Alimentario Argentino, 2014) defines Reggianito cheese as a low-moisture or hard cheese with a moisture content up to 35.9%, a minimum FDM of 32% and a ripening period of at least six months. 'Parmesano', 'Reggiano' and 'Sbrinz' are the other names allowed for this cheese by national legislation (Código Alimentario Argentino, 2014).

1.2.3 Description and Sensory Characteristics

Reggianito has the shape of a cylindrical block with flat sides or is slightly convex (diameter and height are approximately 24 cm and 14 cm, respectively), and the usual weight is around 5–10 kg.

Its body is compact, firm with a brittle texture and fine grain, and it has a slight salty and piquant taste with a mild and well-developed flavour. The colour is white or light yellow and uniform. The rind is smooth and consistent.

1.2.4 Method of Manufacture

Milk preparation: Milk having an initial acidity of 14°D–18°D and a pH of 6.6–6.75 is standardised in its fat content to 2.2%–2.4% (ratio of fat to protein: 0.65–0.75), after which milk is pasteurised at high temperature/short time (HTST). Raw milk may also be used, as cheese ripening exceeds 60 days (Código Alimentario Argentino, 2014). After cooling to 31°C–34°C, $CaCl_2$ is added to give a final concentration of 0.02%.

Starter culture: Natural whey starter cultures have been traditionally used, and are still employed in many mid-sized and small plants. It consists of whey from the previous cheese-making incubated overnight. A volume of natural whey starter is added to milk to produce an increase in its acidity of 4°D (400 mg/L lactic acid). In the last decades, technological packages of commercial starters composed of *Lactobacillus helveticus* strains combined with organic acids or acidogens have been introduced into the market to replace 'natural' whey starters; they are currently used in most large dairy plants.

Rennet: Chymosin produced by fermentation is the milk-clotting enzyme usually used, which has displaced adult bovine coagulant.

Coagulation: Milk should be at 31°C–34°C and 6.3–6.45 of pH, before the addition of the coagulant; the coagulation lasts 12–25 min.

Cutting: The curd is cut into uniforms grains with a size of 1–3 mm (the size of a rice grain).

Cooking: The mixture of particles of curd and whey is cooked under stirring to favour the expulsion of whey from the curd by applying two heating steps: from 31°C–34°C to 45°C at 0.5°C/min and from 45°C to 49°C–51°C at 1°C/min. The curd grains lie under the whey for 15–20 min, causing a pre-pressing.

Moulding/pressing: The curd is manually transferred into the moulds and pressed for 24 hr. A pH of 6.3 is registered after cooking, and a drop of the pH to 5.2 is produced after pressing and airing.

Salting: Salting is performed by immersion in brine solution (19°Bé, 12°C–14°C, pH 4.95–5.1) for 1 day per kg cheese.

Ripening: Cheese is ripened at 10°C–15°C and 80% Relative Humidity (RH) for at least 180 days. Reggianito cheese is characterised by proteolysis mediated mainly by the plasmin/plasminogen system resulting in the production of γ-caseins from â-casein. The residual coagulant was usually considered fully inactivated, but recent research showed that denaturation during cooking is at least partially reversed during ripening; hydrolysis of α_{s1}-casein to give α_{s1}-I is verified in Reggianito cheese during ripening (Costabel et al., 2015; Hynes et al., 2004). In addition, the formation and catabolism of free fatty acids is favoured to occur given the prolonged ripening time (Perotti et al., 2005, 2008; Sihufe et al., 2007; Vélez et al., 2010; Wolf et al., 2010).

1.2.5 Relevant Research

Several studies have been carried out in order to characterise different aspects of Reggianito cheese (Bertola et al., 1995; Descalzo et al., 2012; Hough et al., 1996; Hynes, Bergamini & Suárez, 2003; Lombardi et al., 1994; Perotti et al., 2008; Vélez et al., 2015; Wolf et al., 2010).

Strains of *Lb. helveticus* were isolated from natural whey starters (Reinheimer et al., 1995, 1996) and their technological, biochemical and genetic properties were studied (Quiberoni et al., 1998). These strains re-suspended in sterile whey from the same cheese type were used in experimental Reggianito cheesemaking; proteolysis, peptidolysis, lipolysis profiles and sensorial analyses demonstrated that the use of selected strains of thermophilic lactobacilli is an adequate alternative to replace natural whey starter and achieve uniformity of cheese quality (Candioti et al., 2002; Hynes, Bergamini & Suárez, 2003; Milesi et al., 2011; Perotti et al., 2005).

Different approaches were successfully applied to accelerate Reggianito cheese ripening and diversify flavour. Sihufe et al. (2007, 2010; Sihufe, Zorrilla & Rubiolo, 2010) showed an increase in proteolysis and peptidolysis and in the lipolysis patterns of cheeses ripened at higher temperature than usual. This effect was also detected in the sensory characteristics of cheeses.

Application of milk pre-treatments: Heat treatments, mechanical agitation (Vélez et al., 2010) and homogenisation (unpublished results) had a considerable effect on lipolysis and volatile compound production. Also, different cooking temperatures and coagulant enzymes had an impact on proteolysis, peptidolysis and the ripening rate (Costabel et al., 2015).

The use of wild mesophilic lactobacilli (*Lactobacillus paracasei* and *Lactobacillus plantarum*) as adjunct starter is another strategy which satisfactorily improved cheese flavour (unpublished results).

Acknowledgements

The authors thank Sucesores de Alfredo Williner S.A. for providing cheeses for the photographs.

1.3 Sbrinz PDO – Switzerland

Name: Sbrinz PDO	  Sbrinz PDO and its preparation as nuggets, grated, and planed cheese for consumption (Interprofession Sbrinz and Switzerland Cheese Marketing)
Production area: Cantons of Lucerne, Schwyz, Obwalden and Nidwalden, Zug, the district of Muri (canton of Aargau) and a few municipalities in the cantons of Bern and St. Gallen	
Milk: Cow's, raw	

1.3.1 Introduction

Sbrinz PDO is an extra-hard, full-fat cheese made from raw cow's milk that is produced in the central part of Switzerland. Its annual production is 1,700 tonnes; about 90% is consumed by the domestic market. Sbrinz was registered with PDO status in 2001 (FOAG, 2015). This designation has been recognised by the European Union since December 2011 (Swiss Commission for Denominations of Origin and Geographical Indications, 2011).

Sbrinz was first mentioned in a document dating from 1530. At that time, it was only produced during the summer when cows grazed on the mountain pastures. In the middle of the nineteenth century, Sbrinz was increasingly produced in the lowlands throughout the whole year. By the seventeenth century, Sbrinz was exported to other countries (e.g., Italy). Sbrinz was probably named after the village Brienz in the Bernese Highlands, where the cheese was collected and stored before export.

1.3.2 Type

Sbrinz is a full-fat, extra-hard cheese with a dry rind. According to the specifications, a Sbrinz aged 18 months must fulfil the following criteria: moisture content ≤ 33%, FDM of 45.0%–54.9%, moisture in the non-fat substance (MNFS) MFFB ≤ 50% and salt content ≥ 1.3%. The moisture content was 34.4% at 1 month, 33.3% at 4 months, 30.3% at 12 months and 29.4% at 18 months (Sollberger et al., 1991).

When the additional designation 'Hobelkäse' (planed cheese) or 'Gehobelt' (planed) is used (e.g., 'Sbrinz Hobelkäse'), then the cheese must have been manufactured in an alpine cheese dairy during the summer season from milk produced in the same area.

1.3.3 Description and Sensory Characteristics

The cheese must present a yellow, glossy rind; the cheese body is rather dry, hardly malleable, somewhat crumbly and contains visible crystals. Usually, there are no holes in the cheese body, although a few small holes are tolerated. The colour is ivory to bright yellow. Sbrinz has a fruity aroma with a slightly roasted character of chicory and a salty and slightly sweet taste.

1.3.4 Method of Manufacture

The manufacture of Sbrinz is carried out in small-scale dairies. The milk is processed in copper vats. The use of processing aids other than starter cultures, rennet, non-GMO rennet substitutes and salt is prohibited.

Milk preparation: Only raw cow's milk is processed for Sbrinz. The milk comes exclusively from farms that are located within the indicated geographical area. The preparation, storage and distribution of silage of any kind is forbidden on these farms. A minimum percentage of 70% of the cattle feed (calculated on dry matter basis) must come from the fodder area of the farm. Other fodders such as supplementary and concentrate feeds are defined in detail in the Animal Feed Book Ordinance (FOAG, 2015).

The distance between the milk production plant and the cheese dairy may not exceed 30 km. Milk must usually be delivered to the cheese dairy twice a day, immediately after milking. In this case, the milk is only pre-cooled to about 20 °C prior to transportation. The raw milk is partially skimmed either mechanically or by creaming; it may not be heat-treated or subjected to other treatments such as bactofugation, ultrafiltration or microfiltration. At the beginning of cheesemaking, milk in the vat has to be less than 24 hold.

Starter cultures: For the production of Sbrinz, thermophilic starter cultures derived from traditional whey cultures are used. The cultures are provided as liquid stock cultures by Agroscope. All cultures consist of strains of the species *Streptococcus thermophilus* and *Lactobacillus delbrueckii* ssp. *lactis*. In the cheese dairy, the bulk starter is prepared daily from sterile skim milk. Usually, 1 litre of fresh bulk starter is added to 1000 litres of milk about 30 min before the addition of rennet.

Rennet: Coagulation is carried out at 31°C–32°C, and calf rennet is most often used. However, microbial rennet substitutes derived from *Cryphonectria parasitica* (e.g., Suparen®) are also used. Flocculation should occur 30–35 min after the addition of rennet. Usually, the coagulum is cut 10–15 min after flocculation has occurred.

Cutting/Scalding: The final grain size is the size of a wheat grain. The curd is heated to a temperature of 54°C–57°C without added water (there is no curd washing). After the desired temperature has been reached, the curd is stirred for another 10–20 min until the optimal firmness of the grains is obtained. At the end of this process, the temperature is 52°C–53°C.

Moulding/draining: The curd is usually transferred into moulds by pumping. Attention must be paid to the whey level in the moulds, which should always be high enough to prevent air from being trapped in the cheese mass.

Pressing: Pressing lasts 10–16 hr, and the maximum pressure is usually 0.5 bar. Good cohesion of the grains must be ensured.

Salting: Salting begins immediately after the cheeses have been turned out of the moulds. They are then immersed in a brine bath with a salt content of at least 20°Bé and a temperature of 10°C–15°C for a minimal length of 15 days.

Maturation: The cheese is dry-matured and kept for at least 15 days at 12°C–20°C. A natural film of fat builds on the surface. The wheels are rubbed off weekly with a cloth. Then, the cheese wheels are ripened in an upright position at a temperature of 9°C–14°C and an RH of 60%–75% until they reach the minimal age of 18 months for consumption.

Mature Sbrinz can be stored at any temperature between 2°C and 14°C. Although the cheese preserves well due to its low water activity, care must be taken to avoid the growth of moulds, either through continued care of the wheels or through the exclusion of oxygen in packed cheese.

1.3.5 Relevant Research

Propionic acid fermentation in Sbrinz causes significant financial losses. The monitoring of propionic acid bacteria (PAB) in raw milk with the plate count method is slow (7–10 days) and does not discriminate between the four species of dairy PAB. Recently, Turgay et al. (2016) investigated a total of 51 vat milk samples originating from dairies producing Sbrinz (14), Le Gruyère (29) and Emmentaler (8) with a new procedure that includes an efficient extraction of the bacterial flora from milk and cheese as well as a fast, sensitive and species-specific quantification of *Propionibacterium freudenreichii, Propionibacterium thoenii, Propionibacterium jensenii* and *Propionibacterium acidipropionici* by qPCR. Vat milk from Sbrinz dairies showed the lowest contamination. However, up to 10^3 copies/mL of PAB were detected in individual vat milk samples, with *P. freudenreichii* being the dominant species. In addition, the concentration of PAB and propionic acid was analysed in the corresponding lots of Sbrinz aged for 11 months. Despite the high scalding temperature of 54°C–57°C, counts in the range of 10^6–10^7 copies/g were observed in some of the cheeses. In some cases, *P. freudenreichii* even caused faulty fermentation when present in vat milk in a concentration of 10^1–10^2 copies/mL.

References

Bertola N., Bevilacqua, A. & Zaritzky, N. (1995). Rheological behavior of Reggianito Argentino cheese packaged in plastic film during ripening. *LWT – Lebbensmittel-Wissenschaft und Technologie*, 28, 610–615.

Bottari, B., Santarelli, M., Neviani, E. & Gatti, M. (2010). Natural whey starter for Parmigiano Reggiano: Culture-independent approachNatural whey starter for Parmigiano Reggiano: Culture-independent approach. *Journal of Applied Microbiology*, 108 (5), 1676–1684.

Candioti, M., Hynes, E., Quiberoni, A. Palma, S. B., Sabbag, N. & Zalazar, C. A. (2002). Reggianito Argentino cheese: Influence of *Lactobacillus helveticus* strains isolated from natural whey cultures on cheese making and ripening processes. *International Dairy Journal*, 12 (11), 923–931.

Código Alimentario Argentino (2014). *Cap. VIII: Alimentos lácteos*. http://www.anmat.gov.ar/alimentos/codigoa/CAPITULO_VIII.pdf. Accessed: June 2015.

Costabel L., Bergamini C., Pozza L., Cuffia F., Candioti M. & Hynes E. (2015). Influence of chymosin type and curd scalding temperature on proteolysis of hard cooked cheeses. *Journal of Dairy Research*, 82, 375–384.

Descalzo A., Rossetti L., Páez R., Grigioni G., García P., Costabel L., Negri L., Antonacci L., Salado E., Bretschneider G., Gagliostro G., Comerón E. & Taverna M. (2012). Differential characteristics of milk produced in grazing systems and their impact on dairy products. In Chaiyabutr N. (Ed.), *Milk Production – Advanced Genetic Traits, Cellular Mechanism, Animal Management and Health*. InTech, pp. 339–368, doi: 10.5772/50760. Accessed from: http://www.intechopen.com/books/milk-production-advanced-genetic-traits-cellular-mechanism-animal-management-and-health/differential-characteristics-of-milk-produced-in-grazing-systems-and-their-impact-on-dairy-products

FOAG (2015). *Register of appellation of origin and geographical indications*. Federal Office for Agriculture. An Office of the Federal Department of Economy, Bern, Switzerland http://www.blw.admin.ch/ [last assessed: 28.04.2015]

Gatti, M., Bottari, B., Lazzi, C., Neviani, E. & Mucchetti, G. (2014). Invited review: Microbial evolution in raw-milk, long-ripened cheeses produced using undefined natural whey starters. *Journal of Dairy Science*, 97 (2), 573–591.

Hough, G., Califano, A., Bertola, N., Bevilacqua, A., Zaritzky, N., Martinez, E. & Vega, M. (1996). Partial least squares correlations between sensory and instrumental of flavor and texture for Reggianito grating cheese. *Food Quality and Preference*, 7 (1), 47–53.

Hynes, E., Aparo, L. & Candioti, M. (2004). Influence of residual milk-clotting enzyme on α_{s1} casein hydrolysis during ripening of Reggianito Argentino cheese. *Journal of Dairy Science*, 87 (3), 565–573.

Hynes, E., Bergamini, C. & Suárez, V. (2003). Proteolysis on Reggianito Argentino cheeses manufactured with natural whey cultures and selected strains of *Lactobacillus helveticus*. *Journal of Dairy Science*, 86 (12), 3831–3840.

Lombardi, A., Bevilacqua, A. & Califano, A. (1994). Variation in organic acids content during ripening of Reggianito cheeses in air-tight sealed bags. *Food Chemistry*, 51, 221–226.

Malacarne, M., Summer, A., Franceschi, P., Formaggioni,P., Pecorari, M. Panari, G. & Mariani, P. (2009). Free fatty acid profile of Parmigiano-Reggiano cheese throughout ripening: Comparison between the inner and outer regions of the wheel. *International Dairy Journal*, 19 (10), 637–641.

Melia, S., Losi, G. & Castagnetti, G. B. (2009). The influence of milk κ-casein and β-lactoglobulin phenotypes on fatty acid composition of milk from Reggiana cows. *Dairy Science & Technology*, 89 (1), 115–122.

Milesi, M., Bergamini, C. & Hynes, E. (2011). Production of peptides and free amino acids in a sterile extract describes peptidolysis in hard-cooked cheeses. *Food Research International*, 44, 765–773.

Neviani, E., Bottari, B., Lazzi, C. & Gatti, M. (2013). New developments in the study of the microbiota of raw-milk, long-ripened cheeses by molecular methods: The case of Grana Padano and Parmigiano Reggiano. *Frontiers in Microbiology*, 4, 36.

Pampaloni, B., Bartolini, E. & Brandi, M. L. (2011). Parmigiano Reggiano cheese and bone health. *Clinical Cases in Mineral and Bone Metabolism*, 8 (3), 33–36.

Pellegrino, L. & Resmini, P. (2001). Cheesemaking conditions and compositive characteristics supporting the safety of the raw milk cheese Italian Grana. *Scienza e Tecnica Lattiero Casearia*, 52 (2), 105–114.

Perotti, M. C., Bernal, S. M., Meinardi, C. A. & Zalazar, C. A. (2005). Free fatty acid profiles of Reggianito Argentino cheese produced with different starters. *International Dairy Journal*, 15, 1150–1155.

Perotti, M. C., Bernal, S., Wolf, V. & Zalazar, C. A. (2008). Perfil de ácidos grasos libres y características sensoriales de quesos Reggianito elaborados con diferentes fermentos. *Grasas y Aceites*, 59 (2), 152–159.

Quiberoni, A., Tailliez, P., Quénée, P., Suárez, V. & Reinheimer, J. (1998). Genetic and technological diversities among wild *Lactobacillus helveticus* strains. *Journal of Applied Microbiology*, 85, 591–596.

Reinheimer, J., Quiberoni, A., Tailliez, P., Binetti, A. & Suárez, V. (1996). The lactic acid microflora of natural whey starters used in Argentina for hard cheese production. *International Dairy Journal*, 6, 869–879.

Reinheimer, J., Suárez, V., Bailo, N. & Zalazar, C. (1995). Microbiological and Technological characteristics of natural whey cultures for Argentinian hard cheese production. *Journal of Food Protection*, 58 (7), 796–799.

Santarelli, M., Bottari, B., Malacarne, M., Lazzi,C., Sforza,S., Summer, A., Neviani, E. & Gatti, M. 2013. Variability of lactic acid production, chemical and microbiological characteristics in 24-hour Parmigiano Reggiano cheese. *Dairy Science & Technology*, 93 (6), 605–621.

Sforza, S., Cavatorta, V., Galaverna, G., Dossena, A. & Marchelli, R. (2009). Accumulation of non-proteolytic aminoacyl derivatives in Parmigiano-Reggiano cheese during ripening. *International Dairy Journal*, 19 (10), 582–587.

Sforza, S., Cavatorta, V., Lambertini, F., Galaverna, G., Dossena, A. & Marchelli, R. (2012). Cheese peptidomics: A detailed study on the evolution of the oligopeptide fraction in Parmigiano-Reggiano cheese from curd to 24 months of aging. *Journal of Dairy Science*, 95 (7), 3514–3526.

Sihufe, G. A., Zorrilla, S. E. & Rubiolo, A. C. (2010a). The influence of ripening temperature and sampling site on the proteolysis in Reggianito Argentino cheese. *LWT – Food Science and Technology*, 43, 247–253.

Sihufe, G. A., Zorrilla, S. E., Mercanti, D. J., Perotti, M. C., Zalazar, C. A. & Rubiolo, A. C. (2007). The influence of ripening temperature and sampling site on the lipolysis in Reggianito Argentino cheese. *Food Research International*, 40, 1220–1226.

Sihufe, G. A., Zorrilla, S. E., Sabbag, N. G., Costa, S. C. & Rubiolo, A. C. (2010b). The influence of ripening temperature on the sensory characteristics of Reggianito Argentino cheese. *Journal of Sensory Studies*, 25, 94–107.

Sollberger, H., Glättli, H., Nick, B., Rüegg, M., Sieber, R. & Steiger, G. (1991). Untersuchung über den Reifungsverlauf guter Sbrinz-Käse. *Schweizerische Milchwirtschaftliche Forschung*, 20 (4), 63–69.

Swiss Commission for Denominations of Origin and Geographical Indications (2011). *Annual report 2011. Federal Office for Agriculture FOAG*. http://www.blw.admin.ch/ [last assessed: 08.09.2015].

Turgay, M., Schaeren, W., Wechsler, D., Bütikofer, U. & Graber, H. U. (2016). Fast detection and quantification of propionibacteria in milk samples using real-time quantitative polymerase chain reaction (qPCR). *International Dairy Journal*, 61, 37–43.

Vélez, M. A., Bergamini, C., Ramonda, B., Candioti, M., Hynes, E. & Perotti, M. C. (2015). Influence of cheese making technologies on plasmin and coagulant associated proteolysis. *LWT – Lebbensmittel-Wissenschaft und Technologie*, 64, 282–286.

Vélez, M. A., Perotti, M. C., Wolf, I. V., Hynes, E. & Zalazar, C. (2010). Influence of milk pretreatment on production of free fatty acids and volatile compounds in hard cheeses: Heat treatment and mechanical agitation. *Journal of Dairy Science*, 93, 4545–4554.

Wolf, I. V., Perotti, M. C., Bernal, S. M. & Zalazar, C. A. (2010). Study of the chemical composition, proteolysis, lipolysis and volatile compounds profile of commercial Reggianito Argentino cheese: Characterization of Reggianito Argentino cheese. *Food Research International*, 43, 1204–1211.

Zannoni, M. & Hunter, E. A. (2013). Evaluation of a sensory scorecard for grated Parmigiano-Reggiano cheese. *Italian Journal of Food Science*, 25 (1), 23–34.

Zannoni, M. (2010). Evolution of the sensory characteristics of Parmigiano-Reggiano cheese to the present day. *Food Quality and Preference*, 21 (8), 901–905.

2

Hard Cheeses

Katja Hartmann[1], Giuseppe Licitra[2], Elisabeth Eugster-Meier[3], Marie-Therese Fröhlich-Wyder[4], Ernst Jakob[4], Daniel Wechsler[3,4], Jean L. Maubois[5], Kimon-Andreas G. Karatzas[6], Thomas Bintsis[7], Efstathios Alichanidis[8], Maria Belén López Morales[9], Françoise Berthier[10], İrem Uzunsoy[11], Barbaros Özer[12] and Ylva Ardö[13]

[1] Anton Paar GmbH, Germany
[2] Department of Agriculture, Nutrition and Environment, University of Catania, Catania, Italy
[3] Bern University of Applied Sciences, School of Agricultural, Forest and Food Sciences HAFL, Zollikofen, Switzerland
[4] Agroscope, Research Division Food Microbial Systems, Federal Department of Economic Affairs, Education and Research EAER, Bern, Switzerland
[5] Dairy Research Laboratory, INRA Rennes, France
[6] Department of Food and Nutrition Sciences, The University of Reading, United Kingdom
[7] 11 Parmenionos, 50200, Ptolemaida, Greece
[8] Department of Food Science and Technology, School of Agriculture, Aristotle University of Thessaloniki, Thessaloniki, Greece
[9] Food Science and Technology Department, International Excellence Campus for Higher Education and Research 'Campus Mare Nostrum', Veterinary Faculty, University of Murcia, Spain
[10] Unité de Recherches en Technologie et Analyses Laitières Rue de Versailles, France
[11] Bülent Ecevit University Caycuma Vocational High School, Department of Food Technology, Zonguldak, Turkey
[12] Ankara University, Faculty of Agriculture Department of Dairy Technology, Ankara, Turkey
[13] Department of Food Science, University of Copenhagen, Denmark

2.1 Allgäu Mountain Cheese – Germany

Name: Allgäu mountain cheese	 © www.bayern.by - Gert Krautbauer
Production area: Allgäu region (southern Germany)	
Milk: Cow's, raw	

Global Cheesemaking Technology: Cheese Quality and Characteristics, First Edition.
Edited by Photis Papademas and Thomas Bintsis.

2.1.1 Introduction

Allgäu mountain cheese has held a PDO (Protected Designation of Origin) label since January 1997 (EC, 1997). Allgäu mountain cheese is a traditional and very popular cheese from the Allgäu region in Germany and is manufactured from raw cow's milk. It is mostly produced by small artisanal dairies in rural areas with extensive dairy farming due to the climate and hilly countryside, which has a positive effect on milk quality and thus the produced cheeses. The production of mountain cheese (German: *Bergkäse*) has a long tradition in the German Allgäu region. A document from 1059 reports on the first production of mountain cheese in the Alps in the town of Oberstdorf. Until today, it is mostly produced on so-called 'Sennalpen' (where cows, goats and sheep graze on alpine pastures throughout the summer time) in alpine huts at altitudes between 900 and 1800 m. However, with increased milk production, several dairies developed in villages in the valleys, which produce Allgäu mountain cheese all year long (Hofmann, 2005; Mair-Waldburg, 1974). The geographical regions of Allgäu mountain cheese production include the districts Lindau, Oberallgäu, Ostallgäu, Unterallgäu, Ravensburg and the Lake Constance district as well as the towns of Kempten, Kaufbeuren and Memmingen, located in the districts of Bavaria and Baden-Württemberg in the south of Germany (Deutsche Käseverordnung, 2013).

2.1.2 Type

The specification for Allgäu mountain cheese in the German regulation on cheese requires a fat-in-dry-matter (FDM) content ≥ 45% and a dry matter (DM) content ≥ 62%.

2.1.3 Description and Sensory Characteristics

Allgäu mountain cheese is rind-matured and produced as a cheese wheel with a weight between 15 and 50 kg (Deutsche Käseverordnung, 2013). The cheese texture is firm but elastic and smooth. Eye formation is sparely developed and has the size of peas. The cheese rind is of a dark yellow to brownish colour, whereas the cheese inside is of a dull yellow colour. The cheese has a mild, nutty-like and slightly sweet flavour, which becomes full-bodied and intense with increased maturation (Allgäuer Bergkäse, 2015; Hofmann, 2005; Mair-Waldburg, 1974).

2.1.4 Method of Manufacture

Allgäu mountain cheese is traditionally produced in copper vats in mountain huts and village dairies at altitudes between 900 and 1800 m (Bergkäse, 2015; Weltgenusserbe Bayern, 2015).

Milk preparation: Raw milk is used for the production of Allgäu mountain cheese, which is produced in the permitted manufacturing areas. Feeding of silage is forbidden, and the cheesemilk is not allowed to be heated >40°C. Furthermore, the addition of preservatives or colouring agents is not allowed. The cheesemilk is usually delivered twice a day to the manufacturing dairy. The filtrated milk is tempered to 15°C and stored in copper vats until the next morning. A mixture of morning and evening milk is then adjusted to a fat content of 3.1% to 3.4% (Allgäuer Bergkäse, 2015; Deutsch Käseverordnung, 2013; Mair-Waldburg, 1974).

Starter culture: A mixture of mesophilic and thermophilic bacterial starter cultures is added to the cheesemilk at temperatures between 30°C and 32°C (Kammerlehner, 2009; Mair-Waldburg, 1974).

Rennet: Calf's rennet is added to the cheesemilk at 30°C–32°C, and coagulation lasts between 25 and 30 min (Mair-Waldburg, 1974).

Cutting: The curd is cut with a cheese harp until the curd reaches the size of peas or hazelnuts.

Scalding: The cheese curd is scalded at temperatures between 48°C and 52°C and stirred for another 20–40 min.

Curd draining: The cheese curd is manually transferred to the moulds. The cheese curd is pressed for 20–24 hr with increasing pressure up to 5–8 kg per kg cheese; during the pressing, the cheese is frequently turned.

Salting: The cheeses are brined for 1–3 days at a concentration of 20° Bé–22° Bé and a temperature of 10°C–15°C.

Maturation: The temperatures during ripening depend on the location of the fermentation cellar and are usually between 14°C–16°C. An increased ripening temperature (18°C) leads to increased development of gas opening. The cheeses are stored at 90% Relative Humidity (RH) and are usually turned and treated with brine every other day. Maturation lasts between four months for up to one year (Bergkäse, 2015; Deutsche Käseverordnung, 2013).

Storage: Until marketing, the cheeses are stored at 10°C. The German regulation requires a minimum ripening time of four months. The ideal packaging of Allgäu mountain cheeses is in foil or with a cheese cover (Weltgenusserbe Bayern, 2015).

2.2 Asiago PDO – Italy

<table>
<tr><td>Name: Asiago PDO</td><td rowspan="3"></td></tr>
<tr><td>Production area: Vicenza and Trento and part of the provinces of Treviso and Padua, Veneto region</td></tr>
<tr><td>Milk: Cow's, raw or pasteurised</td></tr>
</table>

2.2.1 Introduction

In 1955, the Asiago cheese obtained its name, and on 21 December 1978, it obtained the Denomination of Origin (DO) under Italian law 125/54. On 6 December 1996, it received the recognition of PDO (EC, 1996). Asiago PDO has been produced since the year 1000 on the Asiago plateau. Between the tenth and fifteenth centuries in the plateau Asiago, fertile in good herbs, sheep production was the predominant activity. It was actually called 'Pegorin' by local residents. In 1500, the local cheesemakers grasped the importance of using cow's milk, and since 1700, thanks to the gradual improvement of cheesemaking techniques, production expanded to other areas surrounding the plateau. In the eighteenth and nineteenth centuries, Asiago d'Allevo was produced after long maturation. Only in the 1920s of the twentieth century was a shorter-maturing Asiago cheese, called Asiago Pressato, produced. It is a technological variant that was already adopted in huts or on mountain pastures, especially in the first period of pasturing cattle. This 'new' product has won the approval of the modern consumer, who prefers sweet flavours and a soft texture. The area of milk collection and production includes the entire territory of the provinces of Vicenza and Trento and part of the provinces of Treviso and Padua, corresponding to their range, the foothills (fascia Pedemontana) of the Veneto region. The production areas, located at an altitude of not less than 600 m, are classified as mountain territory. The Asiago PDO, with 23,000 tonnes, is ranked fourth by the quantity of cheese produced in Italy. Asiago's popularity among Italians and foreign consumers allowed a fast growth that is almost unequalled in the industry and that over the years has generated a profitability of the dairy sector which has allowed, through the phenomenon of aggregated cooperatives (cooperative dairies), the continuation of small farms (family farms).

2.2.2 Type

The Asiago PDO is categorised into two different types of cheese: (1) Asiago pressato, which is a short- and medium-aged (minimum 20 days), semi-hard cheese and (2) Asiago d'Allevo (minimum 60 days of maturing) which, depending on the length of the ripening, is divided into three categories: (1) Asiago d'Allevo Mezzano, for aged cheese 4–6 months, (2) Asiago d'Allevo aged over 10 months and (3) Asiago d'Allevo Stravecchio, aged over 15 months. The cheese produced in mountainous areas (>600 m) may use the term 'Asiago Prodotto della Montagna' and may relate to both the Asiago Pressato (slightly aged for at least 30 days) and the Asiago d'Allevo (also with a minimum more prolonged aging: 90 days from the last day of the month of production). The moisture content of Asiago Pressato is 43.1% and that of Asiago d'Allevo is 31.6%, while the fat and protein content have been respectively recorded as 30.3% and 21.8% for the former and 35.5% and 28% for the latter (Manzi et al., 2007).

2.2.3 Milk

The production regulations of the Asiago PDO list both the fodder that can and cannot be used in animal feeding, (Anonymous, 2006). For the production of 'Asiago Prodotto di Montagna', it is also forbidden to supply the animals with any kind of silage.

2.2.4 Description and Sensory Characteristics

Asiago Pressato cheese has a cylindrical shape with a diameter of 30–40 cm and a height of about 11–15 cm. The average weight of the forms ranges from 11 to 15 kg. The crust is thin and elastic; the body inside is soft, buttery, white or slightly yellow and with noticeably irregular holes. The taste is sweet, delicate and pleasant. Asiago d'Allevo cheese has a cylindrical shape with a diameter of 30–36 cm and a height of about 9–12 cm. The average weight of the forms ranges from 8 to 12 kg. The Vézzena has a hard and grainy paste, yellowish colour, with very few small holes; the taste and flavour are enhanced with recognisable scents of herbaceous pastures and mountain meadows in the area.

2.2.5 Method of Manufacture

The technology of production varies according to the maturation period of the cheese produced: Asiago Pressato (fresh) or Asiago d'Allevo (aged).

2.2.5.1 Asiago Pressato PDO

Milk: Asiago Pressato PDO is produced using raw whole milk or pasteurised at 72°C for 15 s.

Starter cultures/rennet: Commercial starter culture is added in the milk at a temperature of about 35°C or milk culture of the previous processes, and liquid calf rennet. Some cheesemakers add small amounts of sodium chloride to the milk.

Coagulation / curd cutting and scalding: The clotting times range from 15 to 25 min, and the clot is broken down to the size of a walnut shell/hazelnut. Syneresis is stimulated by the addition of hot water, and the curd will reach a temperature of 44 ± 2°C.

Moulding: The curd is then placed in moulds with perforated walls and dry salt, and with a press, usually hydraulic, is pressed for about 4 hr (the most common duration), up to a maximum of 12 hr. Subsequently, the forms are wrapped laterally with the moulds of plastic which imprint the mark 'Asiago' around the entire shape and are put in a cool room, called '*frescura*' (at 10°C–15°C with an RH of 80%–85%), for about 48 hr to dry.

Salting: Salting is performed in brine (18°Bé–22°Bé) for two days to achieve a concentration of sodium chloride of approximately 1.7% (w/w).

Maturation: The maturing processes will last for a minimum period of 20 up to a maximum 60 days.

2.2.5.2 Asiago d'Allevo PDO

Milk: Asiago d'Allevo PDO is produced from cow's milk from two milkings, morning and evening, one of which is skimmed by natural surfacing. It uses raw or thermised milk at 57°C–68 °C for 15 s.

Starter cultures/rennet: See Asiago Pressato PDO.

Coagulation/ curd breaking and scalding: The clotting times range from 15 to 30 min, and the curd is broken down to the size of a hazelnut. For long-aged cheese (Vecchio / Stravecchio), the curd is broken down to the size of a grain of rice.

Moulding: See Asiago Pressato PDO.

Salting: The salting phase will be done in brine (18° Bé–22° Bé) for a variable time depending on the length of the maturation. In any case, a concentration of sodium chloride of approximately 2.5% (w/w) is achieved.

Maturation: The length of maturation will give rise to different types of Asiago d'Allevo PDO: (1) Asiago mezzano (aged for 3–8 months) – the body is compact, although still quite soft, pale yellow in colour rather intense, with small- and medium-sized holes, very tasty but still sweet; (2) Asiago vecchio (aged for 10–18 months) – hard cheese, firm, pale yellow, with medium holes, strong flavour tending towards the piquant and (3) Asiago stravecchio (aged for at least 15 months, but often for more than two years) – smooth and regular rind with a very hard body, grainy, straw-coloured, with small holes and an intense, penetrating and spicy taste.

2.2.5.3 Asiago 'Prodotto di Montagna'

Most of the mountain cheeses are made for long aging. Among them is one of the oldest cheeses in the dairy tradition of Trento province and part of Vicenza that is made exclusively from the milk of cows of the summer pastures of the Plateau Vézzena, which takes the name of this specific product: Vézzena. The request for a specific PDO for Vézzena has been made. Breeders follow a strict discipline of production where the animals must be kept on pasture and supplemented only with high-quality concentrates; no silage, industry by-products or, of course, GMO feed is allowed. The production technique of Vézzena respects the traditional one of semi-fat alpine cheese, but Vézzena is unique to the pasture land principles of the alpine plateau and to the long maturation.

Milk: It is used only raw cow's milk, from two milkings, morning and evening; the last one is skimmed by natural surfacing.

Starter culture/rennet: The milk is heated slowly by adding the milk culture and, then, at 33°C–35°C, liquid calf rennet; coagulation occurs in 20–25 min.

Coagulation/curd breaking and scalding: The coagulum is broken with the 'lyre', to the extent of a grain of corn. During this process, the curd is slowly cooked to 45°C–48°C, the curd mass is left on the bottom and is pulled out of the whey.

Moulding: The curd blocks, placed in wooden moulds, are subjected to pressing. In the evening the weights are taken off, and the forms are placed in a humid and warm room called 'frescura'.

Salting: Dry or brine salting.

Maturation: The cheeses are placed on wooden boards, which are cleaned and treated with linseed oil once a month. The Vézzena is sold after 18 months (old) or 24 months (extra-mature).

2.2.6 Relevant Research

The effects of the period of milk production and ripening on the quality traits of Asiago cheese have been studied (Segato et al., 2007). Another study (Renna et al., 2012) was performed to verify the suitability of using fatty acids (FAs), terpenoids and electronic nose response as potential tracers of the seasonal variations in Asiago d'Allevo PDO cheese. The results showed that coupling FA and terpenoid information could be a suitable method for tracing Asiago d'Allevo PDO cheeses according to their season of production. However, no reliable information at this level seemed to be obtainable from the electronic nose response. The traceability (Favaro et al., 2005) of Asiago mountain cheese was established by analysing samples of herbaceous species, milk and cheese of mountain origin using the headspace solid-phase micro-extraction sampling procedure coupled with gas chromatography–mass spectrometry. As preliminary work had highlighted the characteristic presence of sesquiterpenes in Asiago mountain cheese, these species were considered effective markers of mountain origin. To discriminate the production chain of Asiago d'Allevo cheese, it was also studied through nuclear magnetic resonance spectroscopy and principal component analysis (Schievano et al., 2008), and by near-infrared spectroscopy as an alternative to chemical and colour analysis (Cozzi et al., 2009). Safety aspects of the Asiago cheese have been the subject of extensive studies: from the production of bacteriocin by *Lactococcus lactis* ssp. *lactis* active on lactate-fermenting clostridia (Zezza et al., 1993) through the genetic biodiversity of wild-type *Streptococcus thermophilus* strains (Morandi & Brasca, 2012) and the presence and characterisation of peptides inhibiting *Listeria* spp. growth in Asiago d'Allevo cheese (Lignitto et al., 2012; Thi et al., 2014).

2.3 Berner Alpkäse PDO and Berner Hobelkäse PDO – Switzerland

<table>
<tr><td>Name: Berner Alpkäse PDO and Berner Hobelkäse PDO</td><td rowspan="3">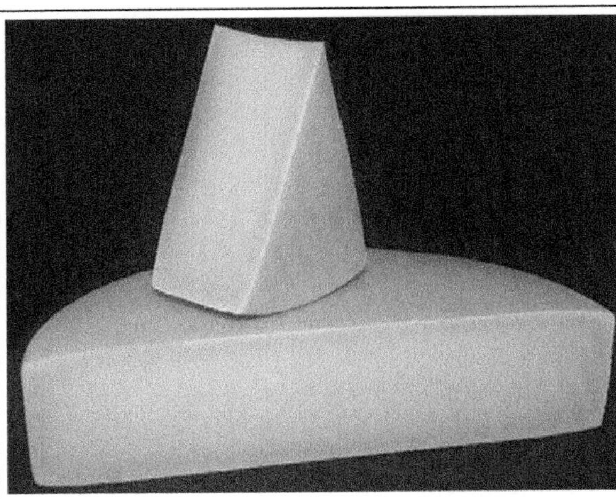
Berner Alpkäse PDO, about 12 months old (Agroscope)

Berner Hobelkäse PDO (Switzerland Cheese Marketing)</td></tr>
<tr><td>Production area: Summer pasture holdings in the eastern part of the canton of Bern, for example, Berner Oberland (Bernese Highlands) and Emmental, and in a few neighbouring districts in the cantons of Lucerne, Vaud and Fribourg</td></tr>
<tr><td>Milk: Cow's, raw</td></tr>
</table>

2.3.1 Introduction

Berner Alpkäse and Berner Hobelkäse, an extra hard variety of Berner Alpkäse, were registered as PDOs in 2004 (FOAG, 2015). This designation has been recognized by the European Union since December 2011 (Swiss Commission for Denominations of Origin and Geographical Indications, 2011).

Berner Alpkäse PDO is a full-fat hard cheese made from raw cow's milk. These alpine cheeses are smear-ripened up to the age of six months. For the production of Berner Hobelkäse PDO cheese wheels, Berner Alpkäse cheeses with good storage quality are washed at the age of 6–8 months and further dry-ripened up to a minimum age of 18 months. As a result of the prolonged dry-ripening, the cheese becomes an extra-hard cheese that is thinly sliced for consumption using a cheese plane ('Hobel' in German). The total annual production of the two cheese varieties is about 1,000 tonnes (Jakob et al., 2007). Berner Alpkäse and Berner Hobelkäse are mainly consumed within their region of production. In the mountain region of the canton of Bern, the production of rennet-curd cheese began early in the sixteenth century. A chronicle dating from 1548 first mentions 'Saanerkäss', a local designation for Bernese alpine cheese that is produced in the area of Saanen. Berner Alpkäse is still manufactured exclusively in alpine cheese dairies during the summer season.

2.3.2 Type

Berner Alpkäse is a full-fat hard cheese that is ripened for at least 4.5 months. The smear-rind of the cheese is usually washed off before sale. The moisture content of the cheese ranges from 27% to 37% and is dependent on its age. Berner Hobelkäse that is ripened for at least 18 months has a moisture content of 20% to 32%. The salt content of Berner Alpkäse and Berner Hobelkäse is not specified but is usually within the range of 1.3%–1.9% and 1.7%–2.1%, respectively (Jakob et al., 2007).

2.3.3 Description and Sensory Characteristics

Berner Alpkäse has the shape of a wheel (diameter: 18–48 cm, height: 9–12 cm, weight: 5–16 kg) and a brownish-yellow rind. In contrast, Berner Hobelkäse has a very firm rind that is covered with a natural fat film. Berner Alpkäse must have only a few or no openings. The texture is medium-hard to hard, slightly crumbly and grainy and not sticky. The flavour is slightly sweet, slightly to moderately sour and salty, milky and slightly animal, mildly to distinctly spicy and slightly roasted, with or without a hint of smoke. Berner Hobelkäse must not have any cracks. The cheese body is hard, slightly dry and slightly to moderately crumbly. It is also not sticky. Crystals are visible. The flavour is slightly sour and moderately salty, spicy and animal. The taste is slightly to distinctly roasted, with or without a hint of smoke.

2.3.4 Method of Manufacture

Berner Alpkäse is manufactured during the alpine pasture season but not before May 10 or after October 10. The milk must be processed in a copper vat that is heated directly or indirectly by a wood fire. The use of processing aids other than starter cultures, rennet and salt is prohibited.

Milk: The milk processed for Berner Alpkäse is obtained exclusively from cows grazing on alpine pastures located within the geographical area of production. The average daily ration of a cow must contain at least 90% alpine forage (on a dry matter basis). On farms in the lowlands, the feeding of silage must be stopped at least two weeks before the start of the alpine season in order to avoid cheese defects caused by *Clostridium tyrobutyricum*. Usually, the alpine cheese dairies process only raw milk that has been produced on their own summer pasture holdings. Transportation of milk hardly occurs, and distances may not exceed 10 km or 30 min from the dairy farm. The fresh evening milk is cooled below 18°C and is usually stored overnight in a copper vat. In the morning, the milk is partially skimmed with a ladle before the fresh morning milk is added to the vat. The oldest milk in the vat may not be older than 15 hr at the beginning of the cheesemaking process. Only raw milk may be processed; pre-treatments of cheese milk (e.g., bactofugation) are forbidden.

Starter cultures/rennet: For the production of Berner Alpkäse, a self-cultured thermophilic whey culture is used. Individual dairies start by using a freeze-dried culture that is specifically produced for the manufacture of alpine cheese by Agroscope. The culture contains *S. thermophilus, Lb. delbrueckii* ssp. *lactis* and *Lc. lactis*. The whey used for the preparation of the whey culture is briefly heated to 61°C–63°C and is thereafter immediately cooled and incubated at 38°C for 20 hr. The acidity of the whey culture should be 25°SH–32°SH. A quantity of 3–7 litres of fresh whey culture is added to 1000 litres of milk about 30 min before the addition of rennet.

Coagulation/curd cutting and scalding: Coagulation is carried out at 31°C–32°C, and usually calf rennet is used. Flocculation should occur 30–35 min after the addition of rennet. Usually, the coagulum is cut 10–15 min after flocculation has occurred. The final grain size corresponds to the size of a wheat grain. The curd is thereafter slowly heated to 50°C–53°C without curd washing. The temperature should not exceed 40°C after the first 15 min, and the final temperature should be achieved 30–50 min after the scalding begins. Then, the curd is stirred for another 5–20 min until the optimal firmness of the grains is obtained.

Moulding: The curd is taken out of the vat with cheesecloth. In larger alpine cheese dairies, the curd is often transferred first into a rectangular basin, where it is pre-pressed into a block. Then, the cheese mass is cut into equal pieces, which are transferred to round moulds where they are wrapped in a cheesecloth before pressing starts.

Pressing: Pressing takes at least 15 hr, and a pressure of 6–10 kg per kg of cheese is applied (with an initial pressure of only 1–2 kg). During pressing, the cheeses are turned several times. For the last 2 hr, the cheeses are pressed without the cheesecloth in order to obtain a smooth surface. It is important to prevent cheeses from cooling too fast. If this occurs, the acidification of the cheese may be too slow, and the conversion of lactose into lactic acid will be incomplete. At the end of pressing, the pH value of the cheese should be in the range of 5.05–5.10.

Salting: Salting begins immediately after the cheeses have been turned out of the moulds. They are immersed for 24 hr in a brine bath of 18°Bé–22°Bé and a temperature of 10°C–18°C. Thereafter, the cheeses are dry-salted twice. On the first day, a handful of salt is rubbed onto the upper side and the edge. On the second day, the cheeses are turned upside down, and the other side is treated.

Maturation: A temperature of 12°C–15°C and an RH of 85%–95% should be maintained in the ripening cellar. During the first two weeks of ripening, the cheese wheels are brushed daily with salt water (approximately 5% NaCl). Usually, no smear culture is used. When a smear has formed and the rind starts to become sticky, the brushing frequency is reduced to about three times a week. Berner Alpkäse must be ripened on spruce shelves for at least 4.5 months before sale. Wheels of Berner Alpkäse selected for refinement to Berner Hobelkäse are washed at the age of 6–8 months in order to remove the smear and then are further dry-ripened in an upright position up to a minimum age of 18 months. Room temperature should be 13°C–15°C with an RH of 60%–75%. Until the rind is sufficiently dry and a natural fat film has been formed (after 1–2 months), the wheels are rubbed weekly with a cloth in order to prevent the growth of mould.

2.3.5 Relevant Research

Jakob et al. (2007) analysed the composition of ten samples of top-rated Berner Alpkäse and Berner Hobelkäse. Concentrations of free n-caproic acid were 0.4 and 1.0 mmol/kg, respectively, indicating that lipolysis is important for the typical flavour of these cheeses. Berner Hobelkäse showed much less variation in acetic acid content than Berner Alpkäse, and concentrations of total biogenic amines were all below 9 mg/100 g. This may reflect the fact that only cheese with perfect preservation quality is refined to Berner Hobelkäse. At the age of 25 months, the cheese contains an amount of free amino acids of 360 mmol/kg (Jakob et al., 2007).

2.4 Cantal PDO – France

Name: Cantal PDO	
Production area: Auvergne	
Milk: Cow's, raw, thermised or pasteurised	

2.4.1 Introduction

Cantal AOC (Appellation d'Origine Contrôlée) was finally granted in 2007 (JORF, 2007). This designation has been recognised at the European level (AOP) since June 1996 (EC, 1996). Cantal cheese is the third AOC cheese in France, with a yearly production of 14,707 tonnes. It is made in the central part of the country (Auvergne) and is mainly consumed in France. Very little is exported despite its similarity with Cheddar. It is commercialised as 35–40 kg, 40 cm height, 36–42 cm diameter round wheels, with a dry crust. The total solids (TS) content and the fat/TS ratio are, respectively, at least 55.7% and 0.45 (De Freitas et al., 2005). Two other varieties produced in the same area of France, 'Laguiol' and 'Salers', are very similar to Cantal. Cantal was first mentioned by Pliny the Elder in his monumental book *Histoires Naturelles*. It was, according this author, brought by Roman soldiers to Roma and then to Great Britain, and so it can be considered as the ancestor of Cheddar. Its making process was then described by Diderot and later detailed by Duclaux (1893). Originally, this cheese was made from May to September in small stone houses (named '*burons*') built in high meadows. Nowadays, except for the 'Salers' variety, Cantal cheese is mainly produced in modern factories located in valleys.

2.4.2 Milk

Milk must be produced by cows grazing at least 120 days per year. It is collected after not more than 48 hr storage in farms and is used either raw or after heat treatment (thermisation 65°C for 20 s or HTST pasteurisation). Protein adjustment is not allowed, but fat skimming is allowed in order to obtain the ratio of fat to TS as 0.45 in the cheese.

2.4.3 Description and Sensory Characteristics

Cantal is a hard cheese made from either raw (16% of the total production) or pasteurised milk. Its crust is initially thin and grey-white in colour. It becomes thicker and shows golden spots as the ripening time increases. Cheese pastry has an ivory colour and becomes browner with age. TS is at least 57% and the fat/TS ratio is at least 0.45. Young Cantal has a sweet and lactic taste with slight nut and vanilla flavours. A medium-aged cheese (3 months; 'Entre Deux') has a creamy taste with some floral flavours. Old ripened cheese (at least 8 months) has a peppery and spicy flavour. Its crust is thick and includes some golden buttons.

2.4.4 Method of Manufacture

Milk preparation: Milk, kept overnight at 10°C–13°C in a conic wooden container ('*gerle*'; see Part I, Chapter 7) where a cold maturation takes place, is heated to 32°C, after which mesophilic lactic starters and yeasts are added.

Coagulation: Rennet is added in order to get a coagulation in 20 min.

Cutting: After 6 min, curd cutting takes place until wheat-sized curd grains are obtained.

Pre-pressing/moulding: The mixture of curd and whey is then directed in a 'pre-pressing' vat in which whey drainage and pressing (about 3500 kg/m^2) take place for 20 min at room temperature. The 'pre-pressed' curd is cut into 2 kg pieces and placed in 50 kg moulds in which the curd is turned.

Pressing: The curd is pressed four to six times in pressure cycles with an increase from 3200 kg/m^2 to 3800 kg/m^2 for 90 min.

Salting: The pressed curd is cut into 15 kg pieces and ripened for 14 hr at 19°C in order to obtain the 'Tome'. This Tome (pH 5.0 to 5.4) is ground and mixed with 2.0% dry salt at 17°C, moulded and pressed for 48 hr (the pressure is increased progressively from 8500 kg/m^2 to 18000 kg/m^2).

Maturation: After removal from the moulds, the cheeses are ripened in adjusted conditions, that is, at 10°C and 95% RH (De Freitas et al., 2007).

2.5 Cheddar – United Kingdom

Name: Cheddar	
Production area: All over the world; West Country Farmhouse Cheddar is produced in the counties of Devon, Cornwall, Dorset and Somerset, in the United Kingdom	
Milk: Cow's, raw or pasteurised	

Cheddar cheese is the most consumed cheese in the world, originating from Somerset around the late twelfth century and taking its name from the caves in the town of Cheddar that were used to store the cheese. The constant temperature and humidity of the caves provided a perfect environment for maturing the cheese. The town also gave its name to a unique part of the cheesemaking process – known as Cheddaring – which is the process of turning the slabs of curd and piling them on top of each other in a controlled way to help drain the whey. Cheddar making in Somerset goes back more than 800 years, with records from the King of England's accounts (the so-called 'Great Roll of the Pipe') noting that in 1170 the king purchased 10,240 lbs

(4.6 tonnes) of Cheddar cheese at a cost of a farthing a pound. The king at the time – Henry II – declared Cheddar cheese to be the best in Britain, and his son Prince John (who reigned between 1199 and 1216) clearly thought the same as there are records of him continuing to buy the cheese for the great royal banquets (Anonymous, 2015a).

Nowadays, Cheddar cheese is still made in Somerset, but it is also made all over the world. It is made on farms in the West Country, and 14 makers are licensed to use the EU PDO 'West Country Farmhouse Cheddar'. The cheese must be made on a farm in the four counties of Devon, Cornwall, Dorset and Somerset from locally produced milk and using traditional Cheddar making techniques – including hand Cheddaring.

During the latter half of the twentieth century, there were a number of significant changes to the way which Cheddar cheese is manufactured: the availability of reliable starter cultures and the development of automated mechanised systems for continuous manufacture (Lawrence et al., 2004). West Country Farmhouse Cheddar is a PDO cheese (EC, 1992) manufactured by members of the Farmhouse Cheesemakers Ltd, Somerset, United Kingdom. The procedure described below is for Cheddar, and the specific requirements for West Country Farmhouse Cheddar can be found on the web site of Farmhouse cheesemakers (Anonymous, 2015a).

2.5.1 Type

Hard cheese, with a maximum moisture content of 39% and a minimum FDM of 48%.

2.5.2 Milk

Artisanal cheesemakers may use raw cow's milk. Industrial manufacturers use pasteurised cow's milk.

2.5.3 Description and Sensory Characteristics

Hard cheese, made from cow's milk – for West Country Farmhouse Cheddar produced predominantly in the designated area, which may or may not have been pasteurised. West Country Farmhouse Cheddar is only sold after it has been stored for at least nine months. West Country Farmhouse Cheddar may be produced in either a traditional cylinder shape (in varying sizes to suit the needs of different customers) or in a block form (again in varying sizes to suit the needs of different customers). It has a close texture and is uniformly free of slits or round holes. Cheddar has a clean flavour, with a hint of sharpness and nutty notes.

2.5.4 Method of Manufacture

Milk preparation: Bovine milk is standardised according to the season, particularly when the natural balance in the milk between fat and casein differs significantly from a ratio of 1:0.7. For the manufacture of West Country Farmhouse Cheddar, the milk should be from farms from the designated area.

Starter culture: When pasteurised milk is used, mixed starter cultures are added at 1.5%–3%.

Coagulation: With rennet (25–30 mL per 100 L of milk according to the season) at 30°C–32°C.

Cutting: Coarse cutting.

Scalding: At 40°C, at the rate of 0.2°C/min, up to 0.3°C/min.

Curd draining: Whey is removed when the curd is dry enough and of the right acidity (pH will typically be 6.05–6.1).

Cheddaring: The compacted curd is cut into blocks and piled up every 15–20 min. According to the rate of acid development, the curd is turned and kept warm or aerated and cooled until acidity of 0.75%–0.85% is reached (pH < 5.3). At this stage, the curd should be dry, firmly mellow and exhibit a 'chicken breast' structure. Wet acid curds should be cooled and spread out before milling. Dry sweet curds are kept warm and piled up until dry and mellow.

Milling: The curd is milled or chopped in a coarse mill to finger-sized pieces. The milled curd is aerated and cooled to 25°C–26°C to cool the enclosed fat before salting.

Salting: Approximately 2% salt is sprinkled on the curd.

Pressing: For small-scale production, pressure is applied gradually at first, and then increased, as the curd cools, to 75 kPa for 12–16 hr. Turn into fresh cheese cloths and re-press at 200 kPa for a further 24 hr. Most industrial pressing is overnight.

Maturation: Store at ambient temperature to dry off, then at temperatures from 7°C to 11°C according to marketing requirements. Cheddar is fully matured at 9–12 months. Block cheese is marketed at 4–6 months.

2.6 Cheshire – United Kingdom

Name: Cheshire	
Production area: North of England	
Milk: Cow's, pasteurised	

The first cheese is believed to have been produced in Cheshire, by the Romans, in the garrison town of Chester. Cheshire cheese is mentioned in the Domesday Book at the end of the eleventh century. It originated around the banks of the River Dee, and due to the fluctuating quality of milk produced from the water meadows near the river, three recipes were needed for (1) a quick-ripening cheese in spring, (2) a medium-ripening cheese in summer and (3) a longer-ripening cheese in autumn (Robinson & Wilbey, 1998). Cheshire cheese was traded across the country. The different varieties of Cheshire cheese were aged to a sufficient level of hardness to withstand the rigours of transport (by horse and cart, and later by boat) to London for trading purposes. This trade started in the early seventeenth century. From 1739, Cheshire cheese was the only cheese bought for consumption by the British Navy. During the nineteenth century, sales of Cheshire cheese grew strongly as it became a firm favourite in the industrial towns and cities of the Midlands and north of England. Cheshire cheese was a cheap form of protein and was sometimes referred to as the poor man's meat. As the towns grew, milk produced on nearby farms was used for liquid consumption, and the main Cheshire-cheese-producing areas spread further out, to the south of Chester and towards the Shropshire borders (Anonymous, 2015b).

2.6.1 Type

Hard cheese, with a crumbly texture.

2.6.2 Description and Sensory Characteristics

It is a hard cheese, has a cylindrical shape and is made from cow's milk. Young Cheshire is naturally bright and white in colour. It is a firm-bodied cheese with a crumbly texture that breaks down gently in the mouth. As Cheshire matures, it becomes firmer in texture and slightly darker in colour. The flavours become more complex, but the cheese remains clean tasting, with no hint of bitterness. The crumbly texture remains, but the cheese has a drier mouth feel. Traditionally, three varieties (spring, summer/winter and autumn cheese) are made according to the chemical composition of the milk produced in these seasons. Cheshire has an open, short silky to granular texture and a clean milky flavour, with acidity and fresh on the palette with a very slightly tangy finish. As Cheshire matures, the flavour becomes more complex, but the cheese remains clean tasting, with no hint of bitterness and a drier mouth feel.

2.6.3 Method of Manufacture

Milk preparation: Cow's milk is pasteurised.

Starter culture: Mixed starter cultures are added (1%–3%) at a minimum of 21°C.

Ripening and colouring: The temperature is raised to 30°C–31°C and held for 40–60 min until the acidity reaches 0.22%–0.24%. Annato may be added according to market requirements, at 0.016% up to 0.02%

Coagulation: With rennet (30–35 mL per 100 L milk according to season) at 30°C–31°C.

Cutting: The curd is cut when it breaks cleanly (usually in 40–60 min).

Scalding: At different temperatures depending on the initial whey acidity: low acid, 0.135% – scald to 32°C in 40 min; medium acid, 0.15% – scald to 35.5°C in 45 min and high acid, 0.16% – scald to 34.5°C in 50 min. The curds are stirred and turned on special cheese tables, making it easier to drain away the whey. The curd is stirred until it is firm and free of whey pockets inside the curd.

Curd draining: Whey is removed when the curd is dry enough and of the right acidity (0.21%–0.23%, and the pH will typically be below 6).

Texturing: The curd is not textured but is kept free and open as far is practicable. The curd is cut into blocks, turned over and broken into blocks by hand every 15–20 min until small in size. Subsequently the curd is disturbed to allow whey drainage until the acidity is 0.65%–0.75%.

Milling: The curd is fine-milled once.

Salting: The curd is then salted by hand using local Cheshire salt. Salting is performed either as the curd passes through the mill or on the milled curd at 1.8%–2.1% salt; then it is re-milled, put into moulds and wrapped in cheese cloth to help drain the excess whey from the curd. Salted curds are then placed in large presses.

Pressing: Cheese is pressed at 35 kPa (about 250 kg per cheese) for 2 hr, then placed in a cheese cloth and pressed at 70 kPa. Next day, it is turned, and the pressure is increased at 140 kPa. After pressing, the cheese has a crumbly texture.

Maturation: Cheese is taken out of the moulds, and the cheese cloth is removed. The cheese is vacuum-packed in special bags, traditionally bound or waxed, and then placed in storage to mature for a matter of weeks or months depending on the maturity required. The cheese is stored at gradually reducing temperatures: 13°C down to 10°C in 1 week, then at 7°C for 1 month or until marketed. The cheese is normally fully matured for 6–9 months. Expert graders check the cheese throughout the maturation process.

2.7 Fiore Sardo PDO – Italy

Name: Fiore Sardo PDO	
Production area: Sardinia region	
Milk: Sheep's, raw	

2.7.1 Introduction

The cheese Fiore Sardo, 'the cheese of the shepherds of the island', received recognition for typicality in 1955, the Denomination of Origin in 1974 and the PDO in 1996 (EC, 1996). Sheep farming in Sardinia has ancient origins dating back to the Nuragic civilisation (before the first millennium BC). It is said to date back to the Bronze Age. It is one of the best-known Sardinian cheeses and bears this name because in the past they used a coagulant plant, possibly obtained from the flower of a thistle. The production area is Sardinia Island (provinces of Cagliari, Nuoro, Oristano and Sassari). Its place of origin is Nuoro, and especially in Barbagia.

2.7.2 Type

Fiore Sardo is a raw, hard, pressed cheese with a minimum FDM of 40%. It is consumed as a table cheese when it matures for up to three months and grated if maturation has exceeded six months. The moisture content was 34.7% at 120 (days), 33.3% at 180 (days) and 30.9% at 240 days (Pirisi et al., 2007).

2.7.3 Description and Sensory Characteristics

The weight varies from 1.5 kg to 4 kg depending on the technical conditions of production. The forms are 12–15 cm high and have a diameter of 12 to 20 cm. The crust varies from bright yellow to dark brown. The body of the cheese is compact with a pale yellow or white colour, while its flavour is more or less spicy depending on its age.

2.7.4 Method of Manufacture

Animal feeding regime: Mainly grazing on wild and cultivated pasture. The sheep receive a balanced diet, consisting of concentrates and/or single feeds, as a function of the availability forage and the animals' needs.

Milk: The raw whole milk of Sarda sheep (predominantly) from the evening milking or from the morning, or mixed, after being filtered is poured into the classic copper vat.

Starter cultures and rennet: The milk can be inoculated with autochthonous lactic acid bacteria (mainly whey culture) from the original area of production. Paste lamb rennet is also used, and more rarely kid paste.

Coagulation: At a temperature of 34°C–36°C after the addition of the rennet (30–40 g in 100 L) milk coagulates in 12–17 min (holding time). The next curd firming occurs in 25–28 min (firming time).

Cutting: The breaking/cutting of the curd is very energetic and continues for about 3 min, so as to reduce the coagulated gel to the size of a grain of millet. A resting phase of about 5 min follows where the curd is left to settle at the bottom of the vat.

Moulding: The curd withdrawn from the copper vat is placed in a suitable mould, a truncated cone, to obtain the shape of the desired dimensions. Once, pierced containers of chestnut wood were used, called 'pischeddas', or even oak or wild pear, on the bottom of which was carved a flower, perhaps the lily or the asphodel, which left on the cheese a real brand which often contained the initials of the name of the manufacturer. Nowadays other types of materials are also used for moulding.

Afterwards, the cheese is turned in the mould and on a table, the 'spersore', for further draining of the whey.

Scalding: In order to encourage the formation of a more resistant crust, the forms are immersed for a few minutes in hot water (>60°C) or in the whey after the production of Ricotta.

Salting: After completing the syneresis, salting of a mixed type (either in saturated brine or dry salting) takes place. The duration varies depending on the size of the forms, but the average is 36–48 hr (from 8 to 12 hr/kg).

Maturation: The maturing of the Fiore Sardo is carried out in three phases. In Phase 1, a first maturation occurs at the house of the shepherd, on trellis canes suspended on the fireplace, where it dries and undergoes a slight smoking for about 2 hr per day at a temperature of 18°C–20°C. To produce the smoke, the shepherds are still using wood fire with fresh twigs of shrubs and trees typical of the area of origin. This stage lasts about two weeks, and during this time the forms are turned over several times. Phase 2 takes place in a room adjacent to that of the first phase and lasts about three months, at a temperature of 10°C–15°C. The cheese forms continue to be turned over frequently. Phase 3 takes place in special cellars of maturation, with temperatures not higher than 15°C and an RH of 80%–85%. The cheeses undergo frequent turnings, and the crust is repeatedly rubbed with an emulsion consisting of olive oil, wine vinegar and cooking salt. Sometimes the fat of sheep is also used. With the approach of the hot season, the cheeses are transferred to special basements usually located in mountainous areas. This phase lasts about three months. In all, the maturation process lasts about six months.

2.7.5 Relevant Research

The Fiore Sardo PDO is one of the oldest and is still a traditional cheese of the Sardinia region. The literature selected mainly deals with the chemical and microbiological characterisation of Fiore Sardo PDO (Ledda et al., 1994; Pisanu et al., 2006; Zazzu, 2011). Further studies have been done on the enzymatic composition of lamb rennet paste (Addis et al., 2005) in traditional sheep milk cheese production (Addis, Piredda & Pirisi, 2008), and its influence on proteolytic and lipolytic pattern and the texture of Fiore Sardo PDO sheep cheese (Pirisi et al., 2007). The volatile fraction under predominantly lipolytic maturation conditions typical of Fiore Sardo PDO is reported by Urgeghe et al. (2012). Furthermore, the food safety aspects of cheeses in the Sardinia region are reported by Zazzu (2011).

2.8 Graviera Kritis PDO – Greece

Name: Graviera Kritis, PDO	
Production area: Island of Crete, Greece	
Milk: Sheep's or mixture with goat's, thermised or pasteurised	

2.8.1 Introduction

Graviera cheese was firstly manufactured in Greece a century ago. Its name is a transliteration of the Swiss Gruyère, the making technique of which was introduced from the Swiss-trained cheesemaker Nikolaos Zygouris (Zygouris, 1952). Initially, Graviera cheese was manufactured in Greece from sheep's milk, after modifications of the original method. Although sheep's milk remains the main raw material, nowadays other milks, like cow's milk or mixtures of sheep's and goat's milks, are also used in different regions of the country. Thus, several varieties of Graviera cheese are produced which are distinguished from the name of the region of their production, for example, Graviera Kritis, Graviera Naxou and Graviera Agrafon, which have been recognised as PDO cheeses, while Graviera from Tinos, Graviera from Epirus and others have not. One of the most popular Graviera cheeses is that of Kriti island, which is manufactured from sheep's milk or with a mixture of it with up to 20% goat's milk, the making technology of which is described below.

2.8.2 Type

Graviera Kritis is a hard cheese, with a maximum moisture (content) of 38%, a minimum FDM of 40% and a maximum salt content of 2%. The gross composition is as follows: moisture 33.8%, fat 34.7%, FDM 52.4, protein 27%, salt content 1.8%, and pH 5.6 (Nega & Moatsou, 2012). The PDO status for Graviera Kritis was recognised by the EC in 1996 (EC, 1996).

2.8.3 Milk

Sheep's, or mixture with goat's (max 20%). Thermised or pasteurised milk can be used.

2.8.4 Description and Sensory Characteristics

Graviera Kritis is a hard cheese having the shape of a wheel with a hard rind, which is formed during maturation from the surface microflora. It has a white-yellowish colour. It is used as a table cheese and has a pleasant flavour. Graviera Kritis has a cohesive and slightly elastic body with small holes from propionic fermentation. It has a sweet and salty flavour with cheesy and fruity notes and a rich aroma from propionic acid fermentation, although in aged cheese it can be sharp.

2.8.5 Method of Manufacture

Milk preparation: Raw or pasteurised or thermised milk is used.

Starter culture: Traditionally, no starter culture is added. When pasteurised or thermised milk is used, thermophilic (i.e. *S. thermophilus, Lb. delbrueckii ssp. bulgaricus* and *Lb. helveticus*) or a blend of thermophilic and mesophilic (homofermentative strains of *Lc. lactis* ssp. *lactis* and *Lc. lactis* ssp. *cremoris*) starter cultures are added. If eye formation is required, propionibacteria are added in the milk for secondary fermentation.

Coagulation: Coagulation takes place at 34°C–36°C for 30 min.

Cutting: The coagulum is cut at the size of rice to corn.

Scalding: After cutting, the curd is stirred for 5–10 min and progressively scalded, under continuous stirring, up to 50°C–52°C. Stirring is continued for about 15 min more, and the curd is left for to settle 5 min.

Curd draining: The curd is transferred into moulds, and the draining is assisted by pressing, in rooms at 14°C–16°C and an RH of 85% for 24 hr.

Pressing: The curd is pressed for 24 hr.

Salting: After pressing, the cheeses are placed on shelves in rooms with temperature 12°C–14°C for 2 days, and then they are immersed in brine 18°Be–20°Be for 2–5 days depending on the size of cheese.

Maturation: Maturation takes place in rooms at 14°C–18°C and a relative humidity of 85%–90%. During maturation, up to 10 dry saltings are carried out, and the cheeses are inverted. Surface microflora develops, resulting in the formation of a hard rind. Graviera Kritis is matured for at least three months.

2.8.6 Relevant Research

Kandarakis et al. (1998) compared cheeses made using mixtures of *S. thermophilus, Lb. helveticus* (1:1) and *Propionibacterium freundreichii* ssp. *shermanii; Lc. lactis* ssp. *lactis, Lc. lactis* ssp. *cremoris, S. thermophilus, Lb. helveticus* (1:1:10:2) and *P. freundreichii* ssp. *shermanii*; and *Lc. lactis* ssp. *lactis, Lc. lactis* ssp. *cremoris, S. thermophilus, Lb. helveticus* (2.5:2.5:1:1) and *P. freundreichii* ssp. *shermanii* with cheese made without a starter culture, and no significant difference was detected in the flavour, body and texture. Moatsou, Moschopoulou and Anifantakis (2004) demonstrated that traditional cheesemaking without lactic acid and propionic acid cultures resulted in the most acceptable organoleptic characteristics for Graviera Kritis. However, the use of a mixture of lactic and propionic acid starters could improve the openness characteristics of the texture, if it was combined with a curd wash to control acidity development in order to obtain a more elastic curd with evenly distributed eyes in the cheese mass and high flavour scores. The ripening at 20°C for three months increased the proteolysis level and resulted in a cheese of inferior quality compared to that ripened at 16°C. More recently, Samelis et al. (2010) investigated the secondary microfloras of Graviera cheese and reported that the LAB species identified (and their prevalence in the cheese samples) were as follows: *Lb. casei/paracasei* (68.8%), *Lb. plantarum* (19.5%), *S. thermophilus* (8.9%), *E. faecium* (2.1%) and *Lc. lactis*

(0.7%). These results were validated using culture-independent FTIR-based methods (Samelis et al., 2011).

2.9 Idiazabal PDO – Spain

Name: Idiazabal cheese PDO	
Production area: Basque Country, Navarre, Spain	
Milk: Sheep's, raw	

2.9.1 Introduction

The PDO status of Idiazabal cheese was recognised by the Ministry of Agriculture, Fisheries and Food in 1993 and at the European level in 1996 (EC, 1996). The cheese is made mainly from raw sheep's milk. Under the PDO regulation, 338 farmers and 122 dairy plants are registered. In 2013, the quantity produced was 1,105,646 kg, which represented a total value of 17,530,000 euros. The cheese is mainly sold on the national market, with an increasing market in European countries. It represents 8.30% of the economic value of Spanish PDO cheese production (Ministry of Agriculture, Food and Environment, 2014). With little variations over 8,000 years, since the Neolithic, shepherds and sheep have follow the same routes determined by the seasons: high mountain pastures in spring, summer and autumn, and valleys during the winter, in order to exploit a natural food that follows an immutable cycle. The result of this and the traditional culture of this region in the processing of milk into cheese makes Idiazabal cheese a product appreciated by the consumer and recognised by the PDO (Ministry of Agriculture, Food and Environment, 2015). The cheese is manufactured in all the areas where the Latxa and Carranzana sheep breeds are found naturally in the Basque Country, and in Navarre, except for the municipal districts which lie in the Roncal Valley (Navarre).

2.9.2 Type

Pressed, hard cheese made from whole, raw sheep's milk from the Latxa and Carranzana breeds, with a minimum ripening period of 60 days. When raw milk from the owners' farm is used, the label may indicate the word 'Artzaigazta' or farmhouse cheese.

The fat and protein content in dry matter has a minimum of 45% and 25%, respectively, with a maximum moisture content of 45% and a pH value of 4.9–5.5.

2.9.3 Milk

This cheese is made from Latxa and Carranzana sheep's milk. The pastures of this region have traditionally been utilised for livestock according to long-standing practice. The milk must be full-fat, without any preservatives, and the cheese produced must have fat content in dry matter higher than 45%.

2.9.4 Description and Sensory Characteristics

The cheese has a cylindrical shape, flat on the top and bottom. The diameter is 10–30 cm with a height in the range of 8–12 cm and a weight ranging from 1–3 kg. The rind is hard and smooth; the colour is pale yellow to whitish grey and dark brown in smoked cheeses. The texture is compact and may have a few, irregularly distributed uneven holes smaller than a grain of rice. The odour is intense and characteristic, sharp, clean and sheepy. The flavour is intense, full and balanced. The taste is clean and consistent, and a marked taste of ripened sheep's milk and a hint of the natural rennet are characteristic. The piquancy is slight, and the sweetness and acidity are very slight to medium. There is no bitterness, the saltiness is of medium intensity and the smoky flavour is light to medium in smoked cheeses with a pronounced aftertaste. The body of Idiazabal is dry and grainy at first but becomes smoother as it is warmed in the mouth, with medium elasticity and fairly firm.

2.9.5 Method of Manufacture

Milk preparation: Milk must be collected from the registered farms, meet the requirements laid down by legislation and be kept under at a temperature of less than 10°C. The milk is gently heated until it reaches the coagulation temperature.

Coagulation: This stage is carried out using rennet, usually lamb rennet paste. The milk must be stored at a temperature between 28°C and 30°C for 20–45 min.

Cutting: The curd must be cut to a size of 0.5–1.0 mm in diameter.

Curd draining: The curd is stirred for about 15 min and heated at 36°C–38°C for 25 min while being gently stirred until the grains reach the appropriate hardness. The curd is placed in cylindrical moulds and pressed for 5–6 hr.

Salting: This is accomplished by dry salting or by placing it in saturated brine at a temperature 8°C–13°C for 24 hr for small cheeses and 48 hr for large ones.

Maturation: The pre-ripening depends on the cheesemaker, but is never more than 30 days, at 10°C–13°C and an RH of 80%. Maturation must last at least 60 days from the date of moulding, and is carried out at 8°C–10°C and an RH of 85%–90%. The cheese remains on trays and is turned and cleaned periodically, as required, so that it gains its special characteristics. The cheese may be smoked using beech or alder wood a few days before sale. The different stages involved in producing the milk and making and maturing this type of cheese must take place in the defined geographical area. The PDO products are identified with a numbered casein label issued and controlled by the Regulatory Council.

2.9.6 Relevant Research

Studies have been carried out related to the influence of different parameters such as seasonality. Barron et al., 2007 and Mendia et al., 2000 stated that cheeses manufactured in February had higher sensory analysis scores mainly for characteristic odour and taste compared with those produced in June, which has a higher score for sweet flavour and bitter taste. Barron et al.

(2001) observed differences in the final properties of industrially produced cheeses made with raw or pasteurised milk. Specifically, they observed that the aroma attributes developed at three months in raw milk cheeses, whereas in pasteurised cheeses they developed later, at six months of ripening.

Some other studies have focused on the influence of

1) the season on casein hydrolysis; mainly in the free amino acids associated with the stage of amino acid release and the free amino acids associated with amino acid catabolism release (Ordoñez et al., 1998);
2) pasteurisation, which caused a decrease in amino acid release that may be directly or indirectly responsible for the atypical sensory characteristics of the cheeses (Ordoñez et al., 1999);
3) the ripening process on the free amino acids profiles (Barcina, Ibαρez & Ordσρez, 1995);
4) brining, with which a positive correlation was observed among most of the microbial groups studied except for *Enterococcus* spp., *Cl tyrobutyricum*, yeast and moulds (Pérez Elortondo, Albisu, & Barcina, 1999, 2002);
5) the volatile composition of Idiazabal/farmhouse Idiazabal and the finding that fresh pasture grazing in spring increased the odour impact ratios of esters and alcohols, which revealed that spring cheeses might be more intense fruity and sweet compared with winter cheeses (Abilleira et al., 2009; Chávarri et al., 1999);
6) the type of rennet (lamb rennet paste or bovine rennet)/starter in the proteolysis profile (Bustamante et al., 2003; Vicente et al., 2001);
7) lipase addition related to the type of lipase used, pre-gastric or fungal (Hernández et al., 2005); and
8) lipolysis, proteolysis and sensory properties observed at the addition of lipase; that is, increase in the content of short-chain FFA, and total partial glycerides in cheeses (Barron et al, 2004; Hernández et al., 2009).

As regards the microbiology, several studies have been carried out on the proteolytic activity of different microorganisms (Arizcun, Barcina & Torre, 1997); indigenous LAB (Pérez Elortondo et al., 1998) and the origin of different strains of *Enterococcus* spp. in farmhouse cheeses (Ortigosa et al., 2008).

2.10 Kefalograviera PDO – Greece

<table>
<tr><td>Name: Kefalograviera, PDO</td><td rowspan="3"></td></tr>
<tr><td>Production area: Western Macedonia, Epirus and Prefectures of Etoloacarnanias and Euritanias, Greece</td></tr>
<tr><td>Milk: Sheep's or mixture with goat's, raw, thermised or pasteurised</td></tr>
</table>

2.10.1 Introduction

Kefalograviera is a hard cheese, the characteristics of which, as the name implies, stand between Kefalotyri (see Part II, Section 2.11) and Graviera (see Part II, Section 2.8) cheeses. It is produced in Western Macedonia, Epirus and the prefectures of Etoloacarnanias and Eurytanias.

2.10.2 Type

Hard cheese, with a maximum moisture of 40% and a minimum FDM of 40%. Gross composition: moisture content 37.8%, fat content 31.2%, FDM 50.2, protein content 24.5%, salt content 3.5, Salt-in-moisture (S/M) 8.5 and pH 5.5 (Nega & Moatsou, 2012). The PDO status for Kefalograviera was recognised by the EC in 1996 (EC, 1996).

2.10.3 Milk

Sheep's or mixture with goat's, up to 10%.

2.10.4 Description and Sensory Characteristics

Kefalograviera is a cheese in the shape of a wheel, with a hard and thin rind. It has a firm, elastic body with few holes or eyes interspersed in its mass, and a colour ranging from whitish to pale yellow. Its flavour is mild to medium piquant and slightly salty.

2.10.5 Method of Manufacture

Milk preparation: Raw or pasteurised or thermised milk is used.

Starter culture: Traditionally, no starter culture is added. When pasteurised or thermised milk is used, thermophilic (i.e. *S. thermophilus, Lb. bulgaricus* and *Lb. helveticus*) or a blend of thermophilic and mesophilic (homofermentative strains of *Lc. lactis* ssp. *lactis* and *Lc. lactis* ssp. *cremoris*) starter cultures are added. If eye formation is required, propionibacteria are added in the milk, for secondary fermentation.

Coagulation: Coagulation takes place at 32°C–34°C for 35 min using commercial rennet or a mixture (50:50) with a traditional rennet paste.

Cutting: When the coagulum is firm enough, it is cut at the size of corn.

Scalding: The curd is heated gradually up to 48°C in 20–25 min, and curd stirring is continued for approximately 20 min more at the final temperature.

Curd draining: After this, it is transferred into moulds (heights 12–13 cm and 31–32 cm in diameter).

Pressing: The curd is pressed at 14°C–16°C and RH 85% approximately for 24 hr.

Salting: The curd is immersed in brine 18°Be–20°Be for 48 hr.

Maturation: Maturation is carried out initially at 14°C–16°C and RH 85%–90% and approximately 10 dry saltings take place with corresponding cheese inversions. Then maturation continues at 6°C for at least three months in all.

2.10.6 Relevant Research

The manufacture of low-fat Kefalograviera-type cheese has been extensively studied. When full-fat and low-fat cheeses were manufactured with different starters, it was observed that the type of starter did not affect the composition of the low-fat cheese, but the latter received significantly higher flavour scores than the low-fat control (Katsiari, Voutsinas & Kondyli, 2002). In addition, the fat content affected the pattern of proteolysis, water-soluble nitrogen was significantly affected by the addition of adjunct cultures and the production of low-molecular-mass nitrogenous compounds was enhanced (Michaelidou et al., 2003). The experimental cheese had significantly higher levels of acetone at 90 days and acetic acid, diacetyl and acetoin at 180 days than the control low-fat cheese, which had significantly higher levels of butan-2-ol and butan-2-one than the former cheeses at both sampling ages. The type of starter also affected the total levels of free FAs (Kondyli et al., 2003).

2.11 Kefalotyri – Greece

Name: Kefalotyri	
Production area: Greece	
Milk: Cow's, sheep's, goat's and/or mixtures, pasteurised	

2.11.1 Introduction

Kefalotyri is a hard cheese, with a high salt concentration, made traditionally from sheep's or goat's milk or their mixtures. Dating back to the Byzantine era, Kefalotyri cheese is believed to be the ancestor of most hard Greek cheeses. Before the twentieth century, all hard cheeses produced in Greece from sheep's and/or goat's milk were called Kefalotyri. The name comes from the words *kefali* which means 'head', and *tyri*, which means 'cheese'. Kefalotyri cheese is manufactured in various parts of Greece, with technologies that differ more or less from place to place, thus giving a great variation in organoleptic properties.

2.11.2 Type

This is a hard cheese, with a maximum moisture content of 38% and a minimum FDM of 40%. Gross composition: moisture content 33.1%–37.2%, fat content 8.9–35.7%, protein content 23.9%–28.5% and salt content 1.8%–3.5% (Andrikopoulos et al., 2003).

2.11.3 Milk

The milk for Kefalotyri cheese manufacture must be of good quality, and sheep's milk is standardised to a fat content of 5.8% to 6.0%, and cow's milk to 3%.

2.11.4 Description and Sensory Characteristics

Kefalotyri has the shape of a wheel with various sizes weighing 5–10 kg. It is used as a table cheese, usually pan-fried or grilled, but mostly for grating. Kefalotyri has a hard body with irregular holes. It has a salty taste with a very strong flavour.

2.11.5 Method of Manufacture

Milk preparation: Traditionally, raw milk was used. For industrial manufacture, pasteurisation usually takes place either at 68°C for 10 min or 72°C for 15 s. Raw milk cheese matures faster and acquires better flavour and taste, but if the milk quality is low, many defects may arise.

Starter culture: Thermophilic starter or ablend of thermophilic and mesophilic (*S. thermophilus, Lb. ssp. delbrueckii bulgaricus, Lb. helveticus* and *Lc. lactis* ssp. *lactis* and *Lc. lactis* ssp. *cremoris*) is used for the manufacture of Kefalotyri cheese.

Coagulation: Renneting takes place at 35°C–36°C with a sufficient quantity of rennet to produce a coagulum ready for cutting within 30–35 min. When a strong piquant flavour is required, traditional rennet is used or a mixture of classical and traditional rennet.

Cutting: When the coagulum acquires the correct firmness, it is cut into particles 0.5–1.0 cm diameter. In some artisanal dairies, instead of standardising the fat content of the milk, the coagulum is further divided so that the curd particles acquire the size of rice and more fat diffuses into the whey. The latter is usually used for the production of high-quality whey cheeses (see Anthotyros, Part II, Chapter 13, Section 13.2).

Scalding: The curd is heated gradually to 43°C–45°C under continuous stirring.

Curd draining: The coagulum is then transferred into moulds.

Pressing: Pressing takes place overnight at 14°C–16°C.

Salting: The cheese wheels, still in moulds, are transferred to the ripening room, and the next day they are immersed in brine 18°Be–20°Be for 1–2 days.

Maturation: Kefalotyri matures at 14°C–16°C for 90 days. Throughout maturation, 20–25 dry saltings take place, with subsequent corresponding cheese inversions. The final cheese is washed with brine and transferred to a cold room at 2°C–4°C and RH 85%, to complete the maturation process.

2.11.6 Relevant Research

Changes in LAB of Kefalotyri from pasteurised cow's milk throughout maturation were studied (Litopoulou-Tzanetaki, 1990). Lactobacilli and enterococci counts were high (10^6–10^8 cfu/g and 10^4–10^7 cfu/g, respectively). Leuconostocs, lactococci and obligatory heterofermentative lactobacilli disappeared early in ripening. *E. faecium, Lb. casei* ssp. *casei* and *Lb. plantarum* became predominant with maturation. After four months of maturation, *E. faecium* was dominant (35.5%) followed by *Lb. plantarum* (18.4%) and *Lb. casei* ssp. *casei* (15.8%). Results on lactic microflora suggested that a starter consisting of both LAB that disappeared early and LAB that survived throughout maturation should be tried in order to improve the quality of Kefalotyri from cow's milk.

During ripening, many chemical, biochemical, microbiological and physical changes occur, which contribute to the development of its organoleptic properties. The quality of the milk and rennet used for cheesemaking as well as the pasteurisation of the milk and the starters used are the main factors which affect the ripening process (Veinoglou et al., 1983).

2.12 Le Gruyère PDO – Switzerland

<table>
<tr><td>Name: Le Gruyère PDO</td><td rowspan="3">
Cheese wheel of Le Gruyère PDO (Interprofession du Gruyère)</td></tr>
<tr><td>Production area: Cantons of Fribourg, Vaud, Neuchâtel and Jura, as well as the districts of Courtelary, La Neuveville, Moutier and the Bernese municipalities of Ferenbalm, Guggisberg, Mühleberg, Münchenwiler, Rüschegg and Wahlern, as well as a few other cheese dairies in the German-speaking part of Switzerland.</td></tr>
<tr><td>Milk: Cow's, raw</td></tr>
</table>

2.12.1 Introduction

Le Gruyère PDO is a hard, smear-ripened cheese made from raw cow's milk. It is probably the best-known cheese in Switzerland. It is also the most produced and, after Mozzarella, the most consumed cheese in the country. Its yearly production is about 29,200 tonnes, and 12,376 tonnes was exported in 2014 (TSM Treuhand GmbH, 2015). The request of the inter-professional organisation of Le Gruyère to include its cheese variety in the register of PDO products was accepted in 2001 (FOAG, 2015). This designation was recognised by the EU in December 2011 (Interprofession du Gruyère, 2014).

Cheesemaking in the Le Gruyère district was mentioned as early as 1249; however, the first traces of cheese production in the region go back to 1115 AD. In 1655, the cheese was officially named after the Gruyère district, which is located in the canton of Fribourg. In 1762, the French Academy included the word 'Gruyère' in its dictionary with the note that it concerns a cheese produced in the Gruyère region. Since the nineteenth century, efforts have been made to protect Le Gruyère, as it has been increasingly copied due to its excellent reputation (Kulinarisches Erbe der Schweiz, 2008). Le Gruyère's inter-professional organisation celebrated the 900-year anniversary of the cheese in 2015.

2.12.2 Type

The specifications for Le Gruyère require that at the time of quality assessment (five months), the cheese should have an FDM of 49%–53%, a moisture content of 34.5%–36.9% and a salt content of 1.1%–1.7%. Le Gruyère is produced in four different varieties: 'Le Gruyère Classic', a rather young cheese of 6–9 months; 'Le Gruyère Réserve', aged for at least ten months; 'Le Gruyère Bio', an organic cheese and 'Le Gruyère d'Alpage', which is produced only during the summer months in the Alps within the designated geographical area.

2.12.3 Description and Sensory Characteristics

The cheese body has a fine surface and is slightly damp; the texture is smooth with a medium firmness and is slightly brittle. With increasing age, crystals develop, which are characteristic for long-ripened Le Gruyère. Its even ivory colouring varies according to the season. Fruity notes dominate the cheese's flavour and are supported by a salty taste of varying intensity. These aroma notes develop from the combined action of the rind and lactic fermentation and can vary according to the terroir. The presence of occasional eyes with a diameter in the range 4–6 mm is desirable but not essential. A few small and isolated cracks are allowed in aged cheeses.

2.12.4 Method of Manufacture

The manufacture of Le Gruyère is carried out in about 170 small-scale dairies. The use of additives is forbidden.

Milk preparation: High-quality cow's milk is Le Gruyère's base raw material. The preparation and distribution of silage of any kind are forbidden on farms that produce milk for this cheese. Regarding the feeding of the dairy cattle, a minimum percentage of 70% of the feed ratio (calculated on dry matter) must come from the fodder area of the farm. Other fodders such as supplementary and concentrate feeds are defined in detail in the Animal Feed Book Ordinance (FOAG, 2015). Furthermore, the use of any type of growth activators is forbidden. Milk processed for 'Le Gruyère d'Alpage' is obtained exclusively from cows grazing on alpine pastures located within the geographical area and is transformed in alpine cheese dairies during the summer season.

The milk can only come from producers assigned to the cheese dairies, and the milk supplier must be located within a maximum radius of 20 km from the cheese dairy. The raw milk must be cooled at the farm to a temperature of 12°C–18°C. With few exceptions, the milk must be delivered to the cheese dairy twice a day, immediately after milking. The raw milk must be transformed in open copper vats of a maximum capacity of 6,600 litres. Natural or mechanical skimming is allowed in order to standardise the fat content.

Starter culture: In-house whey cultures are traditionally added to the vat milk. However, defined liquid starter cultures authorised by the inter-professional organisation may also be added up to a maximum total amount of 49%; the proportion of the in-house whey culture must, therefore, predominate. These cultures consist of strains of thermophilic lactic acid bacteria of the species *S. thermophilus, Lb. delbrueckii* ssp. *lactis* and *Lb. helveticus* (Häni & Keller Möcklin, 2006). The liquid stock cultures are produced by Agroscope.

Rennet: Only calf rennet is allowed. This is added at 31°C, and coagulation lasts 30–50 min.

Cutting: The curd grain size corresponds to the size of a wheat grain.

Scalding: The scalding temperature varies between 54°C and 59°C, but it is most commonly around 57°C. Curd washing is not applied.

Moulding/pressing: The transfer of the curd into the moulds is done manually or mechanically. Each cheese must be marked with the production date and the admission number of the cheese dairy. Pressing lasts at least 16 hr, and a good cohesion of the grains must be ensured.

Salting: The salting starts immediately after the cheeses have been turned out of the moulds, either by dry salting or by immersing the cheeses in brine with more than 20°Bé and a temperature of 12°C–20°C.

Maturation: After salting, the cheeses are stored in cellars at a temperature of 12°C–18°C and an RH of about 92%. The atmosphere of the ripening cellars of Le Gruyère is characterised by a noticeable presence of ammonia. The cheeses are ripened on rough, unpolished red spruce shelves that allow an optimal moisture balance between the cheese rind and the support plate. During the first two weeks of ripening, the cheeses are turned over daily, brushed with smear water (approximately 5% NaCl), and then dry-salted. The initial ripening of Le Gruyère must take place in its area of production and is usually done in local cheese dairies. Before sale, Le Gruyère has to be stored for at least five months in ripening cellars located in Switzerland. Le Gruyère is produced in different categories: 'Le Gruyère Classic' is aged for 6–9 months, whereas 'Le Gruyère Réserve' requires a minimum ripening period of 10 months, but can be found on some markets at the age of up to 18 months. Le Gruyère must have good preservation qualities, allowing it to reach its maturity without any loss in quality.

2.12.5 Relevant Research

NSLAB represent an important component of the ripening flora of Le Gruyère cheese. A study investigating the diversity of non-starter lactic acid bacteria (NSLAB) in Le Gruyère revealed that over 90% of the isolates belonged either to the species *Lb. casei* or *Lactobacillus rhamnosus*. No NSLAB were found in the cheese starter culture, whereas raw milk contained a large number of different genotypes (Casey et al., 2006). In a recent study, the quality of milk delivered by ten farms using an automatic milking system (AMS) and eight farms using a milking parlour (MP) was compared. Milk from the AMS farms had significantly shorter methylene blue reduction times, higher titratable acidity and slightly higher aerobic mesophilic counts. Levels of free butyric acid were much higher in AMS milk than in MP milk. For all parameters except free butyric acid, farm-to-farm variations were more important than variations between the milking systems (Jakob et al., 2013).

2.13 Ossau Iraty PDO – France

Name: Ossau Iraty PDO
Production area: Bearn and French Pays Basque, two adjacent regions of mountains located in the Atlantic Pyrenees (0–2,400 m)
Milk: Sheep's, raw

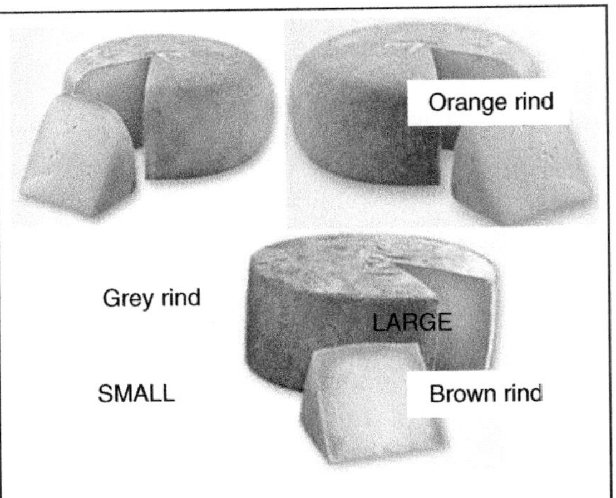

From http://www.ossau-iraty.fr

2.13.1 Introduction

Ossau Iraty is one out of the three PDO French sheep's cheeses and the 187 pressed and uncooked French cheeses. It is the only PDO French cheese made from full-fat sheep's milk that is pressed and uncooked and the only PDO cheese made in the Pyrenees mountains. In 2014, its production represented 22 % of the production of the first PDO French cheese made from sheep's milk in volume, namely, the blue cheese Roquefort. Ossau Iraty was recognised as AOC in 1980 and PDO in 1996 (EC, 1996). Its name comes from the two places located in the two associated regions, namely, the 'Pic du Midi d'Ossau' for the region Bearn and the Iraty forest for the region Pays Basque. The total annual production of Ossau Iraty has steadily increased between 2001 and 2015 (1.5-fold), with 4,200 tonnes produced in 2015 (INAO-CNAOL, 2014). Ossau Iraty is made and matured exclusively in a delimited area included in the French Pyrenees mountains (medium-sized mountains located in southwest France and north Spain). This area was reduced in 2001 and actually covers 600,000 hectares in the two associated regions, namely, Bearn with the 'Pic du Midi d'Ossau' and Pays Basque with the Iraty forest, the largest European beechwood.

Based on an artisanal know-how developed over centuries in Pyrenees, Ossau Iraty cheese is an uncooked and pressed cheese, rennet-coagulated and with a brushed rind, made exclusively from full-fat sheep's milk and in the delimitated area. Some of them (10% in 2013–2014) are made from raw milk at farms (Ossau Iraty 'at farm'), while the rest is made from pasteurised (82.5% in 2013) or raw milk (7.5% in 2013) at dairy plants (sold under the specific commercial label Ossau Iraty 'at plant') (Anonymous, 2015c; INAO-CNAOL, 2014). The cheeses are manufactured throughout the year, except in September and October; in winter and spring, the manufacture is located in plains and hills, while in summer it may be also located at specific places at higher altitudes ('estives'). 'Estive' is mentioned on the cheese label.

Actual Ossau Iraty cheeses are based on ancient cheeses that have been probably manufactured in the Pyrenees mountains since at least the fourteenth century ('Official website for Ossau Iraty', 2016). A traditional practice still in use, the seasonal migration of livestock, called *transhumance*, takes place at the beginning of the warm period. The herds, driven by shepherds and dogs, reach the rich pastures of the highlands (900–2500 m), called 'estives', while hay for winter is harvested from the pastures located in the lowlands. Milking and cheese manufacture are done in the shepherd huts, which are collectively managed by their co-owners. The highland pastures are collective and managed by several associations of owners, in order to locally better distribute herds and work.

2.13.2 Description and Sensory Characteristics

Ossau Iraty is a semi-hard to hard cheese. According to its specifications, it should have a minimum FDM of 50% and a minimum DM content of 58%. On average, its DM is 64%; its fat content 34%, including 25% of saturated fat; its salt content 0.6%; its protein content 24% and its carbohydrate content is 0%.

Ossau's texture is firm (especially for the small cheeses) to creamy (especially for the large cheeses) and smooth, can exhibit small holes, if any, and becomes firmer and slightly breakable as maturation progresses. Ossau Iraty's body is light yellow (ivory) to amber-yellow, depending of the maturation. The rind is orange-yellow to grey, with the rind of large cheeses being generally more orange. The rind is tightly bound to the cheese body, undigested, uncrumbling and unflaking, as well as dry to slightly wet, but not sticky. Small cheeses generally have a thicker (up to 1 cm) and darker under-rind area than large cheeses. At its optimum maturation, Ossau Iraty's flavour is characterised by a subtle fruity (dried-fruit) aroma and a well-balanced flavour; the lactic aroma predominates when it is younger, with a more persistent flavour and strong aroma when it is older. In general, Ossau Iraty 'at plant' has a milder flavour.

2.13.3 Method of Manufacture

Ossau Iraty is manufactured according to rules governing milk production to cheese maturation. All steps, including maturation, should be performed in the delimited area. Some specifications are different, depending on whether Ossau Iraty is 'at farm', 'at plant' or 'd'estive'. For the last cheese type, milk production, milking and manufacture should be done at the same location, which has to be different from the winter location.

Milk production: Milk has to be produced by 'Basco-Béarnaise', 'Manech tête noire' or 'Manech Tête rousse' cows, which are local traditional cows. Milk production should not exceed 300 litres per year and per dairy ewe. The dairy ewes should be six months or older. They are fed fodder, completed with non-GMO cereals in limited amounts; grass (240 days minimum), fresh or dried or dehydrated fodder, straw (without any ammoniac treatment) and fermented feed are the main feeds. Feeding of ewes by any fermented foods during the milking period will be banned from 1 February 2018. More restrictive specifications are associated with milk produced by ewes fed on pasture 'estive', their main fodder being grass pastured in the highlands. The mineral fertilisation for grazing surfaces is limited. Feed has to be produced within the delimited geographical area, except for 280 kg of dry matter per year and per ewe.

Milking should only be done up to 265 days/year and is forbidden in September and October. Depending on the herd, the milking period can be thus restricted to January–June, but can also extend to December–August. Milking is performed twice a day.

Production should be daily only for the cheese 'd'estive'. Cooled milk from the evening is mixed with warm milk of the morning.

Additives on the surface, namely, natamycin and polyvinyl acetate and rind-colouring agents are banned, as are milk concentration and the storage of milk, curd and fresh cheese at subzero temperatures. Thermal treatment of milk and removing of part of its lactose is authorised for manufacturing at dairy plants, while milk filtration on a layer of nettle to eliminate impurities, a traditional practice, is authorised for manufacturing at farms.

Coagulation: The milk is inoculated with starter culture and chymosine (520/10 L of milk) is added. Renneting should be performed 24 hr after milking for cheese made in 'estive', 40 hr for cheeses made at farm ('at farm') and 48 hr for cheeses made at dairy plant ('at plant'), at 28°C–35°C.

Cutting/cooking: The curd is cut to cubes of 1 cm^3 and heated at 38°C–44°C with stirring for 1 hr.

Moulding/labelling: The curd is transferred in moulds (two different sizes, small or large cheese). The cheese should be turned at least once for those made at farms. Between the moulding and de-moulding steps, Ossau Iraty is labelled by a counter-relief fingerprint in the curd or a label on casein put on the rind, which differ in form between 'at plant' (side profile of an ewe's head with its horns) and 'at farm' (full face of an ewe's head with its horns).

Pressing: Pressing is applied.

Salting: Cheeses are brined or dry-salted, just after the mould is removed at temperatures less than 15°C. Salting with dry salt cannot exceed 24 hr and 12 hr per kg of cheese for dry salting and brine salting, respectively. Brine contains water and salt (≤330 g/L), has a pH less than or equal to 5.5, can contain acetic or lactic acid and can be filtered.

Maturation: Cheeses are matured at 6°C–15°C and an RH higher than 75%; they can be brushed with water, salt, maturation microbial strains and mashed red chilli. The rind can be treated with vegetal oil or white vinegar.

Cheeses are matured in one of the maturation cellars, exclusively located in the delimited area, for a minimum of 80 days or 120 days depending on the cheese size, but often longer (12 months or more). After the minimum time of maturation, cheeses from every cheesemaker are sampled twice (at farm) or four times (at plant) per year for scoring their flavour, texture, aftertaste, odour, openness and appearance by a trained panel, and for assessing their gross physico-chemical composition; some of them can then be disqualified as Ossau Iraty (the fingerprint is removed by scratching).

2.13.4 Relevant Research

Very few studies have been dedicated to Ossau Iraty. In 2000, its total and individual amino-acid content was assessed (Izco et al., 2000a) and was found to be similar to the content measured in other related sheep's cheeses varieties; this content was increased in mature Ossau Iraty by adding proteolytic enzymes in the vat milk to accelerate maturation, leading to variation in the sensory scores (Izco et al., 2000b).

In 2010, research has focused on the LAB of Ossau Iraty cheesees manufactured at different farms from raw milk (Feutry et al., 2012a). They were distinct for their taxonomical composition and/or levels of the individual species, depending on the manufacturing stage. Wild lactoccocal strains isolated from Ossau Iraty's milk were characterised in order to use them to formulate new starters for making Ossau Iraty (Feutry et al., 2012b). Some of these strains were further used as a starter to make experimental Ossau Iraty in a very recent study (Feutry et al., 2016) which showed that the presence of *S. thermophilus* in the starter had a more powerful effect than changing the nature of the lactococcal strains present in the starter culture. In addition, powerful tools for diversifying the scores for the sensory characteristics were investigated.

2.14 Tête de Moine PDO, Fromage de Bellelay – Switzerland

<table>
<tr><td>Name: Tête de Moine PDO and Fromage de Bellelay</td><td rowspan="3">
Tête de Moine PDO on a Girolle® that allows the unique preparation of rosettes (Interprofession Tête de Moine)</td></tr>
<tr><td>Production area: The mountainous area and the enclosed summering area of the districts of Franches-Montagnes and Porrentruy, the municipality of Saulcy and the administrative district of Bernese Jura, with the exception of the municipalities of Nods, Diesse, Lamboing, Prêles and la Neuveville</td></tr>
<tr><td>Milk: Raw cow's milk</td></tr>
</table>

2.14.1 Introduction

Tête de Moine PDO (French for 'monk's head') is a small cylindrical smear-ripened, full-fat, semi-hard cheese made from raw cow's milk. Unlike other cheeses, Tête de Moine is only consumed in the form of shaved rosettes. This special kind of preparation for consumption is unique worldwide. Until the 1980s, these rosettes were formed with a knife. In 1981, a special cheese shaver (called Girolle®) was developed for this purpose and allows an easy and fast preparation of the rosette-shaped trimmings. As a result of this invention, the production and export of Tête de Moine increased considerably. In 2014, the production was 2,263 tonnes, and 1,390 tonnes was exported (TSM Treuhand GmbH, 2015). Tête de Moine, also called Fromage de Bellelay, was registered with PDO status in 2001 (FOAG, 2015; Swiss PDO-PGI Association, 2015). This designation has been recognized by the European Union since December 2011 (Swiss Commission for Denominations of Origin and Geographical Indications, 2011).

The monastery of Bellelay was founded in 1136. The manufacture of cheese by the monks of the monastery was first mentioned in 1192. At that time, the monks paid their tithes with cheese made in their abbey. The oldest description of Bellelay cheese dates to 1628 and states that a 'very fatty milk of impeccable quality from the best grasses and herbs of the country is used'. The more familiar name, Tête de Moine, was introduced around 1790, and its origin probably derived from the fact that the number of cheese wheels kept in the abbey depended on the number of monks living there. In 1797, during the French Revolution, the monks were evicted from the monastery, but the cheese continued to be produced in the former abbey. Towards the end of the nineteenth century, several village cheese dairies were established, and cheese manufacture gradually shifted to them (Kulinarisches Erbe der Schweiz, 2008).

2.14.2 Type

The specifications for Tête de Moine include a minimal fat content of 31.5%. The FDM has to be in the range of 51.0%–54.0% (preferably 52.5%), and the moisture content is 32%–36.0%. The salt content should not exceed 2.5%.

2.14.3 Description and Sensory Characteristics

Tête de Moine has the shape of a slightly convex cylinder with a diameter of 10–15 cm (the height is 70%–100% of the diameter) and an average weight of 0.9 kg (range: 0.7–3 kg). The rind is firm, smeared and moist with a reddish brown colour. The unique rosettes that are made exclusively from Tête de Moine with a Girolle® or a similar device must remain compact.

The texture of the cheese is fine and suitable for the preparation of rosettes and for cutting, has a yellow to ivory colour, is homogeneous, slightly moist, sticky and smooth. Tête de Moine may have a few small isolated openings of 1–8 mm. The flavour is pure and aromatic and becomes more intense with maturity. Usually a slight aroma of rind and mushrooms is perceptible. The taste is similar to sour milk, moderately salty, slightly spicy and is somewhat reminiscent of cheese rind and hay.

2.14.4 Method of Manufacture

Tête de Moine is produced in copper vats from raw cow's milk. Water, salt, rennet, lactic acid bacteria and smear cultures are the only permissible processing aids. During ripening, which lasts for at least 75 days, the loaves are placed on planks of spruce and regularly washed and turned.

Milk preparation: Raw cow's milk is the base raw material. On farms whose milk is used in the making of Tête de Moine, the preparation and feeding of silage of any kind is forbidden. Calculated on dry matter, the roughage must constitute at least 70% of the food ration of the dairy cattle and must originate from the region of production. Other fodders such as supplementary and concentrated feed are defined in detail in the specifications (FOAG, 2015). Furthermore, the use of any type of growth activators is forbidden. During the green forage period, the dairy cows must be left on the pasture for at least 120 days.

The milk must be delivered twice a day. In the case of a single delivery, the milk must be stored at the farm at a maximum temperature of 18°C (usually 8°C–12°C) and can be processed no later than 18 hr after the oldest milk has been collected (if the cooling temperature is below 8°C, processing may be no more than 24 hr after the oldest milk has been collected). The only authorised processes for the adjustment of the fat content are natural skimming and centrifugation. Milk treatments such as bactofugation and ultrafiltration are prohibited.

Starter culture: Inoculation of the vat milk is undertaken using liquid bulk starters, which are prepared from specific stock cultures authorised by the inter-professional organisation of Tête de Moine. In addition to yogurt cultures (containing *S. thermophilus* and *Lb. delbrueckii spp. bulgaricus*), thermophilic cultures containing *Lb. delbrueckii* ssp. *lactis* are used (Interprofession Tête de Moine, 2015). The use of GMOs and derivatives thereof is prohibited. Food additives are not allowed.

Cutting: The curd grain size corresponds to the size of a maize kernel. No lactose is removed by curd washing.

Scalding: The cheese curd is heated at a temperature of 44°C–53°C.

Pressing: As soon as the curd grains have attained the desired firmness, the curd is usually pre-pressed and then transferred to the cylindrical moulds or directly filled into the moulds. The pressing lasts for about 1 hr.

Salting: The cylinders are then immersed in a brine bath for at least 12 hr (usually >20°Bé). The duration of brining depends on the desired salinity.

Maturation: The cheeses are regularly treated with salt water and, if needed, natural yeasts and other microorganisms are applied in order to obtain the typical smear-rind. The ideal temperature of the cellar is 13°C–14°C during storage. The RH should be at least 90%. Tête de Moine must mature in the region of production for at least 75 days.

2.14.5 Relevant Research

Recently, a new culture has been developed by Agroscope to protect Tête de Moine against counterfeiting. The culture contains natural LAB with specific gene sequences that can be used to verify the authenticity of the cheese (Haldemann et al., 2013).

2.15 Tulum Cheese –Turkey

<table>
<tr><td>Name: Tulum cheese</td><td rowspan="3"></td></tr>
<tr><td>Production area: Eastern Anatolia, Central Anatolia, South Anatolia and with less frequency in other parts of Turkey</td></tr>
<tr><td>Milk: Sheep's, goat's, cow's, raw</td></tr>
</table>

2.15.1 Introduction

Tulum cheese is a hard cheese traditionally ripened in a goatskin bag. Goatskin bags are stronger than sheepskin bags and are permeable to water and air due to its porous structure

(Hayaloğlu et al., 2007). However, at present, due to the lack of goatskin bags and for hygienic reasons, hardened plastic containers (barrels) are widely used for ripening Tulum cheese. Tulum cheese is the third most economically important cheese variety following Beyaz Peynir and Kashar cheese, respectively. The annual production of Tulum cheese is well over 30,000 tonnes. This variety has a white or cream colour, a high fat content and is crumbly in texture (Kurt et al., 1991). At the industrial level, the goatskin is replaced by special cheese moulds. This cheese variety has a thin, dry and yellowish rind. Its body is homogeneous with no gas holes, and it has a piquant rancid and sharp flavour (Koçak, Aydemir & Seydim, 2005). Savak Tulum cheese is originally ripened in caves for at least three months. Erzincan Savak Tulum cheese was granted a PGI certificate by Turkish Standards Institution (TSE) in 2000. According to the PGI definition, Erzincan Savak Tulum cheese is described as follows: 'A cheese variety which is produced from milks of Karaman sheep breed grazed by 90–100 plants endemic to the hills of Erzincan mountains between the fifth and ninth months of the year'. Tulum cheese is produced practically all over Turkey. However, a majority of traditional Tulum cheese producers are located in eastern and central Turkey. Tulum cheeses produced in eastern and central parts of Turkey are called 'Erzincan Savak Tulum' and 'Divle Tulum', respectively. Tulum-type cheeses with less economic importance are also produced under different names (e.g. Kargi Tulum, Cimi Tulum, Tomas Tulum and Karin Kaymagi) in different parts of Turkey.

2.15.2 Type

Tulum cheese is a hard cheese, with thin, dry and yellowish rind and a pH of 4.8–5.2.

2.15.3 Description and Sensory Characteristics

The manufacturing practices of Tulum cheese vary depending on the region. While in some parts of Turkey (i.e. central Anatolia) Tulum cheese is ripened in caves, in other parts of the country, ripening largely takes place in cold rooms. The yield of Erzincan Savak Tulum cheese is around 12%–14%.

2.15.4 Method of Manufacture

Milk preparation: Sheep's milk, goat's milk, cow's milk (preferably with a far less amount) or a mixture, with sheep's or goat's milk being dominant. The milk is used raw.

Renneting: In the manufacture of Tulum cheese, coagulation is achieved within 60–150 min at 32°C–34°C by means of either home-made rennet or by commercial coagulants of animal origin.

Cutting: Coagulum is cut into 0.5–3.0 cm^3 parts and is poured into a cheesecloth bag. Whey separation is achieved by means of hanging cheesecloth bags for 14–16 hr at the ambient temperature (Koçak & Gürsel, 1992).

Moulding/pressing: Following wheying off, the coagulum is pressed for 6–24 hr until the desired TS level is attained and then cut into small pieces by hand.

Salting: Small cheese pieces are dry-salted with NaCl at concentrations of 2%–3%.

Moulding: Salted cheese curd is transferred into cheese moulds and pressed so that the air inside the mould is removed.

Ripening: The cheese moulds are placed into cold stores with 75%–85% RH, and the cheese is ripened at 4°C–12°C for 120 days or longer (Koçak et al., 1995). In the Southeastern Anatolia region, Tulum cheese is largely produced from milk of the Awassi sheep's breed.

2.15.5 Relevant Research

Hayaloğlu et al. (2007) demonstrated that alanine, valine, leucine and phenylalanine were the most abundant free amino acids in Savak Tulum cheese. Öztürkoğlu-Budak (2014) showed that concentrations of hydrophobic and intermediate peptides were higher than that of hydrophilic peptides in the 120-day-old Divle Tulum cheese. Herbs (i.e. black cumin – *Nigella sativa*) may be added to Tulum cheese, when development of proteolysis is faster (Tarakçı et al., 2005).

Methyl ketones, alcohols and esters constitute the volatile components of Savak Tulum cheese (Hayaloğlu, Özer & Fox, 2008). Hayaloğlu et al. (2007) identified a total of 100 volatile compounds in 90-day-old Savak Tulum cheese (including 11 acids, 16 esters, 12 methyl ketones, 7 aldehydes, 22 alcohols, 7 sulphur compounds, 6 terpenes and 19 miscellaneous compounds). The characteristic flavour of Tulum cheese comes from free FAs, and addition of external lipases in commercial form (i.e. Piccantase) is a common practice in industrial Tulum cheese production (Yılmaz, Ayar & Akın, 2005). The most abundant FAs of Divle Tulum cheese were palmitic, oleic, stearic and myristic acids (Öztürkoğlu-budak, 2014).

2.16 Västerbottensost – Sweden

Name: Västerbottensost	 
Production area: Västerbotten county	
Milk: Cow's, pasteurised	

2.16.1 Introduction

Västerbottensost has been protected by registered trademarks since 1910, which is today owned by the dairy company Norrmejerier, Umeå, Sweden. Västerbottensost is the most famous cheese made in Sweden. It bears the name of the province in the north of Sweden where it has been produced since the 1870s. The manufacturing protocol was developed by dairymaid Ulrika Eleonora Lindström, who was educated at a dairy school in the southwest province of Sweden, Västergötland, which is famous for the high-quality cheeses it has been making for hundreds of years (Magnus, 1555). She introduced and developed a new recipe at the dairy in Burträsk. She was responsible for all production and had to take care of several things by herself at the dairy, and to get around she prolonged the holding times as much as possible. She observed that by increasing the cooking temperature and holding time, the cheese quality improved, and the shelf-life of the cheeses increased. She made detailed notes about the different processing steps and related them to the quality of the final cheese after ripening for a year or more. Her notes are still kept in a safe at the dairy in Burträsk, and the recipe she developed is still used to obtain the unique characteristics of Västerbottensost. (Ränk, 1987)

2.16.2 Type

Västerbottensost is a hard to extra-hard cheese with a moisture content in non-fat substance of around 50%, and it is made with a mesophilic starter. Its unique properties result from a very long cooking time at moderate temperatures, which is unlike most other hard cheeses, which are made with thermophilic starters and cooked at high temperatures for rather short time periods. The texture is open with several irregularly shaped pea-sized eyes, and the fat content is 31%. Västerbottensost is delicious to consume just as it is with bread and is also very nice in cooked food (Ardö, 1993; Sveriges Ostkollegium, 1993).

2.16.3 Milk

Västerbottensost is made of pasteurised cow's milk produced in a defined area around Burträsk in Västerbotten, Sweden.

2.16.4 Description and Sensory Characteristics

The cheese is made in low cylinders with a diameter of 40 cm and height of 14–18 cm and a weight of 18–20 kg. The surfaces are dry and coated with wax, and the colour is yellow. The interior of the cheese is light yellow to yellow. The cheese is ripened for at least one year. The texture is open with several small irregularly shaped eyes. The body is brittle and juicy. The unique flavour is full-bodied, fruity, aromatic and pungent with a sharp note and hints of caramel and truffle.

2.16.5 Method of Manufacture

Milk preparation: The raw milk is standardised for fat content, pasteurised (72°C /15 s) and heated to 30°C before the starter is added.

Starter culture: Mesophilic DL-starter with undefined mixed strains of *Lc. lactis* ssp. *lactis* and *Lc. lactis* ssp. *cremoris* (80%–90%), *Lc. lactis* ssp *lactis* biovar. *diacetylactis* and *Leuc. mesenteroides*.

Rennet: Calf rennet (75/25 of chymosin/bovine pepsin) is used.

Cutting: Grains at the size of a pea.

Cooking: Just above 40°C for several hours to promote exhaustive syneresis. This tough treatment is atypical for cheeses made with DL-starters and creates stressful conditions close to the limit for survival of the mesophilic starter bacteria.

Curd draining: Whey is removed, and the curd is kneaded in the vat before it is put in moulds containing cheesecloths.

Salting: Brine salting for three days at 12°C to reach a salt content of 0.8%–1.2% salt in the ripened cheeses.

Maturation: The cheeses are kept at 12°C for two weeks before they are sent to the ripening unit for keeping at about 18 days at 18°C with turning over each day. Thereafter the cheeses are waxed and further matured at 12°C and 50% RH for 12–14 months or sometimes longer.

2.16.6 Relevant Research

Peptides from starter activity on the chymosin-derived peptide α_{s1}-casein (f1-23) and plasmin-derived peptides in the area of α-casein (f1-107) characterise the peptide profile. The peptide α_{s1}-casein (f29-93) accumulates characteristically during ripening, likely as a result of joint plasmin and chymosin activity and the absence of proteolytic thermophilic bacteria to mediate further degradation of these peptides. The total amount of free amino acids is about 300–400 mmol/kg cheese after a year of maturation. The starter as well as NSLAB influence the composition of individual amino acids and the content of volatile compounds. Secondary alcohols have been shown to decrease between six and twelve months while the corresponding methyl ketones increased, and during the same period branch-chain aldehydes increased (Rehn et al., 2010). The quite elastic fresh cheese becomes softer during the first month as a result of proteolysis and rearrangement of calcium phosphate. Further ripening increases the amount of small molecules, such as amino acids, which contribute to water binding and a shorter and more brittle texture (Rehn et al., 2010).

2.17 Würchwitzer Mite Cheese – Germany

Name: Würchwitzer Mite cheese	 
Production area: Würchwitz (East Germany)	
Milk: Made from low-fat Quark	

2.17.1 Introduction

Mite cheese is a traditional cheese product from Germany which is mainly produced on agricultural farms. In 2006, it became a member of the German Slow Food association, 'Ark of Taste', which aims to maintain and preserve Mite cheese (Milbenkäse, 2015; Slow Food, 2015). Würchwitzer Mite cheese is a registered trademark in Germany (Original Würchwitzer Milbenkäse).

Only scarce information can be found in the literature on Mite cheese, but it supposedly dates back to the Middle Age. Traditionally, Mite cheese was produced on agricultural farms in the border region of the districts of Zeitz and Altenburg in Saxony-Anhalt and Thuringia, respectively. In the village of Würchwitz in Saxony-Anhalt, Mite cheese has been traditionally produced for more than 500 years, and the manufacturing protocol was inherited from generation to generation. In 2006, Helmut Pöschel und his partner Christian Schmelzer built a company supported by a European funding project to produce and distribute Mite cheese in Germany. The *Würchwitzer Milbenkäse Manufaktur* is the only company that is allowed to produce and market Mite cheese in Germany (Milbenkäse, 2015).

Mite cheese has been traditionally produced in the border region of Saxony-Anhalt and Thuringia on agricultural farms and in households. Since 2006, the WürchwitzerMilbenkäseManufaktur produces Mite cheese commercially.

2.17.2 Type

Mite cheese is a low-fat hard cheese. Ripening is performed by enzymes produced from the cheese mites *Tyroglyphus casei* (Storch & Welsch, 2004)

2.17.3 Description and Sensory Characteristics

Mite cheese has the form of a longish roll with 7–9 cm length and a diameter of about 3 cm. These rolls are called 'Quargeln'. Moreover, small balls are formed in which an elderflower panicle is wrapped. The average weight is 100 g with 30%–50% moisture content (Mair-Waldburg, 1974; Milbenkäse, 2015).

The surface is covered with fine grey dust with small holes. With increasing ripening time, the rind becomes amber-coloured until it turns black after approximately one year. The cheese has an acidic smell and a slightly nutty and bitter taste, which increases with ripening.

2.17.4 Method of Manufacture

The WürchwitzerMilbenkäseManufaktur is the only company that produces Mite cheese in Germany under permit of the local food safety office, and HACCP compliance of the product is enforced (Milbenkäse, 2015).

Raw material: Mite cheese is made from low-fat Quark, which is produced according to a standardised process and obtained from dairies located close to the company. After it is drained thoroughly and dried for one day, the Quark is mixed with 2%–3% salt, caraway and elderflowers (Milbenkäse, 2015).

Production: The formed Quark rolls and balls are dried and transferred into a wooden box with cheese mites for at least three months. The size of the wooden boxes depends on the amount of cheeses and varies usually between 5 and 15 litres. The boxes are closed, but gas exchange is possible. The boxes are best stored in a dark, humid place (Mair-Waldburg, 1974; Milbenkäse, 2015; Schmelzer, 2015).

Maturation: The cheese ripens slowly from the outside to the inside. They are turned daily and aired. Periodically, the mites are fed with rye flour in order to avoid too much cheese loss.

Ripening is induced by enzymes and digestive juices produced from the cheese mites. After at least three months of ripening, the cheeses lose 50% of their weight. The optimum conditions for growth of the cheese mites are an RH of 80% to 90% and temperatures from 10°C to 15°C. Ripening time depends on the season; during winter it may be more than three months.

References

Abilleira, E., Collomb, M., Schlichtherle-Cerny, H., Virto, M., Renobales, M. de & Barron, L. J. R. (2009). Winter/spring changes in fatty acid composition of farmhouse Idiazabal cheese due to different flock management systems. *Journal of Agricultural and Food Chemistry*, 57, 4746–4753.

Addis, M., Piredda, G. & Pirisi, A. (2008). The use of lamb rennet paste in traditional sheep milk cheese production. *Small Ruminant Research*, 79, 2–10.

Addis, M., Pirisi, A., Di Salvo, R., Podda, F. & Piredda, G. (2005). The influence of the enzymatic composition of lamb rennet paste on some properties of experimentally produced PDO Fiore Sardo cheese. *International Dairy Journal*, 15, 1271–1278.

Allgäuer Bergkäse (2015). *Allgäuer Bergkäse, Sennerei Schweineberg* http://www.allgaeuer-bergkaese.de/ [Accessed 14 July 2015].

Andrikopoulos, N. K., Kalogeropoulos, N., Zerva, A., Zerva, U., Hassapidou, M. & Kapoulas, V. M. (2003). Evaluation of cholesterol and other nutrient parameters of Greek cheese varieties. *Journal of Food Composition and Analysis*, 16, 155–167.

Anonymous (2006). *Official website of Asiago*. www.asiagocheese.it/it/consorzio/il-disciplinare-di-produzione/; (D.M.03/08/2006 Gazzetta .Ufficiale n. 190 del 17/08/06). Accessed 14 June 2016.

Anonymous (2015a). *Cheddar cheese*. http://www.farmhousecheesemakers.com/ Accessed 30 May 2015.

Anonymous (2015b). *Cheshire cheese*. http://www.cheshirecheese.com Accessed 30 May 2015.

Anonymous (2015c). Le guide 2015 des fromages au lait cru. *Profession fromager, hors-série* 13, 114.

Ardö, Y. (1993). Swedish cheese varieties. In P.F. Fox (ed.), *Cheese: Chemistry, Physics and Microbiology. Vol. 2. Major Cheese Groups*, 2nd edition. Chapman & Hall, London, pp 254–257.

Arizcun, C., Barcina, Y. & Torre, P. (1997). Identification and characterization of proteolytic activity of Enterococcus spp. isolated from milk and Roncal and Idiazαbal cheese. *International Journal of Food Microbiology*, 38, 17–24.

Barcina, Y., Ibαpez, F. C. & Ordσpez, A. I. (1995). Evolution of free amino acids during Idiazαbal cheese ripening. *Food Control*, 6, 161–164.

Barron, L. J. R., Hernandez, I., Bilbao, A., Flanagan, C. E., Najera, A. I., Virto, M., Perez-Elortondo, F. J., Albisu, M. & Renobales, M. de. (2004). Changes in lipid fractions and sensory properties of Idiazabal cheese induced by lipase addition. *Journal of Dairy Research*, 71, 372–379.

Barron, L. J. R., Fernαndez de Labastida, E., Perea, S., Chαvarri, F., de Vega, C., Vicente, M. S., Torres, M. I., Nαjera, A. I., Virto, M., Santisteban, A., Pιrez-Elortondo, F. J., Albisu, M., Salmerσn, J., Mendύμa, C., Torre, P., Ibαpez, F. C. & Renobales, M. de. (2001). Seasonal changes in the composition of bulk raw ewe's milk used for Idiazabal cheese manufacture. *International Dairy Journal*, 11, 771–778.

Barron, L. J. R., Redondo, Y., Aramburu, M., Gil, P., Pιrez-Elortondo, F. J., Albisu, M., Najera, A. I., Renobales, M. de & Fernandez-Garcia, E. (2007). Volatile composition and sensory properties of industrially produced Idiazabal cheese. *International Dairy Journal*, 17, 1401–1414.

Bergkäse (2015). http://www.bergkaese.de [Accessed 14 July 2015].

Bustamante, M. A., Virto, M., Aramburu, M., Barron, L. J. R., Perez-Elortondo, F. J., Albisu, M. & Renobales, M. de. (2003). Lamb rennet paste in ovine cheese (Idiazabal) manufacture.

Proteolysis and relationship between analytical and sensory parameters. *International Dairy Journal*, 13, 547–557.
Casey, M. G., Häni, J. P., Gruskovnjak, J., Schaeren, W. &Wechsler, D. (2006). Characterisation of the non-starter lactic acid bacteria (NSLAB) of Gruyere PDO cheese. *Lait*, 86, 407–414.
Chávarri, F., Bustamante, M. A., Santisteban, A., Virto, M., Barron, L. J. R. & Renobales, M. de. (1999). Changes in free fatty acids during ripening of Idiazabal cheese manufactured at different times of the year. *Journal of Dairy Science*, 82, 885–890.
Cozzi, G., Ferlito, J., Pasini, G., Contiero, B. & Gottardo, F. (2009). Application of near-infrared spectroscopy as an alternative to chemical and color analysis to discriminate the production chains of Asiago d'Allevo cheese. *Journal of Agricultural Food Chemistry*, 57, 11449–11454.
De Freitas, I., Pinon, N., Lopez, C., Thierry, A., Maubois, J.-L., & Lortal, S. (2005). Microstructure, physicochemistry, microbial populations and aroma compounds of ripened Cantal cheeses. *Lait*, 85, 453–468.
De Freitas, I., Pinon, N., Lopez, C., Thierry, A., Maubois, J.-L., & Lortal, S. (2007). In depth dynamic characterisation of French PDO Cantal cheese made from raw milk. *Lait*, 87, 97–117.
Deutsche Käseverordnung, Established 1965, *Revision 1986, Amendment 2010, 2013. German Federal Ministry for justice and consumer protection.*
Duclaux, E. (1893). *Principes de laiterie*, Ed. A. Colin (Paris).
EC (1996). Commission Regulation (EC) No 1107/96 of 12 June 1996 on the registration of geographical indications and designations of origin under the procedure laid down in Article 17 of Council Regulation (EEC) No 2081/92. *Official Journal of the European Union*, L 148, 1–10.
EC (1997). Commission Regulation (EC) No. 123/97 of 23 January 1997 supplementing the Annex to Commission Regulation (EC) No. 1107/96 on the registration of geographical indications and designations of origin under the procedure laid down in Article 17 of Regulation (EEC) No. 2081/92. *Official Journal of the European Union*, L 22, 24.1.1997, 19–20.
Favaro, G., Magno, F., Boaretto, A., Bailoni, L. & Mantovani, R. (2005). Traceability of Asiago mountain cheese: A rapid, low-cost analytical procedure for its identification based on solid-phase microextraction. *Journal of Dairy Science*, 88 (10), 3426–3434.
Feutry, F., Oneca, M., Berthier, F. & Torre, P. (2012a). Biodiversity and growth dynamics of lactic acid bacteria in artisanal PDO Ossau-Iraty cheeses made from raw ewe's milk with different starters. *Food Microbiology*, 29, 33–42.
Feutry, F., Torre, P., Arana, I., Garcia, S., Desmasures, N. & Casalta, E. (2012b). Lactococcus lactis strains from raw ewe's milk samples from the PDO Ossau-Iraty cheese area: levels, genotypic and technological diversity. *Dairy Science & Technology*, 92, 655–670.
Feutry, F., Torre, P., Arana, I., Garcia, S., Elortondo, F. J. P. & Berthier, F. (2016). Suitability of a new mixed-strain starter for manufacturing uncooked raw ewe's milk cheeses. *Food Microbiology*, 56, 52–68.
FOAG (2015). *Register of appellation of origin and geographical indications*. Federal Office for Agriculture. An Office of the Federal Department of Economy, Bern, Switzerland http://www.blw.admin.ch/ [Accessed 28 April 2015]
Häni, J. P. & Keller Möcklin, S. (2006). Nouvelles cultures pour le gruyère AOC. *Alimenta*, 25, 4–5.
Haldemann, J., Berthoud, H., Roetschi, A., von Ah, U., Rollier, D. & Eugster, E. (2013). Tête de Moine AOP: Eine neue Kultur für den Herkunftsnachweis (Tête de moine PDO: A new culture for the proof of origin). *Agrarforschung Schweiz*, 4 (10), 448–450.
Hayaloğlu, A. A., Çakmakçı, S., Brechany, E. Y., Deegan, K. C. & McSweeney, P. L. H. (2007). Microbiology, biochemistry and volatile composition of Tulum cheese ripened in goat's skin or plastic bags. *Journal of Dairy Science*, 90, 1102–1121.
Hayaloğlu, A. A., Özer, B. & Fox, P. F. (2008). Cheeses of Turkey: 2. Varieties ripened under brine. *Dairy Science and Technology*, 88, 224–244.

Hernández, I., Barron, L. J. R., Virto, M., Pırez-Elortondo, F. J., Flanagan, C., Rozas, U., Nαjera, A. I., Albisu, M., Vicente, M. S. & Renobales, M. de. (2009). Lipolysis, proteolysis and sensory properties of ewe's raw milk cheese (Idiazabal) made with lipase addition. *Food Chemistry*, 116, 158–166.

Hernández, I., Renobales, M. de, Virto, M., Pırez-Elortondo, F. J., Barron, L. J. R., Flanagan, C. & Albisu, M. (2005). Assessment of industrial lipases for flavour development in commercial Idiazabal (ewe's raw milk) cheese. *Enzyme and Microbial Technology*, 36, 870–879.

Hofmann, S. (2005). *Edler Käse und seine Geheimnisse*. ReiseArt Verlagsgruppe, Weimar, Germany.

INAO-CNAOL, (2014). *Produits laitiers AOP – les chiffres clés, 2014*. [Online]. Available: https://www.google.fr/search?q=produits+laitiers+AOP++2014&ie=utf-8&oe=utf-8&client=firefox-b&gfe_rd=cr&ei=iI3jV6OgJ8nDaKWRn8AO [Accessed 16 July 2016].

Interprofession du Gruyère AOP (2014). *Pringy – Gruyères*, Switzerland. http://www.gruyere.com/ [Accessed 28 April 2015].

Interprofession Tête de Moine AOP (2015). *St-Imier*, Switzerland. http://www.tetedemoine.ch/ [last accessed: 14.07.2015].

Izco, J. M., Irigoyen, A., Torre, P. & Barcina, Y. (2000a). Effect of added enzymes on the free amino acids and sensory characteristics in Ossau-Iraty cheese. *Food Control*, 11, 201–207.

Izco, J. M., Irigoyen, A., Torre, P. & Barcina, Y. (2000b). Effect of the activity levels of the added proteolytic enzyme mixture on free amino acids in ripening Ossau-Iraty cheese. *Journal of Chromatography A*, 881, 69–79.

Jakob, E., Goy, D., Haldemann, J. & Badertscher, R. (2013). Automatic milking systems in cheese milk production: Potential for improvements. *Agrarforschung Schweiz* 4 (6), 256–263.

Jakob, E., Badertscher, R. & Bütikofer, U. (2007). Zusammensetzung von Berner Alp- und Hobelkäse. *Agrarforschung Schweiz* 14 (3), 96–101.

JORF (2007). *Décret 08 03 2007 relatif à l'appellation d'origine contrôlée « cantal » ou « Fourme de Cantal »*.

Kammerlehner, J. (2009). *Cheese Technology*. Josef Kammerlehner Verlag, Freising, Germany.

Kandarakis, G., Moschopoulou, E., Moatsou, G. & Anifantakis, E. M. (1998). Effect of starters on gross and microbiological composition and organoleptic characteristics of Graviera Kritis cheese. *Lait*, 78, 557–568.

Katsiari, M. C., Voutsinas, L. P. & Kondyli, E. (2002). Improvement of sensory quality of low-fat Kefalograviera-type cheese with commercial adjunct cultures. *International Dairy Journal*, 12, 757–764.

Koçak, C. & Gürsel, A. (1992). *New aspects in Tulum cheese technology. Proceedings of 5th Egyptian Conference for Dairy Science and Technology*, 19–21 October, Egypt, pp. 15–23.

Koçak, C., Aydemir, S. & Seydim, Z. B. (2005). Levels of proteolysis in important types of Turkish cheeses. *Gida*, 30 (6), 395–398.

Koçak, C., Gürsel, A., Avşar, Y. K. & Semiz, A. (1995). Effect of lipase enzyme (Palatase A750 L) on the ripening of Tulum cheese. *Turkish Journal of Agriculture and Forestry*, 19, 171–177.

Kondyli, E., Massouras, T., Katsiari, M. C. & Voutsinas, L. P. (2003). Free fatty acids and volatile compounds in low-fat Kefalograviera-type cheese made with commercial adjunct cultures. *International Dairy Journal*, 13, 47–54.

Kulinarisches Erbe der Schweiz. (2008). Lausanne, Switzerland. http://www.kulinarischeserbe.ch/ [Accessed 28 April 2015]

Kurt, A., Çağlar, A., Çakmakçı, S. & Akyüz, N. (1991). [Erzincan Tulum (Şavak) peynirinin mikrobiyolojik özellikleri. *Doğa-Turkish Journal of Veterinary and Animal Sciences*, 16, 41–50.]

Lawrence, R. C., Gilles, J., Creamer, L. K., Crow, V. L., Heap, H. A., Honoré, C. G., Johnston, K. A. & Samal, P. K. (2004). Cheddar cheese and related dry-salted cheese varieties. In Fox, R. F.,

McSweeney, P. L. H., Cogan, T. M. & Guinee, T. P. (eds.), *Cheese: Chemistry, Physics and Microbiology*, 3rd edition, Vol. 2. Major Cheese Groups, Elsevier Academic Press, London, pp. 71–102.

Ledda, A., Scintu, M. F., Pirisi, A., Sanna, S. & Mannu, L. (1994). Caratterizzazione tecnologica di ceppi di lattococchi e di enterococchi per la produzione di formaggio pecorino Fiore Sardo. *Scienza e Tecnica Lattiero Casearia*, 45, 443–456.

Lignitto, L., Segato, S., Balzan, S., Cavatorta, V., Oulahal, N., Sforza, S., Degraeve, P., Galaverna, G. & Novelli, E. (2012). Preliminary investigation on the presence of peptides inhibiting the growth of Listeria innocua and Listeria monocytogenes in Asiago d'Allevo cheese. *Dairy Science and Technology*, 92 (3), 297–308.

Litopoulou-Tzanetaki, E. (1990). Changes in numbers and kinds of lactic acid bacteria during ripening of Kefalotyri cheese. *Journal of Food Science*, 55 (1), 111–113.

Magnus, O. (1555). *Historia de gentibus septentrionalibus* [History about the Nordic people]. Rom, in protection of Pope Julius III.

Mair-Waldburg (1974). *Handbuch der Käse: Käse der Welt von A-Z: eine Enzyklopädie.* Volkswirtschaftlicher Verlag, Kempten, Germany.

Manzi, P., Marconi, S., Di Costanzo, M. G. & Pizzoferrato, L. (2007). Composizione di formaggi DOP italiani. *La rivista di Scienza dell'Alimentazione, anno*, 36, 9–22.

Mendia, C., Ibapez, F. C., Torre, P. & Barcina, Y. (2000). Influence of the season on proteolysis and sensory characteristics of Idiazabal cheese. *Journal of Dairy Science*, 83, 1899–1904.

Michaelidou, A., Katsiari, M. C., Voutsinas, L. P., Kondyli, E. & Alichanidis, E. (2003). Effect of commercial adjunct cultures on proteolysis in low-fat Kefalograviera-type cheese. *International Dairy Journal*, 13, 743–753.

Milbenkäse (2015). http://www.milbenkaese.de [Accessed 14 July 2015].

Ministry of Agriculture, Food and Environment. (2014). *Secretarva General Tıcnica Centro de Publicaciones.* Datos de las Denominaciones de Origen Protegidas (D.O.P.) e Indicaciones Geogrαficas Protegidas (I.G.P.) de Productos Agroalimentarios, 3, 22–23.

Ministry of Agriculture, Food and Environment. (2015). *Electronic catalogue of Spanish cheeses* http://www.alimentacion.es/es/conoce_lo_que_comes/bloc/queso/default/principal. (Accessed 11 March 2015).

Moatsou, G., Moschopoulou, E., & Anifantakis, E. (2004). Effect of different manufacturing parameters on the characteristics of Graviera Kritis cheese. *International Journal of Dairy Technology*, 57 (4), 215–220.

Morandi, S. & Brasca, M. (2012). Safety aspects, genetic diversity and technological characterisation of wild-type Streptococcus thermophilus strains isolated from north Italian traditional cheeses. *Food Control*, 23, 203–209.

Nega, A. & Moatsou, G. (2012). Proteolysis and related enzymatic activities in ten Greek cheese varieties. *Dairy Science and Technology*, 92 (1), 57–73.

Official website for Ossau Iraty PDO, 2016. [Online]. Available: http://www.ossau-iraty.fr [Accessed 16 August 2016].

Ordoňez, A. I., Ibapez, F. C., Torre, P. & Barcina, Y. (1999). Effect of ewe's-milk pasteurization on the free amino acids in Idiazαbal cheese. *International Dairy Journal*, 9, 135–141.

Ordoňez, A. I., Ibapez, F. C., Torre, P. & Barcina, Y. (1998). Characterization of the casein hydrolysis of Idiazabal cheese manufactured from ovine milk. *Journal of Dairy Science*, 81, 2089–2095.

Ortigosa, M., Irigoyen, A., Urdin, M., Garcνa, S., Ibapez, F. C. & Torre, P. (2008). Sources of enterococci in Idiazαbal-type cheese. *International Journal of Food Microbiology*, 125, 146–152.

Öztürkoğlu-Budak, S. (2014). [*Divle obruk peynirinin bazi ozelliklerinin belirlenmesi.* PhD. Thesis, Ankara University Institute of Natural and Applied Sciences, Department of Food Engineering, Ankara, Turkey.]

Pérez Elortondo, F. J., Albisu, M. & Barcina, Y. (1999). Brining time effect on physicochemical and microbiological parameters in Idiazαbal cheese. *International Journal of Food Microbiology*, 49, 139–149.

Pérez Elortondo, F. J., Albisu, M. & Barcina, Y. (2002). Brining time and smoking influence on acceptability of Idiazabal cheese. *Journal of Food Quality*, 25, 51–62.

Pérez Elortondo, F. J., Aldαmiz Echobarria, P., Albisu, M. & Barcina, Y. (1998). Indigenous lactic acid bacteria in Idiazαbal Ewes' milk cheese. *International Dairy Journal*, 8, 725–732.

Pirisi, A., Pinna, G., Addis, M., Piredda, G., Mauriello, R., De Pascale, S., Caira, S., Mamone, G., Ferranti, P., Addeo, F. & Chianese, L. (2007). Relationship between enzymatic composition of lamb rennet paste and proteolytic, lipolytic pattern and texture of PDO Fiore Sardo ovine cheese. *International Dairy Journal*, 17, 143–156.

Pisanu, M. B., Fadda, M. E., Deplano, M., Corda, A. & Cosentino, S. (2006). Microbiological and chemical characterisation of Fiore Sardo, a traditional Sardinian cheese made from ewe's milk. *International Journal of Dairy Technology*, 59, 171–179.

Ränk, G. (1987). *[From Milk to Cheese]. Nordiska Museets Handlingar 66*, 2nd edition. Berlings, Arlöv, Sweden. ISBN 91-7108-269-7.

Rehn, U., Petersen, M. A., Hallin Saedén, K. & Ardö, Y. (2010). Ripening of extra hard cheese made with mesophilic DL-starter. *International Dairy Journal*, 20, 844–851.

Renna, M., Cornale, P., Lussiana, C., Giordano, M., Belviso, S., Zeppa, G. & Battaglini, L. M. (2012). Efficacy of fatty acids and terpenoids and weakness of electronic nose response as tracers of Asiago d'Allevo PDO cheese produced in different seasons. *Dairy Science & Technology*, 92 (3), 203–218.

Robinson, R. K. & Wilbey, R. A. (1998). *Cheesemaking Practice – R. Scott*, 3rd edition. Aspen Publishers, Maryland, USA.

Samelis, J., Bleicher, A., Delbes-Paus, C., Kakouri, A., Neuhaus, K. & Montel, M.-C. (2011). FTIR-based polyphasic identification of lactic acid bacteria isolated from traditional Greek Graviera cheese. *Food Microbiology*, 28 (1), 76–83.

Samelis, J., Kakouri, A., Pappa, E., MatiJasic, B. B., Georgalaki, M. D., Tsakalidou, E. & Rogelj, I. (2010). Microbial stability and safety of traditional Greek Graviera cheese: Characterization of the lactic acid bacterial flora and culture-independent detection of bacteriocin genes in the ripened cheeses and their microbial consortia. *Journal of Food Protection* 73 (7), 1294–1303.

Schievano, E., Pasini, G., Cozzi, G. & Mammi, S. (2008). Identification of the production chain of Asiago d'Allevo Cheese by Nuclear Magnetic Resonance Spectroscopy and Principal Component Analysis. *Journal of Agricultural Food Chemistry*, 56, 7208–7214.

Schmelzer, C. (2015). *Personal interview led by the author, Berlin*, 30. June 2015.

Segato, S., Balzan, S., Elia, C.A., Lignitto, L., Granata, A., Magro, L., Contiero, B., Andrighetto, I. & Novelli, E. (2007). Effect of period of milk production and ripening on quality traits of Asiago cheese. *Italian Journal of Animal Science*, 6 (1), 469–471.

Slow Food (2015). http://www.slowfood.de/biodiversitaet/die_arche_passagiere/wuerchwitzer_milbenkaese/[Accessed 14 July 2015].

Storch, V. & Welsch, U. (2004). *Systematische Zoologie*. Spektrum Akademischer Verlag, Heidelberg, Berlin, Germany.

Sveriges Ostkollegium. (1993). *Svenska Ostsorter [Swedish cheese varieties]*. Falkenberg, Sweden.

Swiss Commission for Denominations of Origin and Geographical Indications. Annual report (2011). Federal Office for Agriculture FOAG. http://www.blw.admin.ch/ [last assessed: 08.09.2015].

Swiss PDO-PGI Association (2015). Bern, Switzerland. http://www.aop-igp.ch/ [last assessed: 28.04.2015].

Tarakçı, Z., Ekici, K., Sağdıç, O. & Küçüköner, E. (2005). The effect of black cumin on ripening of Tulum cheese. *Archive für Lebensmittelhygiene*, 56, 135–139.

Thi, P. N., Dupas, C., Adt, I., Degraeve, P., Ragon, M., Missaoui, M. F., Novelli, E., Segato, S., The, D. P. & Oulahal, N. (2014). Partial characterisation of peptides inhibiting Listeria growth in two Alpine cheeses. *Dairy Science & Technology*, 94 (1), 61–72.

TSM Treuhand GmbH (2015). *Mehrjahresvergleich ab 2004, Beilage zur Jahresstatiskik 20014.* Bern, Switzerland. http://www.tsmtreuhand.ch/ [Accessed 28 April 2015]

Urgeghe, P. P., Piga, C., Addis, M., Di Salvo, R., Piredda, G., Scintu, M. F., Wolf, I. V. & Sanna, G. (2012). SPME/GC-MS Characterization of the Volatile Fraction of an Italian PDO Sheep Cheese to Prevalent Lypolitic Ripening: the Case of Fiore Sardo. *Food Analytical Methods*, 5 (4), 723–730.

Veinoglou, B., Baltadjieva, M., Anifantakis, E. & Edgaryan, M. (1983). Contribution to the study of Kefalotyri cheese made from sheep's milk. *Food Technology Hygiene Reviews*, 5, 9–14.

Vicente, M. S., Ibαρez, F. C., Barcina, Y. & Barron, L. J. R. (2001). Changes in the free amino acid content during ripening of Idiazabal cheese: Influence of starter and rennet type. *Food Chemistry*, 72, 309–317.

Weltgenußerbe Bayern (2015). http://www.weltgenusserbe.eu/spezialitaeten/allgaeuer-bergkaese.html [Accessed 14 July 2015].

Yılmaz, G., Ayar, A. & Akın, N. (2005). The effect of microbial lipase on the lipolysis during the ripening of Tulum cheese. *Journal of Food Engineering*, 69, 269–274.

Zazzu C. (2011). *Metodi analitici per la caratterizzazione e la valutazione della sicurezza alimentare di prodotti tradizionali della filiera lattiero-casearia ovina della Sardegna.* Tesi di Dottorato in Scienze e Tecnologie Chimiche XXIV ciclo, Università Degli Studi Di Sassari.

Zezza, N., Pasini, G., Lombardi, A., Mercenier, A., Spettoli, P., Zamorani, A. & Nuti, M. (1993). Production of a bacteriocin active on lactate-fermenting clostridia by *Lactococcus lactis* subsp. lactis immobilized in coated alginate beads. *Journal of Dairy Research*, 60 (4), 581–591.

Zygouris, M. P. (1952). *The Dairy Industry*, 2nd edition. Ministry of Agriculture, Athens, Greece.

3

Semi-hard Cheeses

Elisabeth Eugster-Meier[1], *Marie-Therese Fröhlich-Wyder*[2], *Ernst Jakob*[2], *Daniel Wechsler*[2], *Maria Belén López Morales*[3], *Giuseppe Licitra*[4], *Françoise Berthier*[5], *Photis Papademas*[6], *Ylva Ardö*[7,8], *Tânia G. Tavares*[9,10], *F. Xavier Malcata*[9,11], *Zorica Radulovic*[12] *and Jelena Miocinovic*[12]

[1] Bern University of Applied Sciences, School of Agricultural, Forest and Food Sciences HAFL, Zollikofen, Switzerland
[2] Agroscope, Research Division Food Microbial Systems, Federal Department of Economic Affairs, Education and Research EAER, Bern, Switzerland
[3] Food Science and Technology Department, International Excellence Campus for Higher Education and Research 'Campus Mare Nostrum', Veterinary Faculty, University of Murcia, Spain
[4] Department of Agriculture, Nutrition and Environment, University of Catania, Catania, Italy
[5] Unité de Recherches en Technologie et Analyses Laitières Rue de Versailles, France
[6] Department of Agricultural Sciences, Biotechnology and Food Science, Cyprus University of Technology, Limassol, Cyprus
[7] Department of Food Science, University of Copenhagen, Denmark
[8] Department of Food Science, University of Copenhagen, Rolighedsvej, Denmark
[9] LEPABE – Laboratory of Engineering of Processes, Environment, Biotechnology and Energy, Rua Dr. Roberto Frias, Portugal
[10] REQUIMTE/Department of Chemical Sciences, Faculty of Pharmacy, University of Porto, Portugal
[11] Department of Chemical Engineering, University of Porto, Portugal
[12] Department of Food Microbiology, Faculty of Agriculture, University of Belgrade, Serbia

3.1 Appenzeller® – Switzerland

Name: Appenzeller®	 Cheese wheel of Appenzeller® Mildly spicy (Sortenorganisation Appenzeller Käse GmbH)
Production area: Semi-cantons of Appenzell Innerrhoden, Appenzell Ausserrhoden and parts of the cantons of St. Gallen and Thurgau	
Milk: Cow's, raw and thermised	

Global Cheesemaking Technology: Cheese Quality and Characteristics, First Edition.
Edited by Photis Papademas and Thomas Bintsis.

3.1.1 Introduction

Appenzeller® is a semi-hard cheese which is very popular for its unmistakable spicy flavour produced through the application of a herbal brine during ripening. The herbal brine (called 'Kräutersulz') is a mixture of wine, yeasts, salt and approximately 40 different herbs and spices (Kulinarisches Erbe der Schweiz, 2008). The precise recipe of the herbal brine is treated as confidential and has been handed down from generation to generation. Today, Appenzeller® is a well-known brand which is protected worldwide. The yearly production of Appenzeller® is about 9.028 tonnes, and in 2014, 5,188 tonnes was exported (Sortenorganisation Appenzeller Käse GmbH, 2017; TSM Treuhand GmbH, 2015).

As early as the Middle Ages, the monks at the St. Gallen Abbey enjoyed this cheese, which they received as a tithe from the farmers of Appenzell, a mountainous region in the eastern part of Switzerland. The cheese assumed the name of its place of origin and was first cited in a document in the year 1282.

3.1.2 Type

Appenzeller® is produced in six different qualities: 'Mildly spicy', aged for at least three months; 'Strongly spicy', aged for at least four months; 'Extra spicy', aged for at least six months; 'Edel-Würzig', aged for at least nine months; '1/4-fat aromatic spicy', a quarter fat; and 'Bio mildly spicy', which is organic and aged for at least three months. The specifications for full-fat Appenzeller® Mildly spicy when ready for consumption require a minimal fat-in-dry-matter (FDM) content of 48%, a maximum moisture content of 40%, and a salt content of 1.4%–1.7%.

3.1.3 Milk

Raw cow's milk is the basic raw material. On farms whose milk is used for the production of Appenzeller®, the preparation and distribution of silage of any kind is forbidden. Basic fodders such as fresh grass and hay come mainly from the farms of the milk suppliers. Other fodders such as supplementary and concentrate feeds are defined in detail in the Animal Feed Book Ordinance of the Federal Office for Agriculture (FOAG). Fodders originating from genetically modified organisms (GMOs) are forbidden, as is the use of any type of growth activator. In order to prevent propionic acid fermentation during cheese ripening, the absence of propionic acid bacteria is an important quality criterion of the raw milk.

3.1.4 Description and Sensory Characteristics

The texture of the cheese is soft and easy to cut. The cheese curd has an ivory to yellowish colour, which varies according to the season. The salty flavour is accompanied by herbal and spicy notes that develop from the combined action of the lactic acid fermentation and the cheese curing with 'Kräutersulz', a herbal brine preparation that is used for smear-ripening. The typical appearance of Appenzeller® includes a few evenly distributed and pea-sized eyes (3–6 mm) on the cut surface.

3.1.5 Method of Manufacture

The manufacture of Appenzeller® is carried out in small-scale dairies. Water, salt, rennet, lactic acid bacteria and smear cultures are the only permissible processing aids. The use of food additives is forbidden.

Raw cow's milk is the base product. On the farms whose milk is used for the production of Appenzeller®, the preparation and distribution of silage of any kind is forbidden. Basic fodders such as fresh grass and hay come mainly from the farms of the milk suppliers. Other fodders such as supplementary and concentrate feeds are defined in detail in the Animal Feed Book Ordinance of the Federal Office for Agriculture (FOAG). Fodders originating from genetically modified organisms (GMOs) are forbidden, as is the use of any type of growth activator. In order to prevent propionic acid fermentation during cheese ripening, the absence of propionic acid bacteria is an important quality criterion of the raw milk.

Milk preparation: The milk is collected only from suppliers that are affiliated to the cheese dairies. Usually, the milk is cooled at the farm to a temperature between 12°C and 18°C and delivered to the cheese dairy twice a day, immediately after milking. Today, Appenzeller® is made mostly from a mixture of raw and thermised cow's milk. The evening milk is usually thermised and then combined with the fresh morning milk, which is transformed without any heat treatment. Natural or mechanical skimming is allowed in order to standardise the fat content. Cheesemaking is done in traditional open copper vats.

Starter cultures: The cheese dairies prepare the bulk starter and adjunct cultures from authorized liquid stock cultures produced by Agroscope, Institute for Food Sciences IFS (Liebefeld-Bern, Switzerland). The starter cultures consist of strains of thermophilic LAB belonging to the species *Streptococcus thermophilus* and *Lactobacillus delbrueckii* ssp. *lactis.* Adjuncts of *Lactobacillus casei* are added to achieve the typical eye formation. In some dairies, in-house whey cultures are used as well. No propionic acid bacteria are added in the manufacture of Appenzeller®.

Rennet: Calf's rennet is added at 31°C, and coagulation lasts about 40 min.

Cutting/scalding: The curd is then cut into grains of 3–6 mm in diameter. The grain–whey mixture is heated to 41°C–44°C within 15–20 min; 10%–20% of warm water is added to dilute the lactose content in the curd.

Curd draining: The transfer of the curd into moulds is done either manually or mechanically. Each cheese must be marked with the production date and the admission number of the cheese dairy.

Pressing: The pressing lasts for about 60 min, during which time a good cohesion of the grains must be ensured.

Salting: The salting of the cheeses starts immediately after they are removed from the moulds. This is done by immersion for 24–48 hr in a brine bath with >20°Bé and a temperature of 14°C–16°C.

Maturation: After salting, the cheeses are stored in cellars at a temperature of 14°C–15°C and a Relative Humidity (RH) of about 92%–95%. The atmosphere of the ripening cellars of Appenzeller® is characterised by a noticeable presence of ammonia. The cheeses are ripened on rough, unpolished red spruce shelves (*Picea abies*) that are indispensable for an optimal moisture balance between the cheese rind and the supporting plate. During the first ten days of ripening, the cheeses are brushed and turned daily. To induce the forming of a smear-rind, the cheeses are cured with brine (2%–3% salt) inoculated with a surface culture. The surface culture consists of different strains of coryneform bacteria as well as yeasts. Finally, the cheeses are transferred to ripening rooms on the cheese merchant's premises and treated with 'Kräutersulz', a herbal brine used to obtain the characteristic spicy and herbal flavour, at regular intervals for a minimum of three months. Cheeses with excellent storage quality are selected by experts for the manufacture of Appenzeller® with longer periods of ripening such as 'Appenzeller Strongly spicy' or 'Appenzeller® Extra spicy'.

3.1.6 Relevant Research

A long, fine and soft texture is an important feature of Appenzeller®. The switch between summer and winter feeding – from fresh grass to hay – renders the cheese texture more firm and makes the cheese body more prone to the formation of splits and cracks. In a recent study, numerous textural, biochemical, chemical and sensorial properties were determined in two groups of cheeses, with either optimal or insufficient texture (Wechsler et al., 2012). Statistical evaluation of the data revealed that moisture and calcium content are key factors for controlling cheese texture. In order to obtain an optimal texture, a moisture content of 38.2 ± 0.8% should be achieved in cheeses aged three months. Furthermore, the study concluded that a slow acidification rate at the beginning of lactic acid fermentation and a strong acidification at the end increased the withdrawal of calcium during pressing. There are several other factors influencing cheese texture, and multiple interactions between them occur. The most important factors are the content of water, fat and calcium after 24 hr, and the hardness of the milk fat (Wechsler et al., 2013).

3.2 Arzúa-Ulloa PDO – Spain

<table>
<tr><td>Name: Arzúa-Ulloa, PDO</td><td rowspan="3">
</td></tr>
<tr><td>Production area: Galicia, Spain</td></tr>
<tr><td>Milk: Cow, raw or pasteurised</td></tr>
</table>

3.2.1 Introduction

Arzúa-Ulloa cheese is one of the most important Galician cheeses, with different names according to the area or historical period: Arzúa, Ulla, A Ulloa, Curtis, Chantada, Friol or Lugo. The first historical references to livestock in Galicia were made by the Greek historian Polybius (204–122 BC), who spoke of the importance of livestock cattle, sheep and goats in the northwest of the peninsular. This region's farmers have a long tradition of producing this cheese, which can be immediately identified at local markets, promoting a well-deserved reputation among consumers. The product was born out of the rural community's need to preserve the milk, with two purposes: self-sufficiency and improved shelf life at the same time, thus enabling the product to be easily transported for trade. In this area of production, the milk supply is almost guaranteed, which justifies the short ripening time (a week). Cheesemakers solved the problems due to shortages of isolates or a drop in the cheesemaking quality of the milk during the summer by making matured Arzúa-Ulloa cheese from the surplus milk at the end of autumn and winter and subjecting it to a long aging process of at least six months. The production of this type of cheese is adjusted to the circumstances of the place and time: farmers were making cheese from daily milk surpluses which they then sold weekly, fortnightly or monthly at the local fairs. The PDO status of Arzúa-Ulloa cheese was recognised by the Ministry of Agriculture, Fisheries and Food in 1997 and at the European level in January 2010 (EC, 2010). The cheese is made from raw or pasteurised cow's milk. Under the PDO regulation, 1,222 farmers and 21 dairy plants are registered. During 2014, production was 3,209,794 kg representing sales of 19,452,639 euros, mainly on the national market. Together with Manchego cheese, they both represented 62% of Spanish cheese production (Ministry of Agriculture, Food and Environment, 2014). The area for producing the milk and making the cheeses covered by the 'Arzúa-Ulloa' PDO comprises the provinces of Pontevedra, Lugo and A Coruña, which are all located in the basin of the river Ulloa. The soil and weather conditions promote the development of natural meadows and foraging crops.

3.2.2 Type

Semi-hard pressed cheese is made from cow's milk with a predominant enzymatic coagulation. Three types can be identified: Arzúa-Ulloa with at least of 6 days' ripening; farm-manufactured Arzúa-Ulloa, which is made from cows' milk entirely sourced from herds on the farm where the cheese is made is matured for six days, while matured Arzúa-Ulloa has a maturation period of at least six months. The cheeses weigh 0.5–3.5 kg. The minimum fat and protein contents in dry matter are 45% and 35%, respectively, with a minimum moisture content of 55% for Arzúa-Ulloa. Matured Arzúa-Ulloa has a minimum FDM content of 50% but with a minimum moisture content of 35%.

3.2.3 Milk

Arzúa-Ulloa is produced with pasteurised milk, but matured Arzúa-Ulloa cheese can be made with either pasteurised or raw milk. The cheese is only made from whole raw milk from Rubia Gallega, Pardo Alpina and Friesian cows or cross-breeds of the three breeds. According to its PDO regulation, the livestock must be fed mostly on fodder produced on the farm holding itself, feeding by grazing when the weather permits. Concentrated feed of vegetable origin acquired from outside the holding can only be used as a supplement to cover the energy needs of the livestock and must be sourced, so far as possible, from the defined area.

3.2.4 Description and Sensory Characteristics

The shape of Arzúa-Ulloa cheese may be convex or cylindrical, its edges rounded, its diameter 10–26 cm and its height 5–12 cm. The cheese can never be taller than the length of its own radius. It may weigh 0.5–3.5 kg. As regards matured Arzúa-Ulloa, its shape is also convex or cylindrical; its upper surface may be concave; its diameter is 12–20 cm and its height 2–10 cm. It may weigh 0.5–2 kg. In both cheeses, the rind, which is not differentiated, is deep yellow in colour and shiny and greasy in appearance. For Arzúa-Ulloa cheese, the rind is medium to dark yellow, shiny, clean and smooth. The colour is uniform, ranging from ivory white to pale yellow. The texture has no cracks, although it may have a small number of small angular or rounded eyes irregularly distributed through it. Its aroma is milky, reminiscent of the smell of butter and yogurt, with light hints of vanilla, cream and hazelnut. Its taste is essentially of milk, slightly salty and medium to low acidity. The texture is fine, slightly to averagely moist, not too firm and of medium elasticity. In the mouth, it is averagely firm, soft and soluble and of medium springiness. In matured Arzúa-Ulloa cheeses, the rind is similar to the soft one with a greasy appearance. The colour is deep yellow, paler towards the centre, compact and may have a few eyes. Its aroma is intense and milky, smelling strongly of slightly rancid butter. The aroma is pungent and sharp. The flavour is salty, not very acid and of medium to low bitterness. It is mostly reminiscent of butter, with light vanilla and nutty notes which may vary between the centre and the outer edge. The aftertaste is bitter and of butter and vanilla. Its body is hard and very compact to the touch (Ministry of Agriculture, Food and Environment, 2015).

3.2.5 Method of Manufacture

Milk: Milk must be collected from the registered farms and kept after milking at no more than 4°C. The milk used for cheesemaking should not contain colostrum or preservatives and must meet the requirements laid down by legislation.

Starter cultures: When pasteurised milk is used, mesophilic homofermentative starters are usually used although autochthonous recovered microflora is recommended.

Rennet/coagulation: Calf rennet or other authorised coagulant enzymes can be used. Coagulation takes place at 30°C–35°C for 30–75 min, which can vary according to the milk and coagulation process.

Cutting: The curd must be cut to the size of a grain of maize (0.5 to 1 cm in diameter). Washing then takes place with potable water to reduce the acidity of the curd so that the pH of the manufactured product is between 5.00 and 5.55.

Curd draining: The cheese must be moulded in the shape of a cylinder of the defined cheese size.

Pressing: The length of the pressing process varies according to the pressure applied and the size of the pieces.

Salting: Salting can be done directly to the curd in the vat or by dipping the cheeses in brine at 17 °Bé–18°Bé. Cheeses must remain in the brine for no more than 24 hr.

Maturation: Ripening chambers will have a temperature of less than 15°C and an RH between 75% and 90%. The maturation period must be six days from the end of the pressing process, or from the salting if the latter is by soaking in brine. In the case of matured cheeses, the minimum maturation period must be six months. During this stage, turning and cleaning practices should be applied until the cheese reaches its specific characteristics. PDO Arzúa-Ulloa cheese must have a designation label bearing a sequential alphanumeric code and the official logo of the Designation of Origin. The label must indicate whether raw or pasteurised milk was used for cheesemaking. Where 'Arzúa-Ulloa' cheese is of the 'de Granja' and 'curado' types, this may be indicated on the label.

3.2.6 Relevant Research

Halotolerant flora and staphylococci remain practically constant throughout ripening (Centeno, Rodríguez-Otero & Cepeda, 1994b). Centeno, Cepeda and Rodríguez-otero (1996) reported a predominance of *Lactococcus* spp. in the isolates from raw Arzúa-Ulloa cheese, although at the end of ripening, lactococci and lactobacilli isolates were present at similar levels. Most lactococci isolates had weak acidifying and slow milk-coagulating activities. Most *Lactobacillus* spp. isolated were weakly proteolytic. During ripening, α_{s1}-casein was more hydrolysed in pasteurised milk cheeses than in those produced with raw milk between day 1 and day 45, whereas β-casein decreased by only about 30% in raw milk and remained practically constant in cheeses from pasteurised milk (Centeno, Rodríguez-Otero & Cepeda, 1994a). Some research has been done regarding the incidence of toxigenic moulds in Arzúa-Ulloa cheese and the inhibitory effect of eugenol and thymol (Vázquez et al., 1995; Vázquez et al., 2001). Menéndez et al. (2000) recommended the use of selected *Lactobacillus* spp. strains to enhance the desirable characteristics of this cheese, that is, a soft texture due to á$_{S1}$-casein proteolysis but without the bitter taste due to â-casein degradation and a spicy and slightly rancid aroma and taste. Miranda et al. (2009) studied the resistance of Escherichia coli and Staphylococcus aureus isolated from conventional and organic Arzúa-Ulloa cheese. Studies published by Garabal (2007) and Garabal, Rodríguez-Alonso and Centeno (2008) indicated that some changes occurred in the predominant microflora, the most significant being the absence of enterococcal strains among the isolates. Rodríguez-Alonso, Centeno and Garabal (2009) investigated the levels of volatile compounds in Arzúa-Ulloa cheeses manufactured from raw and pasteurised milk, providing useful information for the selection of the most suitable starter and adjunct bacteria for manufacturing pasteurised milk Arzúa-Ulloa cheeses with sensory properties similar to those of traditional raw milk cheeses. The influence of packing (vacuum and modified atmosphere) was evaluated for storing Arzúa-Ulloa cheese by Rodríguez-Alonso, Centeno and Garabal (2011).

3.3 Castelmagno PDO – Italy

<table>
<tr><td>Name: Castelmagno PDO</td><td rowspan="3"></td></tr>
<tr><td>Production area: Some areas of the Valle Grana, in the Hautes Alpes, in Piedmont, in the province of Cuneo, in the municipalities of Castelmagno, Pradleves and Monterosso Grana</td></tr>
<tr><td>Milk: Cow's, sheep's and goat's, raw</td></tr>
</table>

3.3.1 Introduction

The 'Castelmagno' was recognised by the Italian government as a Designation of Origin on 16 December 1982. The PDO was registered by the EU (EC, 1996b). The protected area of origin of 'Castelmagno PDO' includes some areas of the Valle Grana, the Hautes Alpes, Piedmont, the province of Cuneo and the municipalities of Castelmagno, Pradleves and Monterosso Grana. According to the specifications of production, the milk used to produce Castelmagno PDO has to come exclusively from this area. The label of 'Castelmagno' cheese produced at 600 m above sea level includes the wording 'Castelmagno PDO, Prodotto di Montagna.' When milk production and cheesemaking occur at 1,000 metres above sea level, the PDO Castelmagno will be labelled with the word 'Castelmagno PDO, di Alpeggio.' The Castelmagno PDO is produced throughout the year, except for the Alpine pasture variety, which is produced from May to October. The origin of the ancient Castelmagno is perhaps a little later compared to the Gorgonzola cheese, which was already known in 1100. The first official document that records the existence and appreciation of Castelmagno cheese is an arbitration award of 1277, according to which, in a dispute between the municipalities of Castelmagno and of Celle di Macra, it was fixed as the annual fee – to be paid to the Marquis of Saluzzo – a certain quantity of cheese of Castelmagno. In 1722, however, forms of Castelmagno were to be provided to the local feudal lord, by the will of a decree of King Vittorio Amedeo II. In 1800, the cheese became the king of Italian cheeses and appeared in the menu of the most prestigious restaurants in London and Paris. Then with wars and the depopulation of the mountain during the 1960s its decline began, and even its existence was threatened. Production recovered from the early 1980s when the Castelmagno obtained national recognition, the DOC, in 1982 and the European PDO recognition in 1996.

3.3.2 Type

Castelmagno PDO is a pressed semi-hard cheese. The minimum FDM is 34%.

3.3.3 Description and Sensory Characteristics

Castelmagno PDO has a cylindrical shape with flat faces of 15–25 cm diameter. The weight of the cheese ranges between 2 and 7 kg. The rind is 12–20 mm thick, inedible, thin yellow-brown, smooth, tending to be rigid. With maturation it assumes a darker colour, thickens and becomes wrinkled with maturation. The body is pearly white or ivory white to a minimum of aging, yellow-ochre with potential greenish blue veins inside as ripening progresses, also obtained by piercing the cheeses using the traditional technique (manual with long irons). The texture is crumbly when the maturation period is short, becoming denser as ripening progresses. The flavour is subtle, delicate and moderately salty at minimum maturation, and becomes tastier as ripening progresses.

3.3.4 Method of Manufacture

Animal feeding: The feeding base of the cattle and possibly sheep and goats must be composed of green fodder or hay that are derived from the lawn, pasture and hay grass meadows sourced primarily from the municipalities of Castelmagno, Pradleves and Monterosso Grana. The use of silage is not allowed. The feed materials making up the integration of the food ration will be made up of cereals, legumes, mineral salts and vitamins permitted by applicable law.

The particular variety and flavour of the herbs present in the pastures, characterised by a flora composed of grasses of the genera *Poa* and *Festuca*, upper valley Grana, give this cheese a unique flavour.

The cheese 'Castelmagno DOP, di Alpeggio' must meet the following requirements:

- The milk must come exclusively from cattle, goats and sheep grazing in pastures between early May and late October. Animals must be fed at pastures with at least 90% of local flora.
- The entire production process must take place in a pasture.
- Cheesemaking must take place higher than 1000 m above sea level in territories falling in the area of origin.

Milk preparation: The cattle used to produce milk for the production of Castelmagno PDO must be attributable to genetic types Barà Pustertaler, Bruna, Oropa Simmental, Simmental, Montbeliard, Grigio Alpina, Piemontese, Valdostana and their cross-breeds. The 'Castelmagno' cheese is made from raw cow's milk with possible additions of sheep and/or goat as a percentage, ranging from a minimum of 5% to a maximum of 20%. It is allowed to cool the milk for storage to a temperature not lower than 6°C. The milk may be skimmed by means of surfacing of fat by gravity overnight. Subsequently, whole or skimmed, the milk must be heated to a temperature of 30°C–38°C. Pasteurisation and thermisation of milk is not allowed, and thus no starter culture is added.

Coagulation: Coagulation occurs at a temperature between 30°C and 38°C, in 30–90 min with liquid calf rennet (with at least 70% chymosin).

Cutting: When the clot has reached a sufficient degree of hardening, it rises up and subsequently breaks. The break-up is first done coarsely and then it is continued until homogeneous granules with a size between corn and hazelnut are obtained.

The processing in the vat takes place while the mass is continuously stirred for 10 to 15 min in order to facilitate the separation of the curd from the whey, called 'laita'. The broken curd is left to settle in the bottom of the vat, or discharged in containers for resting.

Resting: The curd is placed in a clean, dry cloth called 'risola' in plant tissue or synthetic. The risola is then eventually pressed and left hanging or resting on an inclined plane. It is left to rest for at least 18 hr, which necessary to allow the residual whey to emerge without the pressing action. It is allowed during this rest period to cool the curd.

Acidification, curd mixing and salting: After rest for at least 18 hr, the curd is placed in containers (also made of wood), submerged in the serum of the previous processes that must have a temperature of at least 10°C for a period ranging from 2 to 4 days for regular fermentation. The curd is then broken and finely ground, mixed and salted.

Moulding: The product is then wrapped in a vegetable or synthetic canvas tissue, and introduced in 'fascelle' made of wood or other suitable material, where it remains for at least a day and is subjected to an appropriate manual or mechanical pressing. A matrix bearing the mark of origin in the negative form is positioned on the basis of fascelle.

Second salting: After moulding, one further dry salting is allowed, to give the appropriate colour and consistency to the rind.

Maturation: Maturation occurs in cool, damp natural caves for a minimum period of 60 days on wooden boards or other suitable material. Aging for at least six months enhances the features of this great cheese. Maturation occurs at 5°C–15°C with an RH of 70%–98%, so as to ensure the necessary conditions for correct development of natural and typical Castelmagno PDO moulds. During maturation, the cheeses may be brushed and/or washed using natural substances that have no direct colouring effect, in order to limit the development of unwanted moulds and mites.

3.3.5 Relevant Research

Biochemical, volatile and textural profiles during manufacture and ripening were determined in artisan Castelmagno PDO (Bertolino et al., 2011). The HPLC analysis of organic acids and carbohydrates showed exhaustion of lactose content, while Urea-PAGE Electrophoresis indicated extensive primary proteolysis of both β-casein and α_{s1}-casein. During ripening of Castelmagno PDO cheeses, it is possible to observe high degradation of α_{s1}-casein with an increase of all its degradation products, and evolution of the hydrophilic peptides associated also with the highest concentrations of glutamic acid, valine, leucine phenylalanine and lysine. The volatile profile was characterised by a high level of acids, in particular of hexanoic, octanoic and decanoic acids, which are the primary products of lipolysis metabolism. During ripening, the volatile are characterised by a decrease in acid compounds and an increase in ketones and alcohols as a consequence of free fatty acid metabolism. Texture profiles show increases in hardness, gumminess, chewiness and adhesiveness properties and a decrease in cohesiveness. Further interesting studies have been carried out to analyse the microbial diversity, dynamics and activity throughout manufacturing and ripening of Castelmagno PDO cheese (Dolci et al., 2010), and the microbial dynamics of lactic acid bacteria ecology (Dolci et al., 2008).

3.4 Comté PDO – France

<table>
<tr><td>Name: Comté PDO</td><td rowspan="3">
The green band indicates a high-quality cheese. *Photo © CIGC/Studiovision*</td></tr>
<tr><td>Production Area: French Jura Massif (200–1,500 m)</td></tr>
<tr><td>Milk: Cow's, raw</td></tr>
</table>

3.4.1 Introduction

Comté is the first French PDO cheese from the standpoint of the volume marketed (23% of French PDO cheeses in 2014), representing 87% of pressed and cooked French PDO cheeses (INAO-CNAOL, 2014). It is one of the 37 pressed and cooked French cheeses which include three PDO cheeses, and it represents 23% of the pressed and cooked French cheeses produced in 2014. Comté was recognised as AOC in 1958 and as PDO in 1996. Comté is made, matured, grated and sliced exclusively in the French part of the Jura Mountains (medium-sized mountains located in Eastern France, near Switzerland). Besides Comté, several other cheese varieties, including three PDO cheeses (Morbier, Mont-d'Or and Bleu de Gex) are manufactured from cow's raw milk, which is the first agricultural resource of this area. Actually, Comté's manufacture directly employs around 7,600 persons, and Comté's milk is produced by around 3,000 family farms covering 2,300 km^2 (equivalent to 23% of the French Jura Mountains' surface) and including 150,000 dairy cows. Based on an artisanal know-how developed in the Jura Mountains 1,000 years ago, Comté cheese is a cooked and pressed cheese made, every single day, from raw cow's milk partially skimmed at around 153 small village cheese plants located in the Jura Mountains. Then it is matured in one of the 16 maturation cellars exclusively located in the Jura Mountains, for a minimum of four months, but often more (eight months on average; up to several years). At any time of the year, Comté cheeses manufactured in different seasons are available for sale since their ripening times vary according to the cheese. After four months of maturation (the minimum), each wheel is scored for its flavour, texture, odour and appearance. Wheels scoring over 14 points get a green band indicating excellent quality. Wheels scoring between 12 and 14 points get a brown band. For example, the brown band can indicate a slight appearance defect, while flavour, odour and texture are excellent. Wheels scoring less than 12 do not carry the official label. After maturation for more than four months, a second quality test is performed at the end of maturation in order to definitively validate the band colour. Comté is France's most popular PDO cheese. Comté cheese is bought by more than half of the French households (mean 1.7 kg/year) and is very popular in the East of France. The total annual production of Comté cheese was 64,179 tonnes in 2014 (+29% in 10 years), representing around 1,600,000 wheels. Actually, most Comté cheese is sold in self-service supermarkets (95%). A small part of the production (in 2015, 8% – 4,683 tonnes) is exported, principally to three European countries (Germany, Belgian and England). Actual Comté cheeses are based on ancient cheeses that have been manufactured seasonally in the Jura Mountains for centuries, and which had to be kept for many months in order to ensure a food source for the farmer's family throughout the cold season (Dasen, 2013). The natural resources (pasture, water and firewood) available in the Jura Mountains allowed the production and heating above 50°C of the large quantity of milk (500 litres) required for manufacturing a wheel. For managing milk collection and cheese manufacturing, specific forms of organisation were created in the Jura Mountains at least eight centuries ago and evolved over time into the current organisation (Melo, 2015).

3.4.2 Description and Sensory Characteristics

According to Comté's specifications, the FDM content of cheese is 45%–54%, salt content > 0.6%, moisture content ≤38%, moisture in non-fat substance (MNFS) ≤54% and NPN/TN ratio ≤15.5% (17.5% when FDM > 52%). On average, its dry matter is 64%; fat content 34.6%, including 22.5% of saturated fat; salt content 0.8%; protein content 27% and carbohydrate content 0% (traces). It is especially rich in calcium, phosphorus, zinc and copper and in vitamins A, B_2 and B_{12} (Official Website for PDO Comté, 2016). Comté's rind is solid, granulated and golden yellow to brown, depending on the maturation cellar. Comté's body is strong yellow when the cows are grazing (high-carotene diet, warm season) and light yellow (ivory) when the cows are fed with hay (low-carotene diet, cold season). Comté's texture is smooth and dense, and exhibits a few small eyes (holes), if any. Amino acids accumulate as maturation progresses, leading to the formation, especially in long-matured cheeses, of white crystals of tyrosine, which get crunched under the teeth and are commonly mistaken for salt crystals. Owing to its specifications, Comté cheese is characterised by high flavour diversity. For mature Comté cheeses, 83 different aroma/odours have been frequently cited by a sensory panel specially trained for Comté. They have been grouped into six families (milky, animal, vegetable, toasted, spiced and fruity), and are depicted as a wheel. Any individual mature cheese shows 2–3 families and 5–10 aroma/odours, related to (1) its geographical location (including the farm, the farmer and its practices), (2) its season of manufacture and (3) the specific know-how of both the cheesemaker (who daily adapts his cheesemaking practices to the milk composition) and of the maturation technician (who chooses the maturation conditions and manages the smearing).

3.4.3 Method of Manufacture

Comté is manufactured according to strict rules that apply from milk production to cheese for sale.

Milk production: Milk has to be produced by Montbéliarde or Simmenthal cows (in fact, 95% are Montbéliarde). The minimum surface area of the pasture should be 1 ha/dairy cow, and milk production should not exceed 4,600 L/ha. As a consequence, milk for a wheel is provided by 20 cows, as each Montbéliarde cow produces 20 L per day. Feeding of cows by any fermented foods is banned. Feed bought outside the farm is limited at both the quantitative and qualitative levels. Fodder has to be produced exclusively within the delimitated geographical area. Milking twice a day is mandatory.

Daily, each dairy plant receives milk from dedicated dairy farms located within a 25 km radius around the plant. Milk should be processed raw within 24 hr after the first milking (in the morning). Additives and colouring agents are banned. Use of copper vat is mandatory.

Addition of starter: Great importance is attached to the starter cultures inoculated. The majority of cheesemakers use starter strains previously isolated in the French Jura Mountains area and propagated by a local technical service dedicated to the PDO cheeses of this area. Starters are composed of different mesophilic and thermophilic strains, inoculated at 10^5–10^7 CFU/mL of milk. They include *Lactococcus lactis*, *S. thermophilus* and *Lactobacillus helveticus* and/or *Lactobacillus delbrueckii* strains (Charlet et al., 2009). *Lc lactis* do not survive as a cultivable population to the heating above 50°C occurring in the vat (Jeanson et al., 2003), but its inoculation in the vat at 10^6–10^7 CFU/mL of milk is sufficient to boost the rate of acidification (Jeanson, 2000).

Coagulation: Renneting is performed by adding commercial and/or artisanal rennet preparation.

3.4.4 Relevant Research

After centuries of copper use for vat, the replacement of copper with inox in the 1980s, because of the latter was cheaper, changed the sensory characteristics of Comté dramatically, showing that copper was significant for Comté manufacture. Studies by professionals showed the different impacts of the use or copper versus stainless steel in regard to proteolysis, fat oxidation, volatile compounds profile and lactic acid bacteria fate. It is the reason why copper has become mandatory for Comté manufacture since 1990. In the 2000s, PCR-based methods allowed monitoring the population of thermophilic and mesophilic lactobacilli populations at both species and strain levels during Comté cheese ripening, as well as assessment of the source (raw milk, starters) of the different lactobacilli species (Berthier et al., 2001; Bouton et al., 2002; Depouilly et al., 2004). In general, since Comté cheese has long been referred to as a Swiss-type cheese (when it systematically showed eyes) and still belongs to the same cheese family, some of the results obtained for this cheese category since the 1950s are relevant for Comté cheese.

Recently, the transfer of metal trace elements from soils to Comté milk and cheese was studied (Maas et al., 2011). Concentrations of non-essential elements (cadmium and lead) in cheese largely remained below those considered as dangerous for consumers, while cheese may constitute a useful source of copper and zinc in human diet.

3.5 Flaouna Cheese – Cyprus

Name: Flaouna Cheese	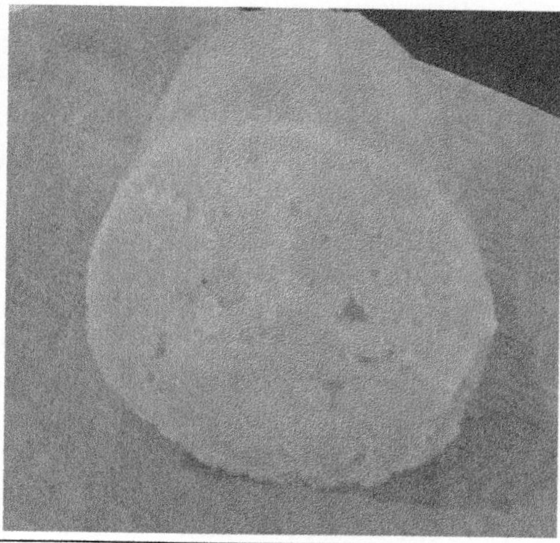
Production area: Cyprus	
Milk: Cow's, sheep's and goat's (mixtures), pasteurised	

3.5.1 Introduction

Flaouna is a Cyprus traditional cheese that is produced during Easter, and is the main ingredient (usually blended with mature Halloumi cheese; see Part II, Chapter 7, Section 7.5) in the traditional 'Flaouna' Easter cakes also, from which the cheese takes its name. Flaouna cheese is rather a modern/industrial version of the 'Paphitiko cheese' made in the area of Paphos and traditionally was left to mature for at least two months before marketing. Cheese from Paphos used for Flaouna cakes is mentioned in references dating back to the eighteenth century (Economides, 2004).

3.5.2 Type

Flaouna cheese is a semi-hard cheese, air-dried in specially designed drying rooms. It has an average moisture content of 37.5%, fat content 29.1% protein content 28.6%, and salt 1.0% (Cyprus State Laboratory, 2013). This is not a standardised product (i.e. milk type, moisture content, fat in dry matter) therefore the chemical characteristics vary according to producer and duration of drying.

3.5.3 Description and Sensory Characteristics

Flaouna cheese is a yellowish cheese that has a cylindrical shape, 15–20 cm height and 8–10 cm width and 1.0–1.5 kg weight. It is compact, with no holes, lightly salted and has the characteristic flavour of the milk type prevailing in the mixture. It is sliceable, and if left to dry for a longer period, becomes harder and more easily grated.

3.5.4 Method of Manufacture

The industrial method of manufacture closely resembles that of Halloumi cheese (see Part II, Chapter 7, Section 7.5).

Milk preparation: Pasteurised milk (sheep/goat, or mixed with cow's milk – the traditional "Paphitiko" cheese is only made with raw sheep's/goat's milk) is heated to reach a temperature of 33°C–35°C.

Rennet/starter cultures: Non-animal commercial rennet is used while no starter cultures are added in the manufacture of the cheese.

Coagulation/cutting: The milk is coagulated within 45–50 minutes, and the cheese curd is cut into cubes of 0.5–1.0 cm. The curds are briefly stirred with no reheating and left to settle to the bottom of the vat.

Moulding/pressing: The curd is then transferred to a plastic cylindrical mould, where it is manually pressed to fill up the mould, fuse the curd particles and drain the whey. In modern dairies, light mechanical pressure is applied on individual cheeses for about 20 min.

Cooking/scalding: The cheese curd is placed in de-proteinated whey (see Anari cheese; Part II, Chapter 13, Section 13.2) and heated to 90°C for at least 30 min. The core temperature of the cheese reaches 80°C. For traditional 'Paphitiko cheese' the scalding temperature is rather short, and the core temperature of the cheese would reach 40°C–45°C.

Salting: Individual pieces of cheese are hand-salted and kept at room temperature until the cheese is cool and the salt is absorbed. The traditional Paphitiko cheese was salted either in brine 18%–20% for 1–2 days or dry-salted for 10–20 days until it reached 5% of the cheese's weight.

Drying: The cheeses are then placed in specially designed drying rooms where an air stream of 18°C–20°C is applied and cheeses are left to dry for about 4–5 days. Traditional Paphitiko cheese was left to mature at 16°C–18°C for two months.

Packing: Flaouna cheeses are individually vacuum-packed and kept at refrigeration temperatures until sold.

3.6 Formaggio di Fossa di Sogliano PDO – Italy

Name: Formaggio di Fossa di Sogliano PDO	
Production area: Emilia Romagna and Marche	
Milk: Sheep's, or cow's, or a mixture of cow's (maximum 80%) and sheep's (minimum 20%)	

3.6.1 Introduction

The Fossa cheese from Sogliano received the PDO recognition in 2009 (EC, 2009). The tradition of the pit ('Fossa') was introduced during the Middle Ages and soon became an integral part of the rural culture of the territory between the Rubicon and Marecchia valleys, to the river Esino, overlapping Romagna and Marche. The use of pits was naturally linked to the need for preservation of the product, as well as the desire to protect it from raids by the tribes and armies which over the centuries tried to occupy the territory. According to legend, the special cheese processing pit was born purely by chance. It appears that in 1486 Alfonso of Aragon, defeated by the French, had obtained the hospitality by Girolamo Riario, lord of Forli. The resources of Forli did not allow long sustenance of the troops, who soon began to despoil the surrounding neighbourhood. The farmers, in order to defend themselves, began hiding provisions in pits. In November, once the armies left and the raids ended, the farmers unearthed supplies and discovered that the cheese had improved organoleptic characteristics. The first documents concerning the pits and the 'infossatura' technique date back to the fourteenth century and belong to the archives of the Malatesta family, which owns the land. The pits were used to preserve various foodstuffs and to age cheese, in the eventuality of siege, epidemic and famine.

3.6.2 Type

Fatty, medium-matured semi-hard cheese.

3.6.3 Description and Sensory Characteristics

The shape of the fresh cheese must have a height varying from 6 to 10 cm and a diameter ranging from 12 to 20 cm. The weight will be between 600 g to 2 kg when placed in the pits, called *infossatura.* Due to the loss of whey and fat during maturation, and by the compression of the weight of the cheeses, they lose much of their round shape. The crust is practically absent, and the colour of the surface of the cheese is ivory with straw yellow-orange shades. It is tough and greasy and may have a little mould. The body is hard, brittle and pale straw in colour, with no holes. The seasoning gives the cheese a strong taste and aroma with a very intense scent; its aroma recalls the smell of brushwood, wood, truffle, musk and hints of herbs, but also the aroma of mushroom, boiled chestnut, the smell of the pit and of the cloths where the cheeses were kept in the pits. Its flavour is distinctive and unique; sweet to slightly spicy.

3.6.4 Method of Manufacture

Animals: Cow breeds for the production of milk are Italian Friesian, Alpine Brown, Simmental and their cross-breeds, while sheep breeds are Sarda, Comisana, Massese, Vissana, Cornella White, Fabrianese Langhe and Pinzirita and their cross-breeds.

Animal feed: The animal's diet is made up of spontaneous forages (rich in flora species with herbaceous plants and shrubs) or cultivated forages comprising grasses and legumes collected from meadows grown in the territories of origin of the PDO cheese. The green, the hay fodder and/or the pastures can be supplemented by simple or compound concentrates and with mineral and vitamins supplements. The concentrates must not exceed 30% of the daily total diet. The use of silage is not allowed.

Milk: The milk used comes from two daily milkings. The cheese must be produced using the following types of milk, exclusively or as a mixture: full-fat sheep milk; full-fat cow milk; or a mixture of full-fat cow milk (up to 80%) and full-fat sheep milk (minimum 20%). The milk can be raw (within 48 hr of the first milking) or pasteurised, in which case the use of starter cultures is required. The use of additives is prohibited.

Starter cultures/rennet: The milk is curdled with natural calf rennet. Starter cultures are used only when pasteurised milk is the raw material.

Coagulation: The milk is heated to 30°C–38°C. The clotting times can vary from 7 to 20 min.

Cutting: The curd is cut to a grain size of about 0.5–1 cm. It is then separated from the whey and placed in suitable forms for further syneresis.

Moulding: The curd is transferred to appropriate containers for further syneresis of the whey. It is subjected to manual pressing.

Salting: Either dry salting of the cheese by distributing the salt in the entire outer surface for 24 hr, or in brine salting (17%–19% NaCl w/v) for about 18 hr, is carried out.

Maturation: During the first phase of maturation, cheese ripening takes place in appropriate locations adjacent to the dairy plant, at a temperature of 10°C–15°C and RH of 70%–85%, for a period of 60–70 days. It is also permitted to mature the cheese in a cellar at a temperature between 6°C and 14°C, at an RH of 75%–92%. The second phase of maturation is considered unique for Fossa cheese from Sogliano, where a pit is used so that the anaerobic maturation phase of the cheeses can take place. The cheeses (collected in varying numbers in bags of cloth that are tightly closed and identified) are laid, by experts called *infossatori* (a unique professional profile among international dairies), in an ancient pit (flask-shaped) carved into the sandstone rock at a depth of 3 m. The bags are stacked up to the entrance of the pit, which is then filled and covered with canvas in order to prevent excessive transpiration. The pits are closed by a wooden lid sealed with plaster. This maturation phase varies from a minimum of 80 days to a maximum of 100 days. Normally the producers are allowed to make two *infossature* per pit in a year, during spring and summer. The final and third phase is when the expert proceeds to take out the cheese from the pits, called *sfossatura*. The traditional opening of the pits takes place on November 25, the day of St. Catherine of Alexandria.

3.6.5 Relevant Research

The microbiology and biochemistry of Fossa (pit) cheese have been studied by Gobbetti et al. (1999), who reported that several common features seemed to be shared by the Fossa cheeses: (1) no hygienic risks, (2) selection of NSLAB during ripening and very low survival of lactic acid bacteria starters, (3) a very high degree of proteolysis, (4) a very high concentration of free amino acids which increased the flavour and (5) moderate lipolysis which varied within samples and was also considerable in cheeses produced from bovine or mixture of bovine and ewes' milks. The safety of Fossa cheese was also considered satisfactory by Massa, Turtura and Trovatelli (1988) in a study on the hygienic quality of Fossa cheese. Babieri et al. (2012) confirmed the absence of hygienic risks to the consumer and attributed the genetic diversity of microbial Fossa cheese to the occurrence of *Lb. plantarum*, *Lb. casei*, *Lb. paracasei*, *Lb. rhamnosus* and *Lb. fermentum*. Avellini et al. (1999) compared different systems of maturing 'formaggio di Fossa' (pit cheese): in factory (cells) and in the pits. The pit-aged cheese showed notable differences in comparison to the cheese aged in the factory. The Fossa cheese was less hard, but moister, saltier, sharper and more acidic, with a more pronounced aroma than the factory-aged cheese. Chemical analysis showed significantly different values of water content, non-casein and non-protein nitrogen, amount of free amino acids and composition of free amino acids and free fatty acid fractions for the pit-aged cheese. The environmental conditions of the pit, together with the presence of moulds on the surface of the pit-aged cheese, probably are responsible for the development of the unique chemical and sensorial characteristics of Fossa cheese.

3.7 Havarti – Denmark

Name: Havarti	
Production area: Denmark	
Milk: Cow, pasteurised	

3.7.1 Introduction

Havarti was included in the Order of the Danish Ministry of Agriculture of 13 March 1952 and may still be found in Order No 2.22 of 4 January 2013 on milk products. It is described in General Standard for Cheese (Codex Alimentarius, 1978). Havarti is made from pasteurised cow's milk. It was first made in the 1920s as an attempt to manufacture a Tilsiter type of cheese by P. Hansen of Ruds Vedby dairy and J. Hansen of Hallebygaard in cooperation with G. Morgentahler from Switzerland, who taught the dairymen how to produce a new pumped-curd cheese. It was a great success, and production of the new cheese type was spread to many other dairies throughout Denmark. The soft texture of the cheese was much appreciated both in Denmark and internationally; however, the very strong flavour from the surface microflora was not, and because large amounts are exported, the flavour has been considerably decreased in strength over the years. The cheese got its name from the Havarti farm, located near Holte in Denmark in 1952. Havarti accounts for about 10% of the Danish cheese production.

3.7.2 Type

Havarti is a quite soft semi-hard cheese made with a moisture content of 42%, 46%, 50%, 52% and 54% and FDM of 60%, 55%, 45%, 40% and 30% (Danish Food Administration, 2013).

3.7.3 Description and Sensory Characteristics

Havarti is pale yellow, has a soft to semi-hard elastic texture and is sliceable with a thin wire. The formation of numerous of rice-grain-sized eyes (mostly 1–2 mm in width and up to 10 mm in length) are typical for Havarti as each eye develops from one of several irregularities, which are evenly distributed all over the body of the cheese as a result of drainage of the whey before forming and pressing. In cheese with a 60% FDM content, the eyes are generally smaller. Havarti is made in different sizes and is commonly square or rectangular in shape with a weight of 4 to 5 kg, but Havarti with a 60% fat content is made in small cylinders with a diameter of 10 cm that weigh about 0.45 kg. Some varieties are made with cumin and other spices. The texture is soft, easily sliced and firm to the bite (*al dente*). The surface microflora gives the otherwise mild, acidic and buttery flavour of Havarti its typical spicy aroma after only a short ripening time of a few weeks.

3.7.4 Method of Manufacture

Milk: Raw milk is standardised and pasteurised and up to 20 g calcium chloride and up to 20 g potassium nitrate per 100 L of milk may be added.

Starter culture/rennet: A mesophilic DL-starter with undefined mixed strains of *Lc lactis* spp. *lactis & cremoris, Lc. lactis* spp. *lactis* biovar. *diacetylactis* and *Leuconostoc mesenteroides* is used. For coagulation, bovine rennet (30–45 mL/100 L of milk) is added.

Cutting/heating of the curd: Curd is cut into grains with a diameter of 10 mm and then heated to 38°C–41°C for about 60 min. 25%–20% whey is removed, and 10%–20% water is added.

Curd draining: Curd is separated from the whey, placed in forms and pressed for 30–60 min very gently in order to ensure that an open structure is formed and maintained. The cheese is then pressed mainly under its own weight. In order to achieve the desired structure and form a coherent block of cheese, the temperature at the beginning of this stage must be kept as high as 41°C.

Salting: The cheeses are chilled for 10–20 hr before they are placed in brine (20–22 Bé) at 12°C for 1–3 days.

Maturation: 1–2 weeks at 16°C–18°C and 90%–95% RH to stimulate the development of smear on the surfaces. This is followed by 2–3 weeks at 13°C–14°C and 85% RH. The surfaces are washed and covered by paraffin and/or aluminium folio and further ripened at 8°C–12°C until sold.

Storage: Cold (2°C–6°C)

3.7.5 Relevant Research

The open texture and relative high water content speed up the biochemical processes, and both the characteristic texture and flavour, which are mainly generated by the smear, appear after three weeks. The internal microflora is dominated by starter bacteria, while a diverse microbial composition is found in the smear Waagner (1993).

3.8 Herrgård – Sweden

Name: Herrgård	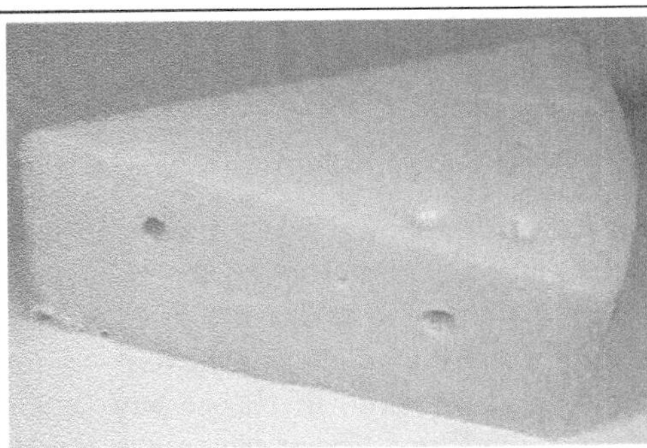
Production area: Sweden	
Milk: Pasteurised cow's milk	

3.8.1 Introduction

Herrgård has been a registered trademark within EU since 2001, owned in fellowship by Swedish dairy companies. Herrgård cheese is the most popular cheese in Sweden. It was developed around 1790 in a manor house dairy in southern Sweden, and Herrgård is Swedish for 'manor house'. A Swiss cheesemaker, Pierre Nicolas Dubas de Rougemont, was invited by count Erik Ruuth, owner of the manor house Marsvinsholm, to teach the art of producing cheese with large round eyes. His recipe, however, depended on a thermophilic, heat-loving microflora from the environment which was not present in the raw milk produced this far north in Europe, and the first trials were complete failures (Ränk, 1987). The mesophilic microflora present died during production and, consequently, the cheeses did not ripen. The temperatures had to be lowered in several steps to stimulate the local mesophilic microflora and thus after several years of adaption of the recipe, a new cheese variety, Herrgård, was born. Herrgård is nowadays produced in modern dairy plants in Sweden using mesophilic undefined DL-starter and moderate cooking temperatures. A part of the whey is replaced with water during manufacture to limit acidification of the fresh cheese, and to obtain a smooth texture.

3.8.2 Type

Herrgård is a semi-hard pressed cheese with round regular eyes (5–15 mm) made in low cylinders of about 12 kg. The surfaces are dried and waxed to prevent development of a surface flora. The fat content is 28% in full-fat and 17% in reduced-fat cheese, and the MNSF is 53%–57%. The salt content in ripened Herrgård cheese is 0.9%–1.3% (Sveriges Ostkollegium, 1993).

3.8.3 Milk

Herrgård is made from pasteurised bovine milk produced in Sweden, mainly at medium-sized farms with an average of 75 cows. The cows have to graze outdoors during summertime. Parameters such as the quality of grass, minimum time to be outdoors each day and length of period are regulated in Swedish law.

3.8.4 Description and Sensory Characteristics

Herrgård is a semi-hard light yellow to yellow cheese with a smooth texture and large round eyes (5–15 mm) evenly distributed in the body and fewer close to the surface. The cheese is shaped like a low cylinder with slightly convex edge sides, a diameter of 35 cm, a height of 10–14 cm and a weight of 12–14 kg. The surfaces are dry and waxed. It is sold after ripening for at least three months and up to about a year. Herrgård has a soft, juicy and slightly rough texture. It is easily sliced and usually consumed by Swedes on bread. Because it melts well, it is also an excellent cheese for cooking. The flavour is mild, slightly acid with sweet and nutty notes and with a long umami aftertaste. Long-duration ripened Herrgård has sharp and aromatic flavour notes.

3.8.5 Method of Manufacture

Milk: Raw milk is commonly microfiltered to remove detrimental clostridia spores, standardised to the intended fat content by mixing cream and milk, and pasteurised (72°C/15 s). The milk temperature is set to 30°C before a mesophilic DL-starter is added.

Starter culture/rennet: A mesophilic DL-starter with undefined mixed strains of *Lactococcus lactis* subsp *lactis* & subsp. *cremoris* (80%–90%), *Lc. lactis* subsp *lactis* biovar. *diacetylactis* and *Leuconostoc mesenteroide* is added. For milk coagulation, bovine rennet with at least 75% chymosin is used.

Cutting: The coagulum is cut after about 30 min into cubes with sides of 5–6 mm.

Heating of the curd: The temperature is raised to 33°C during stirring and about 30% of the whey is removed. During the next 60 min, the temperature is increased stepwise to 36°C. Thereafter 35% water is added, the temperature is raised to 40°C, stirring continues for 25 min, about 45% of the whey–water mixture is removed, stirring continues for another 25 min and the curd is pre-pressed under whey. The cheeses are formed and further pressed at 16°C for about 18 hr.

Salting: The cheeses are salted in saturated brine for three days at 10°C to give a concentration of 0.9%–1.3% salt in the ripened cheeses.

Maturation: After salting, the cheeses are placed in a dry storage at 12°C for one week, and then they are waxed for the first time. It is ripened for another two to three weeks at 16°C to develop its unique appearance and flavour, and thereafter they are ripened at 11°C–14°C until sold. During ripening the cheeses are regularly waxed and turned over. The cheese needs at least three months of ripening and is mainly consumed after four to six months, but also longer ripening periods up to one year or more occur.

3.8.6 Relevant Research

The first biochemical event to occur is catabolism of all lactose into lactic acid within the very first days, which decreases the pH to a minimum of about 5.3–5.4. The pH slowly increases during ripening to reach a value above 5.6 after six months. In reduced-fat Herrgård, the pH is commonly around 0.05 pH units higher during the whole ripening period, which is explained mainly by a higher buffering capacity due to the higher protein content. The citrate is consumed during the first week by the starter bacteria with simultaneous production of aroma compounds such as diacetyl and acetoin, and the carbon dioxide that is responsible for the formation of the round eyes (Ardö, 1993b).

A balanced production of acetic acid, acetaldehyde and ethanol is essential for the quality of Herrgård. Only small amounts of propionic acid and butyric acid are accepted. Hydrolysis of milk fat should be limited because the free fatty acids contribute to rancid off-flavours in quite low concentrations in this cheese; however, small amounts are essential for its typical flavour. After 6 months, around 35% of the casein is degraded into peptides and amino acids (Ardö 1993a, b).

The dominating free amino acids in Herrgård are glutamic acid, leucine, lysine and aspargine (Ardö & Gripon, 1995). The total amount of free amino acids increases during ripening and reach levels around 250 mmol/kg cheese after 10 months (Ardö, Thage & Madsen, 2002). The composition of amino acids changes during ripening as a consequence of bacterial activities. Proline, lysine, isoleucine and glycine increase significantly faster than leucine, phenylalanine, asparagine and glutamine, while the relative amount of glutamic acid remains quite constant during ripening. Proline and glycine contribute to a sweet flavour that is balanced by the high content of the umami-tasting amino acid glutamic acid, which results in a typical background taste of Herrgård. The eyes are formed from accumulation of CO_2 produced by metabolism of citrate performed by *Lc. lactis* subsp. *lactis* biovar *diacetylactis* and *Leuconostoc* spp. in cooperation during the first weeks of ripening.

3.9 Mahón-Menorca PDO – Spain

Name: Mahón-Menorca PDO	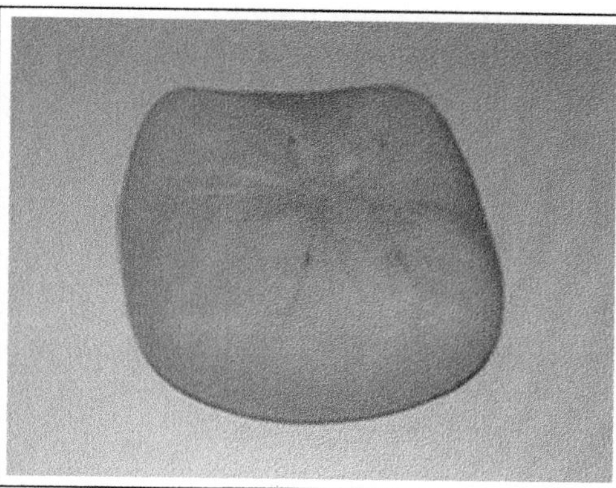
Production area: Balearic Island, Spain	
Milk: Cow's (<5% sheep's), raw	

3.9.1 Introduction

The PDO status of Mahón cheese was recognised by the Ministry of Agriculture, Fisheries and Food in 1985, modified in 1995 and recognised at the European level in 1996 (EC, 1996a). The cheese is made mainly from cow's milk, although 5% sheep's milk is allowed. Under the PDO regulation, 140 farmers and 39 dairy plants are registered. During 2013, production stood at 2,217,964 kg, which represented a total of 14,630,000 euros. The cheese is mainly marketed within Spain with an increasing market in other countries and represents 6.93% of the economic value of Spanish PDO cheese production (Ministry of Agriculture, Food and Environment, 2014). The archaeological finds of ceramic tools used for its production date its origin to 3,000 B.C.

The 'formatjat' or production process of Mahón-Menorca cheese has remained unchanged for a very long time in all the Minorcan 'llocs' (farmhouses) where very old practices are still followed. Mahón cheese is a part of the customs and culture of the Island of Minorca and constituted a key element in the maintenance of its current features. Mahón cheese continues the region's ancient traditions of integrated farming of herds and flocks. Cattle farming is carried out on family owned farms on land divided into fields by a multitude of dry-stone walls. This enables milk to be processed into cheese by the shepherds themselves, maintaining an ecological balance that helped convince UNESCO to declare Minorca a Biosphere Reserve. (Anonymous, 2015a). The production and processing area consists of the island of Minorca (Balearic Islands).

3.9.2 Type

The Mahón-Menorca variety is a half-fat to full-fat pressed cheese which varied from semi-hard to hard cheese according to its ripening time. This cheese has rounded edges and corners and a parallelepiped shape. Pressed cheese is made from whole cow´s milk, occasionally 'with a maximum of 5% of sheep's milk'. It is ripened for a minimum of 21 days. According to the ripening period, different types of cheeses can be described: fresh (ripened for no less than 2 weeks), semi-matured (no less than 2 months) and matured (from 5 to 10 months). Semi-matured and matured cheese can be labelled as farm cheese if raw milk is used. The rind is compact and greasy and not very well developed in the young cheeses, white to yellowish in young cheeses, orange or light brown in the artisan semi-cured cheese, and brown in the cured product. The top of the artisan cheeses is marked by the folds of the cloth used to wrap the cheese during moulding (Ministry of Agriculture, Food and Environment, 2015).

The minimum FDM content is 38%, with a minimum moisture content of 50% and a pH value of 5.0–5.4.

3.9.3 Milk

The cheese is made of milk from Friesian, Mahón or Menorca and/or Brown Alpine cows, with 5% of Menorca sheep's milk. The forage plants of this region have traditionally been used for feeding the cattle.

3.9.4 Description and Sensory Characteristics

The cheese has a parallelepiped shape, flat on the top and bottom. The height is 5–9 cm. The weight ranges from 1 to 4 kg. The natural rind is thin, taking on a reddish-brown colour during smoking, with red, green and blue tinges. The cheese is yellow to ivory in colour, with a few irregular, more or less round holes, in varying sizes but never larger than a pea.

The flavour is generally mildly acidic. The young cheese is mild, the semi-matured has a slight flavour of butter and nuts, especially hazelnuts, and medium persistence in the mouth. The flavour of the matured cheese is complex and intense, with touches of aged wood, tanned leather or the ripening chamber. Its piquancy increases with maturity, and persistence in the mouth is long. The cheese has lactic aromas, with a slight reminiscence of butter and a characteristic acidity, all of which increase during ripening.

The young cheese is soft and elastic, the semi-cured cheese is firmer and easy to cut, and the matured cheese has a firmer, harder, less elastic texture. Semi-matured and matured cheese may be crumbly or flaky when cut.

3.9.5 Method of Manufacture

Milk preparation: Milk must be collected from the registered farms and meet the requirements laid down by legislation.

Coagulation: This stage is carried out by using rennet. The milk must be kept at a temperature of 30°C–34°C for 30 min. This temperature must also be maintained during the cutting and draining of the curds.

Cutting: The curd must be cut to the size of a large pea and left to stand for about 10 min before the whey is drained off.

Curd draining: The curd is placed in special moulds to give the characteristic shape. Farm cheese is moulded by hand, wrapped in a square cotton cloth – the *fogasser* – and hung from the four corners. It is then placed on a table, the whey is removed and the curd is manually pressed and tied. It is then pressed for 6 to 10 hr.

Salting: Salting is done by immersing in brine at saturation point for up 48 hr at 10°C–15°C.

Maturation: After salting, the cheese remains in ventilated rooms for 3–4 days, at which time the surface flora begins to develop. It then remains in ripening chambers until it is ready for sale: 21–60 days for young cheeses, 60–150 days for semi-matured cheeses and over 150 days for matured cheeses. During ripening, the cheese is regularly turned and cleaned, and the rind is rubbed with olive oil and Spanish paprika.

3.9.6 Relevant Research

Studies have been carried out related to the classification of the cheese according to the ripening time and pasteurisation process on the basis of the pattern of amino acid release (García-Palmer et al., 1997; Polo, Ramos & Sánchez, 1985), and a zero-order kinetics was observed (Frau et al., 1997). Mulet et al. (1999) characterised the volatile fraction during ripening and identified the aroma components; only 16 compounds varied with time and exhibited a zero-order kinetics. The use of ultrasonics was studied as a tool for maturity assessment. It was found that the potential of the ultrasonic velocity increased with time, which is perhaps related to water loss and an increase of textural properties (Benedito et al., 2000).

The triacylglycerol composition in Mahón cheeses was determined, and small differences among producers and during ripening are observed due to the low level of lipolysis, although it can be used to determine foreign fat in milk fat (Fontecha et al., 2006). Luna, Juárez and De la Fuente (2007) studied the distribution of conjugated linoleic acid during ripening and confirmed that the effect of ripening was negligible. Simal et al. (2001) developed a diffusion model to identify water and salt effective diffusivity coefficients by using experimental data of ripening experiments and found that moisture and textural parameters can be used as maturity indicators.

3.10 Majorero PDO – Spain

Name: Majorero PDO	
Production area: Canary Island, Spain	
Milk: Goat's, raw	

3.10.1 Introduction

The PDO status of Majorero cheese was recognised by the Ministry of Agriculture, Fisheries and Food in 1996 and at the European level in 1999 (EC, 1999). Cheese is made from raw or pasteurised Majorero goat's milk. Under the PDO regulation, 85 farmers and 24 dairy plants are registered. In 2013, production was 315,557 kg, which represented a total of 2,050,000 euros. The cheese is mainly marketed on the national market and represents 0.97% of the economic value of Spanish PDO cheeses. It was the first Spanish goat's cheese to obtain PDO recognition (Ministry of Agriculture, Food and Environment, 2014).

The Majorero gets its name from the name given to the inhabitants of Fuerteventura (also known as Mahorata, Majorata or Maxorata) because of the shoes worn by the shepherds ('mahos' or 'majos'). They were known by this name from ancient times, and it was extended to all the inhabitants of the island and even to its agricultural products like cheese. The origin of this cheese dates back to the pre-Hispanic period. At this time, cheeses were made from goat's milk, according to documents of fifteenth-century settlers (Anonymous, 2015b). The geographical area comprises all the municipalities (Antigua, Betancuria, La Oliva, Pájara, Puerto del Rosario and Tuineje) on the island of Fuerteventura in the Province of Las Palmas in the Canary Islands.

3.10.2 Type

Full-fat pressed cheese made from Majorero goat's milk which varied from semi-hard to hard cheese, according to its ripening time. Its rind has the imprint of the mould or 'plaits' on the sides. It has a white to yellowish colour as it ranges from fresh to cured cheeses. The surface may be rubbed with paprika, oil or roasted maize meal ('gofio'), which is responsible for its characteristic appearance. Depending on the degree of ripening, it may be young (8–20 days), semi-aged (20–60 days) or aged (over 60 days). The fat and protein content in dry matter are 47%–56% and 16%–3%, respectively, with the moisture content lying between 37% and 50% (Ministry of Agriculture, Food and Environment, 2015).

3.10.3 Milk

This cheese is made from the milk of goats of the Majorero breed, to which may be added up to 15% of Canarian sheep's milk when the cheese is to be ripened. The minimum protein and fat content are 3.2% and 3.8% in goat's milk and 5.3% and 5.8%, respectively, for sheep's milk.

3.10.4 Description and Sensory Characteristics

The cheese has a cylindrical shape, with a height of 6–9 cm and a diameter of 15–35 cm. Cheeses weigh 1–6 kg. Queso Majorero is only produced in dairy plants listed in the Register of Production Plants and Cheese Factories located in the area in which the milk is produced.

The rind, which is almost non-existent in lightly ripened cheeses, bears the markings of the palm leaf bands around the sides and of the mould on the top and bottom. The colour is white to slight ivory when ripened, and the texture is without eyes, but small ones may appear occasionally. The body is firm, compact, of low elasticity and low-medium roughness. The aroma has a low to medium intensity, increasing with ripening, with lactic (whey, yogurt) to goaty hints. Depending on the degree of ripening, some notes of butter, hay and nuts may appear. The taste has a low to high intensity, increasing during ripening but persistent with goaty and lactic notes (goaty, yogurt, whey), slightly sharp. The lactic notes increase during ripening, whey notes being predominant until four months, when they disappear and butter and toasted nuts hints appear. The body is firm, creamy and slightly rough, increasing with ripening. The taste is slightly sharp and piquant in well-ripened cheeses.

3.10.5 Method of Manufacture

Milk preparation: Milk is obtained by mechanical or manual milking and is filtered; it must be collected from the registered farms. The milk used for cheesemaking must meet the requirements laid down by legislation.

Coagulation: Coagulation is carried out by using rennet (preferably rennet paste from dried kid stomachs) at a temperature of 28°C–32°C for approximately 1 hr.

Cutting: The curd is cut in order to obtain a diameter of 0.5–1.5 cm, depending on whether the cheese is to be ripened or not.

Curd draining: The curds are then drained, after initial pressing, to remove as much whey as possible, producing a semi-pressed cheese. The cheese is placed in plaited strands of palm or plastic or in plastic or stainless steel moulds, reproducing the traditional plaiting made with braided palm leaves and of sufficient size to ensure that the cheeses, once ripened, are of the characteristic form, size and weight.

Salting: Salting using sea salt may be wet or dry. When brine is used, the maximum period of salting is 24 hr with a maximum concentration of 20°Bé.

Maturation: The ripening chambers have a temperature of 12°C–18°C and an RH higher than 70%. The rind may be rubbed with paprika or roasted maize meal (gofio) or oiled. During this stage, turning and cleaning procedures are performed until the cheese reaches its specific characteristics.

3.10.6 Relevant Research

Fracturability, hardness, adhesiveness and gumminess were seen to increase from 15 to 90 days of ripening, while elasticity decreased. Furthermore, the ripening time affected most of the sensory parameters analysed and, as the cheeses matured and became drier, there was an increase in roughness and elasticity in addition to odour and aroma intensity (Fresno & Alvarez, 2012).

Different studies have been carried out related to the influence of animal feeding on the final cheese properties as a strategy to improve the quality of Majorero cheese (Alvarez & Fresno, 2004; Alvarez et al., 2007). Special studies have been focused on the use of rennet paste for Majorero cheesemaking to recover the traditional sensory profile and demonstrate the possibility of preparing artisan rennet pastes from commercial-weight kids in an easy way for farmhouse cheesemakers using local resources that would otherwise be destroyed in abattoirs (Calvo et al., 2007, Castillo et al., 2007; Fresno et al., 2014). Studies regarding the influence of supercritical fluid extraction pressure on the final properties of the cheese have also been carried out, the conclusion being that a low-fat goat's cheese, with a lower triglyceride and cholesterol content, can be produced with this technology (Sánchez-Macias et al., 2013).

3.11 Manchego PDO – Spain

Name: Manchego PDO	
Production area: Castilla-La Mancha, Spain	
Milk: Sheep's (pasteurised or raw milk)	

The PDO of Manchego cheese was recognised by the Ministry of Agriculture, Fisheries and Food in 1995 and at the European level in June 1996 (EC, 1996a). This regulation was modified and approved by the European Commission in February 2012 (EC, 2012a)

3.11.1 Introduction

Manchego cheese is a pressed semi-hard cheese made from the milk of the Manchega sheep breed. It is the most well-known Spanish cheese. There has been a continuous increase in the production of the certified product, with a total of 11,176,301 kg being produced in 2013, of which 7,846,305 kg were exported, confirming its export-oriented manufacture. The number of certified dairies is 62, of which 24 make cheeses from raw milk and 38 from pasteurised milk (Ministry of Agriculture, Food and Environment, 2014).

Archaeological finds dating back to the Bronze Age show that the inhabitants of La Mancha made cheese from the milk of sheep that can be considered to be ancestors of the modern-day Manchega breed. The region was named by Arabs as Al Mansha or 'land without water', amply describing the climate and showing that Manchega sheep must have been long adapted to this ecosystem. The cheese is also known as the cheese of Don Quijote, because Miguel de Cervantes mentioned it in his novel, *Don Quijote de la Mancha*.

The PDO area of Manchego cheese covers an area of 44,000 km^2 of Castilla-La Mancha. Nowadays a total of 798 farms and 520,000 Manchega sheep, which produce 60 million litres of milk per year, of which 92.5%, are allocated to the elaboration of Manchego cheese. The production area covers the provinces of Albacete, Ciudad Real, Cuenca and Toledo. The processing and ripening area coincides with the production area.

3.11.2 Type

Pressed semi-hard cheese, made from the milk of sheep of the Manchega breed, aged for a minimum of 30 days for cheeses weighing up to 1.5 kg, and from 60 days up to a maximum of two years for larger cheeses. The minimum fat and protein content in dry matter are 50% and 30%, respectively, with a minimum moisture content of 45% and a maximum salt content of 2.3%.

3.11.3 Milk

Manchego cheese can be made with pasteurised or raw milk. In the latter case, the label may indicate the word 'Artesano'. Manchega sheep graze throughout the year on the natural resources of the area. When in the fold, their diet is supplemented with concentrated feed, hay and by-products. Other specifications regarding sheep farming are defined by the PDO regulation (EC, 1996a).

3.11.4 Description and Sensory Characteristics

It is a pressed semi-hard cheese with a hard rind and a firm and compact paste, with a colour varying from pale yellow to blackish. The shape is cylindrical, with a flat top and bottom surfaces engraved with the typical 'flower' left by the wooden presses. The sides show a zigzag pattern produced by the mat-weed (esparto) of the moulds. Today, industrially produced cheeses have the same engraving, but pre-designed in the new industrial moulds (Ministry of Agriculture, Food and Environment, 2015).

The body is firm and compact with a colour which varies from white to ivory-yellow. It has a smooth surface or an uneven sprinkling of small holes. The smell is intense and persistent, slightly acidic, smell of nuts and hazelnut which evolves to spicy nuances in the most matured cheeses. The taste is slightly acid and strong, which evolves to spicy in mature cheeses. A differentiated residual taste from the Manchega sheep's milk is also noticed. It has low elasticity, and a buttery feel, which can be granulated in the most aged cheeses.

3.11.5 Method of Manufacture

Milk preparation: Milk must be collected from the registered farms and kept after milking at 4°C. The collection and transportation of milk must take place in good hygienic conditions that ensure the milk temperature does not exceed 10°C.

Starter culture: It is made with raw milk, although when pasteurised milk is used, commercial starters containing mainly mesophilic starter cultures are used. Nowadays, the frequency of using thermophilic starters as adjunct cultures is increasing.

Rennet: Calf rennet or other authorised coagulants can be used. Coagulation takes place at 28°C–32°C for 30–60 min.

Cutting: The curd must have the right consistency for ensuring correct draining in accordance with the size of the cheese, as determined by checks during cheese production and the experience of the producer. The curd is stirred and pitched for 15–45 min while being heated at 28°C–40°C.

Moulding/pressing: Curd is transferred to moulds manually or mechanically. A pressing time of between 1 and 6 hr is applied. At this stage, or at moulding, a casein label identifying the number and series (consisting of five digits and two capital letters, respectively) is allocated.

Salting: The time at which the curd must be removed from the mould and moved to the salting stage is determined by measuring the pH. The salting can be done in brine, at 17°C–18° Bé, dry or a combination of both. In brine, the duration of this phase will be a minimum of 5 hr and a maximum of 48 hr.

Maturation: This is done in caves or ripening chambers that are provided with systems to ensure the clear identification and separation of cheeses. These chambers have a temperature of 3°C–16°C and an RH of 75%–90%. During this stage, turning and cleaning procedures are performed until the cheese achieves its specific characteristics.

Manchego cheese can be coated with legally inactive, authorised, transparent films or paraffin, or it can also be dipped in olive oil, while ensuring that the rind maintains its natural colour and the casein label is legible. Substances that turn the rind black are not permitted.

3.11.6 Relevant Research

Over the last ten years, much research has focused on Manchego cheese. The isolation of *Lactobacillus* spp. from artisanal Manchego cheeses (Picon et al., 2010; Sánchez et al., 2006) has been extensively studied. *Lb. paracasei* subsp. *paracasei* exhibits the best technological characteristics, and freeze-drying and storage do not alter its properties (Poveda, Chicón & Cabezas, 2015), which makes it a good alternative for use as an adjunct strain used in industrial cheesemaking conditions.

Twenty-seven *Leuconostoc* spp. isolates from Manchego cheese were characterised by phenotypic and genotypic methods, and their technological attributes were studied in order to test their potential use as dairy starter components (Nieto-Arribas et al., 2009, 2010). *Enterococcus* spp. populations were also isolated from Manchego cheeses (Nieto-Arribas et al., 2011) and representative isolates were assayed for some enzymatic activities that may be considered to have a potential role in cheese ripening. There are also studies related to the late blowing defect (Garde et al., 2012) and the prevalence of *Clostridium* spp. (Garde et al., 2011), which determined that the highest incidence of late blowing is recorded for Manchego cheeses produced in summer. These authors also stated that the only bacteriocins that inhibit the growth of the *Clostridium* spp. isolates are nisin A and Z. Other research focused on the texture profile of Manchego cheese using non-destructive techniques (Benedito et al., 2006a; Conde et al, 2007), ultrasonic and surface probe (Benedito et al., 2006b) and the analysis of the peptidic fraction, which may be associated with the umami and bitter tastes (Gómez-Ruiz et al., 2007).

3.12 Murcia al Vino PDO – Spain

Name: Murcia al Vino PDO
Production area: Murcia, Spain
Milk: Goat's

3.12.1 Introduction

The PDO status of Murcia al Vino cheese was recognised by the Ministry of Agriculture, Fisheries and Food in 2001 and at the European level in 2002 (EC, 2002), and modified in 2013 (EC, 2013). The cheese is made from pasteurised Murciano-Granadina goat's milk. Under the PDO regulation, 171 farmers and 7 dairy plants are registered. In 2013, their production was 377,363 kg, which represented a total of 3,200,000 euros. The cheese is mainly marketed outside the European market (North America) and represents 1.52% of the economic value of Spanish PDO cheese production (Ministry of Agriculture, Food and Environment, 2014).

Murcia al Vino has long been produced in Jumilla and Yecla. In old times, the milk was coagulated by using kids' rennet (rennet paste) macerated in wine and, once the mass had formed, the cheese was placed into small moulds and refined by salting the pieces and humidifying them with wine. This was done several times, thus giving the cheese an excellent taste, which was highly prized. The tradition of making goat's cheese started in goatherds' homes in the province of Murcia, and was not only for domestic consumption but also for sale in the local area (Anonymous, 2015c).

The geographical area comprises all the municipalities of the province of Murcia, situated in the south-east of Spain.

3.12.2 Type

Murcia al Vino is a full-fat, pressed, washed but not cooked, semi-hard cheese made with Murciano-Granadina goat's milk. It is marketed from 45 days after manufacture, although small cheeses may be marketed after 30 days. This cheese is characterised by the deep red colour of the rind, which is obtained by immersing the cheese in wine during ripening. The fat and protein content in dry matter has a minimum of 45% and 32%, respectively, with a minimum moisture content of 45% and a minimum pH of 5.

3.12.3 Milk

The cheese is made only with milk from the Murciano-Granadina goat breed, allowing for 3% morphological defects in the breed due to genetic drift. Feeding is mainly based on direct supply of seasonal pastures, by-products and some traditional high-nutritional-value supplements from the production area, all of which provide the distinctive characteristics that influence the sensory profile of the cheese.

3.12.4 Description and Sensory Characteristics

The cheese has a cylindrical shape, with a maximum diameter and height of 19 and 10 cm, respectively, with a diameter/height ratio between 1.5 and 2.2. The weight is 300 g to 2.6 kg. The different stages involved in producing the milk and making and maturing this type of cheese must take place in the defined geographical area (Ministry of Agriculture, Food and Environment, 2015).

The rind is firm with a maroon-violet colour, smooth without imprints, with slightly curved sides. The body is creamy, uniform with an elastic texture, white to ivory-yellowish in matured cheeses and with a few holes. The aroma has a medium-low intensity with lactic to goaty hints, while the aroma of wine or cellar can be detected in the outer part of the rind. The taste has a low to medium intensity but persistent with an average acidity and salty notes with some hints of fresh lactic aroma and slightly bitter.

3.12.5 Method of Manufacture

Milk preparation: Milk must be collected from the registered farms and kept after milking at less than 6°C. The milk used for cheesemaking must meet the requirements laid down by legislation. The minimum fat content of the milk is 4.7%.

Coagulation: This stage is carried out by using rennet at a temperature of 30°C–34°C for 30–60 min.

Cutting: The curd must have the right consistency for ensuring correct draining in accordance with the size of the cheese, as determined by checks during cheese production and the experience of the producer, and must meet the requirements set out in the specification.

Curd draining: The curd is washed by substituting 15% of whey and is heated to 3°C–5°C above the coagulation temperature. The curd is then placed by manual or mechanical means in a mould of the shape and size required to produce cheeses which fulfil the PDO requirements. The cheeses are subjected to a pressure of 1–2.5 kg/cm^2 for 1–4 hr.

Salting: The time at which the curd must be removed from the mould and exposed to the salting stage is determined by measuring the pH. The salting is done in brine at 17°Bé, and the duration of this phase will be determined by the cheesemaker although the cheeses must comply with the final properties determined by regulation.

Maturation: Cheeses are ripened for a minimum of 45 days for large cheeses and 30 days for small ones in ripening chambers with a temperature of 4°C–12°C and an RH of 70%–90%. The dairy plants must have production systems that prevent crosses with other products. During ripening, the cheeses are bathed in red wine produced in the province of Murcia, including the Monastrell variety, which is the most representative and typical of the area.

The esterified fatty acids are not modified significantly by the ripening time. The most abundant fatty acids in this type of cheese are C_{16} and $C_{18:1}$ (Tejada et al., 2006). Different ripening periods significantly affected the WSN and PTA SN and all FAA, except serine (Boutoial et al., 2013a).

3.12.6 Relevant Research

The use of natural rennet paste produces a cheese with a more hydrolysed protein matrix, which is associated with significant changes in the texture and sensory profile (Ferrandini et al., 2011; Ferrandini et al., 2012). The use of vegetable coagulants in Murcia al Vino cheese accelerates the ripening process as a result of increased cyprosin proteolytic activity (Abellán et al., 2012; Tejada et al., 2006; Tejada et al., 2008a; Tejada et al., 2008b).

Other studies have focused on

- the influence of the size on the calorific value, where it was observed that at 30 days of ripening, large cheeses had higher energy values (Abellán et al., 2007);
- on-line monitoring of coagulation and the syneresis process by using diffuse reflectance backscattering (Rovira, García & López, 2011; Rovira et al. 2012);
- the influence of the manufacturer and season on the physico-chemical properties, where it was demonstrated that cheeses made in spring have the best physico-chemical composition with a lower NaCl content, better fatty acid profile and the same degree of proteolysis (López et al., 2012); and
- the microstructure, in which microstructural and physico-chemical parameters were correlated to predict the porosity, moisture and water-holding capacity (Rovira et al., 2013). Supplementation with aromatic plants and the introduction of distilled and non-distilled thyme/rosemay leaves can be successfully adopted as a strategy to reduce feeding costs (Boutoial et al., 2013b; Boutoial et al., 2013c) or the use of inulin in reduced-fat cheeses to obtain models for the prediction of coagulation and syneresis parameters in milk gels when inulin is added as a fat substitute using a fibre optic light backscatter sensor (Arango, Trujillo & Castillo, 2015).

3.13 Präst – Sweden

Name: Präst	
Production area: Sweden	
Milk: Cow's, pasteurised	

3.13.1 Introduction

Präst has been a registered European trademark since 2001, owned in fellowship by Swedish dairy companies, and it is an important cheese on the Swedish market. The name origins from Präst-ost, which means 'cheese of the vicar/priest'. Präst is made in the lowlands of Sweden, initially in the county of Småland in the south of Sweden since the 1700s, when the farmers paid an annual church tax on their production to their vicar *in natura;* that is, 10% of everything they produced. The farm women or dairymaids came together at cheesemaking parties to make cheese for the vicar, and everybody brought their own milk or sometimes pre-made cheese grains to the party. They usually made high-quality full-fat cheese for the vicar, while their own cheeses were commonly made of half-skimmed milk after the cream had been separated for butter production. Präst is still commonly made with an extra addition of cream.

The cheese is ripened for at least four months (mild); however, cheeses that are eight months (medium), one year (mature) and two years (extra-old) old are on the market with more intense flavours and with a long-lasting aftertaste. (Ränk, 1987)

3.13.2 Type

Präst is a semi-hard, full-fat, sliceable cheese with an open texture containing several small (2–6 mm) irregular eyes. FDM is at least 50%, corresponding to 31% fat in the cheese. Moisture in the non-fat substance (MNFS) is 55%–58%, and the salt content is 1.0%–1.4%. Präst is also made as reduced-fat cheese with 30% FDM (17% fat in cheese), 56%–57% MNFS and 1.4%–1.6% salt content (Sveriges Ostkollegium, 1993).

3.13.3 Milk

Pasteurised Swedish cow's milk is used.

3.13.4 Description and Sensory Characteristics

Präst is made in low cylinders with slightly convex sides with a cloth around it to prevent the cheese from flowing out. The cheese body is evenly light yellow to yellow. The surfaces are dried and covered with wax.

The body melts evenly and is sliceable and flexible. When the ripening time is longer, the body becomes more brittle. The flavour has powerful strength with a distinct saltiness and bitterness that is balanced by creamy, buttery notes, umami, sweetness and fruitiness. The cheese has a long and rich aftertaste.

3.13.5 Method of Manufacture

Milk preparation: The milk is standardised, pasteurised (72°C for 15 s) and the temperature is set to 30°C. Sodium nitrate and calcium chloride may be added to the milk. The milk is pre-ripened for 30 min after addition of a starter.

Starter culture: Mesophilic undefined DL-starter (commonly 0.8% outgrown culture) with undefined mixed strains of *Lc. lactis* subsp. *lactis* and *Lc. lactis* subsp. *cremoris* (80%–90%), *Lc. lactis* subsp *lactis* biovar. *diacetylactis* and *Leuc. mesenteroides.*

Rennet: The milk is coagulated by calf rennet at 31°C for about 35 min.

Cutting: The cheese is cut into cubes with sides of 10–12 mm.

Cooking/curd washing: During cooking at 42°C, a part of the whey (40%) is removed and water (55%) is added.

Curd draining: The curd is pumped with whey, which is removed before the curd is moulded. The cheeses are pressed at 20°C for about 22 hr while the cheeses are turned around several times and the cheesecloths are replaced.

Salting: Brine salting is performed at 15°C for two days.

Maturation: The cheeses are ripened at 18°C–19°C for 20 days and then at 12°C until sold. They are waxed three times during ripening. The cheese has very good storage properties.

3.13.6 Relevant Research

Several peptides have been identified in Präst: 40, 47 and 43 in cheeses at the age of 8, 12 and 24 months, respectively. Of these peptides, at least 21 are suspected to be bioactive in human metabolism, and the highest number was found after eight months of ripening (Rasmusson, 2012).

The texture develops as a result of mineral arrangement, proteolysis and amino acid release. The starter bacteria dominate at the beginning of ripening. Within a couple of weeks, non-starter *Lb casei/paracasei* contribute to ripening; the strain differs between different dairy plants (Ardö, 1993a; Ostlie et al., 2005; Pripp et al., 2006; Rasmusson, 2012).

3.14 Raclette du Valais PDO – Switzerland

<table>
<tr><td>Name: Raclette du Valais PDO</td><td rowspan="3">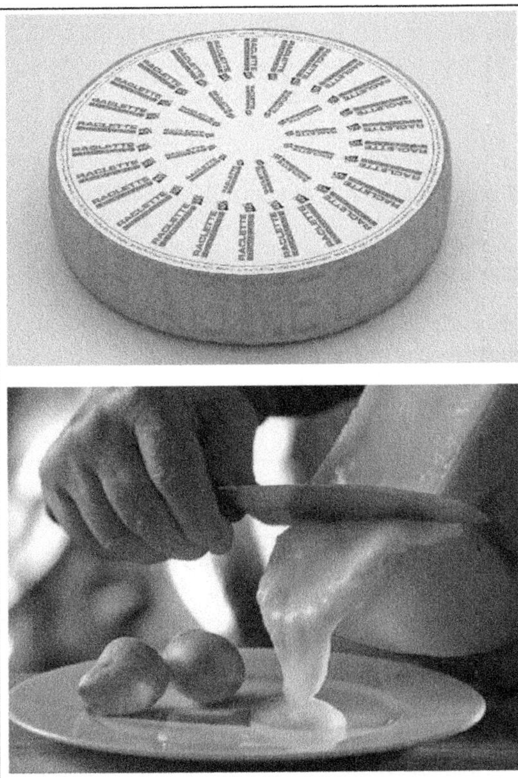
Cheese wheel of Raclette du Valais PDO and its preparation as melted cheese (Interprofession Raclette du Valais).</td></tr>
<tr><td>Production area: Canton of Valais</td></tr>
<tr><td>Milk: Cow's, raw</td></tr>
</table>

3.14.1 Introduction

Raclette du Valais PDO is a semi-hard, smear-ripened cheese produced from raw milk and is mainly consumed in melted form after a ripening time of at least three months. A small part of the production of Raclette du Valais is dry-ripened for at least nine months and is consumed as extra-hard cheese in planed form (similar to Berner Hobelkäse PDO). The regional identification is typically embossed on the hoop side of the cheese. The yearly production is 2,291 tonnes (2014). The product is consumed mainly in local and regional markets and largely during the winter season (TSM Treuhand GmbH, 2015). Raclette du Valais was granted its PDO in 2007 (FOAG, 2015; Swiss PDO-PGI Association, 2015). This designation has been recognised by the European Union since December 2011 (Swiss Commission for Denominations of Origin and Geographical Indications, 2011).

Alpine cheese production in Switzerland has a long tradition dating back to the Middle Ages. In 1574, the physician and pharmacist Gaspard Ambuel from Sion (Valais) first mentioned a 'baked cheese'. From 1875 on, the term 'raclette' has been used in several documents for a cheese melted in front of a fire and afterwards scraped off (*racler* in French). From the beginning of the twentieth century, Raclette became known outside the canton of Valais and is considered a national dish. Until about 1920, Raclette was exclusively an Alpine product although it is now produced also in dairies in the lowland (Kulinarisches Erbe der Schweiz, 2008).

3.14.2 Type

The specifications require an MNFS of 60.0%–64.0% for meltable Raclette du Valais, 57.0%–62.0% for the sliceable variety and ≤50.0% for planed Raclette du Valais. The absolute water content ranges from 37% to 44%. The FDM is 50.0%–54.9%, and the salt content varies between 1.2% and 2.2%.

3.14.3 Description and Sensory Characteristics

Raclette du Valais has a round shape, a diameter of 29–32 cm and a height of 6–7 cm. On the hoop side of the cheese, the regional identification is impressed or embossed. Meltable and sliceable Raclette du Valais has an even, natural brown/orange colour and a slightly damp smear-rind that is removed for dry ripening. The holes should be evenly distributed, not too frequent (2–3 holes per core sample) and have a maximum diameter of 2–3 mm (pea-sized).

The texture of meltable Raclette du Valais is even, smooth and fine; the aroma is reminiscent of fresh butter or cream but is also flowery. The taste is supported by a sour note, also similar to fresh butter or cream, and has a dominating herbal and fruity note. The taste of dry-ripened Raclette du Valais is similar but more salty.

3.14.4 Method of Manufacture

Milk preparation: Raclette du Valais is produced from raw cow's milk, which is usually delivered twice a day to the cheese factory. The forage of dairy cows consists of 75% natural raw feed that grows in the geographical area. The feeding of dairy cows and other animals located in the same building with silage or other fermented feeds, as well as the use of genetically modified feeds, is prohibited. The milk must be cooled to below 8°C within 2 hr of milking. No more than two subsequent milkings may be processed. The oldest milk may not be stored for more than 24 hr between milking and renneting. Treatments such as ultrafiltration, microfiltration and bactofugation, as well as the addition of whey cream, are prohibited. Any thermal treatment of the milk other than cooling is prohibited. The milk is processed in a copper vat, and the maximum processing volume is 5000 litres per batch.

Water, salt, rennet, lactic acid and smear cultures are the only permissible processing aids.

Starter culture: LAB cultures approved by the interprofessional organisation of Raclette du Valais are derived from cultures used in Valais cheese dairies and are further cultured for the manufacture of Raclette du Valais. The milk is inoculated with LAB and then quickly heated to 32°C. A preliminary ripening time of 30–60 min is needed.

Coagulation: Only calf rennet is allowed, which is added when the temperature is 30°C–33°C.

Cutting and curd washing: After cutting the curd into maize-kernel-sized grains, the lactose is usually removed from the curd by adding water (10%–15% of the milk's volume).

Scalding: Scalding is performed at a temperature of 36°C–45°C and lasts for 30 min.

Moulding: The curd is pre-pressed and then transferred to round moulds.

Pressing: The duration of pressing depends on the respective pressing device. The cheese wheels are turned over at least twice during pressing and draining. At the end of this process (≥4 hr), the pH value is in the range of 4.8–5.2.

Salting: The cheese is either dry-salted manually or immersed in a salt bath (8–22°Bé, 22–24 hr, 8°C–15 °C).

Maturation: The cheese is ripened for at least three months in cellars which have a temperature of 7°C–14°C and an RH of 88%–96%. The growth of the surface flora is induced with a smear culture that contains the naturally occurring smear flora of the cheese dairy. After a prolonged break in production (e.g. the summer period), smear cultures from local cheese dairies may be used to restart the production. If this is not possible, commercial cultures of *Brevibacterium linens* may be used. During ripening, which lasts for at least three months, the loaves are placed on planks of red spruce and are regularly washed and turned. The cheeses are stored on raw spruce planks and must be turned over and brushed with a soft brush regularly.

3.14.5 Relevant Research

Research focuses mainly on the cheese's melting properties. Even though most of the research has been conducted for Raclette Suisse®, many of these results are relevant to Raclette du Valais PDO as well. In 2013, the melting properties of the 10 most popular Raclette cheeses were characterised by a trained sensory panel (Foodle.ch, 2014). The results showed that Raclette du Valais PDO had stronger fat separation and a more spicy flavour compared to Raclette Suisse®.

3.15 Raclette Suisse®-Switzerland

Name: Raclette Suisse®	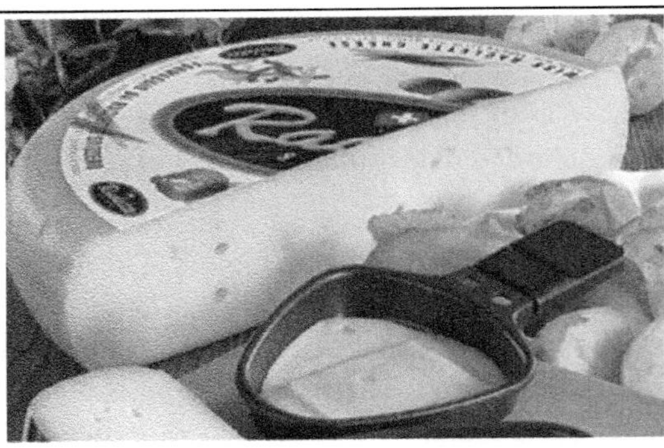 Raclette Suisse® is usually consumed in melted form and prepared directly at the table in small pans (Interprofession Raclette Suisse)
Production area: All over Switzerland	
Milk: Cow's, pasteurised, thermised or raw	

3.15.1 Introduction

Raclette Suisse® is a semi-hard, smear-ripened cheese that is produced from pasteurised or thermised milk and is typically consumed in melted form after a ripening time of at least three months. Raclette Suisse® is a well-known and worldwide protected brand that may be used alone or in combination with manufacturer branding. The yearly production of Raclette Suisse® is about 12,700 tonnes, and 1,559 tonnes were exported in 2014 (TSM Treuhand GmbH, 2015).

In medieval writings dating back as far as 1291, Raclette cheese was described as a particularly nutritious meal of Alpine herdsmen and dairymen. From 1875, the designation 'Raclette' was generically used for the preparation of cheese in front of a fire. As soon as the cut section of the cheese is melted, the surface is scraped off (see Raclette du Valais PDO, Section 3.14). In the German-speaking part of Switzerland, it is referred also as 'Bratchäs' (roasted cheese).

3.15.2 Type

Raclette Suisse® is a semi-hard cheese made from cow's milk. The specifications for Raclette Suisse® require an MNFS of 56.0%–62.0%. The absolute water content ranges from 37% to 44%. The FDM should be in the range 45.0%–52.0%. The salt content varies between 1.7% and 2.4% (Regulations cited by Sieber, 2011).

3.15.3 Description and Sensory Characteristics

Raclette Suisse® has a round or rectangular shape with a diameter of 30–36 cm, a regular height of 6–8 cm and a weight of 4–9 kg. The cheese has an even, natural brown-orange colour and a slightly damp smear-rind, which is often washed off before the cheese is marketed. The few and small-sized eyes should be well shaped and evenly distributed.

The supple body has a long, smooth texture, an ivory to light yellow colour and shows excellent melting properties. The aroma of Raclette Suisse® is mild, milky, the taste is sour, salty and the flavour increasingly spicy with advanced ripening. Raclette Suisse® can also be produced with added herbs and spices or other appropriate ingredients (e.g. garlic, onion or red pepper) and natural extracts thereof in order to obtain specialty cheeses with particular flavours.

3.15.4 Method of Manufacture

Water, salt, rennet, lactic acid bacteria and smear cultures are the only permissible processing aids.

Milk preparation: Cow's milk is the basic raw material. The feeding of dairy cows and other animals located in the same building with silage or other fermented feeds is permitted. However, fodders originating from GMOs are forbidden. Furthermore, the use of any type of growth activators is forbidden.

The milk must be supplied once or twice a day and must be cooled to below 8°C within 2 hr of milking. The milk is usually pasteurised or thermised prior to cheesemaking; however, the use of raw milk is also permitted. Pre-treatment of the milk may include ultrafiltration, microfiltration or bactofugation; the most common pre-treatment is bactofugation. The addition of whey cream to the vat milk is permitted. The milk may be processed in a copper or steel vat.

Starter culture: There are no regulations regarding the type of starter cultures used. The large majority of starter cultures consist of mesophilic LAB (*Lactococcus* spp. and *Leuconostoc* spp.). In addition, thermophilic cultures (*S. thermophilus*, *Lb. delbrueckii* subsp. *lactis* and *Lb. helveticus*) or even in-house whey cultures are sometimes used.

Rennet: Calf rennet or microbial rennet is added when the temperature reaches 32°C. Rennet originating from GMOs is forbidden.

Cutting and curd washing: After cutting the curd to the grain size of a hazelnut, the lactose is removed by curd washing (water addition of up to 30% of the milk's volume).

Scalding: The scalding temperature varies between 35°C and 42°C.

Moulding: The transfer of the curd into the moulds is mostly done mechanically. Each cheese must be marked with the production date and the admission number of the cheese factory.

Pressing: Pressing lasts about 60 min; good cohesion of the grains must be ensured.

Salting: After reaching a pH value of 5.2 at the end of draining, the cheeses are transferred into the brine bath for salting (for 24–48 hr at 12°C–15°C and 20–22°Bé).

Maturation: After salting, the cheeses are stored in cellars at a temperature of 9°C–12°C and an RH of about 92%–96%. The atmosphere of the ripening cellars of Raclette Suisse® is characterised by a noticeable presence of ammonia. During ripening, which lasts for at least three months, the loaves are placed on shelves of red spruce and are regularly washed with smear water and turned. During the first 14 days of ripening, the cheeses are smeared and turned daily. To induce the formation of a rind, the cheeses are rubbed with a surface culture and salted water (2%–5% NaCl). The growth of the surface flora leads to a typical red-orange appearance, contributes to the formation of the flavour, supports the control of moisture and prevents the development of pathogenic bacteria (Roth, 2009). The surface culture consists of different strains of coryneform bacteria as well as yeasts. At the minimal age of three months, Raclette Suisse® becomes ready for consumption. Before packaging, the cheeses are usually put into a water bath for approximately 25–30 min at a water temperature of 22°C–25°C. The cheese rind is then brushed off.

3.15.5 Relevant Research

The melting properties of Raclette Suisse® are particularly important. Several studies carried out by Agroscope have investigated the effect of individual processing factors on its melting properties. The acidification of the cheese curd by lactic acid fermentation is important for the solubilisation of bound calcium and the ensuing loss of calcium into the whey. Good calcium removal from the casein matrix was obtained with a quick decrease in pH during curd making. Similarly, adding citric acid to the cheese curd extracted a lot of calcium from the casein matrix and improved melting. The complexing of calcium by citrate seems to favour the extraction of calcium. Significantly less calcium was extracted by the addition of lactic acid, although the pH values obtained in the whey were comparable. On the other hand, pre-ripening of the milk resulted in better calcium retention compared with standard production. Even in ripened cheese, the ratio of soluble and insoluble calcium is important: a low pH and complexing by citrate increased the proportion of soluble calcium and thus improved the melting properties of Raclette cheese (Fröhlich-Wyder et al., 2007, Fröhlich-Wyder, Guggisberg & Wechsler, 2009).

3.16 San Simón da Costa PDO-Spain

Name: San Simón da Costa PDO	
Production area: Galicia, Spain	
Milk: Cow's, raw or pasteurised	

3.16.1 Introduction

The PDO status of San Simón da Costa cheese was recognised by the Ministry of Agriculture, Fisheries and Food in 2005 and at the European level in 2008 (EC, 2008)

The cheese is made from raw or pasteurised cow's milk. Under the PDO regulation, 105 farmers and 11 dairy plants are registered. In 2013, their production was 370,141 kg, to the value of 2.790.000 euros. The cheese is mainly sold on the national market and represents 1.32% of the economic value of all Spanish PDO cheeses (Ministry of Agriculture, Food and Environment, 2014).

Legend places the origin of this cheese in the Castro culture when people first settled in the hills of *A Carba* and *O Xistral*. During the Middle Ages, all written sources mentioning the cheese were lost, but this cheese was regularly consumed and was also used for tax payments to the nobility or the clergy.

This cheese has always been highly appreciated and formed part of the Chicago Exhibition of 1892 and the Dairy Industry Exhibition held in Madrid in 1913 (Anonymous, 2015d). The milk used for making San Simón da Costa PDO cheeses is produced in the Terra Chá area, in the Province of Lugo.

3.16.2 Type

Smoked semi-hard pressed cheese made from pasteurised or raw cow's milk using predominantly enzymatic coagulation, with a minimum ripening period of 30 days or 60 days. The minimum FDM is 45% and the maximum is 60%, while it has a maximum moisture content of 45% and a pH value between 5.0 and 5.6 (Ministry of Agriculture, Food and Environment, 2015).

3.16.3 Milk

The local dairy herds are mainly composed of Rubia Gallega, Pardo Alpina and Friesian breeds and their cross-breeds.

According to its PDO regulation, the livestock is fed mostly on fodder produced on the holding itself; feeding is by grazing, weather permitting. Cows are fed a high-quality fodder mainly composed of indigenous grasses and pulses. A concentrated feed of vegetable origin acquired from outside the holding must be used only as a supplement to cover the energy needs of the livestock and must be sourced, as far as possible, from within the defined area.

3.16.4 Description and Sensory Characteristics

Cheeses have a height of 13–18 cm, a diameter at the base of 10–15 cm and a weight of 0.8–1.5 kg. Each cheese is pear-shaped and ends with a peak. There are two formats: (1) large, aged for a minimum of 45 days, with a final weight of 0.8–1.5 kg and a height of 13–18 cm and (2) small or 'Bufón' aged for a minimum of 30 days, with a final weight of 0.4–0.8 kg and measuring 10–13 cm in height. Smoking with local birch wood imparts its particular flavour.

The rind is smoked, hard, inelastic, 0.1–0.3 cm thick, yellow-ochre and greasy. The body is uniform between white and yellow, smooth, semi-hard, semi-elastic, firm and compact with a small number of roundish or irregular eyes, varying in size but less than half the size of a pea. Its aroma is milky, reminiscent of the smell of butter with hints of toasted smoked. Its flavour is essentially of milk, slightly salty and sweet. In the mouth, it is firm, soft and soluble and of medium springiness.

3.16.5 Method of Manufacture

Milk preparation: Milk must be collected from the registered farms and kept after milking at no more than 4°C. The milk used for cheesemaking should not contain colostrum or preservatives and must meet the requirements laid down by legislation.

Starter culture: The starter cultures used are various strains of *Lc. lactis* subsp. *lactis* and *Lc. lactis* subsp. *cremoris*. The recovery and use of indigenous strains is promoted.

Coagulation: This process is carried out using rennet at a temperature of 31°C–33°C for 30–40 min, although when raw milk is used the temperature and time are adjusted to 28°C–32°C and 30–35 min, respectively.

Cutting: The curd must be cut to the size of 0.5–1.2 cm in diameter.

Curd draining: The curd is placed in moulds of the shape and size required to produce cheeses with characteristic properties of the certified PDO product.

Pressing: The length of the pressing process varies according to the pressure applied and the size of the pieces. The cheeses are wrapped in cotton cloth to facilitate whey drainage and to produce a smooth rind.

Salting: Cheeses are immersed in brine with a concentration of 14%–17% for a maximum of 24 hr.

Maturation: Ripening chambers have a temperature of less than 15°C and an RH of 75%–80%. A week before the end of ripening, the cheeses are smoked with birch wood without bark, for the time necessary to acquire their characteristic colour, ensuring that the cheeses do not come into close contact with the fire. Large cheeses are aged for a minimum of 45 days after salting and small cheeses (Bufón) for 30 days. The cheeses are turned and cleaned during aging. Optionally, the cheeses can be immersed in an olive oil bath or other authorised product to inhibit the growth of mould.

3.16.6 Relevant Research

San Simón da Costa cheese manufactured from pasteurised milk undergoes only moderate proteolysis and lipolysis during ripening, while the high content of butyric acid (C_4), which represents 18% of the total free fatty acids at the end of the ripening process, is noticeable (García et al., 2001; Nhuch et al., 2008).

Recent studies published by Garabal, Rodríguez-Alonso and Centeno (2008) determined changes in the microflora due to the cheesemaking environments, the most evident being the *Enterococcus* spp. strains. González et al. (2015) studied the microflora of this type of cheese and propose starter cultures and co-cultures for manufacturing San Simón da Costa cheese from pasteurised milk.

Garabal et al. (2010) studied the influence of modified atmosphere packaging on San Simón da Costa cheese and found that this packing system modified the ripening process by altering proteolysis and lipolysis. This may be due to the accumulation of compounds from smoke which were negatively correlated with flavour. Vacuum packaging seems to be the most useful technique to preserve the sensory profile of this type of cheese. There are recent studies on the life cycle inventory and environmental assessment of producing this type of cheese, the results of which can be extrapolated to other Spanish cheeses (González-García et al., 2013).

3.17 Svecia PGI– Sweden

Name: Svecia PGI	
Production area: Lowlands of Sweden	
Milk: Cow's, pasteurised	

3.17.1 Introduction

Svecia has been granted Protected Geographical Indication (PGI) by the EU (EC, 1997). It is a semi-hard, open texture, typical Swedish farm cheese type that has been made from cow's milk using calf rennet for as long as one can follow cheesemaking traditions in Sweden (at least from the 1200s). Most farm cheeses were quite small (1–2 kg), but larger cheeses (7–10 kg or more) were made for special occasions in cooperation between several farms and at small dairy cooperatives from the end of the 1800s. They were produced under different names until an agreement was made around 1920 to use Svecia (from Suecia, Latin for Sweden) for all of them to facilitate trading. Svecia is made with different fat contents. Kryddost is a variety of Svecia spiced with cumin and sometimes also clove, which often are preferred with a reduced fat content (Ardö, 1993a; Ränk, 1987).

3.17.2 Type

Svecia is a semi-hard cheese with an open texture containing several small irregular eyes. Full-fat Svecia has 28% fat, corresponding to 45% FDM and 53%–57% MNFS. Reduced-fat Svecia is made with 30% FDM (17% fat in cheese) and 52%–56% MNFS. The salt content is 1.0–1.5 (Sveriges Ostkollegium, 1993).

3.17.3 Milk

Svecia is produced from Swedish pasteurised cow's milk of two major breeds, the Swedish Red and White Breed (SRB) and the Swedish Holstein Friesian breed (SLB). The cows are kept in barns during the freezing winter months and are grazing outside for large parts of summer. The hygiene in milk production is at a very high level, which means that raw milk is almost free from detrimental bacteria and low in LAB.

3.17.4 Description and Sensory Characteristics

The cheeses are made as low cylinders (diameter, 35 cm; height, 13 cm; 12–15 kg). The colour of the cheese is uniform, light yellow to yellow and the surfaces are dry and waxed.

Svecia is rich in irregular eyes (mainly 1–3 mm, some 3–6 mm) that are evenly distributed in the cheese body. The texture is semi-hard and slightly elastic, creamy, juicy and tender. The flavour of young cheese is acidic, weakly salty and an aromatic flavour develops during further ripening.

3.17.5 Method of Manufacture

Milk preparation: The raw milk is standardised for fat content, pasteurised (72°C/15 s) and the temperature is set to approximately 30°C before the starter is added.

Starter culture: Mesophilic DL-starter with undefined mixed strains of *Lc. lactis* subsp. *lactis* and *Lc. lactis* subsp. *cremoris* (80%–90%), *Lc. lactis* subsp *lactis* biovar. *diacetylactis* and *Leuc. mesenteroides.*

Rennet: The milk is coagulated by calf rennet at 30°C for about 35 min.

Cutting: The curd is cut in cubes with sides of 10–12 mm.

Curd washing/cooking: The curd is stirred for 30 min at 30°C, 25% of the whey is removed and the vat is heated to 40°C and stirred for 25 min; thereafter about 20% water is added to the vat, and stirring continues for another 25 min.

Salting: Salt may be added to the vat at a concentration of 0.3 % (w/w).

Moulding/pressing: The curd grains are separated from the whey and moulded with air between the grains, which gives the characteristic open texture. The cheeses are pressed at 20°C for 20 hr.

Salting: Svecia is salted in brine for three days at 15°C to a salt content of 1.0%–1.5%.

Maturation: The cheese is ripened in dry storage at 16°C for four weeks and then at 12°C for at least another month and up to about a year. The cheeses are waxed about 18 hr after salting and several times during ripening.

3.17.6 Relevant Research

Lactose is converted to lactic acid and citrate to carbon dioxide, diacetyl and acetoin, which contributes to the characteristic open texture and flavour of Svecia. About 30% of the intact casein is hydrolysed in a three-month-old cheese. Proteolysis comprises activity on α_{S1}-casein by rennet and β-casein by plasmin. The peptides formed are further broken down by cell-bound proteases of starter bacteria, and amino acids are released by the intracellular peptidases of starter bacteria. The elastic fresh cheese ripens and softens during the first weeks as a result of proteolysis and rearrangement of calcium phosphate. The starter bacteria dominate the cheese microflora during the short ripening time that most Svecia cheeses undergo (Ardö, 1993a).

3.18 Serpa – Portugal

Name: Serpa	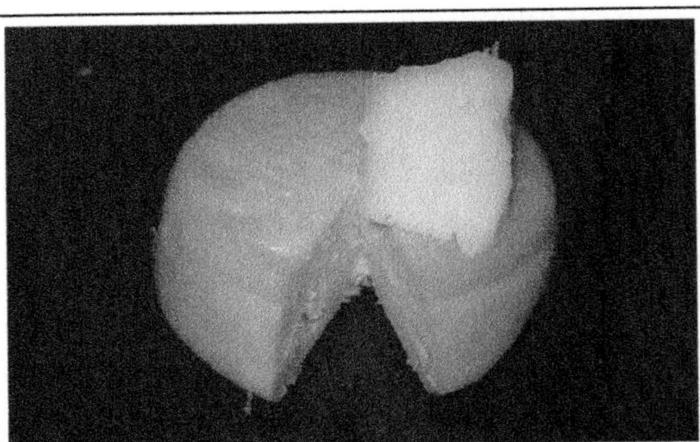
Production area: Alentejo – Southeast of Portugal	
Milk: Sheep's, raw	

3.18.1 Introduction

Serpa is one of the most important and appreciated of Portuguese cheeses. Its production, from the economic point of view, is crucial for local agricultural development, with a sales boom in winter time.

Serpa cheese is produced in the southeastern region of Portugal (Serpa), on the left side of the river Guadiana, and its artisanal production dates back to time immemorial. As the ingredients and manufacturing process are similar to those of Serra da Estrela cheese, it is believed that it was introduced to Alentejo by Serra da Estrela shepherds during their immigration periods. The know-how of traditional cheese manufacture was passed from generation to generation, but remained within the same families for a long time. Cheese milk is produced at the farmhouse level by flocks of 1000–2000 Merino sheep.

Serpa cheese has enjoyed PDO status since 1996 (EC, 1996a). Serpa cheese can only be manufactured from whole raw sheep's milk from Serpa region farms, using an aqueous extract of dried thistle flowers (*Cynara cardunculus L.*) as rennet. Criteria for certification cover compliance with minimum requirements of organoleptic characteristics, shape, chemical and microbial contents of the final product, as well as hygiene during processing. In regard to the region of production, the PDO region of Serpa cheese has been established via a governmental act, and comprises a specific area in Alentejo – the biggest and driest province of inland Portugal, in Beja district (Aljustrel, Almodôvar, Alvito, Beja, Castro Verde, Cuba, Ferreira do Alentejo, Mértola, Moura, Ourique, Serpa, Vidigueira, and Colos and Vale de Santiago from Odemira town); as well as Setúbal district (São Domingos, Alvalade and Abela from Santiago do Cacém town, Azinheira dos Barros and São Mamede do Sádão from Grândola town and Torrão from Alcácer do Sal town) (Anonymous, 1987).

3.18.2 Milk and Rennet

Serpa cheese is traditionally manufactured with raw sheep's milk from the local Merino ovine breed. Most milk producers have gradually been replacing it by dairy breeds improved elsewhere (e.g. Serra da Estrela and Lacaune). The milk quality depends strongly on the seasonal nature of sheep milk production, due to the higher temperatures reached in the Alentejo region, thus limiting the time for production of the cheese from October to June.

The rennet used in Serpa cheese production is the same as that used in Serra da Estrela cheese (see Part II, Section 4.8).

3.18.3 Description and Sensory Characteristics

Serpa PDO has a flat cylindrical shape and is produced in four sizes: (1) weight 200–250 g, diameter 10–12 cm and height 3 – 4 cm; (2) weight 800–900 g, diameter 15–18 cm and height 4–5 cm; (3) weight 1,000–1,500 g, diameter 18–20 cm and height 4–6 cm or (4) weight 2,000–2,500 g, diameter 25–30 cm and height 6–8 cm. Serpa has a ripening period of at least 30 days to ensure food safety. The moisture content ranges from 36% to 41 % and the fat content from 29% to 34%. Serpa cheese is a buttery, full-fat and semi-soft cheese with few or no holes and has a strong, slightly hot and spicy flavour (Anonymous, 1999). This cheese is somewhat similar to Serra da Estrela cheese, but possesses a stronger flavour and aged cheese is rather hot and spicy.

3.18.4 Method of Manufacture

Serpa cheese is made at the farmhouse level, with raw sheep's milk obtained by the cheesemaker on the same day of milking (twice a day). Cheese originates from a slow curd syneresis, after coagulation with the plant rennet. Manufacturing techniques are very similar to those employed with Serra da Estrela cheese (see Part II, Section 4.8).

Milk preparation: Milk is heated, and after it reaches the desired coagulation temperature (28°C–30°C) in a water bath, it is filtered through folded white wool blankets.

Coagulation: The coagulation step comprises preparation of the plant rennet (*Cynara cardunculus L.* flower) infusion. Dry flowers (0.3–0.6 g/L of milk) are soaked and macerated in cold water, and then filtered through a cloth. Half the total salt content (about 9 g/L milk) is added to the milk at this stage, and half is added to the curd, just before moulding. Milk is kept at the coagulation temperature for 1–2 hr, and the end of coagulation is checked by slight agitation in order to evaluate curd consistency.

Cutting: After he intended consistency is reached by the end of coagulation, the coagulum is broken down into a smooth granular paste by vigorous stirring.

Moulding/draining of whey: Small portions of the coagulum are placed in a large mould on a working table. The mould is made of a flexible wooden strip that can be adjusted as the curd becomes drier due to whey draining. This working process continues until the curd is sufficiently dry, and the curd is then distributed into small round open-ended stainless steel moulds.

Pressing: The working and hand pressing process of the curd continues in the stainless steel mould for a while, where it is left to rest for about 1 hr. Afterwards, the moulds are turned over and left for a further 1 hr.

Maturation: After pressing, the cheeses are ripened (30–45 days) in two successive stages: two weeks of ripening in a controlled temperature (8°C–9°C) and RH (92%–97%) room, followed by a second ripening period, at 10°C–13°C and 85%–90% RH. On the second maturation day, the cheeses are demoulded and turned every day for one week. A white cotton bandage is put around the side of the cheeses to prevent breakage of their fine rind. Those two ripening stages lead to a pH reduction due to lactic acid production, and improve whey syneresis, inhibit pathogenic bacteria growth and support development of several microbiological groups that affect cheese aroma and flavour. It is important to ensure regular cheese washing with warm diluted whey or tap water in order to control the natural microflora growth on the surface.

3.18.5 Relevant Research

Physico-chemical, biochemical and microbiological studies were developed by a variety of researchers (Alvarenga et al., 2008; Alvarenga, Canada & Sousa, 2011; Freitas, Macedo & Malcata, 2000; Freitas & Malcata 2000; Roseiro, Wilbey & Barbosa, 2003) in order to evaluate different breed milks, lactation period, effects of ripening, milk status (refrigerated versus non-refrigerated) and cheese location. Although random oscillations are observed during ripening time, the use of refrigerated versus non-refrigerated milk shows that cheese hardness and adhesiveness is higher when refrigerated milk is used – exhibiting the highest values for 28-day-old cheeses, but with a tendency to decrease throughout ripening (Alvarenga et al., 2008). Microbial counts during ripening taken in two distinct seasons (April to May and June to July) do not show significant differences.

Recent studies have attempted to improve dairies, with mechanisation of some cheesemaking steps that clearly improve hygiene and safety. As freezing may be suitable to extend the stability and shelf life of the cheese, some studies have been performed to assess the effect of freezing on physic-chemical Serpa cheese properties at distinct ripening stages (Alvarenga, Canada & Sousa, 2011).

3.19 Sombor Cheese – Serbia

Name: Sombor	
Production area: Northern part of Serbia (around the city of Sombor)	
Milk: Cow's, raw or thermised	

3.19.1 Introduction

Sombor cheese is traditionally produced from raw sheep's milk in the northern part of Serbia (around the city of Sombor) and highly appreciated for its unique flavour and shape. It could be classified as a cheese between soft and semi-hard cheeses (low acidity due to curd washing).

Sombor cheese was certified with a sign of appellation at the national level in 2012 (The Intellectual Property Office, 2013). Nowadays, the cheese is produced in a few households and small craft dairy plants.

The earliest historical record about Sombor cheese dates back to 1748, with a description of its position on the market of Novi Sad city. In the nineteenth century, Sombor tradesmen took Sombor cheese to green markets of all big cities of the Austro-Hungarian Empire, and it was consumed even in the Viennese court. During the 1940s, the production started to slowly die out, and so it completely vanished from the region of Serbia by the end of the twentieth century. During the 1970s, this cheese was still sold in green markets in Novi Sad.

The production of Sombor cheese, adapted to industrial technology, started in 1949 in the 'Somboled' Dairy plant, but it has no longer been produced in this company for the past 20 years. Today, industrial production of these cheeses does not occur, but they are still produced by a few small craft manufacturers, from sheep's and cow's milk.

The cheese from Sombor is produced in the territory of the municipality of the town of Sombor, Republic of Serbia (the northern part of the country).

3.19.2 Type

The moisture content of Sombor cheese is usually within 44%–50% for the upper part and 52%–58% for the soft, inner part of the cheese. The salt content is between 1% and 1.5%. The production yield is about 14%–16%. On the basis of the fat content in dry matter and moisture content in non-fat cheese matter, Sombor cheese can be classified as a full-fat and soft cheese, respectively. The average composition of a typical Sombor cheese is as follows: moisture

50.2%–59%, FDM 45.3%–59.3%, total protein 14.5%–22.7% and MNFS 68.3%–77.2% (Mijačević & Bulajić, 2008).

3.19.3 Milk

Traditionally, raw sheep's milk was used for the production of Sombor cheese. Currently, this cheese type is usually produced from low-heat-treated cow's milk.

3.19.4 Description and Sensory Characteristics

Due to the specific shape and ripening, the lower and upper cheese parts are characterised with different composition and sensory properties. Ripening of one part of the cheese is done inside moulds (approximately 2/3) while the other part is outside (1/3). The upper part has a strong rind yellow colour and a semi-hard texture, while the lower part is characterised by a soft texture, spreadable consistency, higher acidity and white colour (similar to brined cheeses). The specific taste and flavour are the result of complex biochemical changes during ripening mostly due to autochthonous microflora activity.

3.19.5 Method of Manufacture

Milk preparation: For the production of Sombor cheese, raw cow's milk is used. Sombor cheese production is characterised by two specific processes: the addition of water to milk (15%–20%) and the washing of cheese curd in order to prevent extensive acidity development.

Starter culture: In traditional production of Sombor cheese, starter cultures are not used, while in industrial production, mesophilic LAB are used.

Rennet: Addition of calf's rennet occurs at 31°C, and coagulation lasts about 30 min.

Cutting: Rennet-coagulated gel is cut into 4 cm^2 grain size, left to rest for 5–10 min and cut again to the smaller-size cheese grains.

Curd washing and draining: After whey draining, the cheese mass formed is pressed for 20 min (1 kg/kg cheese mass). Then, the cheese block is cut into smaller pieces and placed in water at 30°C–35°C for 20–30 min in order to wash the curd and prevent extensive acidity development.

Salting: Cheese blocks are usually dry-salted and put together in special wooden moulds/buckets ('kačica') where ripening takes place. During pressing, the moulds are upside down, and the next day the upper part is formed. This part is formed by adding a cheeses mass on the top of the moulds (approximately 10–15 cm above), forming a specific shape resembling a mushroom. The upper part is wrapped in cloth.

Maturation: Sombor cheese ripens at 15°C–20°C and 80% RH for 20–30 days. During ripening, the cheesecloths are washed and/or changed in order to prevent mould development.

Storage: Sombor cheese is stored at low temperatures (6°C–8°C).

3.19.6 Relevant Research

The diversity of NSLAB in Sombor cheese was studied. Dominant microflora of Sombor cheese consists of lactococci, lactobacilli and enterococci strains (Mijačević & Bulajić, 2008). The minority of LAB strains isolated from Sombor cheese showed antibiotic resistance to tested antibiotics. However, this small fraction justifies performing the antibiotic susceptibility testing to avoid the microbiological hazards linked with horizontal transfer of antibiotic resistance genes (Bulajić & Mijačević, 2011).

3.20 Tuma Persa PDO – Italy

Name: Tuma Persa	
Production area: Castronovo of Sicily	
Milk: Cow's, thermised	

3.20.1 Introduction

The Tuma Persa cheese is unique for its history and for the cheesemaking technology used. It is produced only by one cheesemaker from Castronovo of Sicily (a small town between Palermo and Agrigento) who revived an old dairy recipe reported in a publication of 1936 Livestock Dairy Institute of Piedmont.

In the eighteenth century, the Tuma Persa was produced in many areas of Sicily, although with names that changed depending on the province in which it was produced: in the province of Palermo it was called 'cheese buffalo' (although it was never made with buffalo's milk, but from cow's and sheep's milk), in the province of Syracuse 'cheese turc' and in the province of Catania 'toma lost' (Romagnoli, 1936).

About the origin of the product and its name, several legends are known, but the most reliable seems the reports of a cheesemaker, who, after producing a classic toma cheese (a pressed cheese to obtain the canestrato cheese), forgot about (or lost) one or a few pieces in a dark corner of the aging room (historically these were places with thick walls, humid and sometimes underground, reminiscent of the French caves). In Italian, the word 'persa' means 'lost', hence the current name 'Tuma Persa'.

After a few weeks, the cheesemaker went back to the aging room and noticed that these pieces, the 'lost' tuma, were covered with mould. After a rough dusting of the surfaces, he vowed to knock them as soon as possible.

After a few weeks, when he went to throw them away, he found that the cheeses were still full of mould. Before throwing them away, he decide to open a form to see what had happened to those cheeses; he was surprised to find the dough inside the cheeses in excellent condition, and from which emerged a smell good; thus, he felt the need to taste a piece. The flavour was excellent enough to make him decide to repeat what had happened accidentally.

As result of the above experience, he began using this 'serendipitous' technology for the production of other cheeses, which achieved great popularity among consumers.

A Sicilian producer, in the early 1990s, having learned, from the literature, of this particular historical cheese, started experiments to reproduce it, and to get the product known today as 'Tuma Persa'.

3.20.2 Type

Tuma Persa is a pressed semi-hard cheese. The moisture content is 34.9%, 33.9% and 35.2% at 4, 8 and 11 months.

3.20.3 Milk

Thermised whole cow's milk is used. The production system most widespread is typically extensive, where the cows are feed mainly on pastures of hilly and mountainous areas, integrated with concentrates and by-products to balance the intake of various nutrients in the diet.

3.20.4 Description and Sensory Characteristics

The final product has a cylindrical shape with a heel 10–12 cm and a diameter of 28–30 cm, with a flat surface and a weight of 7–8 kg. The rind of the cheese is yellow-ochre. It darkens after the application of 'curatina', (i.e. 'cappatura' performed with olive oil and pepper). The body is white/yellow-pale, tender and compact and tends to crumble. It rarely has small holes, which signifies slight abnormal fermentation. The taste of the cheese ranges from sweet to spicy, no saltiness or bitterness, the aftertaste has a very special, long and aromatic persistency due to an excellent combination of flavours that recalls the double natural fermentations, and the mould formed.

Sensory analysis shows an intense flavour profile, characterised by pungent smells and hints of fungi, moulds and wood with a more mellow and smooth taste with greater maturity. The aromatic balance/sensory/taste of the mature product is considered unique and rare in Sicilian medium-long ripening products.

3.20.5 Method of Manufacture

Milk preparation: To produce the Tuma Persa, thermised whole cow's milk, with a starter culture (milk culture) is used.

Coagulation: The evening and morning milk coagulates in a vat at 35°C–36°C, with the addition of a paste of lamb rennet, or at some time of the year, kid rennet, and the clotting time is approximately 50 min.

Cutting: The coagulum is broken down using a tool called 'lira' until the size of a wheat grain is reached.

Moulding: After a few minutes, the caseous mass, which has fallen to the bottom, is picked up by hand, placed on a board, cut into rough pieces and, then placed in a mould 'fascere', in which it is pressed manually to further drain the whey and to give a particular shape to the surface of the cheese.

Scalding: When the mass has settled in the mould, it begins the scalding process, with hot whey (78°C–80°C) coming from the manufacture of Ricotta. The curd is left in the hot whey until it reaches 45°C in the centre of the cheese, usually for about 2–3 hr.

Drainage/pressing: After scalding, the curd is placed for 24–48 hr on a tilted board called the 'tavoliere'. The cheeses are repeatedly rotated inside the mould so that they receive the specific cylindrical shape with a flat surface.

Pre-maturation: The uniqueness of Tuma Persa is related to the early stages of maturation, as well as to the fact that (as the legend tells), once set in shape, it is not touched for about a week (first natural fermentation with development of external moulds), then washed again and allowed to rest for another 10 days of fermentation (the second fermentation). Only after three weeks is it ready to be cleaned from the mould and finally salted.

Salting: The cheese is salted in brine for a few days with the objective of not exceeding 2% salt in the final product. Rarely, and only if necessary, is a subsequent dry salting done on wooden shelves where the cheeses are regularly turned and checked.

Maturation: The aging rooms are of limestone with very thick walls and partially underground. The average maturation time is about eight months.

References

Abellán, A., Cayuela, J. M., Pino, A., Martínez-Cacha, A., Salazar, E. & Tejada, L. (2012). Free amino acid content of goat's milk cheese made with animal rennet and plant coagulant. *Journal of the Science of Food and Agriculture*, 92, 1657–1664.

Abellán, A., Tejada, L., Cayuela, J. M. & Martínez-Cacha, A. (2007). Comparison of various methods for determination of calorific value of Murcia al Vino cheese. *Effect of cheese size. Alimentaria*, 382, 97–99.

Alvarenga, N., Canada, J. & Sousa, I. (2011). Effect of freezing on the rheological, chemical and colour properties of Serpa cheese. *Journal of Dairy Research*, 78, 80–87.

Alvarenga, N., Silva, P., García, J. R. & Sousa, I. (2008). Estimation of Serpa cheese ripening time using multiple linear regression (MLR) considering rheological, physical and chemical data. *Journal of Dairy Research*, 75, 233–239.

Alvarez, S. & Fresno, M. (2004). Effect of the feed of Majorero goats on the cheese-makingprocess employing two types of coagulant. *Alimentación Equipos y Tecnología*, 23, 44–50.

Alvarez, S., Fresno, M., Mendez, P., Castro, N., Fernandez, J. R. & Sanz-Sampelayo, M. R. (2007). Alternatives for improving physical, chemical and sensory characteristics of goat cheeses: the use of arid-land forages in the diet. *Journal of Dairy Science*, 90, 2181–2188.

Anonymous (1987). *Diário da República*. I-Série. N° 1146 de 29-6-1987. Decreto Regulamentar N° 39/87. Imprensa Nacional-Casa da Moeda, Lisbon, Portugal.

Anonymous (1999). DGDR, Produtos Agrícolas e Géneros Alimentícios com Denominação de Origem Protegida, com Indicação Geográfica Protegida ou com Certificado de Especialidade, Divisão da Promoção de Produtos de Qualidade, 26ª Edição, Ministério da Agricultura, do Desenvolvimento Rural e das Pescas, Lisboa, Portugal, 1–21.

Anonymous (2015a). http://www.quesomahonmenorca.com/ (Accessed 16 November 2015).

Anonymous (2015b). http://www.quesomajorero.es. (Accessed 16 November 2015).

Anonymous (2015c). http://www.quesosdemurcia.com/(Accessed 13 May 2015).

Anonymous (2015d). http://www.sansimondacosta.com. (Accessed 15 June 2015).

Arango, O., Trujillo, A. J. & Castillo, M. (2015). Predicting coagulation and syneresis parameters of milk gels when inulin is added as fat substitute using infrared light backscatter. *Journal of Food Engineering*, 157, 63–69.

Ardö, Y. & Gripon, J.-C. (1995). Comparative study of peptidolysis in some semi-hard round-eyed cheese varieties with different fat content. *Journal of Dairy Research*, 62, 543–547.

Ardö, Y. (1993a). Swedish cheese varieties. In Fox, P. F. (ed.), *Cheese: Chemistry, Physics and Microbiology. Vol 2. Major Cheese Groups*, 2nd edition. Chapman & Hall, London, pp. 254–257.

Ardö, Y. (1993b). Characterizing ripening in low-fat, semi-hard round-eyed cheese made with undefined mesophilic DL-starter. *International Dairy Journal*, 3, 343–357.

Ardö, Y., Thage, B. V. & Madsen, J. S. (2002). Dynamics of free amino acid composition in cheese ripening. *Australian Journal of Dairy Technology*, 57, 109–115.

Avellini, P., Clementi, F., Marinucci, M. T., Goga, B. C., Rea, S., Brancari, E., Cavallucci, C., Reali, C. & Di Antonio, E. (1999). Formaggio di fossa: Caratteristiche compositive, microbiologiche e sensoriali | ["Pit" cheese: Compositional, microbiological and sensory characteristics]. *Italian Journal of Food Science*, 11 (4), 317–333.

Barbieri, E., Schiavano, G. F., De Santi, M., Valloni, L., Casadei, L., Guescini, M., Gioacchini, M., Rinaldi, L., Stocchi, V. & Brandi, G. (2012). Bacterial diversity of traditional Fossa (pit) cheese and its ripening environment. *International Dairy Journal*, 23, 62–67.

Benedito, J., Carcel, J., Clemente, G. & Mulet, A. (2000). Cheese maturity assessment using ultrasonics. *Journal of Dairy Science*, 83, 248–254.

Benedito, J., Conde, T., Clemente, G. & Mulet, A. (2006a). Use of the acoustic impulse-response technique for the nondestructive assessment of Manchego cheese texture. *Journal of Dairy Science*, 89, 4490–4502.

Benedito, J., Simal, S., Clemente, G. & Mulet, A. (2006b). Manchego cheese texture evaluation by ultrasonics and surface probes. *International Dairy Journal*, 16, 431–438.

Berthier, F., Beuvier, E., Dasen, A. & Grappin, R. (2001). Origin and diversity of mesophilic lactobacilli in Comte cheese, as revealed by PCR with repetitive and species-specific primers. *International Dairy Journal*, 11, 293–305.

Bertolino, M., Dolci, P., Giordano, M., Rolle, L. & Zeppa, G. (2011). Evolution of chemico-physical characteristics during manufacture and ripening of Castelmagno PDO cheese in winter time. *Food Chemistry*, 129, 1001–1011.

Boutoial, K., Alcantara, Y., Rovira, S., García, V., Ferrandini, E., Abdelkhalek, O. & López, M. B. (2013a). Influence of ripening on proteolysis and lipolysis of Murcia al Vino cheese. *International Journal of Dairy Technology*, 66, 366–372.

Boutoial, K., Ferrandini, E., Rovira, S., García, V. & López, M. B. (2013b). Effect of feeding goats with rosemary (*Rosmarinus officinalis* spp.) by-product on milk and cheese properties. *Small Ruminant Research*, 112, 147–153.

Boutoial, K., García, V., Rovira, S., Ferrandini, E., Abdelkhalek, O. & López, M. (2013c). Effect of feeding goats with distilled and non-distilled thyme leaves (*Thymus zygis* subsp. *gracilis*) on milk and cheese properties. *Journal of Dairy Research*, 80, 448–456.

Bouton, Y., Guyot, P., Beuvier, E., Tailliez, P. & Grappin, R. (2002). Use of PCR-based methods and PFGE for typing and monitoring homofermentative lactobacilli during Comte cheese ripening. *International Journal of Food Microbiology*, 76, 27–38.

Bulajić, S. & Mijačević, Z. (2011). Antimicrobial susceptibility of lactic acid bacteria isolated from Sombor cheese. *Acta Veterinaria (Belgrade)*, 61 (2–3), 247–258.

Calvo, M. V., Castillo, I., Diaz-Barcos, V., Requena, T. & Fontecha, J. (2007). Effect of a hygienized rennet paste and a defined strain starter on proteolysis, texture and sensory properties of semi-hard goat cheese. *Food Chemistry*, 102, 917–924.

Castillo, I., Calvo, M. V., Alonso, L., Juárez, M. & Fontecha, J. (2007). Changes in lipolysis and volatile fraction of a goat cheese manufactured employing a hygienized rennet paste and a defined strain starter. *Food Chemistry*, 100, 590–598.

Centeno, J. A., Cepeda, A. & Rodriguez-Otero, J. L. (1996). Lactic acid bacteria isolated from Arzúa cow's milk cheese. *International Dairy Journal*, 6, 65–78.

Centeno, J. A., Rodríguez-Otero, J. L. & Cepeda, A. (1994a). Changes in the protein profile of Arzúa cheese (NW Spain) during ripening. *Milchwissenschaft*, 49, 319–322.

Centeno, J. A., Rodríguez-Otero, J. L. & Cepeda, A. (1994b). Microbiological study of Arzúa cheese (NW Spain) throughout cheese-making and ripening. *Journal of Food Safety*, 14, 229–241.

Charlet, M., Duboz, G., Faurie, F., Le Quere, J. L. & Berthier, F. (2009). Multiple interactions between Streptococcus thermophilus, Lactobacillus helveticus and Lactobacillus delbrueckii strongly affect their growth kinetics during the making of hard cooked cheeses. *International Journal of Food Microbiology*, 131, 10–19.

Codex Alimentarius (1978). General Standard for Cheese, CODEX STAN 283-1978. Food and Agriculture Organization (FAO), Rome, Italy.

Conde, T., Cárcel, J. A., García-Pérez, J. V. & Benedito, J. (2007). Non-destructive analysis of Manchego cheese texture using impact force–deformation and acoustic impulse–response techniques. *Journal of Food Engineering*, 82, 238–245.

Cyprus State Laboratory (2013). *Cyprus Food Composition Tables*. Press and Information Office (PIO), Nicosia, Cyprus

Danish Food Administration (2013). *Bekendtgørelse om mælkeprodukter m.v.* [Announcement about dairy products etc.]. Notice No 2 of 04/01/2013.

Dasen, A. (2013). *Du Gruyère au Comté, deux siècles d'histoire'* [Online]. Self-published. Available: http://www.comte.com/sinformer/vie-de-la-filiere/lire-larticle/actualite/lecture-du-gruyere-au-comte-deux-siecles-dhistoire.html [Accessed 31 August 2016].

Depouilly, A., Dufrene, F., Beuvier, E. & Berthier, F. (2004). Genotypic characterisation of the dynamics of the lactic acid bacterial population of Comte cheese. *Lait*, 84, 155–167.

Dolci, P., Alessandria, V., Rantsiou, K., Bertolino, M. & Cocolin, L. (2010). Microbial diversity, dynamics and activity throughout manufacturing and ripening of Castelmagno PDO cheese. *Food Microbiology*, 143, 71–75.

Dolci, P., Alessandria, V., Rantsiou, K., Rolle, L., Zeppa, G. & Cocolin, L. (2008). Microbial dynamics of Castelmagno PDO, a traditional Italian cheese, with a focus on lactic acid bacteria ecology. *International Journal of Food Microbiology*, 122, 302–311.

EC (1996a). Commission Regulation (EC) No 1107/96 of 12 June 1996 on the registration of geographical indications and designations of origin under the procedure laid down in Article 17 of Council Regulation (EEC) No 2081/92. *Official Journal of the European Union*, 148, 1–10.

EC (1996b). Commission Regulation (EC) No 1263/96 of 1 July 1996 supplementing the Annex to Regulation (EC) No 1107/96 on the registration of geographical indications and designations of origin under the procedure laid down in Article 17 of Regulation (EEC) No 2081/92. *Official Journal of the European Union, L* 163, 19–21.

EC (1997). Commission Regulation (EC) No 2325/97 of 24 November 1997 supplementing the Annex to Regulation (EC) No 1107/96 on the registration of geographical indications and designations of origin under the procedure laid down in Article 17 of Council Regulation (EEC) No 2081/92. *Official Journal of the European Union, L* 322, 33–35.

EC (1999). Commission Regulation (EC) No. 378/1999 of 19 February 1999 supplementing the Annex to Regulation (EC) No. 2400/96 on the entry of certain names in the "Register of protected designations of origin and protected geographical indications" provided for in Council Regulation (EEC) No. 2081/92 on the protection of geographical indications and designations of origin for agricultural products and foodstuffs. *Official Journal of the European Communities*, L46, 13–14.

EC (2002). Commission Regulation (EC) No 510/2006 of 24 June 2002 supplementing the Annex to Regulation (EC) No 2400/96 on the entry of certain names in the Register of protected designations of origin and protected geographical indications and designations of origin for agricultural products and foodstuffs (Queso de Murcia al Vino-Queso de Murcia). *Official Journal of the European Union*, 166, 8–9.

EC (2008). Commission Regulation (EC) No 1229/2008 of 10 December 2008 registering certain names in the Register of protected designations of origin and protected geographical indications [San Simón da Costa (PDO), Ail blanc de Lomagne (PGI), Steirischer Kren (PGI)]. *Official Journal of the European Union*, L 333, 3–4.

EC (2009). Commission Regulation (EC) No 1183/2009 of 30 November 2009 entering a name in the register of protected designations of origin and protected geographical indications [Formaggio di Fossa di Sogliano (PDO)]. *Official Journal of the European Union, L* 317, 34–35.

EC (2010). Commission Regulation (EC) No 20/2010 of 12 January 2010 entering a name in the register of protected designations of origin and protected geographical indications "Arzúa-Ulloa PDO". *Official Journal of the European Union*, L 20, 8–9.

EC (2012a). Regulation (EU) No 129/2012 of 13 February 2012 approving minor amendments to the specification for a name entered in the register of protected designations of origin and protected geographical indications (Queso Manchego (PDO). *Official Journal of the European Union*, L 172, 6–8.

EC (2013). Publication of an amendment application pursuant to Article 50(2)(a) of Regulation (EU) No 1151/2012 of the European Parliament and of the Council on quality schemes for agricultural products and foodstuffs. *Official Journal of the European Union*, 329, 4–15.

Economides, S. (2004). *Milk and Dairy Products of Cyprus*. Cyprus Milk Industry Organisation, Nicosia, Cyprus.

Ferrandini, E., Castillo, M., de Renobales, M., Virto, M. D., Garrido, M. D., Rovira, S. & López, M. B. (2012). Influence of lamb rennet paste on the lipolytic and sensory profile of Murcia al Vino cheese. *Journal of Dairy Science*, 95, 2788–2796.

Ferrandini, E., López, M. B, Castillo, M. & Laencina, J. (2011). Influence of artisanal lamb rennet paste on proteolysis and texture profile of Murcia al Vino cheese. *Food Chemistry*, 124, 583–588.

FOAG (2015). *Register of Appellation of Origin and Geographical Indications*. Federal Office for Agriculture. An Office of the Federal Department of Economy, Bern, Switzerland http://www.blw.admin.ch/ [Accessed 28 April 2015].

Fontecha, J., Mayo, I., Toledano, G. & Juárez, M. (2006). Triacylglycerol composition of protected designation of origin cheeses during ripening. Authenticity of milk fat. *Journal of Dairy Science*, 89, 882–887.

Foodle.ch (2014). Bern, Switzerland. http://www.foodle.ch/de/raclette-kaese [Accessed 25 January 2016]

Frau, M., Massanet, J., Rosselló, C., Simal, S. & Cañellas, J. (1997). Evolution of free amino acid content during ripening of Mahon cheese. *Food Chemistry*, 60, 651–657.

Freitas, A. C., Macedo, A. C. & Malcata, F. X. (2000) Review: Technological and organoleptic issues pertaining to cheeses with denomination of origin manufactures in the Iberian Peninsula from ovine and caprine milks. *Food Science and Technology International*, 6, 351–370.

Freitas, C. & Malcata, F. X. (2000). Microbiology and biochemistry of cheeses with *Appellation d'Origine Protégée* and manufactured in the Iberian peninsula from ovine and caprine milks. *Journal of Dairy Science*, 83, 584–602.

Fresno, M. & Alvarez, S. (2012). Chemical, textural and sensorial changes during the ripening of Majorero goat cheese. *International Journal of Dairy Technology*, 65, 393–400.

Fresno, M. Alvarez, S., Virto, M. & Renobales, M.de. (2014). Short communication: Sensory profile of raw goat milk cheeses made with artisan kid rennet pastes from commercial-weight animals: Alternative to farmhouse goat cheeses. *Journal of Dairy Science*, 97, 6111–6115.

Fröhlich-Wyder, M. T., Bütikofer, U., Guggisberg, D. & Wechsler, D. (2007). Calcium dans le fromage à raclette: Importance du pH et de la formation de complexes. *Revue suisse Agriculture*, 39 (4), 239–243.

Fröhlich-Wyder, M. T., Guggisberg, D. & Wechsler, D. (2009). Influence of low calcium and low pH on melting characteristics of model Raclette cheese. *Dairy Science & Technology*, 89, 463–483.

Garabal, J. I. (2007). Biodiversity and the survival of autochthonous fermented products. *International Microbiology*, 10, 1–3.

Garabal, J. I., Rodríguez-Alonso, P. & Centeno, J. A. (2008). Preliminary characterization of lactic acid bacteria isolated from raw cows' milk cheeses currently produced in Galicia (NW Spain). *LWT – Food Science and Technology*, 41, 1452–1458.

Garabal, J. I., Rodríguez-Alonso, I. P., Franco, D. & Centeno, J. A. (2010). Chemical and biochemical study of industrially produced San Simón da Costa smoked semi-hard cow's milk cheeses: Effects of storage under vacuum and different modified atmospheres. *Journal of Dairy Science*, 93, 1868–1881.

García, M. C., Franco, I., Prieto, B., Tornadijo, M. E. & Carballo, J. (2001). Microbiological changes in San Simón cheese throughout ripening and its relationship with physico-chemical parameters. *Food Microbiology*, 18, 25–33.

García-Palmer, F. J., Serra, N., Palou, A. & Gianotti, M. (1997). Free amino acids as indices of Mahón cheese ripening. *Journal of Dairy Science*, 80, 1908–1917.

Garde, S., Arias, R., Gaya, P. & Nuñez, M. (2011). Occurrence of *Clostridium* spp. in ovine milk and Manchego cheese with late blowing defect: Identification and characterization of isolates. *International Dairy Journal*, 21, 272–278.

Garde, S., Ávila, M., Gaya, P., Arias, R. & Nuñez, M. (2012). Sugars and organic acids in raw and pasteurized milk Manchego cheeses with different degrees of late blowing defect. *International Dairy Journal*, 25, 87–91.

Gobbetti, M., Folkertsma, B., Fox, P. F., Corsetti, A., Smacchi, E., De Angelis, M., Rossi, J., Kilcawley, K. & Cortini, M. (1999). Microbiology and biochemistry of Fossa (pit) cheese. *International Dairy Journal*, 9, 763–773.

Gómez-Ruiz, J. A., Taborda, G., Amigo, L., Ramos, M. & Molina, E. (2007). Sensory and mass spectrometric analysis of the peptidic fraction lower than one thousand daltons in Manchego cheese. *Journal of Dairy Science*, 90, 4966–4973.

González, L., Fernández Cuadrillero, A., Castro, J. M., Bernardo, A. & Tornadijo, M. E. (2015). Selection of lactic acid bacteria isolated from San Simón da Costa Cheese (PDO) in order to develop an autochthonous starter culture. *Advances in Microbiology*, 5, 748–759.

González-García, S., Hospido, A., Moreira, M. T., Feijoo, G. & Arroja, L. (2013). Environmental life cycle assessment of a Galician cheese: San Simon da Costa. *Journal of Cleaner Production*, 52, 253–262.

INAO-CNAOL (2014). Produits laitiers AOP – les chiffres clés, 2014. [Online]. Available: https://www.google.fr/search?q=produits+laitiers+AOP++2014&ie=utf-8&oe=utf-8&client=firefox-b&gfe_rd=cr&ei=iI3jV6OgJ8nDaKWRn8AO [Accessed 16 July 2016].

Jeanson, S. (2000). La maturation du lait dans la fabrication de fromages à pâte pressée cuite: le rôle des lactocoques'. PhD thesis, Université de Bourgogne, France.

Jeanson, S., Berthier, F., Grappin, R. & Beuvier, E. (2003). Heat resistance of wild Lactococcus lactis strains under a thermal gradient of cooked cheese, in milk and in mini-cheeses. *Lait*, 83, 379–396.

Kulinarisches Erbe der Schweiz. (2008). Lausanne, Switzerland. http://www.kulinarischeserbe.ch/ [Accessed 28 April 2015].

López, M. B., Ferrandini, E., Rodriguez, M., Roca, J. D., Haba, E, Luna, A. & Rovira, S. (2012). Physicochemical study of Murcia al Vino cheese, *Small Ruminant Research*, 106, 154–159.

Luna, P., Juárez, M. & De la Fuente, M. A. (2007). Conjugated linoleic acid content and isomer distribution during ripening in three varieties of cheeses protected with designation of origin. *Food Chemistry*, 103, 1465–1472.

Maas, S., Lucot, E., Gimbert, F., Crini, N. & Badot, P. M. (2011). Trace metals in raw cows' milk and assessment of transfer to Comte cheese. *Food Chemistry*, 129, 7–12.

Massa, S. Turtura, G. & Trovatelli, L. D. (1988). Qualité hygiénique du fromage de " fosse" de Sogliano al Rubicone (Italie). *Lait*, 68, 323–326.

Melo, A. (2015). "Fruitieres comtoises" (Franche-Comte, France) From Village Associations to an Agro-Industrial District: a Mountain Story? *Revue De Geographie Alpine-Journal of Alpine Research*, 103.

Menéndez, S., Centeno, J. A., Godínez, R. & Rodríguez-otero, J. L. (2000). Effects of Lactobacillus strains on the ripening and organoleptic characteristics of Arzúa-Ulloa cheese. *International Journal of Food Microbiology*, 59, 37–46.

Mijačević, Z. & Bulajić, S. (2008). Sensory evaluation and microbiological characterization of autochthonous Sombor cheese. *Acta Veterinaria (Belgrade)*, 58 (5–6), 531–541.

Ministry of Agriculture, Food and Environment (2014). Secretaría General Técnica Centro de Publicaciones. Datos de las Denominaciones de Origen Protegidas (D.O.P.) e Indicaciones Geográficas Protegidas (I.G.P.) de Productos Agroalimentarios 3, 22–23.

Ministry of Agriculture, Food and Environment (2015). *Electronic Catalogue of Spanish Cheeses* http://www.alimentacion.es/es/conoce_lo_que_comes/bloc/queso/default/principal. (Accessed 16 May 2015).

Ministry of Agriculture, Food and Environment (2015). *Electronic Catalogue of Spanish Cheeses* http://www.alimentacion.es/es/conoce_lo_que_comes/bloc/queso/default/principal. (Accessed 14 May 2015).

Miranda, J. M., Mondragon, A., Vázquez, B. I., Fente, C. A., Cepeda, A. & Franco, C. M. (2009). Microbiological quality and antimicrobial resistance of *Escherichia coli* and *Staphylococcus aureus* isolated from conventional and organic "Arzua-Ulloa" cheese. *Journal of Food*, 7, 103–110.

Mulet, A., Escriche, I., Rossello, C. & Tarrazó, J. (1999). Changes in the volatile fraction during ripening of Mahón cheese. *Food Chemistry*, 65, 219–225.

Nhuch, E. L., Prieto, B., Franco, I. & Carballo, J. (2008). Biochemical changes during the ripening of San Simón da Costa cheese (PDO) manufactured from pasteurized milk. *Australian Journal of Dairy Technology*, 63, 68–76.

Nieto-Arribas, P., Poveda, J. M, Seseña, S., Palop, L. L. & Cabezas, L. (2009). Technological characterization of Lactobacillus isolates from traditional Manchego cheese for potential use as adjunct starter cultures. *Food Control*, 20, 1092–1098.

Nieto-Arribas, P., Seseña, S., Poveda, J. M., Chicón, R., Cabezas, L. & Palop, L. L. (2011). *Enterococcus* populations in artisanal Manchego cheese: Biodiversity, technological and safety aspects. *Food Microbiology*, 28, 891–899.

Nieto-Arribas, P., Seseña, S., Poveda, J. M., Palop, L. L. & Cabezas, L. (2010). Genotypic and technological characterization of *Leuconostoc* isolates to be used as adjunct starters in Manchego cheese manufacture. *Food Microbiology*, 27, 85–93.

Official website for PDO Comté (2016). [Online]. Available: http://www.comte.com/ [Accessed 10 July 2016]

Ostlie, H. M., Eliassen, L., Florvaag, A. & Skeie, S. (2005). Phenotypic and PCR-based characterization of the microflora in Prast cheese during ripening. *International Dairy Journal*, 15, 911–920.

Picon, A., García-Casado, M. A. & Nuñez, M. (2010). Proteolytic activities, peptide utilization and oligopeptide transport systems of wild *Lactococcus lactis* strains. *International Dairy Journal*, 20, 156–162.

Polo, C., Ramos, M. & Sánchez, R. (1985). Free amino acids by high performance liquid chromatography and peptides by gel electrophoresis in Mahon cheese during ripening. *Food Chemistry*, 16, 85–96.

Poveda, J. M., Chicón, R. & Cabezas, L. (2015). Biogenic amine content and proteolysis in Manchego cheese manufactured with *Lactobacillus paracasei* subsp. *paracasei* as adjunct and other autochthonous strains as starters. *International Dairy Journal*, 47, 94–101.

Pripp, A. H., Skeie, S., Isaksson, T., Borge, G. I. & Sorhaug, T. (2006). Multivariate modelling of relationships between proteolysis and sensory quality of Prast cheese. *International Dairy Journal*, 16, 225–235.

Ränk, G. (1987). *[From Milk to Cheese]. Nordiska Museets Handlingar* 66. 2nd edition. Berlings, Arloev, Sweden. ISBN 91-7108-269-

Rasmusson, K. (2012). *Bioactive peptides in long-time ripened open texture semi-hard cheese.* MSc thesis made in cooperation by Swedish University of Agricultural Sciences and University of Copenhagen.

Rodríguez-Alonso, P., Centeno, J. A & Garabal, J. I. (2009). Comparison of the volatile profiles of Arzúa-Ulloa and Tetilla cheeses manufactured from raw and pasteurized milk. *LWT Food Science and Technology*, 42, 1722–1728.

Rodríguez-Alonso, P., Centeno, J. A. & Garabal, J. I. (2011). Biochemical study of industrially produced Arzúa-Ulloa semisoft cow's milk cheese: Effect of storage under vacuum and modified atmospheres with high-nitrogen content. *International Dairy Journal*, 4, 261–271.

Romagnoli, A. (1936). *Comitato Nazionale per il latte e suoi derivati*. Roma Società anonima Arte della Stampa.

Roseiro, L., Wilbey, R. A. & Barbosa, M. (2003). Serpa Cheese: Technological, biochemical and microbiological characterisation of a PDO ewe's milk cheese coagulated with *Cynara cardunculus L. Le Lait*, 83, 469–481.

Roth, E. (2009). Control of Listeria contamination on the surface of semi-hard cheeses by natural smear ecosystems and protective cultures. Diss. ETH Nr. 18644.

Rovira, S., García, V. & López, M. B. (2011). Application of a Large Field-of-View sensor during coagulation and syneresis in fresh goat cheese manufacture. *International Journal of Dairy Science*, 65, 51–56.

Rovira, S., García, V., Ferrandini, E., Carrión, J., Castillo, M. & López, M. B. (2012). Usefulness of a large field of view sensor for physicochemical, textural, and yield predictions under industrial goat cheese (Murcia al Vino) manufacturing conditions. *Journal of Dairy Science*, 95, 6320–6331.

Rovira, S., García, V., Laencina, J. & López, M. B. (2013). Microstructure of industrially manufactured goat cheese Queso de Murcia al Vino during syneresis. *International Journal of Dairy Technology*, 66, 382–389.

Sánchez, I., Seseña, S., Poveda, J. M., Cabezas, L. & Palop, L. L. (2006). Genetic diversity, dynamics, and activity of Lactobacillus community involved in traditional processing of artisanal Manchego cheese. *International Journal of Food Microbiology*, 107, 265–273.

Sánchez-Macias, D., Laubscher, A., Castro, N., Arguello, A. & Jiménez-Flores, R. (2013). Effects of supercritical fluid extraction pressure on chemical composition, microbial population, polar lipid profile, and microstructure of goat cheese. *Journal of Dairy Science*, 96, 1325–1334.

Sieber, R. (2011). Zusammensetzung von Milch und Milchprodukten Schweizerischer Herkunft. *ALP Science Nr.* 538, 1–40.

Simal, S., Sánchez, E. S., Bon, J., Femenia, A. & Rosselló, C. (2001). Water and salt diffusion during cheese ripening: effect of the external and internal resistances to mass transfer. *Journal of Food Engineering*, 48, 269–275.

Sveriges Ostkollegium. (1993). *Svenska Ostsorter* [Swedish cheese varieties]. Falkenberg, Sweden.

Swiss Commission for Denominations of Origin and Geographical Indications. *Annual report* (2011). Federal Office for Agriculture FOAG. http://www.blw.admin.ch/ [Accessed 8 September 2015].

Swiss PDO-PGI Association (2015). Bern, Switzerland. http://www.aop-igp.ch/ [Accessed 28 April 2015].

Tejada, L., Abellán, A., Cayuela, J. M. & Martínez-Cacha, A. (2006). Sensorial characteristics during ripening of the Murcia al Vino goat's milk cheese: The effect of the type of coagulant used and the size of the cheese. *Journal of Sensory Studies*, 21, 333–347.

Tejada, L., Abellán, A., Cayuela, J. M. & Martínez-Cacha, A. (2008a). Proteolysis in goats' milk cheese made with calf rennet and plant coagulant. *International Dairy Journal*, 18, 139–146.

Tejada, L., Abellán, A., Prados, F. & Cayuela, J. M. (2008b). Compositional characteristics of Murcia al Vino goat's cheese made with calf rennet and plant coagulant. *International Journal of Dairy Technology*, 61, 119–125.

The Intellectual Property Office of the Republic of Serbia (2013). *Cheese from Sombor – study on the protection of origin names*. No. Reg. 62. http://www.zis.gov.rs [Accessed 21 May 2015].

TSM Treuhand GmbH (2015). Mehrjahresvergleich ab 2004, Beilage zur Jahresstatiskik 20014. Bern, Switzerland. http://www.tsmtreuhand.ch/ [Accessed 28 April 2015].

Vázquez, B. I., Fente, C. A., Franco, C. M., Vázquez, M. J., Rodriguez, J. L. & Cepeda, A. (2001). Inhibitory effects of eugenol and thymol on *Penicillium citrinum* strains in culture media and cheese. *International Journal of Food Microbiology*, 67, 157–163.

Vázquez, B. I., Fente, C. A., Quinto, E., Franco, C. M., Rodriguez, J. L. & Cepeda, A. (1995). Incidence of toxigenic molds in farm level cheese-making units from Arzúa. *Food Science and Technology International*, 1, 91–95.

Waagner, N. E. (1993). Danish cheese varieties. In Fox, P. F. (ed.), *Cheese: Chemistry, Physics and Microbiology. Vol 2. Major Cheese Groups*, 2nd edition. Chapman & Hall, London, pp. 247–250.

Wechsler, D., Winkler, H. & Guggisberg, D. (2012). Vergleich der Eigenschaften von Appenzeller mit gutem und ungenügendem Teig. *ALP forum Nr.* 93, 1–20.

Wechsler, D., Winkler, H. & Guggisberg, D. (2013). Optimierung der Teigeigenschaften von Appenzeller mit festem und zu trockenem Teig. *ALP forum Nr.* 98, 1–20.

4

Soft Cheeses (with Rennet)

Maria Belén López Morales[1], Thomas Bintsis[2], Efstathios Alichanidis[3], Karol Herian[4], Paul Jelen[5], Erica R. Hynes[6,7], Maria Cristina Perotti[6,7], Carina V. Bergamini[6,7], Everaldo Attard[8], Anthony Grupetta[9], Stefania Carpino[10], Tânia G. Tavares[11,12] and F. Xavier Malcata[11,13]

[1] *Food Science and Technology Department, International Excellence Campus for Higher Education and Research 'Campus Mare Nostrum', Veterinary Faculty, University of Murcia, Spain*
[2] *11 Parmenionos, 50200, Ptolemaida, Greece*
[3] *Department of Food Science and Technology, School of Agriculture, Aristotle University of Thessaloniki, Greece*
[4] *Director-Emeritus, Slovak Dairy Research Institute, Slovakia*
[5] *Department of Agricultural, Food and Nutritional Science, University of Alberta, Canada*
[6] *Facultad de Ingeniería Química (Universidad Nacional del Litoral), Argentina*
[7] *Instituto de Lactología Industrial (Universidad Nacional del Litoral – Consejo Nacional de Investigaciones Científicas y Técnicas), Argentina*
[8] *Division of Rural Sciences and Food Systems, Institute of Earth Systems, University of Malta, Malta*
[9] *Veterinary Regulations Directorate, Italy*
[10] *CoRFiLaC – Consorzio Ricerca Filiera Lattiero Casearia, Ragusa, Italy*
[11] *Laboratory of Engineering of Processes, Environment, Biotechnology and Energy (LEPABE), Portugal*
[12] *REQUIMTE/Department of Chemical Sciences, Faculty of Pharmacy, University of Porto, Portugal*
[13] *Department of Chemical Engineering, University of Porto, Portugal*

4.1 Afuega'l Pitu PDO – Spain

Name: Afuega'l Pitu	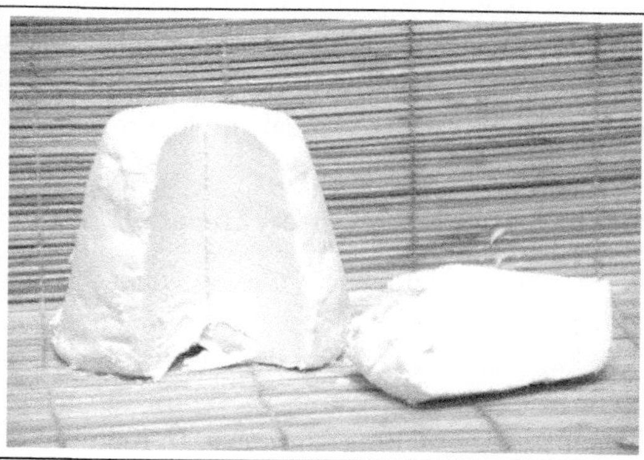
Production area: Asturias	
Milk: Cow's, pasteurised	

Global Cheesemaking Technology: Cheese Quality and Characteristics, First Edition.
Edited by Photis Papademas and Thomas Bintsis.

4.1.1 Introduction

The PDO status of Afuega'l Pitu cheese was recognised by the Ministry of Agriculture, Fisheries and Food in 2004 and at the European level in 2008 (EC, 2008b). The cheese is made from pasteurised cow's milk. Under the PDO regulation, two farmers and nine dairy plants are registered. Production reached 110,793 kg in 2013, representing a total of 1,030,000 euros. The cheese is mainly marketed within Spain and represents 0.49% of the economic value of Spanish PDO cheese production (Ministry of Agriculture, Food and Environment, 2014). The first written references dates to the eighteenth century when this cheese (quesu de puρu or queso de Afuega'l Pitu) was used for tax payment. Literally, 'Afuega'l Pitu' means 'choke the chicken' and is a reference to (1) strangling the neck of the bag ('fardela') in which the cheese is drained, (2) the occasional difficulties encountered in swallowing the cheese and (3) its traditional use as feed for chickens. The defined geographical area of the Afuega'l Pitu PDO covers the municipalities along the rivers Narcea and Nalσn and within the Aramo mountain range. All the cheesemaking stages take place in this area, including maturation and production of the milk used as raw material.

4.1.2 Type

This is a high-fat soft cheese, which may be either fresh or mature, and is made from whole pasteurised cow's milk by lactic acid coagulation. If the ripening stage is more than 60 days, raw milk can be used. The colour is white or reddish-orange, depending on whether paprika is added. According to its shape and colour, four types of cheese can be described: (1) Atroncau blancu: unkneaded, truncated-cone shape, white; (2) Atroncau roxu: kneaded, truncated-cone shape, reddish-orange; (3) Trapu blancu: kneaded, courgette-shaped, white and (4) Trapu roxu: kneaded, courgette-shaped, reddish-orange. The minimum fat and protein content in dry matter are 45% and 35%, respectively, with a minimum moisture content of 70% and a pH value of 4.1-5.0.

4.1.3 Milk

The milk used is obtained from Friesian and Asturiana de los Valles cows and their crossbreeds, and are managed on a semi-stabling basis. The livestock's diet is based on grazing with a supplement during milking of fresh grass, hay and silage obtained from the farms themselves with small quantities of cereal and legume concentrates.

4.1.4 Description and Sensory Characteristics

The resulting product is a cheese shaped like a truncated cone or courgette which weighs 0.2–0.26 kg, has a height of 5–12 cm, a diameter of 8–14 cm measured at the base and a natural rind of variable consistency, depending on its maturation period and whether paprika is added. The cheese is white, shifting to yellow during ripening or reddish-orange if paprika is added, without a rind or with a slight rind in matured cheeses. In fresh cheeses, the body is blind (i.e. no eyes), laminar (flat, in layers), soft and spreadable and short and crumbling in matured cheeses. The taste is mildly acidic, non-salty or slightly salty, creamy and fairly dry, with the red cheeses having a stronger and more piquant taste. It has a mild aroma, which becomes more pronounced as it matures.

4.1.5 Method of Manufacture

Milk preparation: Milk must be collected from the registered farms and kept after milking at no more than 4°C.

Starter cultures: Starter cultures should be used if pasteurised milk is used. Cuesta et al. (1996) selected a specific starter for this type of cheese constituted by the following strains: *Lactococcus lactis* ssp. *lactis* IPLA 947, *Lc lactis* ssp. *lactis* biovar. *diacetylactis* IPLA 838 and *Leuconostoc citreum* IPLA 616.

Coagulation/rennet: A small amount of rennet is added to the cheese vat at a temperature of 22°C–32°C for 15–20 hr.

Curd draining: After coagulation, the curd is transferred to perforated moulds, where the whey drains off over approximately 12 hr. After this time, the partly drained curd is transferred to a smaller mould, and after 12 hr, it is removed from the mould and placed on perforated trays for final draining.

In the case of cheeses made from kneaded (mixed) paste, the curd is drained in larger plastic containers, using gauze, and placed in the kneader (i.e. mixer) after around 24 hr. During kneading, salt is added; approximately 1% of paprika may be added, in which case the variety will be 'trapu roxu'. The kneaded paste, with or without paprika, is then placed in moulds and/or gauze, where it is left to drain for a further 24 hr.

Salting: Salt is added directly in the moulds; in cheeses made from kneaded paste, the salt is added at this stage.

Maturation: Cheeses are left in the maturing chambers for 5 days (fresh cheeses) to 60 days (mature cheeses).The packaged cheeses, bearing labels on the authorised packaging, are kept in cold stores at 4°C–10°C, until sale.

4.1.6 Relevant Research

Lactococcus spp. are the major group in milk, curd and three-day-old cheese but decrease gradually, while *Lactobacillus* spp. are responsible for the pH decrease after 15 days of ripening. *Leuconostoc* spp. is the major group eight-day-old cheese, which may explain the role of these microorganisms in the sensory profile; a decrease in these microorganisms was also observed at the end of ripening (Cuesta et al., 1996).

The ratio of hydrophobic to hydrophilic peptides decreases as the cheese ages (Gonzalez de Llano, Polo & Ramos, 1994). There are some studies on the acidification ability of a mixed strain starter for Afuega'l Pitu cheese using pasteurised cow's and ewe's milk (Cárcoba, Delgado & Rodrvguez, 2000) or related to the growth and metabolic behaviour of *Lc lactis* ssp. *lactis* IPLA 947 (Cárcoba, Pin & Rodrvguez, 2004) or the selection of *Leuconostoc* spp. isolated from artisanal Afuega 'l Pitu (Sánchez, Martvnez & Rodrvguez, 2005).

Other research has focused on the effect of the nisin Z-producing strain *Lc lactis* ssp. *lactis* IPLA 729 on the growth of *Staphylococcus aureus* (Rilla, Martvnez & Rodriguez, 2004) or in the analysis of the whey protein during ripening (Gonzalez de Llano & Santa Maria, 1997).

4.2 Anevato PDO – Greece

Name: Anevato, PDO	
Production area: Prefecture of Grevena and Municipality of Voio (Prefecture of Kozani) in Western Macedonia, Greece	
Milk: Goat's or sheep's, or mixtures, pasteurised or raw	

4.2.1 Introduction

Anevato is a soft and spreadable cheese made from raw sheep's or goat's milk and is produced in the Prefecture of Grevena and Municipality of Voio (Prefecture of Kozani) in Western Macedonia. Its name means 'cheese that is raised' and comes from the word *anevazo*, which means 'rise up', because of the rise of the curd particles on the surface of the whey after cutting the coagulum.

4.2.2 Type

Anevato is a soft cheese with a maximum moisture of 60% and a minimum FDM of 45%. The PDO status for Anevato was recognised by the EC in 1996 (EC, 1996).

4.2.3 Description and Sensory Characteristics

Anevato is a white, soft, grainy, spreadable cheese. It has a grainy and creamy texture and a slightly sour and salty taste with a pleasant milk aroma.

4.2.4 Method of Manufacture

Milk preparation: Whole milk from local herds (sheep and goat) is left at 18°C–20°C to sour (to an acidity of about 35°D) by the activity of the natural microflora.

Starter culture: No starter is added when raw milk is used, but when milk is pasteurised, mesophilic homofermentative strains of *Lc. lactis* ssp. *lactis* and *Lc. lactis* ssp. *cremoris* are added.

Coagulation: Liquid rennet (5 mL/100 L milk) is added to the milk at a pH of about 6.2, and coagulation takes place within 12 hr.

Cutting: The curd is cut into cubes of 2 × 2 × 2 cm and left to rest (usually for 4–5 hr) until the pieces rise up to the surface of the whey.

Curd draining: The curd is put in cheese cloths and drained for 24 hr.

Salting: Dry salt is added, and the curd is thoroughly mixed. Anevato is packed in plastic containers of sizes 250 g, 1 kg or 5 kg, usually under modified atmosphere.

Maturation: Maturation takes place at 4°C for at least two months.

4.2.5 Relevant Research

Lactococci dominated Anevato cheese for 15 days, but lactobacilli became predominant after 30 days. *Lc. lactis* was the most abundant species of lactic acid bacteria found and is suggested as a starter culture to eliminate or suppress the growth of undesirable microorganisms (Hatzikamari, Litopoulou-Tzanetaki & Tzanetakis, 1999).

4.3 Bryndza – Slovakia

Name: Slovak Bryndza	
Production area: Slovakia, especially the northern part, the foothills of the Tatra Mountains	
Milk: Raw or minimally pasteurised sheep's, with or without cows' milk curds added	

4.3.1 Introduction

Bryndza is a traditional Slovak cheese, made from ripened sheep lump (curd) cheese. It is one of the most popular cheeses in Slovakia and is well known also in the neighbouring countries, that is, the Czech Republic, Poland and Hungary. The cheese has PDO status (EC, 2008a). A similar product made in Poland, called Bryndza Podhalanska, is also included in the PDO list.

One of the most important foods in Slovakia, the cheese is used to prepare perhaps the most popular of all Slovak dishes, the 'Bryndzove halusky' ('ryndza gnocc'). This famous Slovak dish is served in virtually every restaurant and in all other food service establishments. A very significant proportion of all sheep's milk produced in Slovakia is used for production of Bryndza.

The history of Bryndza production in the area of today's Slovakia has been documented since the Middle Ages. The artisanal, 'cottage production' of sheep cheese was already thriving in the fourteenth century, as a result of the arrival of Wallachians (Vlachs) and their colonisation of the Carpathian Mountains. Traditionally, sheep cheese was produced either for direct consumption as a moist lump cheese, or as a drier, hard version. The excess lump cheese was left to ferment, then mechanically treated, salted and pressed into either barrels or directly into four-wheel horse-drawn trailers which were used for transporting the cheese to the market. Such hard, salted, well-ripened cheese, called Bryndza even in those early days, was well received in Vienna, Budapest or Bavaria.

About 250 years ago, a Slovak entrepreneur called Jan Vagac developed a method to turn the crumbly lump cheese into a smooth, spreadable soft cheese. The fully fermented cheese was broken up, salty water added was and the mixture kneaded, thus producing the smooth, spreadable and full-flavoured Bryndza as we know it today. The first Bryndza manufacturing plants sprang up, to process the sheep lump cheese that was produced in the high mountain huts (salas) and brought here for reworking into the final product.

However, this was possible only in summer during the sheep milk production period. Thus, the excess sheep lump cheese was crumbled, salted and pressed into barrels, where it was stored for use in the winter months. The salted stored 'raw Bryndza' was mixed with cow's milk curds cheese in a 1:1 ratio, giving rise to the 'winter mixed Bryndza'.

In the Middle Ages, all cheese, not only from sheep's but even from cow's milk, was called Bryndza. In Romanian, *bryndza* is a generic word denoting simply 'cheese'. In Slovakia, where mainly sheep's milk was available, the word *bryndza* always denoted specifically 'sheep's milk cheese'. The contemporary Bryndza cheese, as we know it now, is mainly made in Slovakia, where it is the national cheese, Slovak Bryndza, but it is also made in Poland, parts of Ukraine, Romania and Moldova. The areas where Slovak Bryndza is made are restricted to mountainous regions and the adjoining foothills where sheep have been traditionally reared.

4.3.2 Type

Bryndza is a soft, spreadable cheese made of sheep's milk or a mixture of sheep's and cow's milk. Principally, there are two Bryndza types depending on the type of milk used: (1) Sheep's Bryndza, made exclusively from sheep milk, with a minimum moisture of 52% and a minimum FDA of 48% and (2) Mixed Bryndza, containing at least 51% of fermented or stored salted lump sheep cheese, the rest being acid cow's milk lump cheese. This Bryndza type must have not more than 56% moisture and its FDM is either 48% (full-fat cheese) or 38% (low-fat cheese). Mixed Bryndza can be denoted as either summer or winter. Either raw or pasteurised milk can be used, as long as this information is listed on the label.

4.3.3 Description and Sensory Characteristics

The classical Bryndza is a spreadable cheese, essentially whitish, sometimes with a slightly yellowish/greenish tinge. The shape of the cheese is imparted by the packaging alternatives used. Traditional Bryndza was available in small (1–5 kg) wooden drums. Nowadays, several packaging materials and sizes are being used, including plastic foils or cups. The structure of spreadable Bryndza is soft, slightly mealy with small granules of the milled lump cheese. Bryndza has a strong sheep's milk aroma, with slight acidic/fermented milk characteristics. The cheese taste is likewise reminiscent of pleasantly acid cheese with sheep's milk notes. The so-called 'May Bryndza' has the strongest aroma, as this is the time when the sheep are brought to graze on the mountain meadows.

4.3.4 Method of Manufacture

Milk preparation: The classical sheep's Bryndza is made from freshly obtained milk (32°C–35°C) without any fat standardisation or other processing, except coarse filtration. The milk is turned into the lump cheese in small mountain-cottage-type huts immediately after the morning (sometimes even after the evening) milking. If the raw milk is delivered to a bigger cheese processing facility, it is cooled to 5°C–8°C and subsequently pasteurised at 72°C for at least 15 s. Most of the milk is available in summer when the sheep are grazing on mountain pastures. Only a small amount of the milk is available in winter months, to maintain the year-round production.

Milk for production of sheep's cheeses does not have to be pasteurised according to the legislation. However, when the milk is processed in larger production establishments, it is commonly pasteurised before conversion into either the lump cheese or the final Bryndza. The final product is made from well-fermented, ripened sheep lump cheese, which is milled finely and mixed with table salt and water to achieve the required composition.

Starter cultures/rennet: Traditionally, the lump sheep's cheese was made using only natural microflora containing LAB. Some of the avenues for the microflora to enter the milk included the natural coagulant (rennet enzymes) obtained by extracting pieces of lamb stomach by adding acid whey or by using the cheesemaker's wooden tools. Nowadays, the controlled fermentation in the mountain hut environment is achieved by adding the acid whey from the previous day, or by using selected cultures from reputable suppliers. These (in liquid or powder forms) are also used when the milk is first pasteurised in the larger processing establishments. Starter cultures and soluble calcium salts (10 g $CaCl_2$ per 100 L milk) are added to the milk tempered to 32°C–35°C (immediately after milking or after appropriate warming). The original rennetting by extracts of lamb stomachs has now been largely replaced by industrially produced coagulators.

Coagulation/cutting: The coagulum is first cut to large blocks and, when the whey appears, cut further to small, bean-size particles. The syneresis proceeds for 20–40 min at the minimum constant temperature of 32°C until a firm, non-sticky grain is obtained. The original practice of adding hot water for better grain firming (when controlling the temperature was difficult in the primitive manufacturing conditions) is no longer used.

Moulding: Well-firmed grain, together with the whey, is poured into perforated containers or muslin bags and kept at 18°C–20°C for draining, turning the containers at least once. The curds are kept in these conditions until the next day; the acidity increases and should reach at least a pH of 5.2 or less.

Maturation: Well-acidified lump cheese is not salted but is kept to ripen for 8–19 days at 15°C and a Relative Humidity (RH) of 90%. The cheese is periodically turned.

Salting: Well-ripened lump cheese for the production of the spreadable Bryndza is mechanically disintegrated and milled into a finely granulated mass. Salt is added, followed by vigorous mixing. The paste is immediately filled into retail containers.

To produce the mixed Bryndza, the cow's milk lump cheese is made similarly, with ripening lasting at least seven days to produce the smooth consistency. The milled sheep's lump cheese is then mixed with similarly processed ripened cow's lump cheese in a 1:1 ratio, with the appropriate compositional adjustment of dry matter and salt content as required, and the product is packaged. The keeping quality is usually guaranteed for three weeks provided the storage temperature does not exceed 10°C.

To produce the mixed 'winter Bryndza', the lump cheese is coarsely milled shortly after the manufacture in summer, 4% salt is added and the material is pressed into bags, boxes or drums. These are kept in cold storage and, when needed, the lump cheese is mixed with fermented cow's milk curds in a ratio needed to keep the percentage of the sheep cheese component above 50%. Water is added as needed to achieve the desired dry matter content.

4.4 Cremoso – Argentina

Name: Cremoso cheese	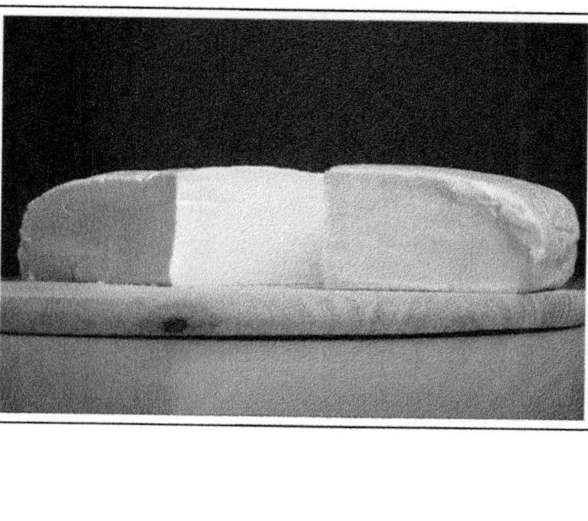
Production area: Northwest/Pampeana region, Argentina	
Milk: Cow's, pasteurised	

4.4.1 Introduction

The dairy area in Argentina is also the region of the country most influenced by European immigration in the nineteenth century. After 1856, when first organised colonies of Europeans immigrants settled, production of cheeses derived mainly from Italian and Swiss cheeses started in Argentina, among them, 'Cremoso', earlier known as 'Cuartirolo'. Even if some sources in the literature and the Argentinean legislation distinguish between the two types of cheese, they are virtually the same, and the name 'Cuartirolo' is much less common nowadays.

The production area is the Pampeana region (provinces of La Pampa, Buenos Aires, Santa Fe, Entre Ríos, Córdoba and San Luis). Cremoso cheese is made also in Entre Ríos province, and a little production may be found in other regions of Argentinean territory, but it is not significant outside Pampeana region.

4.4.2 Description and Sensory Characteristics

Argentinean legislation (Código Alimentario Argentino, 2014) defines Cremoso cheese as a high-moisture cheese (between 46% and 54.9% of moisture content), or very high-moisture cheese (moisture content > 55%). It is a soft, non-cooked cheese, with a minimum fat content in the dry matter of 50%. It has the shape of a cylindrical or parallelepiped block with flat sides or slightly convex (diameter/sides and height approximately of 24 cm and 14 cm, respectively), and the usual weight is around 2.5–5 kg. The ripening time is at least 20 days for 2.5 kg blocks, and 30 days for 2.5 to 5 kg blocks.

Ripened Cremoso cheese, as its name denotes, has a creamy texture. Consumer preferences vary from a slightly elastic texture to a creamy, sticky consistency. Over-ripened Cremoso is difficult to cut and may become fluid, which is normally considered a defect. The flavour is soft and lactic. The colour is white or light yellow and uniform. The presence of holes and/or mechanical openings is undesirable; a limited number of small eyes may be acceptable. A rind is not formed as Cremoso cheese is usually packaged in thermo-contractible plastic bags.

4.4.3 Method of Manufacture

Milk preparation: Cremoso cheese is made with whole milk or milk-fat-enriched milk. The fat content of milk is standardised (ratio of fat to protein > 1), and the milk is HTST (high temperature/short time) pasteurised. After cooling to 37°C, $CaCl_2$ is added to a final concentration of 0.02%.

Starter culture: Cremoso cheese was initially made with 'natural' milk starter cultures of consistently good-quality, thermised milk incubated at 45°C overnight. Natural milk starters contain a complex microbiota dominated by strains of *S. thermophilus* (Reinheimer et al., 1995). They were praised for their phage resistance and the rich sensory characteristics that they imparted to cheeses. However, cheesemaking standardisation and consistency of cheese quality were not easily achieved. Nowadays, milk starters have been completely replaced by commercial starters of mixed strains of *S. thermophilus*, either direct or semi-direct. Commercial starters of *S. thermophilus* may also contain low proportions of mesophilic LAB for aroma development (e.g. *Lc. lactis* ssp. *lactis* biovar. *diacetylactis* or *Leuconostoc* spp.); 'yogurt culture' (*S. thermophilus/Lb delbrueckii* ssp. *bulgaricus*) may also be used. Either way, *S. thermophilus* leads the acidification in Cremoso cheese.

Rennet: Several milk-clotting enzymes are available for Cremoso cheesemaking. This is not an expensive cheese variety; consequently, price is often the main factor affecting coagulant choice. Adult bovine coagulant and microbial coagulant are still used for Cremoso cheese; although they may reduce the shelf life due to extensive proteolysis, when cheeses are expected to be sold within a few days of ripening, this is not a problem. When cheeses have to be on the market over longer periods, the gold standard for coagulation is chymosin. Lately, camel chymosin has been proposed as a technological strategy to increase moisture in cheese (and therefore the yield) without producing excessive proteolysis during ripening (personal communication).

Coagulation, cutting and moulding: The milk-clotting enzyme is added immediately after the starter at the initial pH of the milk and at 37°C; the amount of coagulant is relatively high, and the characteristics of the curd are mostly enzymatic. After coagulation and gel strengthening, the curd is cut into cubes with a size of about 1 cm^3. The mixture of curds and whey is gently stirred to favour the expulsion of some whey, but the temperature is not increased. Other Argentinean cheeses similar to Cremoso but with a firmer body (e.g. 'Port Salut') may include a step of curd–whey heating at 40°C–42°C. Most whey is eliminated from the vat, and then curds and the remaining whey are fed into moulds. Cremoso cheeses are not pressed; moulds are stocked in a warm room or spot, and the cheeses are turned upside down periodically until they reach pH 5.3–5.4, when they are sent to cold brine (Zalazar, Meinardi & Hynes, 1999).

Salting: Cheeses are salted by immersion in brine solution (19°Bé, 5°C, pH 4.95–5.1) for 1 hr per kg, allowed to dry for a day and then packed in thermocontractile plastic bags. Ripening usually proceeds at 5°C, or lower temperatures, to extend the shelf life, for 20 days or more.

Maturation: Maturation of Cremoso cheese is characterised by proteolysis mediated by the residual milk-clotting enzyme. The main event is α_{s1}-casein hydrolysis; cheeses in which the residual coagulant activity was inhibited or diminished showed lower hydrolysis of α_{s1}-casein and a harder texture (Hynes et al., 2001; Hynes, Delacroix-Buchet & Zalazar, 2001; Zalazar et al., 2006).

4.4.4 Relevant Research

Argentinean soft cheese has been successfully proposed as a carrier for probiotic bacteria: *Lb. paracasei*, *Lb. acidophilus* and *Bifidobacterium* spp. were added to a Cremoso-like cheese known and commercialised for a decade as 'Bioqueso' (Vinderola et al., 2000; Vinderola et al., 2009).

Lb. casei, *Lb. paracasei* and *Lb. plantarum* were also isolated from good-quality Cremoso cheeses as NSLAB (Bude Ugarte et al., 2006). These strains were characterised for technological and probiotic properties (Bude Ugarte et al., 2006; Briggiler Marcó et al., 2007). Adjunct cultures obtained were re-inoculated into Cremoso cheese to improve quality, with favourable results, especially regarding increased peptidolysis and bioformation of diacetyl and volatile compounds derived from amino acids (Burns et al., 2012; Milesi, McSweeney & Hynes, 2008; Milesi et al., 2010).

A low-fat Cremoso cheese with increased moisture content and a fat replacer containing 35% of modified whey proteins (Dairy-Lo®) has been obtained; the low-fat cheeses without the fat replacer and slightly increased moisture level were the most similar to the full-fat control (Zalazar et al., 2002). Ultrafiltration techniques are currently applied in some dairy facilities, and also heat-denatured whey protein may be added as a protein concentrate into Cremoso cheese (Portal Lechero, 2015).

Acknowledgements

The authors thank Sucesores de Alfredo Williner S.A. for providing cheeses for the photographs.

4.5 Galotyri PDO – Greece

Name: Galotyri, PDO	
Production area: Epirus and Thessaly, Greece	
Milk: Sheep's or goat's, or mixture, raw	

4.5.1 Introduction

Galotyri is a soft cheese manufactured in Epirus and Thessaly. The name of this cheese is literally translated as 'milk cheese', that is, a cheese which stands between milk and cheese. Galotyri is considered as one of the oldest traditional cheeses in Greece, and about six manufacturing methods (i.e. salting the cheese milk or curd, using rennet or not as well as starter, etc.) are given by Zygouris (1952). The PDO status for Galotyri was recognised by the EC in 1996 (EC, 1996). Since Galotyri cheese has pleasant organoleptic characteristics much appreciated by the Greek consumers, recent years have witnessed a great demand for its production. As a result, both small and large dairies in many regions of the country produce Galotyri-type cheese, which resembles Galotyri, using various procedures different from the one described for the PDO cheese (i.e. curd salting with 2% NaCl and very short or no ripening) (EC, 1996). Thus, today there are several Galotyri-type fresh cheeses in the market, which differ in chemical composition and sensory characteristics.

4.5.2 Type

Soft cheese, with a maximum moisture of 75% and a minimum FDM of 40%.

4.5.3 Description and Sensory Characteristics

Soft, white, creamy, spreadable cheese, characterised by a grainy, curd-like texture. Galotyri cheese has a creamy texture with sourish and brackish flavour and a pleasant milk aroma.

4.5.4 Method of Manufacture

Milk preparation: Milk (sheep's or goat's milk or mixtures of both) is heated to the boil and left for 24 hr at room temperature.

Starter culture: Usually, no starter is added.

Coagulation: The milk coagulates with the addition of rennet at 30°C.

Salting: Salt is added at 3%–4% and is mixed with the curd at room temperature, and the coagulum is left for 48 hr to develop the right acidity.

Draining: The coagulum is then put into leather bags or in wooden barrels for drainage.

Maturation: Maturation takes place in cold rooms (temperature <8°C) for at least two months in case of cheese produced from raw milk.

4.5.5 Relevant Research

Kondyli, Katsiari and Voutsinas (2008) investigated the effects of different manufacturing processes of Galotyri-type cheese currently being used in Greek dairies on its chemical and sensory characteristics, with the aim of helping cheesemakers produce cheese of consistently high quality. In addition, Katsiari, Kondyli and Voutsinas (2009) studied the effect of different starter cultures on the quality of Galotyri-type cheese and concluded that the different starter cultures did not significantly influence the physic-chemical characteristics of the cheese, with the exception of titratable acidity and pH. Moreover, the extent of cheese lipolysis ($C_{4:0}$–$C_{20:1}$ free fatty acids, FFA) was not affected by the starters used, although significant differences for some of the FFA were observed among the cheeses. However, the different starters had a significant impact on the flavour and total quality of Galotyri-type cheese.

4.6 Kopanisti PDO – Greece

Name: Kopanisti, PDO	
Production area: Prefecture of Cyclades	
Milk: Cow's, sheep's or goat's, or their mixture, raw	

4.6.1 Introduction

Kopanisti cheese is a traditional soft cheese which is produced in the Cyclades islands in Greece.

4.6.2 Type

Soft cheese with a maximum moisture of 56% and a minimum FDM of 43%. Gross composition: moisture 44.6%–69.4%, fat 13%–30% and salt 1%–4.7% (Kaminarides, 1986). The PDO status for Kopanisti was recognised by the EC in 1996 (EC, 1996).

4.6.3 Milk

Kopanisti is manufactured from raw cow's, sheep's or goat's milk or their mixture without the addition of any lactic acid cultures.

4.6.4 Description and Sensory Characteristics

Kopanisti cheese is a soft cheese with a creamy texture. It has an intense salty and peppery taste and a strong flavour. The main flavour compound groups found in Kopanisti cheese were alcohols, esters and volatile free fatty acids. Ethanol and several ethyl esters were the main volatile compounds.

4.6.5 Method of Manufacture

Milk preparation: Raw cow's, sheep's or goat's milk is used.

Starter culture: Kopanisti is traditionally manufactured with natural fermentation with no starter added.

Coagulation: Coagulation takes place at 28°C–30°C with rennet in about 2 hr, and the coagulum is left in the cheese vat for 24 hr.

Cutting: The curd is cut to a coarse size and transferred into cheesecloths for drainage.

Salting: Drained curd which called 'petroma' is mixed well by hand with dry salt at 4%–5% (w/w).

Maturation: The curd is transferred into basins or pots with a wide opening, in cool ripening rooms with a high relative humidity, until a rich surface microflora is developed. Then, the curd is kneaded by hand or machine-milled and left for a new layer of surface microflora to be developed. This procedure is repeated 3–4 times, after which the cheese is transferred into plastic barrels or clay containers, pressed by hand so that no empty spaces are left, and its surface is covered with parchment paper and left for 30–40 days to complete the maturation.

4.6.6 Relevant Research

Kopanisti cheese is a cheese with intense lipolysis, with an average total free fatty acid content of 49 mg/kg which contributes to the strong flavour and peppery taste. Acetic, butyric and capric acids were the main volatile acids determined (Karali et al., 2012). Samples (50) of Kopanisti cheese were analysed for total count, coliforms, lactic acid bacteria, micrococcaceae, yeasts and moulds (Tzanetakis, Litopoulou-Tzanetaki & Manolkidis, 1987). *Lb. plantarum* and *Lb. casei* ssp. *casei* were the dominant lactobacillus species. Enterococci and *Pediococcus pentosaceus* were also isolated. Micrococci were found in one sample only. Among yeasts *Pichia membranefasciens* predominated while moulds were characterised as *Penicillium commune*.

4.7 Maltese Ġbejna – Malta

<table>
<tr><td>Name: Ġbejna</td><td rowspan="3">
(A)</td></tr>
<tr><td>Production area: Malta and Gozo</td></tr>
<tr><td>Milk: Sheep's, raw</td></tr>
</table>

(B)

(C)

4.7.1 Introduction

Cow's milk is notably the most popular type with the local consumers. Milk from several dairy farms is collected and then pasteurised at a central unit. Other products are derived from this pasteurised milk. Before the introduction of cow's milk, goat's and sheep's milk were more common. Goat's milk was considered better than sheep's milk for direct human consumption, while sheep's milk was more ideal for cheeselet production. With the advent of Brucellosis, infected animals were removed from herds, and milk pasteurisation became mandatory. Today, sheep's milk is still transformed into cheeselets. The Maltese 'Ġbejna' (plural='Ġbejniet') is a fresh cheese produced with whole raw milk from the 'Maltese' breed sheep (Mason, 1996) and its cross-breeds. It is also known as the Maltese cheeselet. With a total number of 8,326 ewes in 2014 (Tanti, 2014), its yearly production amounts to approximately 205 tonnes. The 'Ġbejna' can be sold as (1) fresh ('Ġbejna friska'), (2) air-dried ('Ġbejna moxxa') or (3) peppered ('Ġbejna tal-bżar') (Cassar, 2010).

Historical records indicate that grazing sheep have been present on the Maltese islands since medieval times (Lattrell, 1975; Wettinger, 1986), while the earliest reports on cheesemaking date back to the fifteenth century and is also mentioned in the seventeenth century (Fiorini, 2006). Agius De Soldanis (1712–1770) states that the 'Ġbejna' used to be exported to Italy, where it was called '*Formaggio Maltese*' or Maltese cheese. The 'Ġbejna' used to be placed in white wine together with pieces of the allergenic wild plant, pellitory of the wall (*Parietaria officinalis*). The arid terrain typical of the Maltese islands favours the grazing of sheep, which can use marginal agricultural areas unsuitable for other agricultural purposes, while cheese-making was a logical way to preserve the milk for longer periods of time, especially since dairy produce could only be consumed on meat-eating days (Cassar, 2010).

The know-how of the cheesemaking process has been passed on from generation to generation, and a recent survey on the production of the 'Ġbejna' further confirmed that most current producers learnt the process from their parents or grandparents. All steps involved in the cheesemaking process can be carried out by the producer, and all ingredients are locally sourced. Moreover, the drying of the cheeselets is carried out naturally taking advantage of the dry and warm climate typical of this geographical area.

The 'Ġbejna' is an integral part of the Maltese culinary heritage, and has also made its way into Maltese expressions and idioms, highlighting the specificity of the word and product and the linkage between the 'Ġbejna' and Maltese culture. The 'Ġbejna' is specific to this geographical area as it is produced with whole raw milk of the Maltese breed sheep; all registered sheep are regularly checked by the national veterinary services to ensure that the herd is free of zoonotic pathogens, such as *Brucella melitensis*. Moreover, most ingredients are sourced locally by the farmers who aim to produce a more genuine product. Whenever possible, the sheep are fed locally produced hays and fodder typical of the Mediterranean climate (e.g. carob and prickly pears).

4.7.2 Type

The 'Ġbejna friska' is a white and glossy cheese, with a fresh and soft core, and is classified as a soft cheese. The 'Ġbejna moxxa' (which is also referred to as 'Ġbejna bajda' or 'Ġbejna t'Għawdex' in certain areas of Malta) is more yellowish in colour, with a firmer consistency. The colour and hardness depend on the period and state of drying and maturation; thus, the 'Ġbejna' becomes yellower and harder as the drying period increases. This may classified as hard cheese. The 'Ġbejna tal-bżar' presents itself with a varying amount of fine to rough ground and crushed black pepper attached to its surface. It may classified as a peppered, hard cheese.

4.7.3 Description and Sensory Characteristics

The 'Ġbejna' appears as a truncated cone of small dimensions with the diameter of the base ranging between 5 and 7 cm, the diameter of the top ranging between 3 and 5 cm, a height of 3–4 cm, and a weight of 60–120 in the fresh form. It has an acidulous taste typical of sheep, and its surface takes the shape of the motifs of the mould in which it is set. Its pH is 5.20, and the total solid content is 68.5%. The total protein is about 23.9%, while the FDM is about 48.5%. The yellowish taint increases, and the parameters vary with the increase of the period of drying and storage (Attard, 2013). The peppered, hard cheese has a characteristic pH of around 4.9 and a total solid content of about 38.5%. The total protein content (determined as total nitrogen content) is approximately 13.2%, while the FDM is about 49.3%. The salt in dry matter is about 3.8%, while the soluble protein determined at a pH of 4.6 and a TCA of 12% are 21.1% and 6.6%, respectively (Carpino et al., 2013).

4.7.4 Method of Manufacture

Milk preparation: The 'Ġbejna' is produced in small-scale dairies. The cheeselet production contributes towards the Maltese cottage industry. Raw sheep's milk originates from herds registered in Malta of the Maltese breed and its cross-breeds. This milk has the following average values: a pH of 6.45, 6% fat, 5.6% protein and 16.6% dry matter (Carpino et al., 2013). The sheep are fed locally sourced hays of leguminous and cereal plants for at least 55% of their intake, and concentrates produced with imported raw materials. Depending on availability, sheep may also be fed locally sourced plants such as carobs, cladodes of prickly pears and vetch. The sheep are milked at the farm, and the milk is then transformed. The milk is first filtered through a very fine strainer to remove any animal hairs that may have fallen in the milk during the milking process, which in most cases is still a hand milking process. The temperature is maintained at 37°C, usually in a stainless steel vat.

Rennet/coagulation: While the milk is still warm (or, if necessary, after re-heating to 37°C), lamb stomach extract or other milk coagulating enzyme is added to the milk. Nowadays, many producers use rennet powder (25 g per 100 L of milk), which is commercially prepared from the inner lining of the fourth stomach of calves, though an effort is being made to re-introduce the use of 'tames' (i.e. the rennet produced on the farm using the stomach lining of a suckling lamb or kid that has not been weaned). Traditionally, one teaspoon of lamb stomach extract is used for every five litres of milk.

Following the addition of curdling agent, the milk is left to coagulate, and after 20 min the curd ('baqta') is formed.

Moulding/drainage: The curd is then collected into small moulds ('qaleb') and sprinkled with thick salt. These moulds are left in an empty stainless steel or plastic container to drain. The moulds were originally made of rushes, but these have been replaced with plastic drain moulds for hygienic reasons. The containers with the Ġbejna still set in the plastic moulds are then placed in a refrigerator, and turned over inside the mould once or twice to allow them to drain well and to give them the time to set in the traditional shape. This is the 'Ġbejna friska'. It is usually marketed within 24 hr of production.

Drying: If desired, the 'Ġbejna' can then be dried in the 'qannić', a wooden or metal frame cupboard covered in wire or nylon mesh with a mesh size ≤ 2 mm. The 'qannić' is placed outside in a ventilated area, normally on a rooftop, to air dry the 'Ġbejna' in a natural environment. The time required for complete drying depends on the wind direction; northerly blowing wind ('riħ fuq') is considered better than southerly blowing wind ('riħ isfel'). Once the 'cheeselet' has hardened sufficiently, it can be sold as ('Ġbejna moxxa'), or it can be pickled ('Ġbejna tal-bżar'). For pickling, the 'Ġbejna moxxa' is placed in white vinegar for as long as 24 hr, after which it is coated with coarsely ground black pepper. Some variation utilising red wine vinegar mostly obtained from Maltese grapes also existed.

Both dried forms can be preserved for months (Cassar, 2010). These are either refrigerated or kept in glass jars or ceramic containers that are tightly covered to prevent the product from becoming humid and mouldy. These jars are frequently stored in ambient conditions in kitchens and pantries.

4.7.5 Relevant Research

The mineral contents of sheep's milk and Maltese Ġbejna were determined using a Microwave Plasma-Atomic Emission Spectrophotometer (Agilent Technologies) (Attard, 2013). The accumulation of certain minerals, such as chromium, in cheese over milk results from the preferential binding of these minerals to caseins (Anastasio et al., 2006). In general, the mineral content is dependent on the fat (globules) and protein (molecules) content of the milk. Therefore, the transfer of nutrients from milk to the cheeselet is dependent in part on the coagulation process.

4.8 Serra da Estrela PDO – Portugal

Name: Serra da Estrela Production area: Serra da Estrela mountains – centre north of Portugal Milk: Sheep's, raw	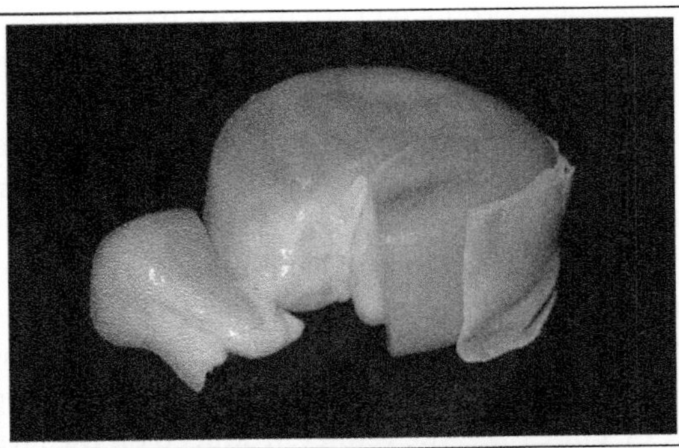
Name: Serra da Estrela Velho Production area: Serra da Estrela mountains – centre north of Portugal Milk: Sheep's, raw	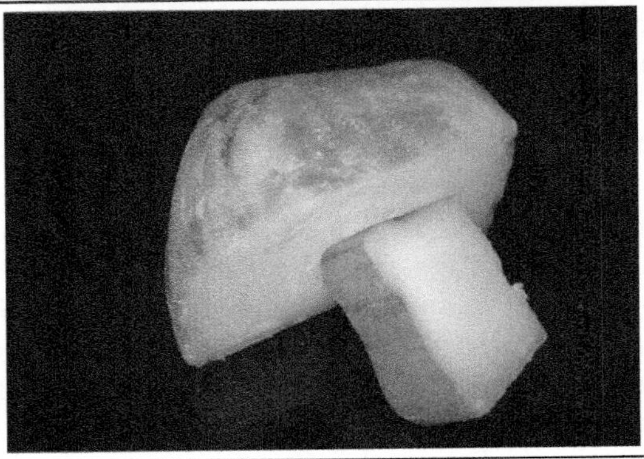

4.8.1 Introduction

Serra da Estrela is economically and organoleptically the most famous and important variety of traditional cheese manufactured in Portugal, owing to its unique organoleptic characteristics.

Serra cheese has been for hundreds of years a major element of the rich cultural heritage of the Beira region (centre of Portugal). Its production (about 1,920 tonnes per year), at the farmhouse level, has become the major source of wealth for Serra's region farmers. Dating from the Roman occupation of the Iberian Peninsula, the traditional techniques of Serra cheese manufacture have been passed through generations of shepherds. This cheese is produced mostly in the neighbourhood of Serra da Estrela mountains, the highest mountains in Portugal (about 2,000 m high), which for environmental protection has the status of a nature reserve. Characterised by a microclimate with long, cold, rainy and snowy winters, but hot and dry summers, this region supports the growth of natural pastures that help give the sheep's milk (and cheese obtained from there) a unique flavour. The exodus of younger generations from the farms to the cities has made it increasing difficult to find true Serra cheese.

Serra da Estrela cheese has had PDO status since 1996. It can only be manufactured from whole raw sheep's milk from Serra da Estrela region farms, using an extract of thistle flower (*Cynara cardunculus L.*) as rennet. Criteria for certification cover compliance with minimum requirements for organoleptic characteristics, shape, chemical and microbial contents of the final product, as well as hygiene during manufacture. In regard to the region of production, the PDO region of Serra da Estrela cheese has been established via a governmental act and includes the following: Celorico da Beira, Fornos de Algodres, Gouveia, Mangualde, Manteigas, Nelas, Oliveira do Hospital, Penalva do Castelo, Carregal do Sal and Seia (Anonymous, 1985).

However, owing to PDO requirements, only about 5% of that cheese is currently certified as Serra da Estrela cheese (Reis & Malcata, 2007). Farmers in these geographical regions are allowed to produce non-PDO sheep's cheese, but only certified cheeses command a premium price in the market: typical prices are 13 and 16 euros/kg for regular and certified cheese, respectively.

4.8.2 Milk

Milk is provided by sheep breeds well adapted to the specific geoclimatic conditions of the Serra da Estrela. A typical flock consists of either black or white varieties of Bordaleira da Serra da Estrela and/or Churra Mondegueira. In order to obtain good milk production, in quantity and quality, the pastures need to be as varied as possible in location; however, they must be ice-free pastures. Milking is done manually twice a day: at sunrise, before the flock is walked to the pasture, and at sunset, after the flock is returned to the stable. The Bordaleira breed is the best Portuguese native breed for milk production; it is able to produce between 150 and 200 litres of milk during the typical lactation season (usually from October to May).

4.8.3 Rennet

Plant species belonging to the *Cynara* genus, family of *Asteracea* – including *Cynara cardunculus L.* and *Cynara humilis L.*, and commonly known as thistle – have been used successfully in the Iberian Peninsula for the manufacture of traditional cheese from sheep's and/or goat's milks. In Portugal, the *C. cardunculus* species is more often used. This is a large plant that grows wild in arid, rocky and untamed areas. Thistle flowers are picked up and then air-dried, without moisture control and under direct sunlight. Since no standard conditions are used for harvesting or drying, the activity of the thistle extracts is extremely variable – and depends on the variety, degree of ripeness, drying time and part of flower used (Silva & Malcata, 2005).

4.8.4 Description and Sensory Characteristics

Serra da Estrela PDO has a flat-cylinder shape; bulging is considerable on the lateral sides, and slight on the upper surface. The diameter is 9–20 cm, height 4–6 cm and weight 0.5–1.7 kg. Serra da Estrela Velho has a flat-cylinder shape; slight or no bulging on the lateral sides and no spine. The diameter is 11–20 cm, height 3–6 cm and weight 0.7–1.2 kg.

Serra cheese exists indeed as two major categories: (1) a buttery one (the genuine 'Serra da Estrela' cheese (soft cheese)) with a ripening period of 30 to 45 days and (2) a ripened one, semi-hard cheese, that requires an ageing process of at least 120 days ('Serra da Estrela Velho' cheese). The moisture content ranges from 43% to 48% and from 35% to 39%, and the fat content ranges from 30% to 35% and is >35%, for the original and the mature variety, respectively.

According to European legislation (EC, 2013), Serra da Estrela cheese bearing the corresponding PDO label should possess a thin, uniform, smooth and well-formed rind and a soft, straw-yellowish paste. It should not have eyes, or have only a few small ones, and its colour should range from ivory to white. Serra da Estrela cheese has a buttery texture, which leads to spontaneous and rapid deformation upon slicing – and possesses a strong aroma, as well as a clean, smooth and slightly acid flavour. From a sensory perspective, odour descriptors associated with typical Serra da Estrela cheese encompass 'acidic', 'sweaty' and 'sheepy-like'; they are a result of the typical composition of sheep's milk, and the relatively short ripening period. Serra da Estrela Velho PDO should possess a smooth to slightly wrinkled surface, and a hard to extra hard rind; it should not have any eyes or have only a few eyes, a slightly dry crumbly body and a colour ranging from yellowish to orange/light brown, becoming darker from the outside towards the centre. Serra da Estrela Velho cheese has a pleasant, lingering, clean and intense nutty flavour – besides a creamy taste, strong to slightly strong and a slightly spicy and salty bouquet.

4.8.5 Method of Manufacture

Serra cheese is manufactured twice a day, mainly at the farm level, following techniques that have been empirically optimised over time. By legal definition, Serra cheese is made only from raw sheep's milk (on average, of 6% fat, 0.26% acidity and pH 6.5), coagulated with thistle flower extract.

Milk preparation: The shepherd does the milking manually. Sheep's milk is accumulated in larger containers, which are kept warm for 30 to 60 min. After that, milk is filtered through a fine clean cloth in order to remove particulates (e.g. hair and dust), and poured into a coagulation vat.

Coagulation: After reaching the desired coagulation temperature (27°C–29°C), plant rennet (thistle) is added (0.2–0.4 g/L of milk). Although the technique of rennet addition may vary, the most commonly used technique by cheesemakers is the immersion of a doll (a cloth with closed ends, containing dry thistle flower, salt and water) in the milk, followed by stirring and squeezing. Milk is kept at the coagulation temperature for 1–2 hr, and the end of coagulation is confirmed by slight agitation to empirically evaluate the consistency of the gel.

Cutting and draining of whey: The work on the curd and drainage of whey are performed on the top of a round, sloped table, which facilitates drainage of whey into an open vessel. The curd is cut and stirred very slowly (up to 1 hr/cheese), since the final texture of the cheese is strongly related to the technique used. At the first stage, cutting is performed by manually stirring the coagulum for about 10 to 15 s, in order to obtain irregular curd pieces. Those are poured into a flexible, perforated metal plate in the form of an open-ended cylinder with an adjustable diameter or in a well-washed cloth bag. In both cases, the curd is pressed to release as much whey as possible. Finally, the curd is placed into fixed-diameter, perforated plastic moulds (about 20 cm wide and 10 cm high), lined with a fine cloth.

Pressing: The drainage is completed via light pressing of the curd while in the mould; this procedure can be performed either by hand or mechanically. Pressing takes place for 3 to 24 hr, with a 6 to 20 kg stone placed on top of the curd, resting in the mould.

Salting: Crude, unrefined kitchen salt is added in two stages; in the first stage, salt is added to milk together with thistle extract at the beginning of the coagulation process. The second salt addition is performed after pressing. Salt is added to rub the top and bottom surfaces of the pressed cheese with crystals of salt, usually with a gap of from 6 to 24 hr between consecutive rubbings. The entire amount of salt used for either process is about 20 g/L of milk, or 0.5 to 0.9 g/cm^2 on the top surface of cheese and 0.5 to 0.6 g/cm^2 on the bottom surface, respectively.

Maturation: After salting and pressing, ripening takes place in a maturation chamber, in the open air, on wooden shelves. This step is usually performed without artificial control of room temperature and RH, and so it is directly dependent on the external weather conditions (which generally ranges within 6°C–12°C and 85%–90% respectively). The minimum ripening time for Serra cheese is 30 days for the original Serra cheese, and 120 days for the aged one. Each cheese is turned upside down once every day. The cheese is washed when a white to reddish viscous smear appears on the exposed surfaces, usually after 8 to 15 days. A band of linen or cotton cloth is wound around the cheese and tied up with a small knot, so as to prevent loss of the cylindrical shape of the cheese at later stages of maturation. At this time, the cheese is moved to a second maturation room, which usually has a cooler temperature than the first, and the same procedure is followed here as in the first room. In the case of Serra da Estrela Velho cheese, the band of linen or cotton cloth is taken off, and the cheese can be rubbed with sweet red pepper.

4.8.6 Relevant Research

Studies showed that the coagulation time does not affect cheese quality significantly, unlike higher coagulation temperatures that lead to losses in cheese softness (Freitas & Malcata, 2000). The rates of proteolysis and lipolysis tend to decrease with ripening time (Partidário, Barbosa & Vilas Boas, 1998; Freitas & Malcata, 2000; Tavaria et al., 2003). The constant temperature and RH in the maturation room are key factors, and ensure better final quality than periodic temperature and humidity fluctuations (Macedo, Malcata & Oliveira, 1993). The RH increases during ripening positively influence microbial growth, lactic acid content, proteolysis, lipolysis and the aroma and softness of the cheese (Freitas, Macedo & Malcata, 2000). Therefore, the best cheeses are produced between December and March. Pressing the curd avoids extensive development of cheese cracks, which is a condition for good final quality (Macedo, Malcata & Oliveira, 1993), and forced ventilation prevents the appearance of smear and crack formation.

Serra da Estrela cheese's microbiological profile is significantly affected by the period within the cheesemaking season, and by the axial location within the cheese (Freitas & Malcata, 2000). The adventitious microflora comprise several LAB and *Enterobacteriaceae* spp. Their contribution to flavour development throughout ripening is crucial, especially due to the production of small volatile molecules with a very low odour threshold (Freitas & Malcata, 2000). The profile of volatile compounds in Serra da Estrela cheese was duly characterised (Freitas, Macedo & Malcata, 2000; Tavaria, Silva-Ferreira & Malcata, 2004) along its regular ripening period – and free fatty acids were found to be the major contributors to the overall aroma. Studies comparing plant and animal rennet showed a lower level of primary proteolysis in cheeses made with the former (Freitas, Macedo & Malcata, 2000). In an attempt to ensure quality and safety standardisation while retaining Serra cheese's typical chemical and sensory attributes, new possibilities have been tested for starter cultures. Therefore, microorganisms previously isolated from the adventitious microflora of Serra cheese were used (Macedo, Tavares & Malcata, 2004; Tavaria & Malcata, 2003; Pereira et al., 2008, 2010).

4.9 Torta del Casar PDO – Spain

Name: Torta del Casar, PDO	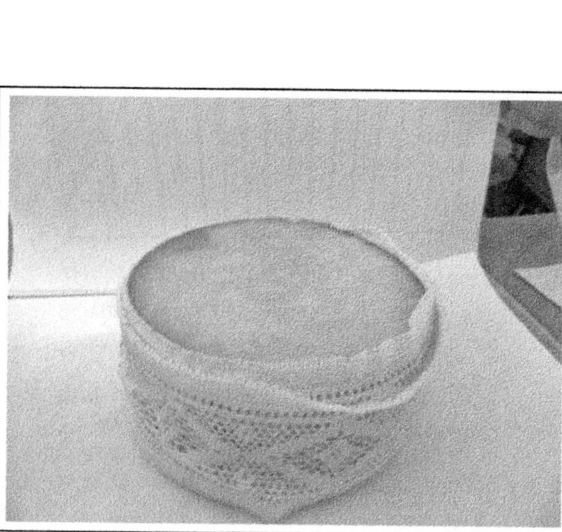
Production area: Extremadura, Spain	
Milk: Sheep's, raw	

4.9.1 Introduction

The cheese is made from raw sheep's milk. Under the PDO regulation, 21 farmers and 11 dairy plants are registered. During 2013, production was 297,536 kg, which represented a total of 3,930,000 euros. The cheese is mainly sold on the national market and represents 1.86% of the economic value of Spanish PDO cheese production (Ministry of Agriculture, Food and Environment, 2014). The PDO status of Torta del Casar was recognised by the Ministry of Agriculture, Fisheries and Food in 2002 and at the European level in 2003 (EC, 2003). Cheeses under the PDO Origin Torta del Casar must have a designation label bearing a sequential alphanumeric code and the official logo of the Designation of Origin.

Transhumance and grazing have since time immemorial been a feature of the districts situated in the geographical area of production. After undergoing a number of changes over the years, the drovers' routes were officially recognised in 1273 as obligatory passageways for flocks under the rules governing the 'Honrado Concejo de la Mesta'. There is evidence of the presence of established flocks in 1291, when King Sancho IV allowed the use of land around the town of Casar as grazing land for flocks. Already then, Torta del Casar was used as a means of payment, but it is not until 1791 that we find the first written reference to sheep's milk cheese from Casar de Cáceres (Anonymous, 2015)

This cheese is made in 36 municipalities in the districts of Los Llanos de Cáceres, Sierra de Fuentes and Montánchez, situated in the province of Cáceres.

4.9.2 Type

It is a soft raw sheep's cheese, coagulated with vegetable coagulant derived from *Cynara cardunculus* spp., which, due to the high proteolysis, imparts to this cheese its particular texture of 'Torta' and has a minimum ripening time of 60 days. The minimum fat and protein content in dry matter are 35% and 50%, respectively, with a maximum moisture content of 50% and a pH value of 5.2–5.9 (Ministry of Agriculture, Food and Environment, 2015).

4.9.3 Milk

The milk used to make the protected cheeses is obtained from healthy sheep of Merino and Entremerino breeds, in accordance with the legislation. The livestock's diet is based wholly or partly on grazing; any supplement fed to the sheep must be subjected to the Regulatory Council controls.

4.9.4 Description and Sensory Characteristics

It is a soft, pressed cheese coagulated with *Cynara cardunculus* flower heads. The cheese has a cylindrical or convex shape with a height of 5–7 cm and a diameter of 11–17 cm. It may weigh 0.5–1.1 kg.

The rind is deep yellow, shiny and greasy in appearance. The body is soft with a white-yellowish colour. It has a strong smell, and its aroma is milky, reminiscent of the smell of yogurt or butter depending on the age, with light hints of animal and plant. Its taste is slightly salty and bitter due to the vegetable coagulant used. The texture is soft and even liquid, spreadable, greasy with moderate or high creaminess in matured cheese, and slightly astringent.

4.9.5 Method of Manufacture

Milk preparation: The milk must be collected from the registered farms and kept after milking at no more than 4°C. The milk used for cheesemaking must meet the requirements laid down by legislation.

Coagulation: This process is carried out by using vegetable coagulant derived from *Cynara cardunculus* spp. at a temperature of 28°C–32°C for 50–80 min.

Cutting: The curd must be cut to the size of rice-like grains.

Curd draining: The curd is placed in moulds of the shape and size required to produce cheeses with PDO characteristic properties of the certified product. The cheeses are subjected to a pressure of 1–2.5 kg/cm^2 for 3–8 hr.

Salting: Brining or dry salting is used. When brine is used, the maximum period of salting is 5-6 hr with a maximum concentration of 16°Bé.

Maturation: Cheeses are ripened for 60 days in chambers at a temperature of 4°C–12°C and an RH of 75%–90%.

4.9.6 Relevant Research

Proteolysis determines the soft and spreadable texture observed in Torta del Casar cheese. Proteolysis is less intense in the first 30 days of maturation than between the 30–60 days of ripening. α_{s1}-Casein degradation is more intense than that of β-caseins, and non-casein proteins/peptides are also hydrolysed at this stage, accompanied by an increase in the free fatty acid levels (Delgado et al., 2010b).

The use of controlled and characterised cardoons in the manufacturing process of Torta del Casar is fundamental to obtaining a homogeneous product, and therefore the most appropriate cardoons for making this cheese are those with higher clotting activities and moderate proteolytic activities, especially on β-casein (Cavalli et al., 2013; Fernández-Salguero & San Juan, 1999; Ordiales et al., 2012; Ordiales et al. 2014). An accurate characterization of *Cynara cardunculus* was carried out by Ordiales et al., 2015 to differentiate it from the wild cardoon by using a developed PCR method. Other studies have focused on the characteristics of proteolysis in Torta del Casar and demonstrated a highly significant correlation between textural parameters, residual caseins levels and nitrogen fractions during maturation, which shows the importance of proteolytic changes for optimal texture formation (Delgado et al., 2010b; Ordiales et al., 2013b), its volatile profile (Delgado et al., 2010a) and free fatty acids and oxidative changes (Delgado et al., 2009). It was observed that changes in free fatty acids, mainly in short-chain fatty acids, could play an important role for cheese final characteristics. On the other hand, oxidative reactions did not play such a role (Delgado et al., 2009).

Although traditionally no starters are used for this type of cheese, several studies have been conducted on the growth and development of fortuitous flora and food pathogens (Ordiales et al., 2013a), characterisation of secondary flora (Caceres, Castillo & Pizarro, 1997) and the influence of microflora on the flavour (Ordiales et al., 2013c). The use of emerging technologies such as high pressure and their influence on microbiological and textural changes has been evaluated, and these authors observed a reduction in the levels of butyric acid in high-pressure-treated cheeses which may help prevent the formation of over-ripened aromas which are rejected by consumers (Rodriguez-Pinilla et al., 2015).

References

Anastasio, A., Caggiano, R., Macchiato, M., Paolo, C., Ragosta, M., Paino, S. & Cortesi, M. L. (2006). Heavy metal concentrations in dairy products from sheep milk collected in two regions of southern Italy. *Acta Veterinaria Scandinavica*, 47 (1), 69.

Anonymous (2015). *Official website of Torta del Casar*. www.tortadelcasar.eu. (Accesed 4 June 2015).

Anonymous (1985). *Diário da República*. I-Série. N° 152 de 5-7-1985. Decreto Regulamentar N° 42/85.Imprensa Nacional-Casa da Moeda, Lisbon, Portugal.

Attard, E. (2013). *Valorisation of the Maltese Ġbejna through Chemometric and Proximate Analysis.* Op Italia-Malta 2007–2013. t-cheesimal project. New technologies supporting traditional cheeses in Sicily and Maltese islands. Final Conference. 4th September, University of Malta, Malta.

Briggiler Marcó, M. M., Capra, L., Quiberoni, A., Vinderola, G., Reinheimer, J. A. & Hynes, E. (2007). Nonstarter *Lactobacillus* Strains as adjunct cultures for cheese making: In vitro characterization and performance in two model cheeses. *Journal of Dairy Science*, 90, 4532–4542.

Bude Ugarte, M., Guglielmotti, D., Giraffa, G., Reinheimer, J. A. & Hynes, E. (2006). Nonstarter Lactobacilli Isolated from Soft and Semihard Argentinean Cheeses: Genetic Characterization and Resistance to Biological Barriers. *Journal of Food Protection*, 69, 2983–2991.

Burns, P. F., Cuffia, F., Vinderola, G., Sabbag, N., Meinardi, C. & Hynes, E. (2012). Technological and probiotic role of adjunct cultures of non starter lactobacilli in soft cheeses. *Food Microbiology*, 30, 45–50.

Caceres, P., Castillo, D. & Pizarro, M. (1997). Secondary flora of Casar de Caceres cheese: characterization of *Micrococcaceae. International Dairy Journal*, 7, 531–536.

Cárcoba, R., Delgado, T. & Rodrνguez, A. (2000). Comparative performance of a mixed strain starter in cow's milk, ewe's milk and mixtures of these milks. *European Food Research Technology*, 211, 141–146.

Cárcoba, R., Pin, C. & Rodrνguez, A. (2004). Growth and metabolic behavior of *Lactococcus lactis* subsp. *lactis* IPLA 947 in anaerobic lactose-limited chemostat cultures. *European Food Research Technology*, 219, 277–281.

Carpino, S., La Terra, S., Ben Younes, R., Farina, G., La Terra, F., Guardiano, C., Ottaviano, M., Rapisarda, T., Licitra, G. (2013). Part 3. Experimental Part. In Attard, E., Carpino, S., Licitra, G. *"New technologies supporting the traditional Cheeses in Sicily and Malta Islands"* T-Cheesimal. Tipografia MODUL MOTTA srl – Ragusa.

Cassar, C. (2010). *Cheeselets in Maltese Food Culture*, April, published in 'Il-Ġens ILLUM'.

Cavalli, S. V., Lufrano, D., Colombo, M. L. & Priolo, N. (2013). Properties and applications of phytepsins from thistle flowers. *Phytochemistry*, 92, 16–32.

Código Alimentario Argentino (2014). *Cap. VIII: Alimentos lácteos.* http://www.anmat.gov.ar/alimentos/codigoa/CAPITULO_VIII.pdf. Accessed 30 June 2015.

Cuesta, P., Fernαndez-Garcνa, E., Gonzαlez de Llano, D., Montilla, A. & Rodrνguez, A. (1996). Evolution of the microbiological and biochemical characteristics of Afuega'l Pitu cheese during ripening. *Journal of Dairy Science*, 79, 1693–1698.

Delgado, F. J., Gonzalez-Crespo, J., Cava, R., Garcia-Parra, J. & Ramirez, R. (2010a). Characterisation by SPME-GC-MS of the volatile profile of a Spanish soft cheese P.D.O. Torta del Casar during ripening. *Food Chemistry*, 118, 182–189.

Delgado, F. J., Gonzalez-Crespo, J., Ladero, L., Cava, R. & Ramirez, R. (2009). Free fatty acids and oxidative changes of a Spanish soft cheese (PDO 'Torta del Casar') during ripening. *International Journal of Food Science and Technology*, 44, 1721–1728.

Delgado, F. J., Rodriguez-Pinilla, J., Gonzalez-Crespo, J., Ramirez, R. & Roa, I. (2010b). Proteolysis and texture changes of a Spanish soft cheese ('Torta del Casar') manufactured with raw ewe milk and vegetable rennet during ripening. *International Journal of Food Science and Technology*, 45, 512–519.

EC (1996) European Commission Regulation (EC) No 1107/96 of 12 June 1996 on the registration of geographical indications and designations of origin under the procedure laid down in Article 17 of Council Regulation (EEC) No 2081/92. *Official Journal of the European Union* L 148, 1 – 10.

EC (2003). Commission Regulation (EC) No 1491/2003 of 25 August 2003 supplementing the Annex to Regulation (EC) No 2400/96 (Ficondindia dellÉtna, Monte Etna, Colline Romagna, Pretuzziano delle Colline Teramane, Torta del Casar, Manzana de Gerona or Poma de Girona). *Official Journal of European Union*, L214, 6–7.

EC (2008a). Commission Regulation (EC) No 676/2008 of 16 July 2008 registering certain names in the Register of protected designations of origin and protected geographical indications (Ail de la Drôme (PGI), Všestarská cibule (PDO), Slovenská bryndza (PGI), Ajo Morado de Las Pedroñeras (PGI), Gamoneu or Gamonedo (PDO), Alheira de Vinhais (PGI), Presunto de Vinhais or Presunto Bísaro de Vinhais (PGI)). *Official Journal of the European Union*, L189, 19–20.

EC (2008b). Commission Regulation (EC) No 723/2008 of 25 July 2008 registering certain names in the Register of protected designations of origin and protected geographical indications (Afuega'l Pitu (PDO), Mazapαn de Toledo (PGI), Agneau de Lozθre (PGI), Oignon doux des Cιvennes (PDO), Butelo de Vinhais or Bucho de Vinhais or Chouriηo de Ossos de Vinhais (PGI), Chouriηa Doce de Vinhais (PGI)). *Official Journal of the European Union*, L 198, 28–29.

EC (2013). Council of Europe, Commission implementing Regulation (EU) No 562/2013 of 14 June 2013 approving a minor amendment to the specification for a name entered in the register of protected designations of origin and protected geographical indications [Queijo Serra da Estrela (PDO)]. *Official Journal of the European Union*, L 167, 10–14.

Fernández-Salguero, J. & San Juan, E. (1999). Influence of vegetable and animal rennet on proteolysis during ripening of Casar de Caceres cheese. *Food Chemistry*, 64, 177–183.

Fiorini S. (2006) Documentary sources of Maltese History Part I. Notarial Documents No 1. *Notary Giacomo Zabara* R 494/1(1): 1487-1488. Malta. Pp. 249-250, 291-292, 330.

Freitas, A. C., Macedo, A. C. & Malcata, F. X. (2000). Review: Technological and organoleptic issues pertaining to cheeses with denomination of origin manufactures in the Iberian Peninsula from ovine and caprine milks. *Food Science and Technology International*, 6, 351–370.

Freitas, C. & Malcata, F. X. (2000). Microbiology and biochemistry of cheeses with *Appellation d'Origine Protégée* and manufactured in the Iberian peninsula from ovine and caprine milks. *Journal of Dairy Science*, 83, 584–602.

Gonzalez de Llano, D., Polo, M. C. & Ramos, M. (1994). Study of proteolysis in liquid chromatography artisanal cheeses: high performance of peptides. *Journal of Dairy Science*, 78, 1018–1024.

Gonzalez de Llano, D. & Santa Maria, G. (1997). Analysis of whey proteins during ripening of Afuega'l Pitu cheese by reversed phase HPLC. *Food Science and Technology International*, 6, 445–449.

Hatzikamari, M., Litopoulou-Tzanetaki, E. & Tzanetakis, N. (1999). Microbiological characteristics of Anevato: A traditional Greek cheese. *Journal of Applied Microbiology*, 87, 595–601.

Hynes, E., Candioti, M. C., Meinardi, C., Sabbag, N., Cattaneo, T. & Zalazar, C. A. (2001). Influence of milk clotting enzyme concentration in the αs1 hydrolysis during soft cheeses ripening. *Journal of Dairy Science*, 84, 1335–1340.

Hynes, E., Delacroix-Buchet, A. & Zalazar, C. A. (2001). Influence of milk-clotting enzyme in proteolysis during ripening of Cremoso Argentino cheese. *Australian Journal of Dairy Technology*, 56, 208–212.

Kaminarides, E. S. (1986). Quality of the Kopanisti cheese. *Deltio Ethnikis Epitropis Gallaktos Ellados*, 3, 1.

Karali, F., Georgala, A., Massouras, T. & Kaminarides S. (2012). Volatile compounds and lipolysis levels of Kopanisti, a traditional Greek raw milk cheese. *Journal of the Science of Food and Agriculture*, 93 (8), 1845–1851.

Katsiari, M. C., Kondyli, E. & Voutsinas, L. P. (2009). The quality of Galotyri-type cheese made with different starter cultures. *Food Control*, 20, 113–118.

Kondyli, E., Katsiari, M. C. & Voutsinas, L. P. (2008). Chemical and sensory characteristics of Galotyri-type cheese made using different procedures. *Food Control*, 19, 301–307.

Lattrell, A. (1975). *Approaches to Medieval Malta*. London, British School at Rome, pp. 36–40.

Macedo, A. C., Malcata, F. X. & Oliveira, J. C. (1993). The technology, chemistry and microbiology of Serra cheese: A review. *Journal of Dairy Science*, 76, 1725–1739.

Macedo, A. C., Tavares, T. G. & Malcata, F. X. (2004). Influence of native lactic acid bacteria on the microbiological, biochemical and sensory profiles of Serra da Estrela cheese. *Food Microbiology*, 21, 233–240.

Mason, I. L. (1996). *A World Dictionary of Livestock Breeds, Types and Varieties*, 4th edition. CAB International.

Milesi, M. M., McSweeney, P. L. H. & Hynes, E. (2008). Viability and contribution to proteolysis of an adjunct culture of *Lactobacillus plantarum* in two model cheese systems: Cheddar cheese-type and soft-cheese type. *Journal of Applied Microbiology*, 105, 884–892.

Milesi, M. M., Wolf, I. V., Bergamini, C. V. & Hynes, E. R. (2010). Two strains of non-starter lactobacilli increased the production of flavour compounds in soft cheeses. *Journal of Dairy Science*, 93 (11), 5020–5031.

Ministry of Agriculture, Food and Environment. (2015). *Electronic Catalogue of Spanish Cheeses* http://www.alimentacion.es/es/conoce_lo_que_comes/bloc/queso/default/principal.(Accessed 27 May 2015).

Ministry of Agriculture, Food and Environment. (2014). *Secretaría General Técnica Centro de Publicaciones. Datos de las Denominaciones de Origen Protegidas (D.O.P.) e Indicaciones Geográficas Protegidas (I.G.P.) de Productos Agroalimentarios, 3*, 22–23.

Ordiales, E., Benito, M. J., Martin, A., Casquete, R., Serradilla, M. J. & Guia-Cordoba, M. de G. (2013a). Bacterial communities of the traditional raw ewe's milk cheese 'Torta del Casar' made without the addition of a starter. *Food Control*, 33, 448–454.

Ordiales, E., Benito, M. J., Martin, A., Fernández, M., Hernandez, A. & Cordoba, M. de G. (2013b). Proteolytic effect of *Cynara cardunculus* rennet for use in the elaboration of 'Torta del Casar' cheese. *Journal of Dairy Research*, 80, 429–438.

Ordiales, E., Fernandez, M., Benito, M. J., Hernandez, A., Martin, A. & Cordoba, M. G. (2015). Differentiation of wild cardoon quality used in the elaboration of traditional cheeses by DNA typing analytical methods. *Food Analytical Methods*, 8, 7–17.

Ordiales, E., Martin, A., Benito, M. J., Hernandez, A., Ruiz-Moyano, S. & Guia Cordoba, M. de G. (2012).Technological characterisation by free zone capillary electrophoresis (FCZE) of the vegetable rennet (*Cynara cardunculus*) used in 'Torta del Casar' cheese-making. *Food Chemistry*, 133, 227–235.

Ordiales, E., Martin, A., Benito, M. J., Fernández, M., Casquete, R. & Córdoba, M. G. (2014). Influence of the technological properties of vegetable rennet (*Cynara cardunculus*) on the physicochemical, sensory and rheological characteristics of 'Torta del Casar' cheese. *International Journal of Dairy Technology*, 67, 402–409.

Ordiales, E., Martin, A., Benito, M. J., Hernandez, A., Ruiz-Moyano, S. & Cordoba, M. de G. (2013c). Role of the microbial population on the flavor of the soft-bodied cheese Torta del Casar. *Journal of Dairy Science*, 96, 5477–5486.

Partidário, A. M., Barbosa, M. & Vilas Boas, L. (1998). Fatty acids, triglycerides and volatile compounds in Serra da Estrela cheese – changes throughout ripening. *International Dairy Journal*, 8, 873–881.

Pereira, C. I., Gomes, E. O., Gomes, A. M. P. & Malcata, F. X. (2008). Influence of bacterial dynamics upon the final characteristics of model Portuguese traditional cheeses. *Food Microbiology*, 3, 862–868.

Pereira, C. I., Neto, D. M., Capucho, J. C., Gião, M. S., Gomes, A. M. P. & Malcata, F. X. (2010). How three adventitious lactic acid bacteria affect proteolysis and organic acid production in model Portuguese cheeses manufactured from several milk sources and two alternative coagulants. *Journal of Dairy Science*, 93, 1335–1344.

Portal Lechero (2015). *Ultrafiltración para fabricación de queso*. http://www.portalechero.com/innovaportal/file/4698/1/2-ultrafiltracion-para-fabricacion-de-queso.pdf. Accessed 30 June 2015.

Reinheimer, J., Bailo, N., Binetti, A., Quiberoni, A. & Renzulli, P. (1995). Fermentos naturales de leche empleados en quesería. Características tecnológicas y microbiológicas. *Revista Argentina de Lactología*, 7 (12), 21–34.

Reis, P. M. J. & Malcata, F. X. (2007). Improvements in small scale artisanal cheesemaking via a novel mechanized apparatus. *Journal of Food Engineering*, 82, 11–16.

Rilla, N., Martνnez, B. & Rodriguez, A. (2004). Inhibition of a methicillin-resistant *Stahphylococcus aureus* strain in Aguega´l Pitu cheese by the nisin Z-producing strain *Lactococcus lactis* subsp *lactis* IPLA 729. *Journal of Food Protection*, 67, 928–933.

Rodriguez-Pinilla, J., Marquez, G., Tabla, R., Ramirez, R. & Delgado, F. J. (2015). Microbiological and lipolytic changes in high-pressure-treated raw milk cheeses during refrigerated storage. *Dairy Science and Technology*, 95, 425–436.

Sánchez, J. I., Martνnez, B. & Rodrνguez, A. (2005). Rational selection of *Leuconostoc* strains for mixed starters based on the physiological biodiversity found in raw milk fermentations. *International Journal of Food Microbiology*, 105, 377–387.

Silva, S. V. & Malcata, F. X. (2005) Studies pertaining to coagulant and proteolytic activities of plant proteases from Cynara cardunculus. *Food Chemistry* 89, 19-26.

Tanti, R. (2014). *Sheep and Goats Census: 2014. External Cooperation and Communication Unit, National Statistics*. https://nso.gov.mt/en/News_Releases/View_by_Unit/Unit_B3/Agricultural_and_Environment_Statisti cs /Documents/2015/News2015_024.pdf

Tavaria, F. K. & Malcata, F. X. (2003). Enzymatic activities of non-starter lactic acid bacteria isolated from a traditional Portuguese cheese. *Enzyme and Microbial Technology*, 33, 236–243.

Tavaria, F. K., Franco, I., Carballo, F. J. & Malcata, F. X. (2003). Amino acid and soluble nitrogen evolution throughout ripening of Serra da Estrela cheese. *International Dairy Journal*, 13, 537–545.

Tavaria, F. K., Silva-Ferreira, A. C. & Malcata, F. X. (2004). Volatile free fatty acids as quality indicators for consumption of Serra da Estrela cheese. *Journal of Dairy Science*, 87, 4064–4072.

Tzanetakis, N., Litopoulou-Tzanetaki, E. & Manolkidis, K. (1987). Microbiology of Kopanisti, a traditional Greek cheese. *Food Microbiology*, 4, 251–256.

Vinderola, C. G., Prosello, W., Ghiberto, D. & Reinheimer, J. A. (2000). Viability of Probiotic (*Bifidobacterium, Lactobacillus acidophilus* and *Lactobacillus casei*) and nonprobiotic microflora in Argentinian Fresco cheese. *Journal of Dairy Science*, 83 (9), 1905–1911.

Vinderola, G, Prosello, W., Molinari, F., Ghiberto, D. & Reinheimer, J. (2009). Growth of *Lactobacillus paracasei* A13 in Argentinian probiotic cheese and its impact on the characteristics of the product. *International Journal of Food Microbiology*, 135 (2), 171–174.

Wettinger, G. (1986). 'The Arabs in Malta', second and much lengthened edition. In *Malta: Studies of its Heritage and History*. Malta, pp. 87–104.

Zalazar, C. A., Meinardi, C. & Hynes, E. (1999). *Quesos típicos argentinos. Una revisión general sobre producción y características*. Ediciones UNL, Santa Fe, Argentina.

Zalazar, C. A., Zalazar, C. S., Bernal, S., Bertola, N. Bevilaqua, A. & Zaritzky, N. (2002). Effect of moisture level and fat replacer on physicochemical, rheological and sensory properties of low fat soft cheeses. *International Dairy Journal*, 12 (1), 45–50.

Zalazar, C. A., Candioti, M. C., Mercanti, D., Bergamini, C. V. & Meinardi, C. A. (2006). Maduración de quesos y su control. In Reinheimer, J. A. & Zalazar, C. A. (eds.), *Advances en Microbiología, Bioquímica y Tecnología de Quesos*. UNL Ediciones, Santa Fe, Argentina, pp. 117–243.

Zygouris, M. P. (1952). *The Dairy Industry*, 2nd edition. Ministry of Agriculture, Athens, Greece.

5

Dutch-Type Cheeses

Eva-Maria Düsterhöft, Wim Engels and Thom Huppertz

NIZO Food Research, Ede, The Netherlands

5.1 Edam Cheese – The Netherlands

Name: Edam cheese	
Production area: The Netherlands	
Milk: Cow's, pasteurised	

5.1.1 Introduction

Edam cheese is a Dutch semi-hard cheese; as a generic type, it is produced in many European countries and even worldwide. For the Edam cheese produced in the Netherlands, different authenticity certifications apply; in 2010, the designation 'Edam Holland' (PGI, Protected Geographical Indication) was recognised at the European level for Edam cheese which is produced in the Netherlands, from Dutch milk and which is matured under the traditional 'natural ripening' regime (under 'drying conditions') (EC, 2010a). The designation 'Noordhollandse Edam' as PDO was recognised in the EU in 1996 for Edam cheese which is produced in the Netherlands, from milk deriving exclusively from the Dutch province of 'Noord Holland', and which is matured under the traditional 'natural ripening' regime (EC, 1996). Edam cheese is a semi-hard cheese primarily made from pasteurised cow's milk. The production volume of Edam cheeses, with the characteristic fat content of 40% in dry matter, was 107×10^3 tonnes in the Netherlands and 188×10^3 tonnes in Germany in 2012. Edam cheese production has developed since the Middle Ages and reached its full growth in the seventeenth century. The cheeses, with a spherical shape, were originally produced in the northern part of the Dutch province of Noord Holland and were sold and sent for overseas shipment to the town Edam (situated in this same area), from which the cheese derives its name. Due to their particular shape and size

Global Cheesemaking Technology: Cheese Quality and Characteristics, First Edition.
Edited by Photis Papademas and Thomas Bintsis.

(~1.7 kg) and quite firm consistency, Edam cheeses were well suited for export. To enhance their shelf life, they were originally coated with a red colour, which was later replaced by red paraffin. Edam cheese was originally prepared from a blend of evening milk (skimmed) and full-fat morning milk, but since the nineteenth century, the skimmed version with a fat content of ~40% in dry matter has taken over as the predominant Edam cheese. The original production area of Edam cheese was the northern part of the Dutch province of Noord Holland. However, nowadays Dutch Edam is produced in a several factories situated across the country. For the PDO specialty 'Noordhollandse Edam', it is mandatory that the milk should be derived exclusively from Noord Holland. Cheeses carrying the generic name 'Edam' are, however, produced all over the world.

5.1.2 Type

To comply with the PGI/PDO criteria, Dutch Edam cheese contains 40%–44% FDM, has a maximum moisture content of 46% and a salt content of 2.4%–2.9% (w/w) at 14 days, for Edam Holland (PGI), and a maximum of 2.1% for the PDO type 'Noordhollandse Edam'. Both are characteristically higher than for Gouda cheese. The typical pH is 5.1–5.3 (depending on the cheeses' size and form), which is thus similar or somewhat lower than the pH of Gouda cheese. For specific types (e.g. 'brittle Edam'), the pH is as low as 4.9–5.0. The compositional criteria for Edam cheese is a minimum fat-in-dry-matter (FDM) of 30% and a maximum moisture of 53% (Codex Alimentarius, 1966). Naturally ripened Dutch Edam cheese is available in several maturation stadia, starting with 'young' cheese with a minimum of four weeks of ripening and ranging – for a small part – to over a year.

5.1.3 Description and Sensory Characteristics

The typical and unique shape of Edam cheese is spherical (with a small flat bottom and top side). It is produced in various sizes: 1 kg (Baby-Edam, about 10 cm diameter), 1.7–2 kg (standard sphere size), 4 kg (so-called Commissiekaas), but also in block size (16–20 kg) and as a loaf (2–5 kg). Naturally ripened Edam cheese has a moisture-permeable, usually transparent, plastic coating, giving the intact cheese a shining, dark yellowish appearance. In particular, the cheeses for export are coated in (red or yellow) wax, often after having undergone the desired period of natural ripening. Foil-ripened Edam cheese does not undergo rind development. Edam cheese normally has a semi-hard consistency and a smooth texture, with or without a few pea-sized holes in the cross section. Edam cheese is generally well sliceable (non-sticking). After prolonged natural ripening, the consistency becomes increasingly firm and short. The latter characteristics apply even more so – and at all maturation stages – for the 'Brittle Edam' variant, due to its low pH. The colour of the body is usually deeper yellow/orange than that of Gouda cheese. In naturally ripened Edam cheese, a darkish, somewhat glassy rind (1 cm) region develops upon maturation. The flavour of Edam cheese varies with the stage of maturity and the type of starter culture used. A young cheese is generally mild, somewhat acidic/tart and salty. The flavour becomes more pronounced with piquant notes with maturation.

5.1.4 Method of Manufacture

The process typically includes the following steps.

Milk preparation: The same procedure is used as for Gouda cheese, except for standardisation to the required fat:protein ratio (approx. 1.05 for Edam cheese with 48% FDM) by mixing full-fat milk and skimmed milk or membrane concentrated skim milk. Whey cream (fat recovered centrifugally from whey) may also be used in standardisation. After standardisation, the milk is cooled to 5°C before being used, generally within 48 hr. Prior to cheesemaking, the milk is pasteurised (HTST, 70°C–75°C for 10–20 s). Milk for Gouda cheese production is generally in-line bactofugated using specific self-desludging centrifuges, to remove the spores of *Cl. tyrobutyricum*, a feed-derived microorganism that may cause late blowing defects in Edam-type cheeses.

Other ingredients: Calcium chloride is added (see Gouda cheese). Sodium or potassium nitrate is traditionally added to prevent the germination of *Clostridium tyrobutyricum*, which leads to the 'late blowing' defect. It is not nitrate but its reduced form, nitrite and/or degradation products of nitrite, which affects spore germination. Xanthine oxidase, an enzyme present on the milk fat globule membrane, catalyses the reaction (Stadhouders, 1990). As nitrate is undesirable in whey, in particular for infant nutrition applications, it is sometimes added to the whey–curd mixture after removal of the first whey (which then remains free of nitrate), rather than to the milk. Carotenoid-based colourings like Annatto or beta-carotene are added to cheesemilk to obtain the typical yellow/orange colour of Edam cheese and to maintain uniform colour intensity throughout the year. For Commissiekaas, an about 25× higher colouring dosage is applied to obtain the intensely dark orange colour of this particular variant.

Starter culture (see Gouda cheese): Cultures used in the Netherlands for Edam-type cheese, for example, 'BOS' and 'A', thus have a long history of use in the dairy environment and can be considered as domesticated cultures (Smid et al., 2014).

Rennet: Traditionally, for Edam cheese, calf's rennet is used. Alternative rennet sources, for example, microbial rennet (such as from *Mucor miehei*) or fermentation-produced chymosin (FPC) may be used for particular purposes.

Coagulation: Coagulation and curd preparation in large industrial factories take place in closed OST or Double-O vats with 15,000–30,000 litres capacity. Rennet addition and coagulation take place at 30°C–31°C, and the coagulation process takes approximately 20–30 min.

Cutting: The same procedure as for Gouda cheese is used.

Washing and scalding: The same procedure as for Gouda cheese is used, with the difference that the scalding temperature ranges from 31°C to 33°C, and stirring is continued for another 30 min.

Curd draining and pressing: The same procedure as for Gouda cheese is used.

Salting: When the pH of the cheese has reached 5.5–5.8, which may be after a short holding period after pressing, the cheeses are immersed in a brine bath. The brine (18°Bé–20°Bé) contains ~0.2% calcium and has a pH of 4.5 and a temperature of 13°C. Brining times vary from 48 to 96 hr for the small 2 kg sized and large 16 kg blocks, respectively.

Maturation: After brining, Edam cheese is either naturally ripened (this process is obligatory for the PGI and PDO varieties 'Holland Edam' and 'Noordhollandse Edam') at 13°C–15°C and 85%–88% RH, or vacuum-packed and foil-ripened (typically at <8°C). The latter process probably accounts for the majority of the Edam-type cheeses produced outside the Netherlands. (For natural ripening, see Gouda cheese.) The intensive treatment (transport, turning, coating and cleaning of shelves) takes place in well-insulated, highly mechanised maturation rooms. Naturally ripened Edam cheese is typically matured for four weeks (young) until up to over a year. During this period, its moisture content gradually decreases. Foil-ripened Edam cheeses are produced with starter cultures having a low CO_2-forming capacity (e.g. O-starter, or L-starters), to prevent loosening of the wrapping (which has low moisture and oxygen permeability). The rectangular cheese blocks are piled and stored in larges boxes during maturation. Maintenance of the cheeses is unnecessary. The different ripening conditions (low temperature, O- or L-starter and anaerobic conditions at the cheese surface) all contribute to a generally less pronounced, rather flat and bland flavour profile of foil-ripened Edam cheese.

5.2 Gouda – The Netherlands

Name: Gouda	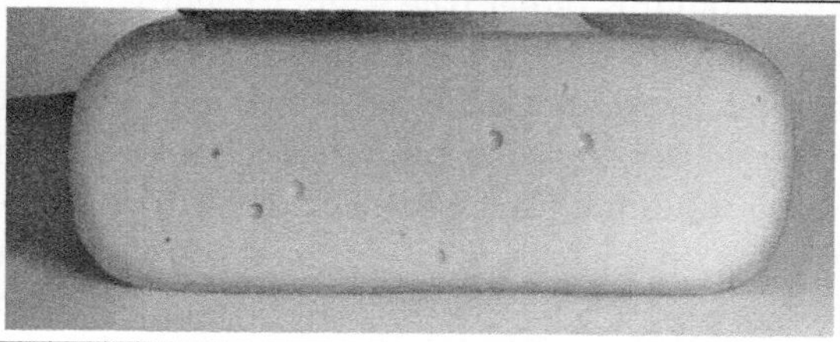
Production area: The Netherlands	
Milk: Cow's, pasteurised	

5.2.1 Introduction

Gouda cheese is a Dutch semi-hard cheese; as a generic type it is produced worldwide. For Gouda cheese produced in the Netherlands (~50% of total Dutch cheese production), different authenticity certifications apply. That is, in 2010, the designation 'Gouda Holland' PGI was recognised at the European level for Gouda cheese produced from Dutch milk in the Netherlands, and matured under the traditional 'natural ripening' regime (under 'drying conditions') (EC, 2010b). The designation 'Noordhollandse Gouda' PDO was recognised at the European level in 1996 for Gouda cheese produced in the Netherlands from milk derived exclusively from the Dutch province 'Noord Holland', and matured under the traditional 'natural ripening' regime (EC, 1996). The major producing countries of Gouda cheeses are the Netherlands (455×10^3 tonnes in 2012) and Germany (265×10^3 tonnes in 2012), (Bundesanstalt für Landwirtschaft und Ernährung, 2014; Productschap Zuivel, 2012). Outside Europe, a semi-hard cheese called Gouda is produced in, for example, Australia, New Zealand, Mexico, Brazil and the United States. Gouda cheese production has developed since the Middle Ages and reached full growth in the seventeenth century. While the cheese was produced in different Dutch provinces, trading took place in the Dutch town of Gouda. The cheese, with its typical flat-cylindrical form, traded on the Gouda cheese market and received the designation 'Gouda' in the eighteenth century. The original production areas of Gouda cheese were located in the Dutch provinces South Holland and Utrecht. However, nowadays, Dutch Gouda cheese is produced in factories situated in various parts of the country. For the PDO specialty 'Noordhollandse Gouda', the milk is derived exclusively from the coastal Dutch province of Noord Holland. Cheeses carrying the generic name 'Gouda' are, however, produced all over the world.

5.2.2 Type

Gouda cheese has a minimum FDM content of 30% and a maximum moisture content of 52% (Codex Alimentarius, 1966). Most Dutch Gouda cheeses contain 48% to 52% FDM, which corresponds to the minimum and maximum criteria for the PGI denominated 'Holland Gouda'. The moisture and salt content of the PGI Holland Gouda are 42.5% (max) and 1.4%– 2.0% at 14 days, respectively. Naturally ripened Gouda cheese is marketed in the Netherlands at several typical maturation stages, starting with 'young' cheese, with a minimum of four weeks of ripening, and ranging to 'old' Gouda , which is generally ripened for over 52 weeks. Sometimes, herbs or spices (e.g. mustard seeds, chive and clove) are added to Gouda cheese.

5.2.3 Description and Sensory Characteristics

Gouda cheese is produced in a flat-cylindrical or rectangular block shape, the weight typically ranging from 10 to 16 kg (cylinder shape: diameter 39 cm, height approx. 10 cm; block size 50 cm × 30 cm × 10 cm) although small Gouda cheeses of 0.2–1.0 kg (called 'baby Gouda') are produced as well. Naturally ripened Gouda cheese has a moisture-permeable, usually transparent, plastic coating, giving the cheese a shiny, dark yellowish appearance. Foil-ripened Gouda cheese's colour is rather pale white to light yellowish. Gouda cheese normally has a semi-soft to semi-hard consistency and a smooth texture, with a few pea-sized holes in the cross section. It is generally well sliceable (non-sticking). After prolonged natural ripening, the consistency becomes increasingly hard and short. The formation of amino acid crystals in matured Gouda cheese is common. The colour of the body may vary from light yellowish to orange-yellow, depending on the season, use of colouring and maturity. In naturally ripened Gouda cheese, a distinct, somewhat darker and translucent rind region (~1 cm thickness) develops upon maturation. The flavour of Gouda cheese varies widely with the stage of maturity, the type of starter and adjunct culture used. A young cheese is generally mild, with some fresh acidic, buttery, sulphury and brothy notes. The latter become more pronounced with maturation, and the flavour becomes more complex, with savoury (umami), salty and sweetish flavour notes developing, while creamy/buttery notes diminish.

5.2.4 Method of Manufacture

Milk preparation: Gouda cheese is usually made from fresh pasteurised partially skimmed cows' milk. Only a small proportion (<2% of the total Dutch Gouda cheese production) is farm-made from raw milk. Standardisation is carried out to the required fat: protein ratio (approximately 1.05 for Gouda cheese with 48% FDM). Prior to cheesemaking, the milk is pasteurised (HTST, 10–20 s at 70°C–75°C). Milk for Gouda cheese production is generally in-line bactofugated using specific self-desludging centrifuges, to remove the spores of *Cl. tyrobutyricum*, a feed-derived microorganism that may cause late blowing defects in Gouda-type cheeses.

Other ingredients: Calcium chloride is added to cheese milk (15–35 g/100 L) to aid rennet-induced coagulation. The addition directly enhances the aggregation of the para-casein micelles and indirectly enhances the enzymatic activity of the rennet (via a pH reduction). The rate of coagulation and firmness of the coagulum can be steered by balancing the dosages of rennet and $CaCl_2$. As goat's milk is usually not bactofugated, nor is nitrate added (being ineffective in goat's milk, due to the much lower activity of xanthine oxidase compared to cow's milk), alternative measures need to be taken to control the risk of butyric acid fermentation. The addition of lysozyme to the cheesemilk is one such option that can be applied. The starters for Dutch goat cheese are basically similar to those for Gouda cheese: traditionally undefined mixed-strain cultures of mesophilic LAB. They were selected for their specific flavour formation, acidification, gas formation and bacteriophage resistance.

Typically, these cultures consist of consortia of acid-producing *Lactococcus lactis* ssp. *lactis* and *Lactococcus lactis* ssp. *cremoris* strains and citrate-fermenting (and carbon-dioxide-producing – eye-forming) *Leuconostoc lactis* and/or *Leuconostoc mesenteroides* ssp. *cremoris* (L-starters), *Lc. lactis* ssp. *lactis* biovar *diacetylactis* (D-starters) or *Lc. lactis* ssp. *lactis* biovar *diacetylactis* and *Leuconostoc* spp. strains (DL-starters). The fermentation of citric acid and CO_2 production are higher and faster for DL-starters, and they are preferably used when more extensive eye formation is desired. Fresh bulk starters grown by the cheese producers themselves or frozen concentrates (DVI) are used, depending on the size of the operation. The inoculation dosage of fresh bulk starter is in the range of 0.5% to 1%.

Bacteriocin-producing starter cultures are sometimes used as alternatives to the usual starters (or as adjunct culture). Bacteriocin (e.g. nisin) produced in situ (or other, non-defined anti-clostridial factors) contribute to the control of butyric acid fermentation.

By the use of (attenuated) adjunct cultures (often, but not exclusively, thermophilic lactobacilli), Gouda cheese with widely varying flavour profiles can be produced, and the time to achieve an appropriate degree of maturation can be shortened

Starter cultures/rennet: The starters for Gouda cheese traditionally are undefined mixed-strain cultures of mesophilic LAB (). Cultures used in the Netherlands for Gouda-type cheese, for example, 'BOS', 'Ur' and 'A', thus have a long history of use in the dairy environment and can be considered as domesticated cultures (Smid et al., 2014). In-house starter propagation is strictly controlled to ensure a uniform bacterial composition, to control the rate of acidification and for complete protection from bacteriophages. Such systems include a bacteriophage-free mother starter concentrate and a propagation tank operated with overpressure of bacteriophage-free air. The inoculation dosage of fresh bulk starter is in the range of 0.5% to 1% (m/m). By using (attenuated) adjunct cultures (often, but not exclusively, thermophilic lactobacilli), Gouda-type variants with widely varying flavour profiles can be produced, and the time to achieve an appropriate degree of maturation can be shortened (van den Berg and Exterkate, 1993). Reduced-fat variants may also benefit from the use of adjunct cultures.

Coagulation: Coagulation and curd preparation in large industrial factories takes place in closed OST vats (TetraTebel, horizontal vat for curd making) or Double-O vats (Tetra Damrow) with 15,000–30,000 litres capacity. Rennet addition and coagulation take place at 30°C–31°C, and the coagulation process takes ~20–30 min.

Cutting: When the desired firmness is reached, the coagulum is cut to a curd grain size ~8–15 mm within approximately 15 min. At the end of curd preparation (see cutting and scalding), the curd grains typically are 5 mm size.

Washing and scalding: When the required curd particle size is achieved, the curd–whey mixture is stirred. By cutting and stirring, a considerable amount of whey is expelled, driven by syneresis. After a short settling time, about 40% of the whey (the so-called first whey) is removed. Warm washing water is then added to obtain a typical scalding temperature ranging from 33°C to a maximum of 38°C, and stirring is continued for another 30–45 min. By whey removal/addition of washing water, the lactose content of the whey (and consequently, by diffusion, of the cheese serum) is reduced. The concentration of lactose and the extent of syneresis of the curd (its moisture content) are the two main determinants of the pH of the final goat cheese. At the end of curd preparation in the vat, the pH has typically dropped to about ~6.3–6.4. Further acidification of the curd block (pH 5.5–5.8) takes place during the subsequent steps of cheesemaking before brining.

Curd draining and pressing: The concentrated curd–whey mixture, after draining part of the second whey, is pumped into continuously working vertical drainage columns (e.g. Casomatic®), or, in smaller-sized production sites, into open drainage vats with a moving perforated belt, where compact curd blocks are quickly formed. They are transferred into perforated moulds, a unique cheese-mark/number made from casein is marked on top of the curd, followed by a short pre-pressing. The cheeses are then pressed for about 1–1.5 hr with stepwise increasing pressure (final pressure 0.2–0.3 kg/cm2) to give the typical shape and achieve the formation of a homogeneous closed rind.

Salting: When the pH of the cheese has reached 5.5–5.8, which may be after a short holding period following pressing, the cheeses are immersed in a brine bath with a strength of 18–20°Bé, a calcium concentration of around 0.2%, a pH of 4.5 and a temperature of 13°C. Brining times vary from 72 to 96 hr for the large 10–15 kg loaves.

Maturation: After brining, Gouda cheese is either naturally ripened (this process is mandatory for the PGI and PDO varieties 'Holland Gouda' and 'Noordhollandse Gouda') at 13°C–15°C and 85%–88% RH, or vacuum-packed and foil-ripened (typically at temperatures <8°C). The latter process probably accounts for the majority of the Gouda-type cheeses produced outside the Netherlands.

For natural ripening, the cheeses are repeatedly coated with a polyvinyl-acetate-based aqueous dispersion which upon drying forms a moisture- and gas-permeable plastic film of hydrophilic nature. The plastic coating provides mechanical stability and protection against mould growth (both by a physical barrier function and by fungicidal components, which may be added to the plastic dispersion, as approved by the respective producing country). The use of wooden shelves in natural cheese ripening is still common. Strict control of ripening conditions (temperature, humidity and air velocity) and an adequate balance between cheese microbial stability and yield is required to ensure high quality. Naturally ripened Dutch semi-hard goat's cheese is typically matured for 3–4 weeks (young) until up to >one year. During this period, its moisture content gradually diminishes, from approximately 43% at 14 days to 28% after 12 months. Foil-ripened goat cheeses are stored at <8°C. As with Gouda cheese, due to the constant, relatively high moisture content throughout the entire ripening period, the maximum maturation time for the foil-ripened cheeses is limited. Prolonged ripening periods (e.g. >3 months) may lead to decreasing quality (sticky and soft consistency, flavour imbalance, off flavours) due to excessive proteolysis.

Foil-ripened Gouda cheeses are also produced The rectangular cheese blocks are piled and stored in larges boxes during maturation. Maintenance, that is, coating and turning of the cheeses as done for natural ripening, is unnecessary. The different ripening conditions (low temperature, O- or L-starter and anaerobic conditions at the cheese surface) all contribute to the generally less pronounced, rather flat and bland flavour profile of foil-ripened Gouda cheese. Due to the relatively high moisture content, the maximum maturation time for foil-ripened Gouda cheeses is limited. Prolonged ripening periods (>three months) may lead to reduced quality (sticky consistency, flavour imbalance and off flavours); thus, the majority of foil-ripened Gouda cheese is marketed at a rather young age.

5.2.5 Relevant Research

The application of new (adjunct) cultures has widened the diversity of flavour profiles of Gouda-type cheeses. Use of an (attenuated) adjunct starter to accelerate the ripening of Gouda cheeses and to improve the quality of fat-reduced variants has been successfully implemented in Gouda production. The steadily increasing knowledge regarding flavour formation pathways enables the design of (adjunct) starters and rational development of new Gouda-type cheese variants. For this purpose, targeted selections of starter bacteria are made for example, on the basis of proteolytic activity, amino acid converting activities and/or lysis sensitivity, using recent technological breakthroughs in the field of automated screening (Bachmann et al., 2009). Small peptides and amino acids produced during proteolysis are responsible for the important desired savoury, brothy, sweet and salty background flavours in a matured cheese. Recently, specific kokumi peptides have been related to the so-called long-lasting taste in Gouda cheese (Toelstede, Dunkel & Hofmann, 2009). The actual composition of mixed complex starter cultures, governed by for example, growth and lysis, appears to be highly dynamic during the process of dairy fermentation and ripening (Smid and Lacroix, 2013).

5.3 Hollandse Geitenkaas (Dutch Goat's Cheese) PGI – The Netherlands

<table>
<tr><td>Name: Hollandse Geitenkaas PGI</td><td rowspan="3"></td></tr>
<tr><td>Production area: The Netherlands</td></tr>
<tr><td>Milk: Goat's, pasteurised</td></tr>
</table>

5.3.1 Introduction

Hollandse Geitenkaas (Dutch goat's cheese) is a semi-hard Dutch-type cheese. In May 2015, the designation 'Hollandse Geitenkaas' (Dutch goat's cheese) was registered as PGI in the EU (EC, 2015). This specialty is produced in the Netherlands from the milk of the Dutch White goats or Dutch White cross-breeds by similar processes as for Gouda cheese. In contrast to the PGI variety 'Gouda Holland', goat's cheese may be naturally ripened or foil-ripened. It is mandatory that the former process takes place in the Netherlands. Professional breeding (initially, Dutch goats and Swiss Saanen) began around the start of the twentieth century and resulted in the so-called Dutch White goat, which exhibits high productivity and a characteristic constant mild taste in its milk (and of the cheese produced with it). The production of Dutch Goat cheese has evolved as a specialty, but also as an economic necessity, from traditional Dutch dairy farming and cheese production. The long tradition of Gouda cheese production was transferred to this goat's milk cheese. The first mention of 'Hollandse Geitenkaas' dates back to 1946.

5.3.2 Type

Hollandse Geitenkaas has a typical minimum fat in dry matter (FDM) of 50% and a maximum of 60%. The maximum moisture content is 44% (at 14 days), and the salt content is 2.3% in the cheese (4.1% dry matter). Similar to natural ripened Gouda, natural ripened Hollandse Geitenkaas is marketed in different maturation stages ranging from 'young' (more than 28 days ripened) to 'old' (up to one year ripened). Herbs or spices, for example, chive, fenugreek and coriander, can be added to the basic plain cheese to create a wide diversity of tastes and visual appearances in semi-hard goat cheese.

5.3.3 Milk

The cheese is made from fresh pasteurised goat's milk of the Dutch White goat and Dutch White goat cross-breeds. The raw milk's fat content typically varies between 3.8% and 4.5% fat and 2.6% and 4.4% protein. Milk from a maximum of eight milkings is collected at the farms in cooled farm tanks, and then transported to the cheese factory.

5.3.4 Description and Sensory Characteristics

The prevailing form of 'Hollandse Geitenkaas' is flat-cylindrical, with the weight ranging from 5 to 10 kg; rectangular blocks are produced as well. Natural ripened Dutch goat cheese has a moisture-permeable plastic coating, while blocks are mainly wrapped in moisture-impermeable foil. Hollandse Geitenkaas has a semi-soft to semi-hard consistency and a smooth, yet typically short texture. The cross section may be blind or have a few pea-sized eyes. It is generally well sliceable (non-sticking). With prolonged natural ripening, the consistency becomes increasingly firm and shorter and amino acid crystals may develop. The colour of the body when young is plain white. It changes to an ivory colour with increasing maturity in naturally ripened cheeses. In the latter, a darker, somewhat glassy rind (1 cm) region develops upon maturation. The flavour and flavour development of Hollandse Geitenkaas is basically similar to that of Gouda; however, the typical aroma components originating from the goat's milk fat are superimposed. The intensity of the goaty flavour in the Hollandse Geitenkaas is generally low; young Hollandse Geitenkaas has a mild flavour profile. However, the flavour intensity increases with maturity and may also vary considerably with the use of adjunct cultures (e.g. sweetish notes).

5.3.5 Method of Manufacture

Hollandse Geitenkaas production takes place in small- to mid-sized industrial cheese factories. The process typically comprises the following steps.

Milk preparation: The cooled milk is collected at the farms by trucks. On receipt at the factory, the milk is thermised (e.g. 65°C for 15 s) to reduce the bacterial load and extend the shelf life during storage prior to further use. The next step is standardisation to the required fat:protein ratio (approx. 1.05 for goat cheese with 50% fat in dry matter). Bactofugation is usually not practised in Dutch Goat's cheese production (due to economic and technological reasons). Prior to cheesemaking, the milk is pasteurised (HTST, 71.8°C/15 s).

Ingredients added: See Gouda cheese.

Rennet: Traditionally, calf's rennet is used. Alternative rennet sources, for example, microbial rennet (such as from *Mucor miehei*) or FPC may be used for particular purposes.

Coagulation: Coagulation and curd preparation via rennet addition and coagulation take place at 30°C–31°C, and the coagulation process takes approximately 25 min.

Cutting: When the desired firmness is reached, the coagulum in goat cheese production is cut to a curd grain size of about 8–15 mm within approximately 15 min. At the end of curd preparation (see 'Cutting and scalding'), the curd grains are typically 5 mm size.

Washing and scalding: See Gouda cheese.

Curd draining and pressing: See Gouda cheese.

Salting: The cheeses are immersed in a brine bath, optionally after a short resting period in the mould, when the pH has reached a pH of 5.5–5.8. The brine typically has a strength of 18–20°Bé, a calcium concentration of around 0.2%, a pH of 4.5 and a temperature of 13°C. Brining times vary between 2 and 3 days (for the 5–10 kg cheeses).

Maturation: After brining, Hollandse Geitenkaas cheese is either naturally ripened at 13°C–15°C and 85%–88% RH, or vacuum-packed and foil-ripened (typically at 4°C–7°C). For natural ripening, see the description of Gouda cheese.

References

Bachmann, H., Kruijswijk, Z., Molenaar, D., Kleerebezem, M. & van Hylckama Vlieg, J. E. T. (2009). A high-throughput cheese manufacturing model for effective cheese starter culture screening. *Journal of Dairy Science*, 92 (12), 5868–5882.

Bundesanstalt für Landwirtschaft und Ernährung (2014). Internet link: http://www.ble.de/SharedDocs/Downloads/01_Markt/09_Marktbeobachtung/02_MilchUndMilcherzeugnisse/JaehrlicheErgebnisse/01_Deutschland/Dt_Herstellung/130910_Herstellung_Kaese_nach_Sorten.html. Accessed 30 June 2015.

Codex Alimentarius (1966). *Codex Standard for Edam*, CODEX STAN 265-1966 and Codex Standard for Gouda, CODEX STAN 266–1966. Food and Agriculture Organization (FAO), Rome, Italy.

EC (1996). Regulation (EC) No. 1107/96 of 12 June 1996 on the registration of geographical indications and designations of origin under the procedure laid down in Article 17 of Council Regulation (EEC) No. 2081/92 with corrections. *Official Journal of the European Communities*, L 148, 1–10.

EC (2010a). Commission regulation (EU) No 1121/2010 of 2 December 2010 entering a designation in the register of protected designations of origin and protected geographical indications [Edam Holland (PGI)]. *Official Journal of the European Union*, L 317, 14–21.

EC (2010b). Commission regulation (EU) No 1122/2010 of 2 December 2010 entering a designation in the register of protected designations of origin and protected geographical indications [Gouda Holland(PGI)]. *Official Journal of the European Union*, L 317, 22–29.

EC (2015). Commission Implementing Regulation (EU) 2015/745 of 4 May 2015 entering a name in the register of protected designations of origin and protected geographical indications (Hollandse geitenkaas (PGI)). *Official Journal of the European Union*, L 199, 1–2.

Gauna, A. (2005). Elaboración de quesos de pasta semidura con ojos. Cuaderno Tecnológico N° 3 Lácteos. Instituto Nacional de Tecnología Industrial. Proyecto Mejora de la Eficiencia y de la Competitividad de la Economía Argentina. Retrieved from https://www.inti.gov.ar/lacteos/pdf/cuadernotecnologico3.pdf. (Accessed: 30 June 2015).

Productschap Zuivel (2012). *Zuivel in cijfers 2012-Zuivelindustrie-update 26 juni 2013*, The Netherlands . Internet link: http://www.zuivelnl.org/. (Accessed: 30 June 2015).

Smid, E. J. & Lacroix, C. (2013). Micro-microbe interactions in mixed-culture food fermentations. *Current Opinion in Biotechnology*, 24, 148–154.

Smid, E. J., Erkus, O., Spus, M., Wolkers-Rooijackers, J. Alexeeva, S. & Kleerebezem, M. (2014). Functional implications of the microbial community structure of undefined mesophilic starter cultures. *Microbial Cell Factories*, 13 (Suppl. 1): S2.

Stadhouders, J. (1990). Prevention of butyric acid fermentation by the use of nitrate. *Bulletin of the IDF*, 251, 40–45.

Toelstede, S., Dunkel, A. & Hofmann, T. (2009). A series of Kokumi peptides impart the long-lasting mouthfulness of matured Gouda cheese. *Journal of Agricultural and Food Chemistry*, 57 (4), 1440–1448.

Van den Berg, G. & Exterkate, F. A. (1993). Technological parameters involved in cheese ripening. *International Dairy Journal*, 3, 485–507.

6

Swiss-Type Cheeses (Propionic Acid Cheeses)

Katja Hartmann[1], Elisabeth Eugster-Meier[2], Marie-Therese Fröhlich-Wyder[3], Ernst Jakob[3], Daniel Wechsler[3], Ylva Ardö[4], Eva-Maria Düsterhöft[5], Wim Engels[5], Thom Huppertz[5], Erica R. Hynes[6,7], Maria Cristina Perotti[6,7] and Carina V. Bergamini[6,7]

[1] Anton Paar GmbH, Germany
[2] Bern University of Applied Sciences, School of Agricultural, Forest and Food Sciences HAFL, Zollikofen, Switzerland
[3] Agroscope, Research Division Food Microbial Systems, Federal Department of Economic Affairs, Education and Research EAER, Bern, Switzerland
[4] Department of Food Science, University of Copenhagen, Denmark
[5] NIZO food research, P.O. Box 20 6710BA, Ede, the Netherlands
[6] Facultad de Ingeniería Química (Universidad Nacional del Litoral)
[7] Instituto de Lactología Industrial (Universidad Nacional del Litoral – Consejo Nacional de Investigaciones Científicas y Técnicas), Santa Fe, Argentina

6.1 Allgäu Emmental PDO – Germany

Name: Allgäu Emmental PDO	
Production area: Allgäu region (southern Germany)	
Milk: Cow's, raw	

6.1.1 Introduction

Allgäu Emmental cheese is a well-known and popular cheese in Germany which is consumed throughout the year. Approximately 10 dairies are allowed to produce Allgäu Emmental cheese. They are small to medium-sized dairies, mostly organised as cooperative associations owned by the milk producers (Bayerische Landesanstalt für Landwirtschaft, 2015; Regierungspräsidium Karlsruhe, 2015). Allgäu Emmental cheese has held a PDO since January 1997 (EC, 1997).

In 1827, Josef Aurel Stadler, a cheesemaker from the Allgäu and a member of a wealthy regional family, brought the Swiss cheesemaker Johan Althaus from the Emmen Valley in Switzerland to Germany to teach the production of high-quality Emmental cheese. In the beginning, the equipment was simple; the milk was heated over an open fire, and the temperature was measured with

Global Cheesemaking Technology: Cheese Quality and Characteristics, First Edition.
Edited by Photis Papademas and Thomas Bintsis.

an elbow (i.e. dipping the elbow in the heated milk and judging whether the temperature was correct for adding the rennet). However, the manufacturing procedure improved, and by around 1840 most dairies in the Allgäu produced Emmental cheese of high quality. A major improvement was the introduction of fermentation cellars based on a knowledge of the interactions between the ripening temperature and the development of the characteristic gas holes (Mair-Waldburg, 1974; Wachter, 1955). Allgäu Emmental cheese is produced in the districts Lindau, Oberallgäu, Ostallgäu, Unterallgäu, Ravensburg, the Lake Constance district and the towns of Kaufbeuren, Kempten and Memmingen.

6.1.2 Type

Allgäu Emmental cheese belongs to the group of hard cheeses and is produced as a full-fat cheese (45% fat-in-dry-matter (FDM)) with a dry matter of ≥62%. It has a salt content of 0.5%–0.7%, and the pH of the cheese lies in the range 5.67–5.78 (Pillonel et al., 2002, 2003).

6.1.3 Description and Sensory Characteristics

Allgäu Emmental cheese is traditionally produced as a rind-matured cheese wheel with a minimum weight of 60 kg. Today, cheese blocks are also available with a minimum weight of 40 kg (Deutsche Käseverordnung, 2013). The yellow to brownish rind is even and slightly convex. The texture is elastic, supple and sliceable with walnut-sized, evenly distributed gas openings (1–3 cm). The aroma is nutty and slightly sweet, which becomes more intense with increasing ripening time (Deutsche Käseverordnung, 2013; Spezialitätenland Bayern, 2015).

6.1.4 Method of Manufacture

Allgäu Emmental cheese is produced according to German cheese regulations in a few dairies located in the Allgäu in Germany (Deutsche Käseverordnung, 2013).

Milk: Allgäu Emmental cheese is processed from raw cow's milk, which is produced in the region of cheese manufacture. The milk needs to meet the quality parameters set out for the production of Allgäu Emmental cheese: silage feeding is not permitted, and the milk is not allowed to be heated to >40°C prior to cheese manufacture (Deutsche Käseverordnung, 2013). Usually, a mixture of evening and morning raw milk is used for Allgäu Emmental cheese production, which is stored at 8°C–12°C until manufacture.

Starter culture: The milk is heated in a fermentation tank to 29°C–31°C, and a specified mixture of mesophilic and thermophilic lactobacilli and streptococci (typically *Lactobacillus helveticus* and *Streptococcus thermophilus*) is added for acidification of the cheesemilk. Furthermore, propionic acid bacteria (PAB) bacteria (typically *Propionibacterium freundenreichii* ssp. *shermanii*) are added. Rapid acidification is important in order to achieve even eye formation, storage stability and development of the typical taste (Kammerlehner, 2009).

Rennet: After acidification (pH = 6.60–6.55), calf's rennet mixed with whey is added to induce coagulation. Curdling time lasts about 30 min.

Cutting: The milk gel (coagulum) is cut, with a cheese harp, to the size of peas (2–3 mm).

Scalding: The curd pieces are heated to 50°C–52°C and for 20 to 35 min.

Curd draining/pressing: The curd pieces are allowed to sink to the bottom of the cheese vat and are collected in steel sieve baskets while the whey is drained. The curd is pressed with a pneumatic press (up to 6 bar) for one day while it is repeatedly turned (4–5 times).

Salting: The raw cheeses are transferred into a salt bath with an NaCl concentration of 20%–22% for 2–3 days until a rind has developed.

Maturation: The cheeses ripen for 2–4 weeks at 9°C. The second ripening stage takes place at temperatures between 20°C and 23°C for 4–5 weeks. During this warm ripening phase, the characteristic gas openings develop. Afterwards, the cheeses are stored at 8°C for 3–6 months. During maturation, the cheeses are washed and salted at least twice a week to support rind and flavour development.

Storage: Allgäu Emmental cheeses are stored at temperatures between 7°C and 10°C, wrapped in parchment paper or perforated aluminium foil.

6.2 Emmentaler PDO – Switzerland

<table>
<tr><td>Name: Emmentaler PDO</td><td rowspan="3">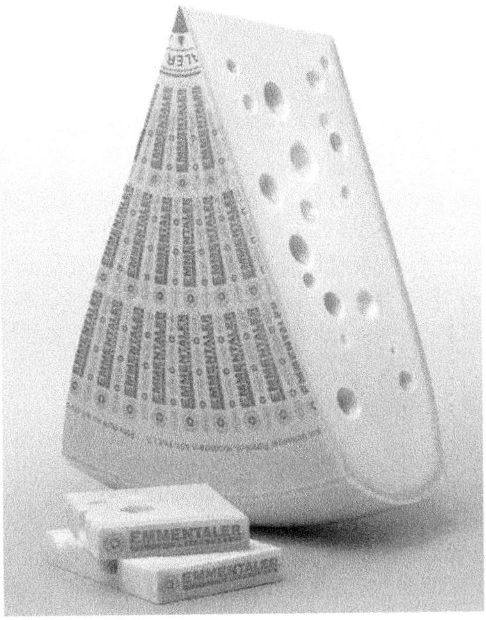

Emmentaler PDO (Emmentaler Switzerland)</td></tr>
<tr><td>Production area: Cantons of Aargau, Bern (except the district of Moutier), Glarus, Lucerne, Schwyz, Solothurn, St. Gallen, Thurgau, Zug and Zurich, as well as the lake district and Singine districts of the canton of Fribourg</td></tr>
<tr><td>Milk: Cow's, raw</td></tr>
</table>

6.2.1 Introduction

Emmentaler PDO is a full-fat hard cheese made from raw cow's milk, which has been subjected to propionic acid fermentation. It is characterised by its regular round eyes and its sweet and nutty flavour. After Mozzarella and Gruyère PDO, Emmentaler is the third most produced cheese in Switzerland (20,259 tonnes in 2014) and is the most exported (13,994 tonnes in 2014). Emmentaler was granted PDO status in 2006 (FOAG, 2015).

The name 'Emmentaler' refers to its area of origin, which is the 'Emmental', a valley ('Tal' in German) located in the canton of Bern. In 1542, 'Emmenthaler Käss' (Emmental cheese) was mentioned for the first time in the 12th dialect dictionary. At that time, the cheese was a popular gift item. Until the beginning of the nineteenth century, the cheese was produced in alpine farms of the Emmental. Thereafter, its production area and trade expanded, and around 1715, production was moved to the lowlands. At that time, Emmentaler was developed into its present form (Kulinarisches Erbe der Schweiz, 2008).

6.2.2 Type

The specifications for Emmentaler require at the time of consumption a FDM of 45.0–54.9%, a fat content of at least 27.5%, an MFFB of 51.0%–54.0%, and a maximum water content of 38.0% (FOAG, 2015). The salt content usually ranges between 0.2% and 0.5%.

6.2.3 Description and Sensory Characteristics

The texture of the cheese must be elastic and non-adhesive with fine to medium-fine curd grains. The colour of the body is ivory in winter and light yellow in summer. The eyes are round to slightly oval, mostly 2–4 cm, clean and regularly formed. A fully mature cheese has eyes with some water and crystals, which are slightly sunken. The characteristic flavour is created by propionic acid fermentation and is acidic, sweet, nutty and slightly salty and spicy.

6.2.4 Method of Manufacture

The manufacture of Emmentaler is carried out in small-scale dairies. The use of any kind of additives is forbidden.

Milk: No silage may be used at the milk production site. At least 70% (on a dry matter basis) of the fodder ration must be based on roughage. Mixtures with vetch, rape, turnip rape and similar annual fodder plants may be used. Feeds of animal origin other than milk as well as feed-grade urea are prohibited. No feedstuffs declared to be 'genetically modified' may be used either. No sewage sludge may be present on productive land or on the summer pastures of the milk production site. The distance between the cheese dairy and the milk production site may not exceed 30 km. The milk must be processed in a copper vat or copper cauldron; daily production is limited to three lots. Only raw milk may be used, and it must be processed within a maximum of 24 hr after the last milking. No bactofugation, ultra- or microfiltration or any similar procedure may be used to process the milk.

Starter cultures/rennet: Cultures of lactic acid bacteria (LAB) and propionic acid bacteria, which originate from the area, are used. The lactic cultures consist of undefined mixtures of strains of thermophilic lactic acid bacteria of the species *S. thermophilus* and *Lb. delbrueckii* ssp. *lactis*; direct inoculation of the vat milk is banned. The propionic culture consists of defined strains of *P. freudenreichii* spp. *shermanii*. Cultures of NSLAB containing strains of *Lb. casei* may also be applied (Fröhlich-Wyder and Bachmann, 2004). The use of genetically modified bacteria is prohibited. The liquid stock cultures are produced and delivered by Agroscope.

Recombinant chymosin or rennet substitutes produced by genetically modified microorganisms may not be used. The rennet is added at 32°C, and coagulation lasts 35–45 min. The curd grain size corresponds to the size of a wheat grain or a maize kernel. Scalding occurs between 52°C and 54°C and within 30–60 min. Lactose is diluted through curd washing (water addition of 10%–20% of the milk's volume).

Moulding: The temperature of the curd at the time of moulding is 49°C–52°C. Pressing lasts up to 20 hr with a press force of 500–2,000 kg per cheese wheel.

Salting: Salting is done by immersing the cheese in brine of 20°Bé–22°Bé and lasts for 24–72 hr.

Maturation: Initially, the cheeses are matured for 30–70 days in a warm room (19°C–24°C, Relative Humidity (RH) 80%) to promote propionic acid fermentation. The wheels must be turned weekly and, if needed, rubbed dry or moist. A ripening in the cold room follows (11°C–14°C, RH 70%–90%) until the cheeses are ready for consumption. Extended ripening in humid storage (RH > 90%) is sometimes also applied (e.g., cave ripening). Emmentaler has to be ripened up to at least four months within the area of production and is sold at different ripening stages, usually at four, eight or twelve months. Emmentaler must have good storage qualities, allowing it to reach its optimal maturity without any loss in quality.

6.2.5 Relevant Research

PAB are naturally present in raw milk at low levels. To evaluate the prevalence and significance of wild strains of propionibacteria in Emmentaler, three cheeses were manufactured from the same raw milk using three different commercial *P. freudenreichii* spp. *shermanii* starters. Strains with high aspartase activity or with a fast growth at 11°C dominated during the whole ripening period, irrespective of whether the strain originated from the culture or from the raw milk. In all three cheeses, the wild strains exhibited high aspartase activity, indicating that this strain-specific property is a key factor in the control of propionic acid fermentation and in the prevention of late fermentation in Swiss-type cheeses (Turgay et al., 2011).

When bactofuged or microfiltrated milk is used in the production of Swiss-type cheeses, eye formation may be unsatisfactory. In an experiment with Emmental cheese, the addition of hay microparticles proved to be an efficient measure for controlling the number and size of the eyes in the cheese in a dose-dependent manner (Guggisberg et al., 2015). Nowadays, the terms 'Emmental' or 'Swiss' cheese are used worldwide for industrial cheeses with typical propionic acid fermentation. Bisig et al. (2010) compared the characteristics of Emmentaler PDO with generic Emmental cheese made in Europe. Significant differences, for example, in the composition of volatile compounds or of the applied cultures are discussed. Pillonel et al. (2002) searched for parameters that would enable a geographic classification of Emmental cheeses. Using chemometry, they were able to distinguish between Emmentaler PDO from Switzerland and that from different European regions. Casey et al. (2008) developed a new method for verifying the authenticity of Emmentaler PDO cheese, which includes the addition of natural LAB to the vat milk that can be traced back even in grated or processed cheese by means of rapid and sensitive qPCR tests.

6.3 Grevé – Sweden

Name: Grevé	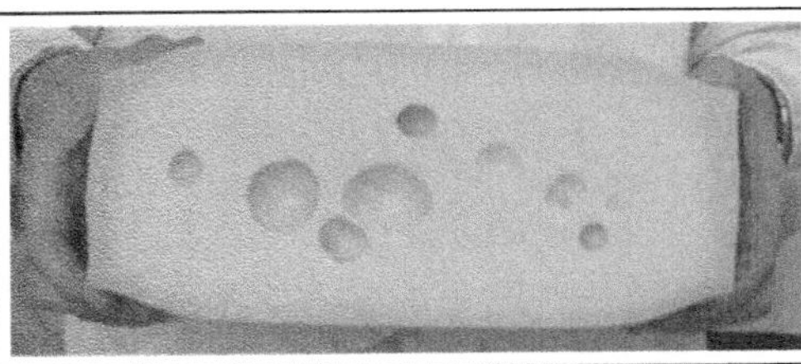
Production area: Sweden	
Milk: Cow's, pasteurised	

6.3.1 Introduction

Grevé has been a registered trademark in Sweden since 1964 and in the EU since 2000, which is owned in fellowship by Swedish dairy companies. Grevé cheese gets its typical characteristics from a combination of mesophilic DL-starter and a culture of PAB. It was developed at the experimental cheese dairy in Örnsköldsvik, Sweden, during the late 1950s by two dairy engineers, Åke Berglöf and Yngve Johansson, who were inspired by the Norwegian Jarlsberg.

6.3.2 Type

Grevé is a semi-hard cheese with moisture in non-fat substance of 54%–57 % and a fat content of 28% fat in cheese corresponding to 45% FDM. Salt content is 1.0%–1.4 %. (Sveriges Ostkollegium, 1993)

6.3.3 Description and Sensory Characteristics

Grevé is a semi-hard cheese with many characteristic large, round and regular eyes (10–40 mm) enlarged by the activity of PAB. The body is evenly light yellow to yellow. The cheese is pressed and produced in low cylinders with somewhat convex surfaces and a weight of 12–14 kg. It has hard, dry and waxed surfaces. The texture is somewhat elastic and sliceable, and it becomes brittle during a longer ripening time. The flavour is creamy and buttery with distinct sweet and nutty notes, and the aftertaste is long and rich.

6.3.4 Method of Manufacture

Milk: Raw milk undergoes microfiltration, fat standardisation and pasteurisation (72°C/15 s). Calcium chloride may be added in amounts up to 20 g per 100 kg milk to facilitate coagulation, and sodium nitrate may be used to prevent detrimental clostridia spores from germinating.

Starter culture/rennet: Mesophilic DL-starter with undefined mixed strains of *Lc. lactis* ssp. *lactis* & *cremoris* (80-90%), *Lc. lactis* ssp. *lactis* biovar. *diacetylactis* and *Leuc. mesenteroides* as well as a single strain of *P. freudenreichi* ssp. *shermanii*. Calf rennet is used for coagulation.

Cutting: The curd is divided into cubes with sides of 5–6 mm.

Heating of the curd: Temperatures are around 37°C–40 °C for about 2 hr.

Curd washing/draining: About 40% of the whey is replaced by water after about 40 min of stirring at 30°C, and about 15% is drained off for 40 min before the cheeses are pressed under whey.

Salting: The cheeses are brine-salted to a salt content of 1.2 % (w/w).

Maturation: Maturation conditions vary among producers, but there are three main stages; cold storage (1–2 weeks/8°C–10°C), warm room (about two weeks at warmer temperatures up to 20°C) and finally ripening storage (8°C–10°C) until consumption. Grevé is mainly sold at a ripening age between four and nine months, but longer storage does occur. The surface is dry and waxed.

Storage: Cold (2°C–6°C)

6.3.5 Relevant Research

The starter LAB convert all lactose to lactic acid and the citrate into carbon dioxide and buttery aroma compounds such as diacetyl and acetoin within a few days of ripening. The Propionibacteriaceae use the lactic acid formed via different metabolic pathways to produce propionate, acetate, carbon dioxide and energy in the form of ATP. This is further enhanced in the presence of the amino acid aspartame and succinate. Proteolysis is mainly performed by the LAB; however, PAB contribute and release more leucine and proline than LAB, of which the latter contribute to a characteristic sweet flavour. PAB also contribute to lipolysis and ester formation. Ethyl propionate is characteristically found in Grevé. Texture changes during maturation and the biochemical changes responsible were also studied (Rehn et al., 2011).

6.4 Maasdammer – The Netherlands

Name: Maasdammer cheese	
Production area: The Netherlands	
Milk: Cow's, pasteurised	

6.4.1 Introduction

Maasdammer is a semi-hard Dutch cheese made from cow's milk. The cheese was created in the 1970s as an alternative to (Swiss) Emmental cheese. Although similar to Emmental, it differs in regard to loaf size (traditional Maasdammer wheels are ~13 kg), moisture content (typically 40%) and the duration of ripening (e.g. 4–6 weeks). Approximately 20% of Dutch cheese production is made up of Maasdammer cheese (150,000 tonnes in 2012). The most well-known Maasdammer cheese brand name is Leerdammer®.

The Maasdammer (Swiss-style) cheese concept was developed for the Dutch dairy companies by NIZO in the 1970s, by combining Gouda cheese cultures and technology with Emmental cultures and technology. The brand Leerdammer was launched in 1974 and was named after the Dutch town of Leerdam, located close to the main production site of the cheese. The production of Maasdammer cheese is essentially in the Netherlands. Related varieties are produced in Scandinavia, for example, Jarlsberg and Samsoe.

6.4.2 Type

Maasdammer cheese is ready for consumption at 4–5 weeks of age. The flavour at the age is mild sweet, buttery and nutty. The cheese can be kept for several months, provided the storage temperature is low (<7°C), and the flavour will become more pronounced with age. Maasdammer cheese has a minimum FDM of 45% (typically 45%–47%). The moisture content is 40.5% (38.5%–41.5%) and the salt content in cheese 1.5%–1.6% (2.5%–2.6% dry matter).

6.4.3 Description and Sensory Characteristics

The most characteristic features are its large 'eyes' (holes), which are prominent especially in the centre of the cheese, the firm elastic texture and the mild sweet, buttery and nutty taste. Naturally ripened Maasdammer cheese has the shape of a round wheel (flat cylindrical form) with a yellow-coloured (by Annatto) clean, dry and smooth rind. The wheel has a diameter of 36–39 cm, a height of ~15 cm and a mass of 13 kg. So-called 16 kg Euroblock rectangular Maasdammer cheeses are also produced, which are ripened in foil. The inner cheese colour is pale yellow. In particular, the centre part of Maasdammer cheese contains 10–40 mm sized oval or round eyes.

6.4.4 Method of Manufacture

Milk preparation: Maasdammer cheese is produced from bactofuged and pasteurised high-quality cows' milk. By bactofugation, using specific self-desludging centrifuges, the spores of *Cl. tyrobutyricum*, a feed-derived microorganism that may cause late blowing defects in some cheeses, can be effectively removed. The milk is standardised to reach the desired fat content in the cheese. Cheese production generally is in large-scale highly automated factories.

Ingredients added: Calcium chloride and sodium or potassium nitrate are added. The starter culture for Maasdammer cheese production consists of (undefined) mixed-strain mesophilic starters (see Part II, Section 5.2) and of PAB. Specific thermophilic lactobacilli may also be applied. The PAB are *P. freudenreichii* species cultivated in-house or applied as a concentrate. PAB are responsible for the second fermentation in Maasdammer cheese. During the second fermentation, the bacteria ferment the lactate developed by the LD culture and produce propionic acid, acetic acid and CO_2. These components determine the characteristic flavour and eye formation. Traditionally, for Maasdammer cheese, calf's rennet is used. Alternative rennet sources, for example, microbial rennet (such as from *Mucor miehei*) or fermentation-produced chymosin (FPC) may be used for particular purposes.

Coagulation: Coagulation and curd preparation in large industrial factories takes place in closed OST or Double-O vats. Rennet addition and coagulation take place at 30°C–31°C and the coagulation process takes ~25 min.

Cutting: The curd grain size corresponds to the size of 6–8 mm cubes.

Curd washing/scalding: After cutting, stirring and a short settling time, up to 40% of the whey is removed, and warm washing water is added to raise the temperature for scalding to 35°C (34°C–37°C) and to reduce the lactose concentration in the whey (and consequently in the curd) by ~30%.

Curd drainage: The transfer of the curd into the moulds is done mechanically, generally using Casomatic equipment for whey drainage and curd block portioning. The next phase is pressing, which lasts for 1.5 hr under stepwise increasing pressure; a good cohesion of the grains must be ensured.

Salting: The salting starts directly after pressing, after the cheeses have been turned out of the moulds, by immersion of the cheese in a brine with a strength of typically 18°Bé–20°Bé and a temperature of approximately 12°C. The duration of brining is typically about two days.

Maturation: After brining, Maasdammer cheese can be either naturally ripened (the wheel cheeses) or foil-ripened (packed in semi-permeable foil). Natural ripened cheeses are stored at a temperature of 12°C–13°C and an RH of about 85% for two weeks. At this temperature, initial proteolysis by rennet and starter enzymes takes place, which is important for flavour development. The wheel cheeses rest on wooden shelves, whereas the foil-packed Maasdammer cheeses are stored stacked in boxes. The surface of the naturally ripened wheel cheeses is regularly treated with a new layer of polyvinylacetate-based coating material during maturation. After two weeks, the cheeses are transferred to a so-called warm room for maturation at ~20°C for a period of about two weeks. During warm room maturation, propionic acid fermentation takes place, yielding the desired typical flavour and eye formation.

Storage: After ripening, Maasdammer cheeses are preferably stored at a low temperature, for example, below 7°C, to prevent undesirable further fermentation and consequently possible defect formation (late fermentation). The cheese can be stored for several months.

6.4.5 Relevant Research

The main (key) flavour compounds in Maasdammer cheese are those that yield the typical sweet and nutty flavour notes: pyrazines, branched-chain aldehydes and benzaldehyde, methylketones, lactones and ethyl esters seem important (Thierry et al., 2006). Research in Switzerland showed that the capability of PAB strains to utilise aspartate may be a very important factor for flavour formation. Very high aspartase activity will, however, also increase CO_2 formation and the risk of late fermentation. However, moderate aspartase activity will have a positive effect on the quality of Swiss-type cheese in regard to both eye formation and flavour intensity (Turgay et al., 2011).

6.5 Pategrás Cheese – Argentina

Name: Pategrás	
Production area: Pampeana region	
Milk: Cow's, pasteurised	

6.5.1 Introduction

Pategrás Argentino cheese is the most popular semi-hard cheese produced in Argentina (Zalazar, Meinardi & Hynes, 1999). The technology of cheesemaking was developed by European immigrants in the late nineteenth and early twentieth centuries, and was initially based on that of Dutch cheeses, but then it was modified and adapted during the last century to Argentinean raw materials and environmental conditions, and nowadays it has its own characteristics. Several types of Pategrás exist. The main Pategrás cheese varieties are Pategrás for table, to which propionic bacteria can occasionally be added to obtain a hybrid cheese, which retains nevertheless the name of Pategrás, and 'Pategrás sandwich' ('Barra' cheese), which forms a block to be sold sliced and does not have holes. The production area of Pategrás cheese includes several provinces pertaining to the Pampeana region of Argentina: Buenos Aires, Santa Fe (south and centre), Córdoba (south and east), La Pampa (centre, north and south) and San Luis (centre and south). It is also produced in Río Negro province, which is in the Patagonia region.

6.5.2 Type

Argentinean legislation (Código Alimentario Argentino, 2014) defines 'Pategrás' cheese as a medium-moisture or semi-hard cheese, with moisture content between 36.0% and 45.9%. The name of Pategrás and Gouda being equivalent for the legislation, the choice is often a marketing matter. According to national legislation, Pategrás cheese is classified as a full-fat cheese because the fat content in the dry matter is between 45.0% and 59.9%. It has the shape of a flattened cylinder, a convex profile (diameter and height approximate of 25 cm and 15 cm, respectively), and the usual weight is around 4 kg, with a range between 1 to 10 kg. The ripening time is at least 1 month for 1 kg blocks, 1.5 month for 1 to 5 kg blocks and 2 months for 5 to 10 kg blocks.

6.5.3 Description and Sensory Characteristics

'Pategrás sandwich' cheese (Código Alimentario Argentino, 2014) has a parallelepiped shape (sides and length approximate of 12 × 12 cm and 40 cm, respectively). In the Argentinean legislation, some cheeses with similar cheesemaking technology are included, such as 'Holanda', similar to Pategrás for table, or 'Tybo' (Código Alimentario Argentino, 2014), similar to 'Pategrás sandwich'. The differences are mainly in the fat content of the cheesemilk. Pategrás cheese has a compact and firm texture with elastic consistency; it can have some sweet eyes, small (1 to 5 mm) and medium sized (5 to 10 mm), which are well spread in the cheese mass, or have no eyes. The typical Pategrás cheese colour is yellowish-white and uniform. The rind is smooth and closed, generally covered with a natural paraffin coating or coloured with a red or yellow plastic emulsion. Pategrás has a characteristic sweet taste and a soft, clean, pleasant and well-developed aroma. The variety 'Pategrás sandwich' is usually packaged in thermocontractible plastic bags and has a more elastic texture than traditional Pategrás, which is aimed at slicing easily and prolonging shelf life. In addition, it has no eyes, and the intensity of the aroma is lower than that of Pategrás. During 2000–2010, the use of propionic cultures and changes in ripening technology of Pategrás cheese were introduced, mainly by starter suppliers, which resulted in a hybrid cheese. The cheese conquered the market, as pungent and intense flavours are appreciated by Argentineans, and the price was lower than that of Le Gruyere and Swiss cheeses. However, the fact that the hybrid cheese retained the original name ('Pategrás') creates confusion about the cheese type

6.5.4 Method of Manufacture

Milk preparation: Pategrás cheese is made with HTST pasteurised cow's milk, in which the fat content is standardised to 2.8% to 3% (ratio of fat to protein: 0.9–1). After cooling to 35°C–37°C, $CaCl_2$ is added at a final concentration of 0.02%.

Starter culture: Natural milk starters, obtained by incubating good-quality thermised milk at 45°C overnight, and containing *S. thermophilus* and heterofermentative citrate-positive lactic acid bacteria, were applied for Pategrás up to 1980. This microbial biodiversity determined the cheese characteristics and imparted a taste similar to that of Dutch cheeses and a flavour resembling that of raw milk cheeses. However, with the aim of obtaining cheeses with a uniform and constant quality, the starter has been completely replaced by commercial selected cultures of *S. thermophilus* for acidification, and *Leuconostoc* spp. or citrate-positive lactobacilli, for aroma and eye formation, since the 1990s (Zalazar, Meinardi & Hynes, 1999).

Coagulation: Coagulant enzyme has traditionally been calf rennet or adult bovine coagulant, but FPC obtained from genetically modified organisms is used almost exclusively today. Milk-clotting enzyme is added to milk at 35°C–37°C, approximately 10 min after the addition of the starter.

Cutting: After coagulation and gel strengthening, the curd is cut in successive steps until it reaches the size of a corn grain (at 37°C – approximately 20 min).

Scalding: The mixture of curd particles and whey is gently stirred, and a step of scalding up to 45°C (0.5°C–1°C/min) is performed and then maintained for 10–15 min approximately with the aim of slightly reducing the content of moisture of the curd.

Curd washing: A curd wash step is often included, especially in the 'Pategrás sandwich' variety, in order to improve the physical stability of this cheese, the commercialised form of which is mainly cut in slices. A pH of 6.3–6.5 is reached at the end of cheesemaking, and further acidification takes places during pressing.

Moulding/pressing: Curds and whey are unloaded by gravity or pumped to draining equipment, where curd is pre-pressed, under whey, before moulding. The curd is then transferred into the moulds and pressed for 12 hr. After that, the cheeses are immersed in cold brine.

Salting: Salting of Pategrás cheese is done by immersion in brine solution (19°Bé, 12°C–14°C, pH 4.95–5.1) for 12–16 hr per kg of cheese. After salting, the cheeses dry for a day and are then covered with a plastic emulsion in order to avoid external mould contamination.

Maturation: Ripening of Pategrás cheese takes place at 12°C and 85% RH, for 1 to 2 months depending on the size of cheese: the minimum ripening time is 1 month for cheeses of 1 kg, 1.5 month for cheeses from 1 to 5 kg and 2 months for cheeses of 5 to 10 kg.

6.5.5 Relevant Research

Ripening of Pategrás cheese is characterised by proteolysis mediated by the residual milk-clotting enzyme, partially inactivated by scalding, which mainly acts on the α_{s1}-casein, followed by the activity of microbial enzymes (proteases and peptidases). In cheeses with the addition of heterofermentative or citrate-positive adjunct cultures, the ripening is also characterised by the production of lactate, acetate, ethanol, diacetyl, acetoin and CO_2 (eye formation) from the fermentation of lactose and citric acid (Bergamini, Hynes & Zalazar, 2006; Vélez et al., 2015; Zalazar et al., 1985, 1988; Zalazar, Meinardi & Hynes, 1999). In the hybrid cheese containing propionic bacteria, propionic fermentation takes place with formation of big holes and development of the typical flavour (Gauna, 2005).

Pategrás Argentino cheese showed excellent performance as a carrier for different commercial probiotic cultures: they were easily added in cheesemaking and maintained high counts during ripening (Bergamini, Hynes & Zalazar, 2006, 2009). On the other hand, the wrapping of the non-matured cheeses in plastic films has been studied because it confers many economic benefits, such as reduced weight losses and fewer handling problems. Ripening of the packaged cheese forms did not modify proteolysis, textural changes and organic acid profiles during ripening, these cheeses being similar to those ripened by traditional procedures (Bertola et al., 2000; Califano and Bevilacqua, 2000).

Acknowledgements

The authors thank Sucesores de Alfredo Williner S.A. for providing cheeses for the photographs.

References

Bayerische Landesanstalt für Landwirtschaft (2015). www.lfl.bayern.de/mam/cms07/iem/dateien/2015_04_07_hersteller_allg%C3%A4uer_emmentaler2.pdf [last assessed: 14. July 2015].

Bergamini, C. V., Hynes, E. & Zalazar, C. (2009). Proteolytic activity of three probiotic strains in semi-hard cheese as single and mixed cultures: Lactobacillus acidophilus, Lactobacillus paracasei and Bifidobacterium lactis. *International Dairy Journal*, **19**, 467–475.

Bergamini, C. V., Hynes, E. & Zalazar, C. A. (2006). Influence of probiotic bacteria in the proteolysis profile of a semi-hard cheese. *International Dairy Journal*, **16**, 856–866.

Bertola, N., Califano, A. N., Bevilacqua, A. E. & Zaritzky, N. E. (2000). Effects of ripening conditions on the texture of Gouda cheese. *International Journal of Food Science & Technology*, **35** (2), 207–214.

Bisig, W., Fröhlich-Wyder, M. T., Jakob, E. & Wechsler, D. (2010). Comparison between Emmentaler PDO and generic emmental cheese production in Europe. *The Australian Journal of Dairy Technology*, **65** (3), 206–213.

Califano, A. N. & Bevilacqua, A. E. (2000). Multivariate analysis of the organic acids content of gouda type cheese during ripening. *Journal of Food Composition and Analysis*, **13**, 949–960.

Casey, M. G., Isolini, D., Amrein, R., Wechsler, D. & Berthoud, H. (2008). Naturally occuring genetic markers in lactobacilli and their use to verify the authenticity of Swiss Emmental PDO cheese. *Dairy Science and Technology*, **88**, 457–466.

Código Alimentario Argentino (2014). *Cap. VIII: Alimentos lácteos*. http://www.anmat.gov.ar/alimentos/codigoa/CAPITULO_VIII.pdf. Accessed 30 June 2015.

Deutsche Käseverordnung, Established 1965, *Revision 1986, Amendment 2010, 2013. German Federal Ministry for justice and consumer protection*.

EC (1997). Commission Regulation EC No 123/97 supplementing the Annex to Commission Regulation EC No 1107/96 on the registration of geographical designations of origin under the procedure laid down in Article 17 of regulation EEC No 2081/92. *Official Journal of the European Union*, **L 22**, 19–20.

FOAG (2015). *Register of appellation of origin and geographical indications*. Federal Office for Agriculture. An Office of the Federal Department of Economy, Bern, Switzerland http://www.blw.admin.ch/ [Accessed 28 April 2015].

Fröhlich-Wyder M. T. & Bachmann H. P. (2004). Cheeses with propionic acid fermentation. In Fox, P. F., McSweeney, P. L. H., Cogan, T. M., Guinee, T. P. (eds.), *Cheese Chemistry, Physics and Microbiology, Vol. 2, Major Cheese Groups*, 3rd edition. Elsevier, Academic Press, Amsterdam, the Netherlands, pp. 141–156.

Guggisberg, D., Schuetz, P., Winkler, H., Amrein, R., Jakob, E., Fröhlich-Wyder, M. T., Irmler, S., Bisig, W., Jerjen, I., Plamondon, M., Hofmann, J., Flisch, A. & Wechsler, D. (2015). Mechanism and control of the eye formation in cheese. *International Dairy Journal*, **47**, 118–127.

Kammerlehner, J. (2009). *Cheese Technology*. Josef Kammerlehner Verlag, Freising, Germany.

Kulinarisches Erbe der Schweiz. (2008). Lausanne, Switzerland. http://www.kulinarischeserbe.ch/ [Accessed 28 April 2015]

Mair-Waldburg (1974). *Handbuch der Käse: Käse der Welt von A-Z: eine Enzyklopädie*. Volkswirtschaftlicher Verlag, Kempten, Germany.

Pillonel, L., Albrecht, B., Badertscher, R., Bütikofer, U., Chamba, J. F., Tabacchi, R. & Bosset, J. O. (2003). Analytical methods for the determination of the geographic origin of emmental cheese. Parameters of proteolysis and rheology. *Italian Journal of Food Science*, **15** (1), 49–62.

Pillonel, L., Badertscher, R., Bütikofer, U., Casey, M., Dalla-Torre, M., Lavanchy, P., Meyer J., Tabacchi, R. & Bosset, J. O. (2002). Analytical methods for the determination of the geographic origin of Emmentaler cheese. Main framework of the project; chemical, biochemical, microbiological, colour and sensory analyses. *European Food Research and Technology*, **215** (3), 260–267.

Regierungspräsidium Karlsruhe (2015). https://rp.baden-wuerttemberg.de/Seiten/Results.aspx?k=emmentaler [Accessed 14 July 2015].

Rehn, U., Vogensen F. K., Persson, S.-E., Hallin Saedén, K., Nilsson, B. F. & Ardö, Y. (2011). Influence of microflora on texture and contents of amino acids, organic acids and volatiles in semi-hard cheese made with DL-starter and *Propionibacterium*. *Journal of Dairy Science*, **94**, 1098–1111.

Spezialitätenland Bayern (2015). http://www.spezialitaetenland-bayern.de/?id=164&an=3&L=0&no_cache=1 [Accessed 14 July 2015].

Sveriges Ostkollegium. (1993). *Svenska Ostsorter [Swedish cheese varieties]*. Falkenberg, Sweden, pp. 22–23.

Thierry, A., Maillard, M. B., Richoux, R. & Lortal, S. (2006). Ethyl ester formation in enhanced by ethanol addition in mini Swiss cheese with and without added propionibacteria. *Journal of the Agricultural and Food Chemistry*, **54**, 6819–6824.

Turgay, M., Irmler, S., Isolini, D., Amrein, R., Fröhlich-Wyder, M.-T., Berthoud, H., Wagner, E. & Wechsler, D. (2011). Biodiversity, dynamics, and characteristics of *Propionibacterium freudenreichii* in Swiss Emmentaler PDO cheese. *Dairy Science & Technology*, **91** (4), 471–489.

Vélez, M. A., Bergamini, C. V., Ramonda, M. B., Candioti, M. C., Hynes, E. R. & Perotti, M. C. (2015). Influence of cheese making technologies on plasmin and coagulant associated proteolysis. *LWT – Food Science and Technology*, **64** (1), 282–288.

Wachter, C. (1955). *Die Allgäuer Milchwirtschaft im 19. Jahrhundertbis zur Gründung des Milchwirtschaftlichen Vereins. Geschichte der Allgäuer Milchwirtschaft*, Milchwirtschaftlicher Verein im Allgäu, Kempten, Germany.

Zalazar, C., Meinardi, C., Bernal, S. & Candioti, M. (1988). Características de los quesos argentinos tipo Pategrás. *Revista Argentina de Lactología*, **1** (1), 27–43.

Zalazar, C., Meinardi, C. & Hynes, E. (1999). *Quesos Típicos Argentinos. Una revisión general sobre producción y características*. Santa Fe, Argentina: Centro de Publicaciones de la Universidad Nacional del Litoral.

Zalazar, C., Meinardi, C., Reinheimer, J., Candioti, M. & Bernal, S. (1985). La maturazione del formaggio "Pategras Argentino". 3a Parte: Influenza della coltora dei fermenti lattici sull'evoluzione della composizione chimica e della flora microbica. *Scienza e Tecnica Lattiero Casearia*, **36** (5), 369–377.

7

White-Brined Cheeses

Thomas Bintsis[1], Efstathios Alichanidis[2], İrem Uzunsoy[3, 6], Barbaros Özer[4], Photis Papademas[5], Zorica Radulovic[7] and Jelena Miocinovic[7]

[1] *11 Parmenionos, 50200 Ptolemaida, Greece*
[2] *Department of Food Science and Technology, School of Agriculture, Aristotle University of Thessaloniki, Thessaloniki, Greece*
[3] *Bülent Ecevit University Caycuma Vocational High School, Department of Food Technology, Turkey*
[4] *Ankara University, Faculty of Agriculture Department of Dairy Technology, Ankara, Turkey*
[5] *Department of Agricultural Sciences, Biotechnology and Food Science, Cyprus University of Technology, Cyprus*
[6] *Bülent Ecevit University Caycuma Vocational High School, Department of Food Technology, Zonguldak, Turkey*
[7] *Department of Food Microbiology, Faculty of Agriculture, University of Belgrade, Serbia*

7.1 Batzos PDO – Greece

Name: Batzos, PDO	
Production area: Western and Central Macedonia and Thessaly	
Milk: Goat's or sheep's or mixtures, raw	

7.1.1 Introduction

Batzos PDO is a low-fat, semi-hard, white-brined cheese with a large number of 'holes' in its body, which are formed by fermentation within few days of manufacture. In the past, Batzos was made mostly when cheesemakers wanted to prepare a whey with high fat content in order to make high-quality Manouri cheese (see Part II, Section 13.3). Thus, the real intention of the cheesemaker was to obtain, in the first place, high-fat whey (for the manufacture of Manouri). This was obtained with special manipulation of the curd. Thus, Batzos can be characterised as a secondary product from the manufacture of Manouri cheese,

Global Cheesemaking Technology: Cheese Quality and Characteristics, First Edition.
Edited by Photis Papademas and Thomas Bintsis.

whereas, in most cases, whey cheeses are the secondary products. The traditional method for the manufacture of Batzos has been described by several authors (Anifantakis, 1991; Nikolaou et al., 2002; Psoni et al., 2003; Psoni et al., 2006). However, nowadays, since Manouri cheese is also produced from the whey of sheep's and/or goat's milk cheeses with the addition of whole milk and cream, the production of Batzos is not a prerequisite. Batzos is mostly manufactured according to the following method (which is quite different from the traditional one).

7.1.2 Type

Semi-hard white-brined cheese with a maximum moisture content of 45% and a minimum fat in dry matter (FDM) of 25%. The PDO status for Batzos was recognised by the EC in 1996 (EC, 1996).

7.1.3 Description and Sensory Characteristics

White cheese in brine, with a slightly acidic, piquant, very salty flavour. It has no rind and a body with many small holes.

7.1.4 Method of Manufacture

Milk preparation: Traditionally, unpasteurised goat's or sheep's milk or their mixture was used, but nowadays, partly skimmed milk is used.

Starter culture: No starter culture is added.

Coagulation: Coagulation is carried out at 28°C–32°C with rennet in amounts that produce coagulation within approximately 50 min.

Cutting: As soon as the coagulation occurs (10 min after addition of the rennet), the coagulum is cut in small pieces and is left to rest for 30 min. A resting period of 30 min follows to complete the coagulation. Afterwards the curd is stirred for 15–20 min, and then is left still to precipitate on the bottom of the cheese vat.

Draining: The curd, with appropriate manipulation, is collected at one side of the vat, cut and put in a cheesecloth where it is left to drain. The whey thus produced has a high fat content and can be used for the manufacture of high-quality Manouri cheese.

Salting: The next day, the curd is cut into slices and salted. Usually four surface saltings are done with a coarse-grained salt, and then the cheese is put in a tin or lacquered metal container, covered with brine at 10%–12% and transferred into a cold room.

Maturation: Maturation takes place in cold rooms for at least three months.

7.1.5 Relevant Research

Lactic acid bacteria (LAB), Enterobacteriaceae and coliforms were the major components of the microflora of Batzos cheese during ripening; LAB grown on M17 agar was the most abundant microbial group in the curd and cheese during storage for one month at 4°C. *Enterobacteriaceae* and coliforms declined throughout maturation, and they declined faster in cheeses made in summer, possibly due to the lower pH values of cheeses made in summer and the higher number of LAB present in curds of summer cheeses (Nikolaou et al., 2002). *Lactobacillus paracasei* ssp. *paracasei* and *Lactobacillus paraplantarum* were isolated frequently from ripened Batzos cheese in summer, while *Enterococcus faecium* was the most frequently isolated LAB species in spring. *Lactococcus lactis* ssp. *lactis* was found only in the curds of spring cheeses. The main lactococci species were *Lc. lactis* ssp. *lactis*, found more frequently in winter cheeses. In spring and summer cheeses, facultatively heterofermentative lactobacilli predominated. *Lb. plantarum* and *Lb. paraplantarum* were the principal species of lactobacilli isolated. *Enterococcus durans* was the predominant *Enterococcus* spp. in cheese made in winter. In an attempt to improve the microbiological quality of the cheese from raw milk, selected *Lc. lactis* ssp. *lactis* strains were used as the starter (Psoni et al., 2006).

7.2 Beyaz Peynir – Turkey

Name: Turkish Beyaz Peynir (white cheese)	
Production area: Turkey	
Milk: Cow's, sheep's, goat's or an appropriate mixture, pasteurised	

7.2.1 Introduction

Turkish Beyaz Peynir was traditionally produced by small-scale dairies for centuries. However, along with developments in dairy technology, the incorporation of mechanisation and automation in cheese manufacturing made large-scale production possible. Today, except for the practices in remote rural areas, the production of Beyaz Peynir is carried out by medium- or large-scale dairy factories. Traditionally, Beyaz Peynir was manufactured from sheep's or goat's milk. However, more readily available cow's milk is used by large-scale dairy processors to meet the growing demands for these cheeses in the markets. Today, almost 95% of the cheeses, which amounts to 574,138 tonnes, are made from cow's milk (Anonymous, 2014). On the other hand, cow's milk is not ideal for Beyaz Peynir because it imparts a yellowish colour and a characteristic 'cowy' odour to the cheese. The main objection to goat's milk is that it produces a hard, dry cheese which is atypical for this type of cheese. Beyaz Peynir is mainly, but not limited to, native to the western part of Turkey.

7.2.2 Type

The minimum total moisture levels of fresh and ripened Beyaz Peynir should be 65% and 60%, respectively, and according to the Turkish Food Codex Cheese Communiqué, this cheese variety is classified as a semi-hard cheese, and is categorised as (1) full fat – having 45% or higher FDM, (2) half fat – having an FDM between 25% and 45%, (3) low fat – having an FDM between 10% and 25% and (4) light – having less than 10% FDM (Anonymous, 2015).

7.2.3 Description and Sensory Characteristics

Beyaz Peynir is a rindless, white-coloured, close-textured (no evidence of holes) variety with a salty acid taste; it may have a slight piquant flavour, especially when made from sheep's milk (Özer, 2014). It is matured for a period of three months or longer. Industrial Beyaz Peynir is usually made from milk pasteurised at 78°C–80°C for 30–60 s using plate heat exchangers. Medium- or small-scale dairies usually prefer to pasteurise milk at 65°C–68°C for 15–30 min (open vat system). In both cases, addition of a starter culture is essential. In some practices, the milk is thermised at 61°C–63°C for 30–60 s, and cheese is manufactured without a starter culture. In this case, the ripening relies on the activity of indigenous flora.

7.2.4 Method of Manufacture

Milk preparation: In large-scale cheesemaking plants, clarified raw milk is usually standardised with respect to the casein to fat ratio (0.8–0.9) (Toufeili & Özer, 2006). The milk is heat treated as described above.

Coagulation: Milk is then cooled to 32°C and inoculated with a starter culture, usually at a level of 1–2 g/100 g. In order to restore the ionic calcium balance after heat treatment, calcium chloride is added at a level of 0.2 g/L cheesemilk. Inoculated milk is rested for about 30 min to allow development of acidity for stimulating rennet activity. The coagulation of milk is mostly achieved by commercial calf rennet. Other suitable coagulants, including proteinases from microorganisms, are becoming more popular in the manufacture of Beyaz Peynir. Most popular are aspartic proteases from *Rizomuchor miehei, Rizomuchor pusillus* and *Cryphonectria parasitica.* The use of microbial proteases has no significant effect on the yield of cheese, but the development of proteolysis is usually faster than with animal rennet. On the other hand, the sensory profile of the product made with calf rennet is superior to a cheese made with the microbial coagulant.

Cutting/moulding/pressing: The gelation of cheesemilk commences within 30–45 min, and the gel is sufficiently firm to be cut after about 90 min. The coagulum is cut into small cubes (1–2 cm^3) and rested for about 5–10 min. Then the coagulum is pressed by employing 20–40 kg weight for each 100 kg of cheesemilk. Pressing is ended when whey separation stops completely. Afterwards, the cheese mass is cut into portions of $7 \times 7 \times 7$ cm^3, weighing ~350–500 g.

Salting: Turkish Beyaz Peynir is kept in dense brine solution (14%–16% NaCl) for about 6–12 hr at 15°C–16°C and then ripened in brine solution (10%–12% NaCl) at 6°C–8°C. The final product should contain a maximum of 6.5% NaCl.

Maturation: Beyaz Peynir must be ripened for ≥ 90 days before being introduced into the market.

7.2.5 Relevant Research

The predominant LAB in Beyaz Peynir are *Lc. lactis* ssp. *lactis, Lc. lactis* ssp. *cremoris, Lactobacillus casei, E. faecalis* var. *liquafaciens* and *Leuconostoc paramesenteroides.* Determination of the technological performance of the natural flora of cheese is of critical importance in the selection of a balanced combination of starter bacteria to obtain the best texture, aroma/flavour and body in cheese (Hayaloğlu et al., 2002). Although white-brined cheese traditionally contains a relatively high level of salt, it is thought that, with moderate modifications in the manufacturing technology of this cheese variety, it may be possible to produce cheese containing probiotic strains at levels higher than the threshold for a probiotic effect without impairing the quality of the final product. Yılmaztekin et al. (2004) monitored the survival of *Bifidobacterium bifidum* and *Lactobacillus acidophilus* used as an adjunct culture in Beyaz Peynir. After 90 days of ripening at 4°C, the numbers of both probiotic bacteria were around 10^6 cfu/g of cheese. In order to improve the viability of *B. bifidum* and *Lb. acidophilus* in Beyaz Peynir, the microencapsulation technique was successfully employed by Özer et al. (2009).

A number of volatile compounds are produced by defined or wild-type lactococcal bacteria used in the manufacture of Turkish Beyaz Peynir. Although it may vary depending on the type of starter bacteria and manufacturing practices, the predominant groups of volatiles are methyl ketones (mainly 2-pentanone, 2-butanone and 2-heptanone) and alcohols (mainly ethanol, 2-pentanol, 2-heptanol and 3-methyl-1-butanol) (Özer et al., 2011).

7.3 Feta PDO – Greece

<table>
<tr><td>Name: Feta, PDO</td><td rowspan="3"></td></tr>
<tr><td>Production area: Macedonia, Thraki, Epiros, Thessalia, Sterea Ellada, Peloponnissos and Lesvos island</td></tr>
<tr><td>Milk: Sheep's or a mixture with goat's (up to 30%), pasteurised</td></tr>
</table>

7.3.1 Introduction

Feta is the most important cheese variety produced in Greece in terms of both production and consumer acceptance (Moatsou & Govaris, 2011). It is known worldwide and is registered as a PDO in the EU (EC, 2002). The word *feta* means 'slice' in Greek, and the name comes from the cutting of the curd pieces after elaboration.

Cheesemaking technology is based on the traditional method and has been described in detail many times, more recently by Abd El-Salam and Alichanidis (2004), Anifantakis and Moatsou (2006), Alichanidis and Polychroniadou (2008) and Alichanidis (2007).

7.3.2 Type

It is a white-brined cheese with a maximum moisture of 56% and a minimum FDM of 43%. Its gross composition is as follows: moisture 53.2%, fat 25.8%, FDM 55.1%, salt 2.3%, salt-in-moisture (S/M) 4.2 and pH 4.7 (Nega & Moatsou, 2012). The ripening period must be at least 60 days.

7.3.3 Milk

Pasteurised sheep's milk is used or a mixture of sheep's with goat's, with the latter not exceeding 30%.

7.3.4 Description and Sensory Characteristics

It is a semi-soft cheese, with a snow-white colour, a moist surface with no rind and a sliceable body. It has small, almond-shaped mechanical openings, while the presence of small round holes is considered a defect. Feta cheese is matured either in tinned or lacquered metal containers holding 15–18 kg each, or in wooden barrels of 40–50 kg, always immersed in brine; it is sold either as a piece taken out from a bulk package, or in packages of 0.5 or 1 kg in plastic boxes filled up with brine.

The body and texture is firm, smooth and creamy, which makes the cheese sliceable. No gas holes should be present, but irregular small mechanical openings are desirable. The flavour of typical Feta is mildly rancid, slightly acid and salty and is frequently described as flavourful and appetising.

7.3.5 Method of Manufacture

Milk preparation: Cheesemilk, after standardisation to a fat content of 5.8%–6%, is pasteurised at 72°C for 15 sec or thermised at 65°C for 5–10 min and is cooled at 32°C–34°C.

Starter culture: Traditionally, Feta cheese was manufactured from raw sheep's milk without the addition of starters (artisanal cheese), and development of the typical flavour and texture was based on the natural lactic acid microflora (Zygouris, 1952), as well as contaminants from the cheesemaking environment and equipment. Nowadays, the milk is pasteurised, and yoghurt culture (*Streptococcus thermophilus* and *Lactobacillus delbrueckii* ssp. *bulgaricus*, 1:1) or *Lc. lactis* ssp. *lactis* and *Lb. delbrueckii* ssp. *bulgaricus* (1:3) or *Lc. lactis* ssp. *lactis* and *Lc. lactis* ssp. *cremoris* culture is added to the cheesemilk at the level of 0.3%–0.5% and incubated for 20–30 min before the addition of rennet.

Coagulation: Coagulation takes place with rennet at 32°C–34°C for ~50 min. Traditionally, the rennet used was home-made from the abomasa of unweaned lambs and kids, and, besides chymosin, it contained several enzymes (e.g. pregastric lipase) which contribute to the development of the pleasant, piquant flavour of traditional Feta (Anifantakis, 1991). Cheeses made using such rennets have a strong, spicy taste, which is mainly due to the accumulation of short-chain free fatty acids in the cheese. In some modern dairy plants, the traditional rennet has been partially or totally substituted by commercial rennets. However, most of the factories use a mixture of these two rennets in a ratio of 1:3 (traditional/commercial), in order to achieve the traditional flavour of Feta cheese.

Cutting: The coagulum is cut into cubes of 2–3 cm and left for about 10 min for partial whey exudation.

Curd draining: The curd is gradually transferred into perforated moulds with a rectangular or cylindrical shape 20–30 cm in height and diameter according to the size of the barrel in which the cheese will be packed. This gradual transfer leads to the formation of small mechanical openings in the cheese mass, which is a characteristic of Feta cheese. The curd is left to drain in the moulds at 14°C–16°C without pressing for 2–3 hr, and the moulds are then inverted and left for another 2–3 hr to complete draining. While still in the moulds, the drained curds are cut to the desired final shape, usually rectangular (11 × 11 or 11 × 22 cm) or sphenoid for wooden barrels.

Salting: The curd pieces are removed from the mould and placed on a salting table, side by side, to retain their shape. The surface of the table has already been sprinkled with coarse sea salt. The curd is then dry-salted on the surface, left for 12 hr, then inverted and dry-salted again. This procedure is repeated until the curd contains about 3%–3.5% salt. During dry salting, the coarse salt (the size of a rice grain or bigger) penetrates slowly into the curd mass and contributes to the normal drainage. The role of dry salting is very important in the development of the secondary microflora (e.g. slime formation in traditional manufacture), which is a major parameter in the development of the desired cheese flavour. The reduction of the time of dry salting or even replacement by brining, have possible implications for the mild flavour of the industrial Feta and Feta-type cheeses; in fact, early brining interrupts the biochemical activities during the maturation, and the lactic acid produced by the starter is accumulated, resulting in high acidity cheese.

Maturation: Cheese pieces are tightly packed in tins or barrels, and brine (7%–8% salt) is added to the container to fill the space between them and cover the cheese surface completely. Cheese is kept at 14°C–16°C until the pH reaches 4.4–4.6 and the moisture decreases to less than 56%. This is the pre-maturation step, which usually lasts for 2–3 weeks. After this pre-maturation period, the tins or barrels are sealed and transferred to a cold room (4°C–5°C) to complete the maturation process, which takes not less than two months.

7.3.6 Relevant Research

Feta cheese has been extensively studied, and scientific papers can be found concerning the microbiology (Bintsis & Papademas, 2002; Manolopoulou et al., 2003; Rantsiou et al., 2008; Vassiliadis et al., 2009) and technology of Feta cheese (Abd El-Salam et al., 1993; Abd El-Salam & Alichanidis, 2004; Alichanidis, 2007; Alichanidis & Polychroniadou, 2008; Anifantakis & Moatsou, 2006).

Proteolysis is not very intense in Feta cheese (Abd El-Salam & Alichanidis, 2004). The low pH of Feta does not favour the proteolytic activities of enzymes other than chymosin. Nega and Moatsou (2012) concluded that the low pH and high moisture in Feta cheese resulted in high residual chymosin and low plasmin and plasminogen-derived activities. In addition, the ripening period (pre-maturation step, 2–3 weeks at 14°C–16°C), after which, subsequently, the cheese is transferred to the cold room, is short. In the cold room, all biochemical reactions, especially proteolysis, are slowed down. It is worth noting that a considerable amount of water-soluble peptides and amino acids produced from the proteolysis, together with some whey proteins which remain in the curd after draining, diffuse into the brine throughout the maturation (Abd El-Salam & Alichanidis, 2004; Michaelidou et al., 2005).

As has been already mentioned, typical Feta cheese is characterised by a slightly rancid flavour and the extensive and distinctive lipolysis that can be attributed to the use of traditional rennet. FFAs are also released by the enzymes of its secondary microflora (mainly yeasts and bacteria), which grow during dry salting and are washed off when the cheese is put in brine (Zygouris, 1952). The biochemistry of the maturation of Feta and other white-brined cheeses has been reviewed by Moatsou and Govaris (2011). The microbiological quality and the most common defects in Feta and other white-brined cheeses have been reviewed (Bintsis & Papademas, 2002).

7.4 Halitzia – Cyprus

Name: Halitzia	
Production area: Tilliria	
Milk: Goat's, sheep's may be added, raw or pasteurised.	

7.4.1 Introduction

Halitzia were originally made at Pyrgos (a village of Tylliria found on the northern foothills of the Troodos mountain region) and Tsakkistra (a village of the province of Nicosia). The milk used was goat's or a mixture with sheep's milk, and there are accounts mentioning that the cheese was made in the past during May–June (Economides, 2004). Nowadays the cheese is also produced in October to December, when the milk quality is good and the weather conditions (i.e. 20°C–25°C) are favourable for the maturation of that cheese. *Halijia* means gravel or small stone; this name is derived from the shape of the finished product, which looks like a small stone.

7.4.2 Type

Halitzia is a white-brined cheese that will mature for at least 40 days in whey-brine.

7.4.3 Description and Sensory Characteristics

Halitzia has a characteristic fresh, sour taste that is moderately salty. It is slightly crumbly, with mechanical holes and it has a rather smooth texture.

7.4.4 Method of Manufacture

Milk preparation: Fresh, raw goat's (and sheep's) milk is filtered with a cloth and then heated to reach a temperature of 33°C–35°C.

Rennet/starter cultures: Non-animal commercial rennet is used, while no starter cultures are added in the manufacture of the cheese.

Coagulation/cutting: The milk is coagulated within 45–50 min, and the cheese curd is cut into cubes of 1.5–2.5 cm. The curds are briefly stirred with no reheating and left to settle to the bottom of the vat.

Moulding/pressing: The curd is then transferred to rectangular stainless steel moulds (5 cm height) with holes lined with cheesecloths to assist the whey drainage. No pressure is applied, and the curd is left to drain for 2 hr at room temperature or for 1 hr if placed in refrigeration temperatures, to a curd height of about 3 cm.

Salting: Individual pieces of cheese (squares of 8–10 cm) are hand-salted and placed in non-pasteurised salted whey (13% NaCl w/v).

Maturation: The cheese is matured in containers of 13% NaCl (w/v) of whey-brine for at least 40 days at 25°C.

Packing: After 40 days of brining, the product can be vacuum-packed and placed in the fridge. However, Halitzia can be kept and be preserved in brine for several months or up to a year. However, it is recommended that Halitzia are kept in brine for no more than 3–4 months, otherwise the product will be extremely salty. This type of cheese is usually consumed within a year from the day of production.

7.5 Halloumi - Cyprus

Name: Halloumi	
Production area: Cyprus	
Milk: Sheep's or goat's or mixtures with the addition (or not) of cow, pasteurised	

7.5.1 Introduction

Halloumi is the traditional cheese of Cyprus and the major agricultural product of the island. The first references that link Halloumi with Cyprus date back to 1554, when Florio Bustron refers to the sheep and goats of Cyprus, and a cheese called Halloumi (in Italian 'calumi') made from a mixture of sheep's and goat's milk. In 1778, Kyprianos of the Cyprus Church, in his historical review, describes Halloumi cheese as 'delicious' and that 'quantities were sold abroad' (Papademas, 2006). Halloumi is a versatile cheese and is used as an ingredient in many different recipes and served in various ways. When grilled, it retains its shape and does not melt, owing to the high pH of the cheese, and most of the calcium is still present in its colloidal form, that is, as calcium phosphate, which holds the protein network intact. Its colour turns golden (due to the Maillard reaction) as the cheese retains much of its initial lactose, because no starter cultures are used in the manufacture of the cheese. Traditionally, Cypriots consume Halloumi cheese fresh accompanied with fruit and especially watermelon, or grilled in pitta bread. Mature Halloumi is usually grated over hot pasta sprinkled with mint, or sliced accompanying traditional deli meats and strong local spirits. Halloumi exports have considerably increased over the last few years, with an annual market value of 80 million euros.

7.5.2 Type

Halloumi is commercially available in two forms: fresh (semi-hard) and mature (hard) when it ripens for at least 40 days in salted whey-brine. The fresh cheese has a maximum moisture content of 46%, a minimum FDM of 43% and a maximum salt content of 3% (Anonymous, 1985a). Mature Halloumi cheese has a maximum moisture content of 37%, a minimum FDM of 40%, a maximum salt content of 6% and acidity of 1.2% expressed as lactic acid (Anonymous, 1985b).

7.5.3 Milk

The milk used in the manufacture of Halloumi cheese is a mixture of sheep's, goat's and cow's. By law, the percentage of sheep's/goat's milk in the mixture must not be lower than 20%, the remainder being cow's milk. On the other hand, the PDO application that was filed with the European Commission states that the milk mixture must comprise 51% sheep's/goat's milk and 49% cow's milk. If Halloumi is granted PDO status, then milk producers will have to considerably increase the quantities of sheep's and goat's milk within a period of eight years from today.

7.5.4 Description and Sensory Characteristics

Fresh Halloumi cheese is white, with a compact, elastic texture that is easily sliced. It has a characteristic flavour owing to the milk mixture used and the addition of mint in the final product. It is a rather salty cheese with very low acidity. On the other hand, mature Halloumi cheese is yellowish, hard, slightly crumbly, easily grated, salty and acidic. Halloumi has a rectangular shape (industrial) or 'half-moon' folded (traditional) in 250–300 g pieces.

7.5.5 Method of Manufacture

The method of manufacture for fresh and mature Halloumi cheese is described in the Cyprus Standards

Milk preparation: Milk is pasteurised (72°C/15 s). The addition of milk powders or any other additives are prohibited.

Rennet/starter cultures: Non-animal commercial rennet is used, while no starter cultures are added in the manufacture of the cheese.

Coagulation/cutting: The milk is coagulated within 45–50 min, and the cheese curd is cut into cubes of 1 cm.

Moulding/pressing: For industrial manufacture, curd pieces are transferred in rectangular moulds with dimensions (7.5 × 11.5 × 3.3 cm). The moulds are stacked on top of each other, and pressure is applied to facilitate the expulsion of whey. When traditional Halloumi cheese is manufactured, the curd is transferred manually to slightly concave plastic moulds with the following dimensions: bottom diameter 9.5 cm, top diameter 12 cm and height 8 cm. No pressure is applied. The cheese is the moulds is turned twice.

Cooking/scalding: The cheese curds are placed in de-proteinated whey (Anari cheese; see Section II, Chapter 13, Section 13.1) and heated to 90°C for at least 30 min. The core temperature of the cheese reaches 80°C.

Cooling/salting/folding: Halloumi cheese pieces are left to briefly cool, a mixture of salt and dry mint is sprinkled on them and they are manually folded.

Salting: Fresh Halloumi cheese is salted overnight (approximately 16 hr) in whey-brine (12% NaCl, w/v).

Packing: Fresh Halloumi cheese is vacuum-packed and kept under refrigeration until it is sold.

Maturation: The cheese is matured in containers of 12% NaCl (w/v) of whey-brine for at least 40 days at 20°C. Mature Halloumi cheese is either sold as individual vacuum-packed pieces or in containers of 1–5 kg filled with whey-brine.

7.5.6 Relevant Research

The cheese's characteristics (i.e. chemical, microbiological, sensory) both as fresh and mature product have been extensively studied by Papademas (2006). Kaminarides et al. (2007) also examined the changes in organic acids, volatile aroma compounds and sensory characteristics of sheep's milk Halloumi cheese kept in brine. More recently, Osorio et al. (2015) looked at the possibility of using trace element analysis as potential biomarkers in milk and Halloumi cheese for distinguishing goat feeding regimes, that is, extensive, semi-extensive and free grazing Interestingly, a study of the volatile fraction (free fatty acids) from the same set of samples and the non-volatile fraction (i.e. polyphenols) illustrated that they could also serve as potential biomarkers for distinguishing feeding regimes and season of production of Halloumi cheese (Papademas, unpublished data).

7.6 Mihalıç – Turkey

<table>
<tr><td>Name: Mihalıç cheese</td><td rowspan="3"></td></tr>
<tr><td>Production area: Northwestern Turkey</td></tr>
<tr><td>Milk: Sheep's, goat's, cow's or their mixtures. Ratio of sheep's milk should not be lower than 60% in mixtures, raw or pasteurised</td></tr>
</table>

7.6.1 Introduction

In terms of production capacity, Mihalıç cheese is the fourth-ranked economically important Turkish cheese variety, coming after Beyaz Peynir, Kashar and Tulum cheeses. This cheese variety was first produced in Mustafakemalpaşa province, and its name comes from the previous name of the province 'Mihalıç'. The history of Mihalıç cheese dates back to 200–250 years ago (Hayaloğlu et al., 2008).

7.6.2 Type

Mihalıç cheese can be produced from goat's milk or appropriate mixtures of sheep's, goat's and/or cow's milks. In the latter case, the share of sheep's milk should not be less than 60% (Çakmakçı et al., 2012). This cheese variety is classified as a hard-type white-brined cheese with high fat and a characteristically high salt level (up to 8%).

7.6.3 Description and Sensory Characteristics

Mihalıç cheese is ripened in brine and characterised with regular openings (2–4 mm) and 3–4 mm rinds (Aday and Karagül-Yüceer, 2010). Mihalıç cheese has a unique colour, aroma and flavour and owing to its high total solids level, it is a nutritious cheese variety. The size of the cheese blocks may vary depending on the manufacturer, but its weight is usually between 3 and 5 kg. This cheese is native to the northwest part of Turkey (specifically, the region of Marmara).

7.6.4 Method of Manufacture

Milk preparation: In the manufacture of Mihalıç cheese, raw, thermised or pasteurised milk may be used. Since the curd is scalded after cutting and pressing the curd, the use of raw milk in the production of Mihalıç cheese is a common practice.

Coagulation: Coagulation of milk is achieved by animal rennet in wooden barrels locally called '*polim*' or '*taari*' of 100–250 litres capacity (Hayaloğlu et al., 2008). Coagulation is complete within 60 to 90 min at 32°C–35°C.

Cutting/cooking: Following coagulation, the curd is cut into very small pieces (i.e. the size of a rice grain) and cooked at 45°C by adding hot water or cheese whey for 10–15 min. The cooked curd is placed into cheesecloths, and whey is expelled by gravity drainage. Whey separation is complete within 3–8 hr depending on the amount of curd. In industrial practice, curd is pressed using 50 kg per 250 kg of milk. Regular openings are achieved by piercing the cheesecloths periodically.

Salting: The drained curd is cut into 3–5 kg of portions and put into brine solution with an NaCl level of 18% for three days. The cheese blocks are put in a denser brine solution (20% NaCl) for the next two days and finally placed in brine with 22% NaCl for another 15–20 days. After the pre-brining period, the cheese blocks are placed on the bottom of a wooden barrel containing a layer of coarse salt. Afterwards, brine solution (20%–22% NaCl) is added onto the cheeses, and the barrel is closed with a wooden lid.

Maturation: The usual ripening period of Mihalıç cheese at <10 °C (preferably at 4°C) takes about 90 days (Hayaloğlu et al., 2008). Mihalıç cheese is usually introduced to the markets as vacuum-packed portions of 200–300 g. The yields of Mihalıç cheese made from sheep's and cow's milks are 20–22% and 10%–12%, respectively (Özcan, 2000; Tekinşen, 1997).

7.6.5 Relevant Research

The rate of lipolysis in Mihalıç cheese increases significantly throughout the ripening period of 90 days. The use of commercial lipase (i.e. Piccantase A) and/or protease (i.e. Fermizyme B500) results in increased acid degree values from 2.30 to 6.16 KOH/g cheese (Özcan, 2000). Similarly, proteolysis increases at a much faster rate when exogenous lipases and/or proteases are added to the cheesemilk. Degradation of α-casein and β-casein is usually faster in the cheese made from pasteurised milk with starter plus added commercial lipase than that made from raw milk. The degradation of these two major casein fractions coincides with the increased level of γ-casein. In general, the use of lipase and/or protease increases the yield; however, selection of the protease level is of critical importance for the establishment of a balanced aroma/flavour as excessive proteolysis may lead to bitterness in Mihalıç cheese. In this respect, Özcan (2000) recommends the use of proteases to accelerate the ripening of Mihalıç cheese at a level lower than 0.002%.

In the manufacture of Mihalıç cheese, a mesophilic-aromatic starter combination containing *Lc. lactis* ssp. *cremoris*, *Lc. lactis* ssp. *lactis*, *Leuc. mesenteroides* ssp. *cremoris* and *Lc. lactis* ssp. *lactis* var. *diacetylactis* is used at a level of 1.0% (Özcan, 2000). *Leuc. mesenteroides* ssp. *cremoris* is primarily responsible for the formation of holes as this bacteria is able to produce carbon dioxide. *Propionibacterium* spp. play a major role in both the ripening and characteristic hole formation of Mihallıç cheese.

7.7 Sjenica - Serbia

Name: Sjenica	
Production area: The Sjenicko-Pesterska plateau, in the southwest part of Serbia	
Milk: Cow's, raw	

7.7.1 Introduction

White-brined cheeses are the most widely produced and consumed cheeses in Serbia (accounting for about 60% of the total cheese consumption in Serbia). There are many different types of Serbian brined cheeses named according to production regions as follows: Sjenica cheese, Zlatar cheese, Svrljig cheese and Homolj cheese. These cheeses may be very similar but also rather different with respect to the type of milk, region of production, manufacturing protocols, composition and sensory properties.

Some of the white-brined cheeses, including Sjenica cheeses, are recognised as products with geographical indication at the national level of protection. Nowadays, Sjenica cheeses from sheep's and cow's milk are protected and have been certified with a sign of appellation of origin at the national level (The Intellectual Property Office of the Republic of Serbia, 2011, 2014).

Reliable data on the beginning of the production of white-brined cheeses is not available. It is assumed that their production in the Balkans began with arrival of the Slavs in the ninth to tenth centuries. The available historical data of organised livestock breeding and cheesemaking dates back to the twelfth century. Although the first written records on cheese trading in this region are from the fourteenth and fifteenth centuries, detailed data on white-brined cheese production in the currently Serbian area date back to the nineteenth century (Djurić, 1906; Dozet et al., 1996; Trojanović, 1896).

The geographical region of Sjenica cheese production is in the Sjenicko-Pesterska plateau, which is located in the southwestern part of Serbia and represents the biggest plateau in the Balkan, which includes the municipalities Sjenica and Tutin. Sjenicko-Pesterska plateau is located at an altitude of 1150 m and covers an area of 63 km^2.

7.7.2 Type

The specifications of the ripened Sjenica cow cheese (60 days old) are as follows: FDM 53%–58%, moisture content 54%–58% and salt content 2.2%–3.1%.

7.7.3 Milk

Traditionally, Sjenica cheeses were made from cow's, sheep's and goat's milk, but now they are produced mainly from cow's milk. Sjenica cheeses are home-made cheeses, manufactured at farms or produced in small dairy craft plants. Traditionally, Sjenica cheeses were produced from raw milk, but thermisation at the 63°C/10–15s is a common practice in small dairy craft plants. On average, 7–8 litres of milk is used for 1 kg of cheese.

7.7.4 Description and Sensory Characteristics

Sjenica cheeses are white-brined cheeses cut into slices, without a rind. The texture/consistency is firm and brittle. Depending on the dry matter content and ripening period, holes may be absent or there may be a few small mechanical holes and openings. Slices are white or yellowish, weighing 200–500 g, in block shape, having dimensions of 10–15 × 10–15 × 3–5 cm or with a triangular shape. The colour of the cheeses is white/plain yellowish with a smooth surface. The odour is mild, typical milky and clearly expressed. The flavour is slightly acidic and distinctive for white-brined cheeses while the cheese is moderately salty. More mature cheeses are characterised by a very pungent odour and flavour.

7.7.5 Method of Manufacture

The production of Sjenica cheeses is carried out at farmhouses and small-scale dairies. The use of additives is forbidden.

Milk preparation: Traditionally, at the farms, milk for cheese production is used immediately after the milking. When mixing of evening and morning milk is necessary, evening milk must be cooled to 7°C–10°C. In small craft dairy plants, milk must be delivered twice a day, immediately after the milking.

Starter culture: For the production of Sjenica cheese, no starter cultures are used. The microflora of milk and airborne bacteria are responsible for the cheese fermentation process.

Coagulation/pressing: Addition of calf's rennet occurs at 30°C–32°C while coagulation lasts 40–60 min. After coagulation, the coagulum is cut and transferred into cheesecloth. Spontaneous whey drainage is carried out for 1 hr after which pressure is applied (0.5–2 kg/kg cheese mass) during the next 2–3 hr.

Cutting: Cutting is carried out manually, in block shape with dimensions 10–15 × 10–15 × 3–5 cm, or in a triangular shape (wedge).

Salting: The salting starts after pressing, by dry salting, and the slices are placed in a wooden vat. In small-scale dairy plants, plastic vats are usually used.

Maturation: After salting, the cheeses are stored in vats at a temperature of 14°C–18°C. During maturation, whey is gradually drained from cheese, which contributes to formation of brine, in which cheese is immersed. All cheese must be covered with brine, and during maturation it is necessary to ensure hygienic conditions and remove any mould which may appear on surface. Ripening lasts a minimum of 60 days in closed vats (anaerobic conditions) in brine with 6%–8% NaCl at 14°C–18°C.

Storage: Ripened cheeses are stored in cold storage at 4°C–10°C. Sjenica cheese can be sold on markets in vats of 3, 5 or 10 kg, depending on the buyer requirement.

7.7.6 Relevant Research

The diversity of autochthonous LAB in Sjenica cheeses was studied. Over 47% of the isolates were lactococci, over 35% were lactobacilli, over 10% were enterococci and over 3% were leuconostocs (Radulović, 2010). From autochthonous microflora, starter cultures are selected for application in standardised Sjenica cheese production in semi-industrial conditions. Sensory evaluation showed no significant differences in comparison with traditionally produced cheeses (Radulović et al., 2011).

7.8 Urfa – Turkey

Name: Urfa cheese	
Production area: Southeastern Anatolia (specifically Urfa province)	
Milk: Sheep's, cow's or mixture (sheep's milk being dominant), raw or pasteurised	

7.8.1 Introduction

The manufacture of Urfa cheese from the milk of Awassi sheep intensifies between February and early May. After mid-May, most of the sheep's milk is used for yoghurt and butterfat production. Therefore, cow's milk is used in the manufacture of Urfa cheese after May. The annual production of Urfa cheese is estimated to be approximately 35,000–40,000 tonnes (Özer et al., 2002). The cheese is characterised by the absence of a rind, white colour, closed texture and a salty acid taste. In the production of Urfa cheese, whey drainage is achieved by leaving the curd hanging in a special cheesecloth (known locally as '*parzın*') (Toufeili & Özer, 2006). Urfa cheese is native to the southeastern part of Turkey. It is commonly produced in the areas between the rivers Euphrates and Tigris.

7.8.2 Type

Urfa cheese is a white-brined, semi-hard cheese variety (Özer et al., 2003). In some parts of southeastern Turkey, the scalding stage is excluded, and cheese is kept in hot water for a short while just before use. As a more general practice, cheese blocks are scalded and then put into brine solution.

7.8.3 Description and Sensory Characteristics

This traditional cheese variety is mainly produced from raw sheep's milk (mainly milk of the Awassi sheep breed) or an appropriate mixture of sheep's and cow's milks. Urfa cheese is increasingly gaining nationwide popularity, and it is also exported to Middle Eastern and Central Asian countries. During the past 5–10 years, attempts have been made to give local cheese varieties an industrial identity. In this context, studies on the characterisation of the chemical, physical and organoleptic properties of traditional Urfa cheese and adaptation of modern technologies to the production of this cheese have been intensified. In this context, ultrafiltration has been successfully employed to produce Urfa cheese with better texture and higher yield (more details are given in the following section).

7.8.4 Method of Manufacture

The production period of sheep's milk in Turkey is fairly short (approximately 6–7 months) so that it is not always possible to extend the Urfa cheese production over the year. For this reason, the possibilities of using cow's milk in the manufacture of Urfa cheese were studied (Özer et al., 2002). Traditional Urfa cheese is made from raw milk. However, since the average daytime temperature in southeastern Turkey, where Urfa cheese production is prevalent, is well over the national average, raw milk has largely been replaced by thermised or pasteurised milk in industrial production. No starter culture is used in traditional production.

Milk preparation: In the case of use of pasteurised milk, a mesophilic (*Lc. lactis* ssp. *lactis* and ssp. *cremoris*) or thermophilic (*S. thermophilus* and *Lb. delbrueckii* ssp. *bulgaricus*) starter culture can be employed. Atasoy et al. (2008) demonstrated that Urfa cheese made from high-heat-treated cow's milk (at 72°C for 5 min) had lower sensory scores than that made from low-heat-treated milk or from raw milk. The authors recommended to inoculate Urfa cheese milk with mesophilic starter bacteria rather than with thermophilic ones.

Coagulation: Pasteurised milk (in industrial practice) or raw milk (in traditional practices) is coagulated with animal rennet at a level sufficient to coagulate milk within about 60 min at 32°C.

Cutting: The coagulated milk is then cut into small pieces of 3–5 cm^3 and transferred into parzins (a conical shape cheese cloth with about 0.5 kg capacity). The whey separation is achieved by gravity drainage in a cold room, say 4°C–8°C. The whey separation takes about 6–10 hr depending on the parzin's capacity and the total solids level of the coagulum as well as the drainage temperature, the separation being faster at higher temperatures.

Salting: Following wheying off, the cheese blocks are dry-salted overnight, with each side being salted by turning the blocks at 4 to 6 hr intervals.

Scalding: The cheese blocks are then scalded in boiling cheese whey for 1–3 min or by pouring boiling cheese whey onto the cheese blocks.

Maturation: Scalded cheeses are kept in brine with 16–20 g/100 g NaCl, w/v, at temperature less than 10°C, for more than 90 days.

7.8.5 Relevant Research

Biochemical changes during ripening of Urfa cheese were studied extensively by Özer et al. (2002 and 2003) and Kırmacı et al. (2014). Kırmacı et al. (2015) investigated the volatile profile of artisanal Urfa cheese made from sheep's milk; in all, 70 volatile compounds were recovered from the cheeses including 10 alcohols (mainly ethanol, 2-methyl-1-propanol and 3-methyl-1-propanol), 20 aldehydes and ketones (mainly 2-pentanone and 2-heptanone), 11 esters (mainly ethyl acetate), 10 acids (mainly acetic acid and 2-hydroxy-4-methyl pentanoic acid), 6 terpenes (mainly α-pinene) and 13 miscellaneous compounds (mainly 3,7-dimethyl-1,6-octadiene). Atasoy (2004) found no significant difference between mesophilic and thermophilic starter cultures with respect to fat hydrolysis in Urfa-type white cheeses, but products made from raw milk had remarkably higher concentrations of volatile fatty acids (14.55 mL NaOH/100 g cheese) than the cheeses made from pasteurised milk (2.90 mL NaOH/100 g cheese).

The scalded cheese has a firmer texture than the unscalded cheese, and the unscalded UF cheese has a more 'springy' body than the unscalded traditional cheese. Overall, scalding produces a more homogeneous structure, but the unscalded UF cheese has a closed texture that resembles the scalded cheeses. With respect to texture and structure, cheeses made with UF milk do not need to be scalded after production.

Özer et al. (2003) showed that the cheese made from UF concentrated milk had a more compact structure than the cheese manufactured from unconcentrated milk. On the other hand, a finer and continuous structure was evident in the latter sample. Surprisingly, fat particles were noticeable in the cheese produced from unconcentrated milk but not in the sample manufactured from UF concentrated milk. Cheeses prepared from milks concentrated by UF became progressively coarser as the concentration factor of the milk increased. The less developed structure of the curd from UF concentrates therefore tends to make it fragile. Scalding seemed to produce a more homogeneous structure in both traditional and UF Urfa cheeses. At the temperatures of ripening, the fat has its own peculiar rheological properties, behaving like a plastic material. As a result of scalding, distribution of fat globules over the casein matrix may contribute to the development of a more springy structure in the scalded cheeses.

In the traditional method of manufacture of Urfa cheese, microbiological safety was provided practically by keeping the cheese in a very dense brine solution (e.g. 20%–23%). However, considering the negative effects of salt on the activities of starter organisms and/or the natural microflora, the brine concentrations have been reduced to 12%–13%, as high brine concentrations greatly hindered the development of proteolysis (Özer et al., 2003). Alternatively, scalding of fresh cheese blocks in boiling whey for about 2–3 min is another practical way of reducing microbial counts in Urfa cheese. In this case, relatively lower brine concentrations are preferred, and the scalded cheese blocks have a springier and harder texture compared to the unscalded ones. However, even this treatment is not satisfactory enough to provide full microbial safety since the efficiency of scalding is dependent upon the initial microbiological load of the milk, the size of the cheese block and the period of scalding. The temperature gradient between the outer and inner layers of the circular cheese blocks is wide, and the temperature at the centre of the cheese block is not high enough to eliminate pathogens completely. Therefore, scalding should not be considered as a method that provides full microbiological safety in Urfa cheese.

References

Abd El-Salam, M. H. & Alichanidis, E. (2004). Cheese varieties ripened in brine. In: Fox, P. F., McSweeney, P. L. H., Cogan, T. M. & Guinee, T. P. (eds.), *Cheese: Chemistry, Physics and Microbiology. Major Cheese Groups*, Vol. **2**. Elsevier Academic Press, London, pp. 227–249.

Abd El-Salam, M. H., Alichanidis, C. & Zerfiridis, G. K. (1993). Domiati and Feta cheese. In Fox, P. F. (ed.), *Cheese: Chemistry, Physics and Microbiology*, 2nd edition, Vol. **2**. Chapman & Hall, London, pp. 301–335.

Aday, S. & Karagül Yüceer, Y. (2010). [Mihalıç peyniri üretimi ve karakteristik özellikleri. pp 312. *1. Uluslararası "Adriyatikten Kafkaslar'a Geleneksel Gıdalar Sempozyumu.*15–17 Nisan, Tekirdağ, Turkey].

Alichanidis, E. (2007). Cheeses ripened in brine. In McSweeney, P. L. H. (ed.), *Cheese Problems Solved.* CRC Press, Cambridge, pp. 330–342.

Alichanidis, E. & Polychroniadou, A. (2008). Characteristics of major traditional regional cheese varieties of East-Mediterranean countries: A review. *Dairy Science and Technology*, **88**, 495–510.

Anifantakis, E. M. & Moatsou, G. (2006). Feta and other Balkan cheeses. In Tamime, A. Y. (ed.), *Brined Cheeses.* Blackwell Publisher, Oxford, UK, pp. 43–71.

Anifantakis, E. M. (1991). *Greek Cheeses: A Tradition of Centuries.* National Dairy Committee of Greece, Athens, Greece.

Anonymous (1985a) Cyprus Standard 94. *Part 1: Fresh Halloumi cheese.*

Anonymous (1985b) Cyprus Standard 94. *Part 2: Mature Halloumi cheese.*

Anonymous (2014). *National Milk Council of Turkey.* USK report on dairy sector in the world and in Turkey.

Anonymous (2015). *Turkish Food Codex, Cheese Coomunique, Communique No: 2015/6*, Published by the Ministry of Food, Agriculture and Livestock of Turkey.

Atasoy, A. F., Yetişmeyen, A., Türkoğlu, H. & Özer, B. H. (2008). Effects of heat treatment and starter culture on the properties of traditional Urfa cheeses (a white-brined Turkish cheese) produced from bovine milk. *Food Control*, **19** (3), 278–285.

Bintsis, T. & Papademas, P. (2002). Microbiological quality of white-brined cheeses: A review. *International Journal of Dairy Technology*, **55** (3), 113–120.

Çakmakçı, S., Çetin, B., Gürses, M., Dağdemir, E. & Hayaloğlu, A. A. (2012). Morphological, molecular, and mycotoxigenic identification of dominant filamentous fungi from moldy civil cheese. *Journal of Food Protection*, **75** (11), 2045–2049.

Djurić, M. Đ. (1906). *Naši mlečni proizvodi.Težak-ilustrovani glasnik Srpskog poljoprivrednog društva, Beograd*, br.32-33.

Dozet, N., Adžić, N., Stanišić, M. & Živić, N. (1996). *Autohtoni mlječni proiozvodi*. Poljoprivredni institut, Podgorica, Silmar, Beograd.

EC (1996). Commission Regulation (EC) No 1107/96 of 12 June 1996 on the registration of geographical indications and designations of origin under the procedure laid down in Article 17 of Council Regulation (EEC) No 2081/92 *Official Journal of the European Union*, **148**, 1–10.

EC (2002). European Commission Regulation No. 1829, Amending to the Annex to Regulation (EC) No. 1107/96 with regard to the name 'Feta'. *Official Journal of the European Union, L* **277**, 10–14.

Economides, S. (2004). *Milk and Dairy Products of Cyprus*. Cyprus Milk Industry Organisation, Nicosia, Cyprus.

Hayaloğlu, A. A., Güven, M. & Fox, P. F. (2002). Microbiological, biochemical and technological properties of Turkish White cheese "Beyaz Peynir". *International Dairy Journal*, **12**, 635–648.

Hayaloğlu, A. A., Özer, B. & Fox, P. F. (2008). Cheeses of Turkey: 2. Varieties ripened under brine. *Dairy Science and Technology*, **88**, 224–244.

Kaminarides, S., Stamou. P. & Massouras, T. (2007). Changes of organic acids, volatile aroma compounds and sensory characteristics of Halloumi cheese kept in brine. *Food Chemistry*, **100**, 219–225.

Kırmacı, H. A., Hayaloğlu, A., Özer, B., Atasoy, F. & Türkoğlu, H. (2014). Evolution of proteolysis in Urfa cheese made from ewe's milk by wild-type starter culture. *Small Ruminant Research*, **119**, 120–129.

Kırmacı, H. A., Hayaloğlu, A. A., Özer, B., Levent, O. & Atasoy, A. (2015). The effects of incorporating wild type starter culture systems (artisanal strains) on volatile profile of Urfa cheese made using sheep milk. *International Journal of Food Properties*, **18**, 1915–1929.

Manolopoulou, E., Sarantinopoulos, P., Zoidou, E., Aktypis, A., Moschopoulou, E., Kandarakis, I. G. & Anifantakis, E. M. (2003). Evolution of microbial populations during traditional Feta cheese manufacture and ripening. *International Journal of Food Microbiology*, **82**, 153–161.

Michaelidou, A., Alichanidis, E., Polychroniadou, A. & Zerfiridis, G. (2005). Migration of water soluble nitrogenous compounds of Feta cheese from the cheese blocks into the brine. *International Dairy Journal*, **15**, 663–668.

Moatsou, G. & Govaris, A. (2011). White brined cheeses: A diachronic exploitation of small ruminants milk in Greece. *Small Ruminant Research*, **101**, 113–121.

Nega, A. & Moatsou, G. (2012). Proteolysis and related enzymatic activities in ten Greek cheese varieties. *Dairy Science and Technology*, **92** (1), 57–73.

Nikolaou, E., Tzanetakis, N., Litopoulou-Tzanetaki, E. & Robinson, R. K. (2002). Changes in the microbiological and chemical characteristics of an artisanal low-fat cheese made from raw ovine milk during ripening. *International Journal of Dairy Technology*, **55**, 12–17.

Osorio, M. T., Koidis, A. & Papademas, P. (2015). Major and trace elements in milk and Halloumi cheese as markers for authentication of goat feeding regimes and geographical origin. *International Journal of Dairy Technology*, (**67**), 1–9.

Özcan, T. (2000). [*Starter, proteaz ve lipaz kullaniminin Mihalıç peynirinin olgunlaşma surecine etkisi.* PhD. Thesis, Uludag University, Institute of Natural and Applied Sciences, Department of Food Engineering, Bursa, Turkey.]

Özer, B. (2014). Microflora of white-brined cheeses. In Batt, C. A. & Tortorello, M. L. (eds.), *Encyclopedia of Food Microbiology,* 2nd edition. Elsevier Ltd., Academic Press, pp. 402–408.

Özer, B. H., Atasoy, A. F. & Akın, M. S. (2002). Some properties of Urfa cheese (a traditional white-brined Turkish cheese) made from bovine's and ovine's milk. *International Journal of Dairy Technology,* **55** (2), 94–99.

Özer, B. H., Kırmacı, H. A., Hayaloğlu, A. A. S., Akçelik, M. & Akkoç, N. (2011). The effects of incorporating wild-type strains of *Lactococcus lactis* into Turkish white-brined cheese (Beyaz peynir) on the fatty acid and volatiles content. *International Journal of Dairy Technology,* **64** (4), 494–501.

Özer, B. H., Kırmacı, H. A., Şenel, E., Atamer, M. & Hayaloğlu, A. A. (2009). Improving viability of *Bifidobacterium bifidum* BB-12 and *Lactobacillus acidophilus* LA-5 in white-brined cheese by microencapsulation. *International Dairy Journal,* **19,** 22–29.

Özer, B. H., Robinson, R. K. & Grandison, A. S. (2003). Textural and microstructural properties of Urfa cheese (A white-brined Turkish cheese). *International Journal of Dairy Technology,* **56** (3), 171–176.

Papademas, P. (2006). Halloumi cheese. In Tamime, A. Y. (ed.), *Brined Cheeses,* Chapter 4. Blackwell Publishing Ltd, Oxford, pp. 117–138.

Psoni, L., Tzanetakis, N. & Litopoulou-Tzanetaki, E. (2003). Microbiological characteristivs of Batzos, a traditional Greek cheese from raw goat's milk. *Food Microbiology,* **20** (5), 575–582.

Psoni, L., Tzanetakis, N. & Litopoulou-Tzanetaki, E. (2006). Characteristics of Batzos cheese made from raw, pasteurized and/or pasteurized standardized goat milk and a native culture. *Food Control,* **17,** 533–539.

Radulović, Z., Miočinović, J., Pudja, P., Barać, M., Miloradović, Z., Paunović, D. & Obradović, D. (2011). The application of autochtonous lactic acid bacteria in white brined cheese production. *Mljekarstvo,* **61** (1), 15–25.

Radulović, Z. (2010). Autochthonous lactic acid bacteria as starter cultures. *Monograph, Faculty of Agriculture,* University in Belgrade, Serbia.

Rantsiou, K., Urso, R., Dolci, P., Comi, G. & Cocolin, L. (2008). Microflora of Feta cheese from four Greek manufacturers. *International Journal of Food Microbiology,* **126,** 36–42.

Tekinşen, O. C. (1997). [*Süt Ürünleri Teknolojisi. Selçuk Üniversitesi Veteriner Fakültesi Yayın Ünitesi,* Konya.]

The Intellectual Property Office of the Republic of Serbia (2011). *Sheep Cheese from Sjenica Study on the Protection of Origin Names.* No. Reg. **59.** http://www.zis.gov.rs [Accessed 21 May 2015].

The Intellectual Property Office of the Republic of Serbia (2014). *Cow Cheese from Sjenica Study on the Protection of Origin Names.* No. Reg. 65. http://www.zis.gov.rs [Accessed 21. May 2015].

Toufeili, I. & Özer, B. (2006). Brined cheeses from the Middle East and Turkey. In Tamime, A. Y. (ed.), *Brined Cheeses.* Blackwell Publishing, Oxford, UK, pp. 188–210.

Trojanović, S. (1896). *Starinska srpska jela i pića.* Srpski etnografski zbornik, Knj. 2, Izd. Srp. Kralj. Akademije, Beograd.

Vassiliadis, A., Psoni, L., Nikolaou, S., Arvanitis, L. Tzanetakis, N. & Litopoulou-Tzanetaki, E. (2009). Changes in microbial populations, kinds of lactic acid bacteria and biochemical characteristics of Greek traditional feta cheese during ripening. *International Journal of Dairy Technology,* **62,** 39–47.

8

Pasta-Filata Cheeses

Giuseppe Licitra[1], Zorica Radulovic[2], Jelena Miocinovic[2], İrem Uzunsoy[3], Barbaros Özer[4], Thomas Bintsis[5], Efstathios Alichanidis[6], Karol Herian[7] and Paul Jelen[8]

[1] Department of Agriculture and Food Science (DISPA), University of Catania, Italy
[2] Department of Food Microbiology, Faculty of Agriculture, University of Belgrade, Serbia
[3] Bülent Ecevit University Caycuma Vocational High School, Department of Food Technology, Zonguldak, Turkey
[4] Ankara University, Faculty of Agriculture Department of Dairy Technology, Ankara, Turkey
[5] 11 Parmenionos, 50200, Ptolemaida, Greece
[6] Department of Food Science and Technology, School of Agriculture, Aristotle University of Thessaloniki, Greece
[7] Director-Emeritus, Slovak Dairy Research Institute, Slovakia
[8] Department of Agricultural, Food and Nutritional Science, University of Alberta, Canada

8.1 Caciocavallo Podolico PDO – Italy

Name: Caciocavallo Podolico PDO	
Production area: From Abruzzo to the southern tip of Calabria through the regions Molise, Campania, Puglia and Basilicata	
Milk: Cow's, raw	

8.1.1 Introduction

Caciocavallo Podolico belongs to the general category of 'Caciocavallo' cheeses and obtained DOC recognition by the Italian government in 1955. Caciocavallo Podolico was also recognised by the Ministry of Agriculture as a 'Traditional Italian Product' (MIPAAF, 2005). It is a symbolic cheese of southern Italy and has several varieties related to its production areas: Caciocavallo Podolico Gargano (Puglia Region), Caciocavallo Podolico Lucano (Basilicata Region) and Caciocavallo Podolico Campano (Campania Region).

The history of cheese is strongly related to the milk's origin. The main common denominator of the different territorial varieties of Caciocavallo Podolico is the exclusive use of milk from Podolica cows. The Podolica breed originates from *Bos primigenius Podolicus*, a large size of cattle with long horns and that is supposed to have been domesticated in the Middle East in the

Global Cheesemaking Technology: Cheese Quality and Characteristics, First Edition.
Edited by Photis Papademas and Thomas Bintsis.

fourth millennium BC. On the origin of the cattle Podolica, there are two theories: (1) Podolian cattle came to Italy in 452 AD in the wake of the Huns from Mongolia and then passed through the Ukrainian steppes, which can be considered the true cradle of Podolica or (2) they are ancestors of the long-horned cattle from Crete, known from Minoan times as *Bos primigenius*. Originally, the Podolica animals, thanks to the robust morphology, were bred especially for land work, but are now used for the production of milk to cheese processing and for meat. Caciocavallo Podolico is produced along the Apennines from Abruzzo to the southern tip of Calabria through the regions Molise, Campania, Puglia and Basilicata.

8.1.2 Type

Caciocavallo Podolico is a pasta-filata cheese that is semi-hard to hard depending on the degree of maturation.

8.1.3 Description and Sensory Characteristics

It has a pear shape, is rounded (diameter 22 cm and height 30 cm) with a head (enlarged upper neck) to which natural fibre cords are applied. The weight of the Caciocavallo Podolico varies between 1 and 2.5 kg. Larger-size cheeses, even 10 kg, are also produced, although more rarely, and are left to mature for years. The cheese with only one month of ripening (which occurs externally) has a thin crust, and it is smooth, uniform and pale yellow. The most aged cheese has instead a hard crust and a deep yellow colour. The compact texture, soft and buttery, and the absence of holes are characteristic of early ripening. The texture later becomes hard with pale yellow overlapping layers that evolve into a darker yellow as it matures. The taste is sweet and delicate in the young cheese, slightly spicy in the aged cheese and very spicy in cheeses matured for years.

8.1.4 Method of Manufacture

Animal feeding: The Podolian cattle live in the natural environment in the geographical area mentioned earlier, despite the poor grazing and scarce water supply. This animal is, in fact, able to make its home in the bushy pastures, stubble and spots, using the leaves of shrub species, the sprouts of the trees and the production of herbaceous undergrowth. The pasture is the staple food for the cows for most of the year (they feed on wild oat grass spigarola, mint, lavender, thyme, oregano and rosemary vegetation typical of the region. Only if a heavy snowfall is expected are animals moved into the barn and fed with feeds and hay produced in the area. Even today they are bred in the wild and raised through the ancient technique of transhumance, which involves a search for natural pastures. This peculiarity gives the Caciocavallo Podolico its typical flavour, which can vary depending on the pasture. In spring and summer, the mountain pasture is rich in aromatic essences, and if it is plain, in autumn and winter the cheese flavours are more pronounced.

Milk preparation: Milking of the herd is largely manual and performed once a day, in the morning. The milk collected is stored at room temperature in the cheese making room, filtered with a metal sieve on which is placed a fine mesh cloth (silk) and manufactured within 1–3 hr of milking.

Starter cultures/rennet: Whey culture is used to produce the Caciocavallo Podolico. The cheesemakers leave the residual whey of the cheesemaking of the previous day in the wood coagulation vat called the 'tinaccio' at night. Next morning, the whey is drained (to be used for the production of Manteca cheese), leaving the walls of the tinaccio wood wet and soaked in acidified serum that will constitute a kind of natural whey starter. Raw milk is heated to a temperature of about 35°C–40°C. The desired temperature is obtained by heating a third of the milk (about 100 litres) at about 70°C in a boiler of tinned copper, called the 'caccavo,' with wood fire, and it is then added to the rest of the milk in the tinaccio. Historically, kid rennet paste (40 g/100 L) has been used, but because it is not always available, an alternative is the use of lamb rennet. In some cases, liquid calf rennet in amounts of about 25 mL/100 L of milk is used. In this phase, the temperature is maintained constant (35°C–40°C).

Coagulation: The coagulation time is about 35–40 min.

Acidification: The curd is then left to ripen for 50 min at room temperature and subsequently is covered with the serum and heated in the caccavo brought to a temperature of about 80°C. In such conditions, the curd ripens in about 3 hr, reaching temperatures that exceed 50°C, with a final cooling to about 40°C.

Stretching: The cheesemaker, before starting the real stretching phase, tests if the curd is ready to stretch by taking small portions of the curd from the tinaccio and trying to stretch it with hot whey in a hemispherical vessel made of elm wood ('coccio') with a capacity of 2–3 litres. When the paste reaches the desired degree of elasticity (evaluated empirically by subjecting it to tension), it is extracted from the tinaccio. The curd, extracted manually, is left to dry for 15–20 min on a wooden table ('tompagno') located in the same processing room. At this point, the curd is pale yellow with a chewy texture. Afterwards, it is cut with a knife, first in thick slices and then in thin slices in a container in wood, low and wide, to which 'white water' (aqueous solution at 20% acid whey from the previous day's processing), previously heated to a temperature of 85°C, is added. The paste is then moulded manually by the cheesemaker, initially assuming a semi-spherical shape and thus the characteristic pear shape with a small head.

Moulding: The Caciocavallo Podolico cheeses are then immersed in water at a temperature of about 15°C, for a time varying between 30 min and 3 hr, according to their size.

Salting: Salting is carried out in brine having a concentration ranging from 20% to 30% NaCl, at a temperature of 15°C–20°C. Caciocavallo remains immersed in the brine for a variable time depending on the size, generally 8–9 hr per kg of cheese.

Maturation: After salting, the Caciocavallo is bound with natural fibre rope, hung to dry riding on a beam (hence the name 'Caciocavallo,' which means 'cheese on horseback') in the cheese plant environment for 2–4 days, after which they are subjected to surface cleaning of the mould by means of a cloth. Maturation takes place in ventilated rooms (sometimes in caves) at a temperature below 15°C, for a period between 2 and 12 months. In the case of very large sizes (8–10 kg), the maturation process can last years.

8.2 Kachkaval (Kačkavalj) – Serbia

Name: Kachkaval	
Production area: Southern part of Serbia	
Milk: Cow's, sheep's or mixture, raw or pasteurised	

8.2.1 Introduction

Kachkaval (Serbian name: Kačkavalj) is a hard pasta-filata cheese made from cow's, sheep's or mixed raw milk and probably the best-known cheese from Serbia. The most popular cheese is a variety named after the town of Pirot (the southern part of Serbia), near the Stara Planina Mountain, known as 'Pirotski Kačkavalj'. Pirot Kachkaval, which is produced by the traditional method, was certified with a sign of appellation at the national level in 2012. This kind of cheese is produced in the Balkan Peninsula, Romania and Italy. Descriptions of making cheese from sheep's milk are mentioned in very old historical records, but nobody knows its real origin. Probably the first production began in huts of Geta in Dacia, after which it was adopted by the Illyrians and Thracians and later knowledge about cheese production was transferred from Balkans to the Apennine peninsula. That cheese nowadays is known as 'caciocavallo'. After the conquest of the Balkans by the Romans, this cheese got the famous name 'Kachkaval' in this region. However, in other countries of the Balkans, there are many similar varieties of Kachkaval cheese with different names (Kindstedt et al., 2004).

8.2.2 Type

According to National Standard for Quality Requirements of Kačkavalj Cheese, the Institute for Standardization of Serbia (Anonymous, 1997) defines two types of cheese: (1) Kačkavalj, with a weight of 5–10 kg and minimum levels of dry matter (DM) and fat-in-dry-matter (FDM) of 56% and 45% after eight weeks of ripening and (2) Kačkavalj Krstas, with a weight up to 3 kg and minimum levels of dry matter and FDM of 54% and 45% after eight weeks of ripening, respectively.

8.2.3 Description and Sensory Characteristics

Kachkaval cheese has a flat cylindrical shape with a strong bright yellow rind, depending on the milk type. The texture is elastic, very close with visible layers, occasionally with random slots but without gas holes, while the taste is salty and very piquant, especially mature cheese produced from sheep's milk.

8.2.4 Method of Manufacture

Milk preparation: Raw cow's, sheep's or mixed milk is used for the traditional production, while in industrial production milk heat treatment is usually done (63°C/30 min. or 72°C/15 s). In artisanal production, whole milk is used in the making of Kachkaval cheese.

Starter culture: A starter culture is not used in traditionally produced Kachkaval cheese. However, in industrial production of this type of cheese, thermophilic lactic acid bacteria are commonly used (*Streptococcus thermophilus*, *Lactobacillus delbrueckii* ssp. *bulgaricus*, *Lactobacillus helveticus*).

Rennet: Addition of calf's rennet occurs at 30°C–32°C, and coagulation lasts 30–45 min.

Cutting: The curd grain size corresponds to a size of 5–10 mm.

Scalding: This takes place between 40°C and 42°C with stirring at this temperature.

Curd draining: The curd is transferred into the pressing vat, and cheese curds are formed.

Ripening of cheese curds: Cheddaring (fermentation of cheese curds) is done at 18°C–20°C for 19–20 hr until the pH drops to 5.1–5.2. In industrial production, when starter cultures are used, acidification takes a shorter time (3–4 hr). After acidification, the cheese curd is cut into small pieces with a knife or mechanically.

Texturising: Texturising of the acidified curd involves heating, kneading and stretching by soaking in hot water or in brine. Traditionally, texturising is done in wooden baskets with hot water (85°C–95°C) where the cheese mass is kneaded with a wooden stick until a homogenous compact structure (5–6 min) is obtained (the temperature of cheese mass is at least 65°C). Wooden basket and sticks are traditionally made from the hazelnut tree. The hot stretched cheese mass is then transferred onto a table and kneaded by hand with addition of dry salt. In industrial conditions, the cheese mass is texturised and kneaded in special equipment designed for this kind of cheese. The cheeses are left in moulds and are cooled for 24 hr (Kindstedt et al., 2004).

Salting: Salting starts during kneading with addition of dry salt to the hot cheese mass. Addition of salt is repeated during the first 10 days of ripening. In industrial production, Kachkaval cheeses could be salted during texturising in hot brine (6%–10% NaCl) or by immersion of the cheese in brine.

Maturation: After salting, the cheeses are stored in cellars with a 22°C–26°C and 65%–75% of relative humidity (RH) for drying, after which ripening is done at 14–18°C/75%–85% RH. At the beginning of ripening, for 5 to 10 days, the cheeses are dry-salted and turned daily. Kaschkaval cheese must be at least two months old, while Kaschakavl Krstas must be ripened for at least one month.

8.2.5 Relevant Research

The diversity of Non starter lactic acid bacteria (NSLAB) in traditional produced Kaschkaval cheese was studied, and the most frequent species within NSLAB belong to lactobacilli, lactococci and enterococci (Radulović et al., 2006). Kaschkaval Krstas cheeses were more chewy and springy, melted at lower temperatures during heating than the Kaschkaval cheeses, which have low levels of cohesiveness and chewiness. The differences in textural, cooking and viscoelastic properties of cheeses were associated with differences in composition and level of protein breakdown (Guinee et al., 2015).

8.3 Kashar (Kaşar Peyniri) – Turkey

Name: Kashar	
Production area: Turkey (especially eastern Turkey and Trakia region)	
Milk: Cow's, sheep's or a mixture, raw or pasteurised	

8.3.1 Introduction

Kashar cheese is a pasta-filata cheese that is commercially the second most important cheese variety in Turkey. Traditional Kashar cheese is usually produced from sheep's milk, but due to the shortage of sheep's milk in some seasons, sheep's milk has been largely replaced by cow's milk in industrial practice. Traditional Kashar cheese is mainly, but not limited to, produced in far eastern part of Turkey. Kashkaval-type Kashar cheese is common in the Trakia region (the northwest part of Turkey close to the border with Bulgaria and Greece).

8.3.2 Type

Kashar cheese is a hard cheese with no holes but a rind, and it is yellowish in colour. It is slightly salty, has a sharp flavour and is springier than brined-type cheeses. Moisture and salt-in-moisture levels of mature Kashar cheese must be equal to or higher than 40% and 2.5%, respectively (Anonymous, 2015).

8.3.3 Description and Sensory Characteristics

As stated above, sheep's or cow's milk is widely used in the manufacture of Kashar cheese. Cheeses produced from sheep's milk have higher levels of proteolysis and lipolysis, and better melting properties (Temizkan et al., 2014). Kashar cheese requires a rather longer ripening period than the cheeses ripened in brine. Therefore, several attempts have been made to reduce the ripening period by the addition of individual and mixed lipase and proteases (Akın, 2012). The addition of a commercial lipolytic enzyme preparation from *Muchor miehei* 'Palatase M 200 L' to Kashar cheesemilk resulted in considerably better sensory scores and reduced ripening time compared with untreated control cheese. Akın et al. (2012a,b) investigated the efficiencies of different encapsulating materials (i.e. κ-carrageenan, gellan and sodium alginate) on the acceleration of cheese ripening. They demonstrated that all three gelling agents could successfully be used as lipase and protease carrier systems to accelerate Kashar cheese ripening. The type of coagulating agent (i.e. calf rennet, chymosin derived by fermentation and proteases from *Rhizomucor miehei* and *Cryphonectria parasitica)* also affect the level of proteolysis in Kashar cheese (Yaşar & Güzeler, 2011).

8.3.4 Method of Manufacture

Although raw milk is mostly preferred by cheesemakers, heat-treated milk (thermised or pasteurised) can also be used in the production of Kashar cheese.

Milk preparation: Cheese milk is pasteurised at 72°C–74°C for 15 s (with plate heat exchangers) or 65°C for 30 min and then cooled to coagulating temperature of 32°C–34°C.

Starter culture: Milk is inoculated with starter culture (*S. thermophilus* and *Lb. delbrueckii* ssp. *bulgaricus*; *S. thermophilus* and *Lb. delbrueckii* ssp. *bulgaricus* and *Lb. casei*; *Lactococcus lactis* ssp. *lactis* and *Lb. casei*) at a level of 0.5% (w/v). In order to restore the ionic calcium balance after heat treatment, calcium chloride is added at a level of 10–15 g/100 L of milk. Milk is acidified by a starter culture to pH 6.40–6.45 within 30–40 min.

Coagulation: Following acidification, coagulating enzyme preparation is added at a level to coagulate milk within 45 min.

Cutting: The curd (pH 6.30–6.35) is cut into small pieces (1.5–2.0 cm^3 parts), left to rest for 10–15 min and then cut again into smaller pieces (6–7 mm^3).

Scalding: The curd is heated up to 38°C–40°C by continuously stirring and kept at this temperature for 15 min. The temperature of the curd is increased stepwise to 38°C–40°C with a 1°C increment for every 3–4 min. This process helps expel whey effectively. The whole process is ideally finished within 30 min.

Pressing: The curd is transferred to the pressing machine. The pH of the curd at the stage of pressing is about 5.90–6.15. The curd is pressed mechanically (i.e. 1 kg weight for 1 kg of curd at the beginning and then increased gradually up to 15 kg weight for 1 kg of curd). The pH of the pressed curd is around 5.25–5.30.

Stretching: The pressed curd is cut into portions of 25–30 cm × 15–20 cm in size and left for incubation at 15°C–20°C until the pH becomes 5.0. Afterwards the curd is stretched at 72°C–75°C (in some cases up to 85°C) for a few minutes (Kurultay et al., 2004). The pH of the curd before stretching should not be below 4.8, otherwise textural problems in the resulting product are likely to occur. The stretching solution contains NaCl at concentrations of 5%–6%. The acidity of the stretching solution should be 10°SH.

Moulding/pressing: The stretched curd is then moulded and pressed. The moulded Kashar cheeses are rested for 12–24 hr, and during this period the cheeses are turned upside down at every 2 hr. Fresh cheeses are kept in an incubation room for about a week and then washed with potassium sorbate before vacuum packaging. If the cheeses are ripened, the cheeses are placed into ripening room at 16°C–18°C with 85%–90% RH.

Salting: The cheeses are dry-salted on both sides.

Maturation: Salted and pre-ripened cheeses are washed in pasteurised whey, left for 1–2 hr and then transferred to the ripening room again at 12°C–16°C with 85% RH for 30–60 days. Afterwards, the temperature of the ripening room is decreased to 5°C–6°C. During the ripening period, the cheese blocks are turned upside down once a week, and if mould is grown on the surface of the cheese, the cheeses are washed with brine solution containing 5% NaCl. Then the ripened cheeses are washed with warm brine solution (with an NaCl level of 10%–12%), packaged and stored in a cold room (2°C–4°C) for 4–10 months. The yield of the cheese varies depending on the milk species used. In the case of use of sheep's milk and cow's milk, the yields of the end products are 16%–18% and 9%–10%, respectively.

8.3.5 Relevant Research

The textural, melting and sensory properties of low-fat fresh Kashar cheeses produced by using two protein-based fat replacers (1.0% w/w Simplesse®D-100 and 1.0% w/w Dairy-Lo(TM)) and one carbohydrate-based fat replacer (5.0% w/w Raftiline®HP) were examined by Koca and Metin (2004). As expected, the use of fat replacers reduced the hardness, springiness, gumminess and chewiness, and increased cohesiveness of the cheeses.

Hayaloğlu (2009) analysed commercial Kashar cheese samples for their volatile compounds and identified a total of 113 volatiles, and Avşar (2009) identified 131 volatile compounds from mature Kashar cheese. Diacetyl, 3-methyl-1-butanol, 2,3-butandion, hexanal and methianal are the most important compounds that determine the aroma of Kashar cheese.

In the case of reduced-fat Kashar cheese production, it is recommended to use *Brevibacterium linens* as an adjunct culture to compensate the aroma/flavour losses. Kashar cheese may be a suitable carrier food for probiotics into the human body. Although scalding may lead to drastic declines in the counts of probiotic bacteria, this handicap may well be overcome by microencapsulation of probiotics, as shown by Özer et al. (2008).

8.4 Kasseri PDO – Greece

Name: Kasseri PDO	
Production area: Macedonia, Thessaly, Lesvos and Prefecture of Xanthi	
Milk: Sheep's or mixture of sheep's with goat's, with the latter not exceeding 20%, pasteurised	

8.4.1 Introduction

Kasseri is a traditional Greek cheese of the pasta-filata type. Traditionally, it was produced in the mountains of Pindos and Olympos. Because milk collection was difficult, shepherds carried out the coagulation of milk and the drainage of cheese curd in the mountains, in order to preserve the milk and reduce its volume, thus decreasing the cost of transportation. Several drained cheese curds (called 'baskies') were gathered together and carried for processing to cheese plants. In the meantime, the pH of the baski dropped (to about pH 5.20) by the activity of the native microflora of the milk. Mature baski was sliced and kneaded in hot water in nearby factories in order to obtain a pasta-filata texture. This procedure was well adapted to the mountainous character of the country because transport of milk over long distances was avoided (Zygouris, 1952). According to traditional practice, Kasseri cheese was made from raw milk, because it was considered that kneading eliminated the pathogenic bacteria and controlled the native microflora (Moatsou et al., 2001).

8.4.2 Type

Semi-hard, pasta-filata cheese with a maximum moisture of 45% and a minimum FDM of 40%. Its gross composition is as follows: moisture 40.9, fat 28.7, FDM 48.5, protein 25.7, salt 1.6, S/M 3.9 and pH 5.7 (Nega and Moatsou, 2012). The PDO status for Kasseri was recognised by the EC in 1996 (EC, 1996a) and amended in 2000 (EC, 2000).

8.4.3 Milk

Sheep's milk, or a mixture of sheep's with goat's, with the latter not exceeding 20%.

8.4.4 Description and Sensory Characteristics

It is a semi-hard pasta-filata cheese, with a closed texture, no holes, cylindrical in shape (wheel) with a diameter of 25–30 cm and a height of 7–10 cm, or rectangular. No rind is formed. Kasseri cheese may be covered with paraffin or some other coating, while rectangular cheeses are wrapped in airtight plastic films. Kasseri has a yellowish colour, and no colouring, antibiotic or preservative is permitted.

Kasseri has a closed texture with no holes and a mild, mellow, faintly sweet flavour. The flavour is more pronounced when cheeses are well matured.

8.4.5 Method of Manufacture

Milk preparation: Traditionally, raw milk was used; however, in industrial manufacture, the milk is either thermised or pasteurised.

Starter culture: Usually, no starter is added when raw or thermised milk is used. When milk is pasteurised, a starter culture (~1%–1.5%) of thermoresistant *S. thermophilus* and *Lb delbrueckii* ssp *bulgaricus* is added. In addition, $CaCl_2$ up to 20 g/100 kg of milk may be added when pasteurised milk is used.

Coagulation: The milk coagulates with rennet at 32°C for 35–40 min.

Cutting: The coagulum is cut to the size of corn and left for 10 min.

Scalding: At 38°C–40°C under continuous stirring.

Curd draining: Curd is moulded with cheesecloths and left for ripening (baski) until the pH drops to 5.2.

Kneading: Ripened curd (baski) is cut in strips and inserted in hot water at 70°C–80°C, where kneading continues until the strips can be easily extended. Then it is moulded and left in the moulds for two to three days.

Salting: Dry salting is done (12–14 times), with medium-size salt.

Maturation: For at least three months.

8.4.6 Relevant Research

Moatsou et al. (2001) studied the effect of technological parameters on the characteristics of Kasseri cheese made from raw or pasteurised sheep's milk and concluded that the pasteurisation of the cheesemilk had a negative effect on the rate of baski acidification. In addition, mature cheeses ripened and stored at 4°C gained higher organoleptic scores than their pairs ripened at 15°C, probably due to the different temperature dependencies of the various enzymes involved in the proteolysis. *Pediococcus* spp. followed by *Enterococcus* spp. were the dominant microbial groups in matured (120 days) Kasseri cheese (Bintsis, unpublished data). Anastasiou et al. (2007) suggested the use of *Streptococcus macedonicus* ACA-DC 198 as an adjunct culture in Kasseri cheese production.

8.5 Mozzarella di Bufala Campana PDO – Italy

Name: Mozzarella di Bufala Campana PDO	
Production area: Mainly in the Campania region	
Milk: Buffalo's, raw	

8.5.1 Introduction

The Mozzarella di Bufala Campana PDO was recognised by the Italian government with the 'denomination of typical' in 1979, obtained the Designation of Origin in 1993 and the PDO in 1996 (EC, 1996a).

Mozzarella di bufala has medieval origins, although historical documents are not always available that certify with certainty the true origin and who invented this unique and delicious product. It is said that the Saracens would transport buffalo first in Sicily and then in the marshy plain of the Garigliano (between Lazio and Campania).

The Lombards defeated the Saracens in the ninth century, and the monks of the place learnt the techniques of farming and dairy processing practised by the Saracens.

The first certain historical information is recorded in a Lombard document. According to these sources, already in the eleventh century the princess Aloara, widow of the Prince of Capua Testadiferro Pandolfo, distributed a 'mozza' (Mozzarella) with a piece of bread to the Abbazia of San Lorenzo at Septimum on the outskirts of Aversa. According to others, it is assumed that it was the Normans who invented Mozzarella; their county town was Aversa, where many dairies are still active which produce and sell the typical Mozzarella Aversana. The use, processing and consumption of products derived from buffalo milk *(the casicaballus, the butyrus, the recocta* and *the provaturo)*, are supported by claims in documents from the twelfth century kept in the episcopal archive of Capua. The first official document that speaks of Mozzarella from Aversa is recent and dates from the early fifteenth century.

Originally, Mozzarella was born as a by-product of the preparation of other type of stretched cheeses, for example, Provola, as it was difficult to store and market, given the unique characteristics of freshness and perishability, and perhaps for these reasons was produced in very small quantities. The spread of the Mozzarella, however, follows the development of the transport road system. With the unification of Italy (1861), between Naples and Caserta, Aversa, the famous 'Taverna', was created, a kind of wholesale market of buffalo Mozzarella and Ricotta which established with daily quotes a relationship between production and demand. The advent of large retailers further allowed the spread and expansion of fresh produce including buffalo Mozzarella campana.

It is produced mainly in the Campania region, in the provinces of Caserta and Salerno, and the provinces of Naples and Benevento. It is also produced in Lazio in the provinces of Frosinone, Latina and Rome; Puglia in the province of Foggia and Molise in the province of Isernia.

8.5.2 Type

It is a stretched (pasta-filata) fresh cheese, also available smoked. The minimum FDM content is 52% and 65% moisture content.

8.5.3 Description and Sensory Characteristics

It has a typical round shape. However, other forms characteristic of the production area are permitted, such as titbits, braids, beads, cherries, knots and 'ovolini'. The weight ranges between 10 and 800 g, according to the mould. To form a braid, a weight up to 3 kg is allowed. It resembles white porcelain in appearance, has a thin crust of approximately 1 mm, a smooth surface and is never slimy or curled. The internal texture made of thin leaves, is elastic in the first 8 to 10 hr after the production and packaging and subsequently tends to become more flowing. It is free of defects such as eyes caused by gaseous or unusual fermentation. No preservatives, inhibitors or colouring agents are permitted. The taste is characteristic and delicate.

8.5.4 Methods of Manufacture

Milk preparation: The Mozzarella di Bufala Campana PDO is produced from the milk of buffaloes of the Italian Mediterranean breed. The buffaloes are reared in a stabled or semi-stabled environment consisting of paddocks with artificial ponds and sheds necessary to protect them from the summer heat. Buffaloes graze in a regime of semi-freedom, and strict sanitary controls keep the herds from diseases such as brucellosis and tuberculosis, so the raw milk need not be subjected to pasteurisation.

Whole fresh milk with at least 7.2% fat and 4.2% of protein must be delivered to the dairy plant before the 60th hour after the first milking.

Traditionally raw milk was used, but the specification (disciplinary) for the PDO production also permits the use of thermisation or pasteurisation of the milk.

Whey starter culture and rennet: The use of natural whey from previous processing of buffalo milk occurring in the same cheese plant or in the area of origin is allowed. It uses natural liquid calf rennet (18–20 mL/100 L of milk).

Coagulation: The coagulation, after heating the milk to a temperature of 33°C to 39°C following addition of the whey culture and rennet, occurs in about 30 min.

Cutting: Breaking the curd is usually done manually with a 'ruotolo' wooden stick (to the ends of which is fixed a wooden disc with the outer face convex) or with a metal tool called the 'spino', which is turned energetically until cheesy clots reach a size between a hazelnut and a walnut (3–6 cm). The curd is broken in two phases: the first break reduces the curd into cubes and, after a break of half an hour, the master cheesemaker proceeds with the second break with the ruotolo or the spino. The extraction of the curd usually takes place manually.

Scalding: After breaking, the curd is left to acidify under the whey for a variable time in relation to the concentration of the starter culture present in the added natural whey.

About 60% of the serum is extracted from the coagulation vat, and a part of it (approximately 5% of the total serum) is heated and added after about 5–10 min in the vat so as to maintain the temperature of the curd at around 46–50°C. This step is important to lower the pH so as to be ready for stretch (5.2–5.6), to select the right dairy microflora and for the health safety of the final product in synergy with the temperatures that will be reached in the stretching phase.

The cycle of artisan acidification lasts 3–4 hr, and rarely up to 8 hr.

A short resting phase of the curd in a table 'spersorio' follows, where it continues to ferment, drying and allow further loss of whey.

Stretching: At the end of curd ripening, with the curd resting on the spersoio table, starts the phase of stretching the curd, which in the traditional processing still performed manually. The curd is cut into thin slices with a mince-curd and placed in a wood container called the 'compecina' in which it is melted by the addition of boiling water. The curd reaches temperatures of over 60°C, representing in fact the equivalent of pasteurisation for raw milk products, contributing to food safety for consumers. In the traditional practice, the master cheesemaker uses a bowl and a wooden stick to work the curd until it stretches well and becomes homogeneous.

Shaping the Mozzarella: Mozzarella di Bufala Campana PDO is produced in different shapes and sizes: in addition to rounded, other forms characteristic of the production area are allowed, such as titbits, braids, beads, cherries, knots and ovolini; the size and therefore the weight can vary from 10 g to 800 g, according to the form. To form a braid, a weight up to 3 kg is allowed. The desired shape and size can be obtained using dividers with predefined moulds.

The artisanal way to form Mozzarella di Bufala Campana is manually, by two workers (called a 'four hand') performing characteristic movements that lead to the 'mozzatur' (cutting off pieces of curd) of the stretched curd: an operator supporting the another operator firmly holds a globular mass of curd (about 2–3 kg), allowing another master cheesemaker to cut it with the thumb and index finger ('mozza', hence the name 'Mozzarella') into pieces of the desired size. The 'braid' forms are traditionally handmade, a stretched segment of the curd being deftly weaved until the final shape is obtained.

In specific areas, such as the territory of Aversa, in accordance with local traditions, the Mozzarella Aversana, whose weight is about 500–600 g, is produced through a handmade process. This specification may be stated on the label as 'Mozzarella di Bufala Campana PDO, Aversana.'

The freshly produced Mozzarella are immersed in cold potable water (5°C–6°C), for times that vary depending on the size, until firming.

Salting: Salting is performed in brine (10%–18%) for times that vary according to the size and the concentration of the salt brine. It follows the packaging (which must be done in the same dairy plant) in its liquid, which is sour and slightly salty (1%–1.5%).

Smoking: The product can be smoked only by natural and traditional means, and is then labelled 'Mozzarella di Bufala Campana PDO Smoked.' The artisanal smoking is accomplished by exposing the cheeses, held in a cylindrical container whose top is closed by a thick cloth wet, to the smoke of wheat straw: the smoke passes through the cylinder and darkens the crust of Mozzarella, which turns from white porcelain to a dark yellow, while the dough takes on a very characteristic pleasantly smoky taste.

8.5.5 Relevant Research

Gianferri et al. (2007a and 2007b) investigated the physical structure and metabolic profile of Mozzarella di Bufala Campana cheese, the water dynamic states and the age-related changes.

Textural modifications have been also studied by Costanzo et al. (2015) using sonoelastrography. The chemical and nutritional characteristics of buffalo Mozzarella are reported in the papers of Manzi et al. (2005, 2007). The sensory profile is described by Pizzolongo et al. (2007). The lipid fractions (trigliceridic, fatty acids and CLA) of buffalo Mozzarella campana are presented by Romano et al. (2008). Tenore et al. (2015) reported the results of an investigation of the in vitro intestinal protection, bioavailability and anti-haemolytic capacity of antioxidant peptides from Mozzarella di Bufala Campana PDO.

8.6 Parenica – Slovakia

<table>
<tr><td>Name: (Slovak) Parenica</td><td rowspan="3"></td></tr>
<tr><td>Production area: Foothills and mountainous regions of Central and Eastern Slovakia</td></tr>
<tr><td>Milk: Sheep's, raw</td></tr>
</table>

8.6.1 Introduction

Slovak Parenica is one of the most traditional and ever-popular Slovak sheep's milk cheese specialties. It is a pasta-filata-type cheese, where the final form is a filamentous ribbon, rolled into a short cylinder. Very often the Parenica is slightly smoked. The cheese has been registered as a Protected Geographical Indication (PGI) in the EU (EC, 2008). Parenica is a typical Slovak cheese specialty, very popular with tourists especially in its more striking versions ('korbacik'). It is well known also in the neighbouring countries, especially the Czech Republic. The name 'Parenica' is derived from the Slovak expression for 'scalding'.

The development of sheep farming in Slovakia in the fourteenth century resulted in production of simple sheep's milk cheeses, predominantly the lump cheese variety (Bryndza). The idea of scalding the sheep cheeses and subsequently the manufacture of the smoked Parenica type came into being in the eighteenth century, in the families of sheep farmers, who could control the high quality of the raw milk and proper acidification of the sheep cheese. The required elevated temperature for proper lump cheese acidification was often found in the family smokehouses. In the nineteenth century, Parenica cheese became popular in

Vienna and other neighbouring cities. In an important Viennese publication 'Codex Alimentarius Austricus' (1917), the following note is found: 'Parenica is a cheese scalded in hot water, pulled into strips, rolled and smoked'. One of the legendary Czech cheese science experts, Prof. Otakar Laxa, pointed out in 1908 that 'no other cheese has a more typical Slovak characteristics than Parenica, which must be considered a unique phenomenon in the cheese world'.

Slovak Parenica, made from untreated sheep's milk, is produced only in the foothills and mountainous regions of Central and Eastern Slovakia, where sheep husbandry is well established. These areas are the least contaminated by industrial pollution, and thus the quality of the sheep's milk is outstanding, which makes it suitable for manufacturing this cheese specialty. The high quality of the fresh raw sheep's milk is very important, especially with respect to the technological implications of the heat treatment process. Artisanal manufacturing has been limited mainly to the sheep farms, as only a few industrial manufacturers have access to fresh sheep's milk. Nowadays, however, the technology of the Parenica manufacture is being used more and more also for production of similar products from cow's milk. These can have a very similar appearance and textural properties but are lacking the specific sheep's milk flavour.

8.6.2 Type

Parenica is a pasta-filata-type cheese, made from full-fat sheep's milk that is not heat-treated. Thus, the cheese is available with uniform quality characteristics. This unripened (but often smoked) semi-hard cheese has a maximum moisture content of 47%, a maximum FDM of 50% and the salt content is 2%.

8.6.3 Description and Sensory Characteristics

The most characteristic feature of Parenica is its unique form of the double ribbon, as well as its stringy structure and attractive, softly acidic flavour with the sheep's milk overtones. The initial long ribbon (up to 4 m) is 5–8 cm wide and 2–4 mm thick, with a mass of 400–500 g. The long ribbon is cut as needed and rolled into a single or a double cylinder, tied by a thin cheese thread so that the cylinder(s) will not unfurl. The cheese may be lightly smoked, which produces a slightly golden brown tinge. The cheese is also offered without smoking, in which case the colour is creamy yellowish. Nowadays, the cheese is made not only in the traditional Parenica shape, but also in various other striking forms, including rope-like woven chains ('korbacik'), or even various figurines, flowers and other decorative assemblies. The texture of all Parenica types is fine, fibrous but elastic. The individual fibres can be easily taken apart, which gives the product its particular appeal for children. The aroma of the unsmoked Parenica is slightly acidic with slight cooked flavour overtones. The cheese taste is likewise reminiscent of pleasantly acid cheese with sheep's milk notes and slight saltiness.

8.6.4 Method of Manufacture

The classical Parenica is made from freshly produced raw unpasteurised full-fat sheep's milk, obtained directly at the alpine meadows. Most of the traditional production occurs in the summer months. However, nowadays sheep farming is expanding, with some large farms offering a year-round availability of the milk. Initial storage of the raw milk is critical for proper technological behaviour. The milk should not be kept in cold storage, as temperatures below 5°C lead to difficulties in the later stages of processing. Such milk is no longer suitable for the scalding and stretching.

Milk preparation: The most important aspect of Parenica manufacturing is the availability of high-quality fresh sheep's milk, resulting in curds with good acidifying characteristics. Before renneting, the milk is pre-filtered and brought to the required 32°C temperature. Traditionally, Parenica production in the mountain cottages relied on the presence of natural microflora, with no starter cultures added.

Coagulation: Extracts from lamb stomachs, made using the residual whey, were added for both renneting and also a source of the microflora needed for the acid production in the lump cheese. The wooden tools used in the cheese manufacture were another source of the microflora. At the present time, the starter cultures are being added in the form of the whey from the previous day, added to the milk at 1%–3%. The largest industrial manufacturers nowadays use lyophilised dairy cultures produced using sheep's milk. Similarly, the original renneting by extracts of lamb stomachs has now largely been replaced by industrially produced coagulators, for example, Fromase and Hannilase. The properly diluted rennet is added to the milk at 32°C after adding the starter cultures. The clotting is expected to start after 15–20 min, and the firming of the coagulum continues for additional 15–20 min. No calcium salts are added to the milk when producing Parenica.

Cutting: After the proper acidity is reached, the coagulum gel is first cut into large blocks and, when the whey appears, cut further to small, bean-size particles. The curd grains are carefully mixed, to facilitate the syneresis, for 20–40 min at the minimum constant temperature of 32°C until a firm, non-sticky grain is obtained.

Moulding: Well-firmed grain, together with the whey, is poured into perforated containers or muslin bags and kept at 18°C–20°C for draining, turning the containers at least once. This lump cheese is kept in these conditions until the next day; the acidity increases and should reach at least a pH of 5.2 or less. Precise pH control is very important for the optimal outcome of the scalding and ribbon stretching.

Cooking: The lumps are cut into smaller pieces corresponding to the desired size of the finished product, and these are introduced into the scalding bath. As soon as a piece of the fully acidified cheese becomes plastic in water kept at 60°C–70°C, the scalding process begins. In the traditional artisanal processing, each piece was kneaded with a wooden spoon to produce a homogeneous cheese mass. This was further kneaded by hand, and finally a ribbon was produced on a flat forming surface. From about 0.5 kg of the cheese mass, a ribbon about 4 m long, 6 cm wide and 2–3 mm thick can be produced. In the modern industrial production of Parenica, the scalding, kneading and ribbon forming are all mechanised, and the machinery is interconnected, forming a continuous process. The modern machinery for the stretching and ribbon forming can produce up to five ribbons simultaneously, discharging these via a shallow trough into the cooling and salting bath. The ribbons move through a cold salting bath for 2–5 min and proceed to a flat surface table where the ribbons are rolled into the characteristic cylinders. To avoid unfurling, the finished cylinders are wrapped by well-scalded and stretched cheese 'threads'. Such thin threads can be also produced separately and weaved into various striking formations, such as chains, flowerettes and many similar forms.

Smoking: After drying, the Parenica final products can be smoked in a smokehouse using cold smoke from hard wood for about 2 hr.

The finished product (whether smoked or not) is individually wrapped in suitable foils. The keeping quality of the Slovak Parenica is, as a minimum, three weeks in the cold storage.

8.7 Provolone Valpadana PDO – Italy

Name: Provolone Valpadana PDO	
Production area: Lombardy, Veneto, Emilia Romagna and the province of Trento	
Milk: Cow's, raw	

8.7.1 Introduction

The Provolone Valpadana PDO was born in the Po Valley in the second half of the nineteenth century, after the unification of Italy in 1861, which made it possible to overcome the barriers between the different areas of the peninsula and thus facilitated the establishment of entrepreneurs from the south. They brought with them the culture and the tradition of 'pasta-filata', stretched cheeses. The name 'Provolone' appears in the literature for the first time in 1871, in the 'Canevazzi-Mancini's Agriculture Vocabulary' (Cappelli, 1871).

Provolone Valpadana received the PDO in 1996 (EC, 1996a), whose area of origin falls in some regions of the Po Valley: Lombardy, Veneto, Emilia Romagna and the province of Trento.

8.7.2 Type

It is a semi-hard, stretched cheese, with short, medium and long ripening times. It is produced in two types: sweet and spicy.

8.7.3 Description and Sensory Characteristics

The rind is smooth, thin, light yellow, golden, to yellow-brown. The rind is absent for the sweet type, which is destined for subsequent portioning and relative packaging.

The body is generally compact and may have a few holes. Slight flaking (peeling) in cheese is permitted at an early age, while a more marked exfoliation is seen in long-aged cheese. The colour is generally pale yellow.

The flavour is delicate for the sweet cheeses with short aging, pronounced for longer maturation and the spicy type where lamb or kid rennet was used.

8.7.4 Methods of Manufacture

Milk preparation: The feed of dairy cattle is based on forage (green or preserved, including silage), integrated with concentrates. It is applied to lactating cows, dry cows, and heifers over seven months old. At least 75% of the dry matter of the fodder in the daily ration must come from feed produced in the production area. Production takes place throughout the year, starting from whole raw cow's milk, collected in the origin area in less than 60 hr after the first milking.

The heat treatment of the milk for the sweet type is to the maximum extent of pasteurisation; for the spicy type, the heat treatment of thermisation.

Starter culture and rennet: The starter cultures used in processing must be natural acidified whey, left over from previous cheesemaking. The rennet used varies depending on the type of cheeses: for the sweet type, calf rennet (a small percentage of lamb and/or kid is allowed) is preferred; for the spicy type, goat rennet and/or lamb are used.

Coagulation: Coagulation occurs at a temperature of 36°C–39°C. The coagulated gel will be broken in two phases in order to obtain curd granules of the dimensions adapted to the different type of cheeses to be produced, and then the curd is cooked at 52°C–53°C.

Acidification: The curd is laid in tables and left for the natural acidification to reach the pH suitable (pH 4.7 to 5.2) for the stretching phase of Provolone.

Stretching: Stretching is carried out using boiling water, producing a variety of shapes and weights.

Shaping: Shaping is carried out manually or with the aid of special moulds. There are distinguished official shapes: salami (cylindrical), melon, truncated conical, tangerine and pear topped by spherical little head (flask); the outer surface can have small inlets determined by the passage of the support ropes. The moulds have a weight ranging from 0.5 kg to over 100 kg. The cheese thus obtained is placed in cold water or cooled for firming.

Salting: The salting of Provolone PDO takes place in brine with a salt concentration up to 22°Bè, for a variable period of time that depends on the weight of the cheeses, from a few hours (12 hr) up to 25 days, depending on the weight of the cheese. Once pulled out of the brine, the cheeses are washed in cold water, dried and tied.

Binding and stewing: The 'binding' of the cheeses with suitable ropes is another special touch applied by the cheesemakers, another old custom that has never been lost. It follow the stewing phase for a week at 22°C–24°C to form the rind, where the cheeses are tied in pairs (horseback) with ropes in special supports.

Maturation: The aging period can vary depending on the weight of the cheese, as follows: up to 6 kg, 10 days; over 6 kg, 30 days; over 15 kg and only for the spicy type, 90 days and over 30 kg of spicy type with branding for developing countries, aging for over 8 months which may reach 16 months. The cheese can be smoked, during which process it is allowed to be covered with paraffin.

8.8 Ragusano PDO – Italy

Name: Ragusano PDO	
Production area: Ragusa province and municipalities in the Province of Syracuse	
Milk: Cow's, raw	

8.8.1 Introduction

Ragusano is one of the oldest cheeses of the island of Sicily, and the name is derived from the area of production (Ragusa). Since the fourteenth century, this cheese, which has a pleasant and distinctive taste, has been the focus of a thriving commerce beyond the boundaries of the Kingdom of Sicily. In *Ferdinando il Cattolico e Carlo V*, Carmelo Trasselli writes that in 1515 there were 'duties' exemption' also for the Ragusano, which was already at the time part of an important trade. In addition, in 'Note sui Ragusei in Sicilia' Trasselli quotes some documents of 'Notaio Gaetano, F.106' that indicate a trade by ship of this cheese. In 1808, the Abbot Paolo Balsamo cites in his work 'the good quality of Modica's cattle' and the 'cheese and ricotta's products, were fifty percent superior to the regular ones, and twenty-five percent to the best of Sicily.' In 1856, Filippo Garofalo mentions the fame and the deliciousness of the cheeses and ricottas of the area of Ragusa.

The name 'Ragusano' was recognised by the Italian government as 'typical production' in 1955. The PDO was registered by the European Union in 1996 (EC, 1996b). The area of origin protected for Ragusano PDO includes the entire territory of the province of Ragusa and some municipalities in the province of Syracuse.

8.8.2 Type

Ragusano PDO is a stretched-curd cheese made with cow's raw whole milk. It is a high in fat and protein cheese, and it has a low moisture content when aged. The chemical composition mandated by PDO regulations limits the FDM to not less than 40%; not less than 38% for cheese aged for more than 6 months and the maximum moisture content is 40%.

8.8.3 Description and Sensory Characteristics

Ragusano PDO has a parallelepiped shape with a square section and rounded corners. You may encounter on the surface (in the central part of the cheese) furrows made by the passage of the cords used in the maturing process. The sides of the square section are between 15 and 18 cm, and the length of the parallelepiped is 43 to 53 cm. The weight ranges between 10 and 16 kg, and the most common weight is 13 kg. The crust is smooth, thin (max 4 mm) and compact; the colour is golden yellow and becomes pale brownish if the cheese is further aged for grating. A film of olive oil can be present on the surface. The structure is compact, with any cracks that occur with further aging sometimes combined with little holes; on cutting, the colour is white to pale yellow, more or less intense. The flavour is very pleasant, sweet, delicate and slightly spicy in the early months of maturation in the table cheeses, and piquant and savoury with advanced aging for cheese grating. The cheese has a pleasant flavour, resulting from the wild plants grazed by the animals and the traditional systems of production.

8.8.4 Methods of Manufacture

Milk preparation: Ragusano cheese is produced exclusively from raw whole cow's milk coming from farms located in the area of origin, in accordance with the PDO regulations. The cows must be fed mainly wild plants on the pasture and green forage produced in Iblean (the old name of Ragusa) territory, eventually preserved as hay.

The milk of one or more milkings will be coagulated at a temperature of about 34°C–35°C, exploiting the spontaneous development of lactic microflora present (biofilm) in the wooden vat used to make the cheese.

Coagulation: Coagulation is achieved by the use of lamb or kid rennet paste, dissolved in an aqueous solution of sodium chloride. The clotting time varies from 60 to 80 min, depending on the type and amount of rennet used.

Cutting: This process takes place in two phases, using a wooden stick called the 'ruotula': in the first step, the coagulated gel mass will be broken through rotational movements of the wooden stick to reduce them to the size of lentils; in the second step, water is added (8–10 litres per 100 l of milk) at a temperature of about 80°C to further reduce the size of the curd pieces to that of a rice grain. The addition of hot water (first cooking) will help the syneresis.

Pressing: The caseous mass obtained by sedimentation and separation from the whey is subjected to pressing to facilitate further whey drainage.

Second cooking: To stimulate acidification, the curd is left to rest for about 85 min in a wooden vat filled with 'scotta' (the liquid resulting from the processing of Ricotta) or with water at a temperature of about 80°C, and covered with a cloth. The temperature of the curd will not exceed 40°C–42°C.

Acidification of the curd: So that the curd reaches a suitable pH for stretching (5.2 to 5.4), it is left to ferment at room temperature, or covered in the colder periods, in a wooden box, the 'mastredda,' for about 20–24 hr. During this period, the curd will also continue, slowly, to drain off whey.

Pre-stretching: The curd, properly acidified, is cut into a small round container made of tinned copper or wood and used for the stretching, called the 'staccio.' The sliced curds are thin and long and covered with water at a temperature of about 80°C for about 8–10 min. The cheesemaker checks whether the curd is ready to stretch, picking up with his or her hands a piece of curd and subjecting it to gentle traction; the piece should not break, but have adequate elasticity for stretching.

Stretching: Traditionally stretching is entrusted to the expert hands of the cheesemakers, who, immersing the curd in a very hot liquid, starts working it using a wooden stick called the 'manuvedda' (a special stick with one side extended and flat, 8–10 cm wide and convex on the back). The aim is to press the curd to eject as much whey as possible, and stretch it until it becomes a single compact piece and with a spherical shape. The outer surface should be smooth with no cracks and welded to a pole.

Shaping: The cheesemaker will mould the sphere of curd in the 'mastredda,' pressing the sphere in one of its corners, and by using wooden tables, 'muolitu,' and tablets, 'cugni,' of different sizes, will form the desired size of the parallelepiped with a square section. Patience and skill are required because the cheese, still hot, must be continually rotated carefully (every 15 min in the early hours), and then at longer intervals until evening, to obtain the perfect flat sides of the parallelepiped with rounded corners.

Salting: The cheeses will be immersed in the saturated brine for a period of time that varies according to the weight of the cheeses; care must be exercised to ensure that 6%/total solids of salt is not exceeded. In reality, the cheesemakers/ripeners, to satisfy the tastes of modern consumers, tend to produce Ragusano PDO with a concentration of sodium chloride not higher than 3% per total solids.

Affineurs apply dry salting only if abnormal fermentation (early swelling) occurs or if cracks form on the surface of the cheeses, when the cheese form is salted externally and perforated with a thin needle in order to let out the excess gas; the salt blocks further fermentation. These products can no longer reach high quality standards, and are often intended for consumption as a grated cheese.

Maturation: Ripening takes place in fresh (14°C–16°C), humid and breezy rooms called 'maize,' sometimes underground, in cellars and natural caves with natural geological walls. The cheeses are hung in pairs on a wooden beam (horseback) with ropes 'liama' or strings of 'Cannu,' of 'Zammarra' or cotton. All the cheeses hanging on the wooden beams are suspended vertically and form a stack with the cheeses inclined at an angle of about 45°, but in such a way as to ensure perfect air circulation over the entire surface of the forms. This system is particular and unique, because the affineurs do not need to turn the forms during ripening, but at the same time they can control them and intervene only for the cheeses that present defects (such as late swelling and rifts in the crust).

The maturation period for obtaining the PDO is four months but can exceed a year. The degree of ripeness preferred by consumers is at about six months of aging.

Cappatura/smoking: Cappatura with olive oil is applied to cheese destined for long aging. The product can be smoked only in a natural and traditional way; in this case, the designation of origin must be followed by the word 'smoked'.

8.8.5 Relevant Research

The technology to produce Ragusano cheese has been described (Licitra et al., 1998), as well as its composition during aging (Licitra et al., 2000), especially the proteolysis profile during ripening (Licitra, 2010). The influence of the cows' diets (native or cultivated pasture, total mixed ration) on the sensory properties and volatile profile have been studied (Carpino et al., 2002; Carpino et al., 2004a; Carpino et al., 2004b, Rapisarda et al., 2014).

8.9 Vastedda della Valle del Belìce PDO – Italy

<table>
<tr><td>Name: Vastedda della Valle del Belìce PDO</td><td rowspan="3">
</td></tr>
<tr><td>Production area: Sicily, Valle del Belìce, several municipalities in the provinces of Agrigento, Trapani and Palermo</td></tr>
<tr><td>Milk: Sheep's, raw</td></tr>
</table>

8.9.1 Introduction

The Vastedda della Valle del Belìce was conferred the PDO by the EU in 2010 (EC, 2010). The first document, among those found, that relates to the sale of cheese made in the Belice Valley dates back to the mid-fifteenth century. In 1497, the territory produced a huge amount of cheese: fresh and aged Pecorino cheese, Ricotta cheese, Caciocavallo and the 'Vasteddadella Valle del Belìce', which was traditionally produced only in the summer months, when the high temperature favoured natural acidification and the milk became richer in intense aromas.

The Vastedda della Valle del Belìce is a unique cheese because it is a stretched cheese from sheep's milk.

The older cheesemakers evoke a unique origin, and narrate that once, the shepherds during a particularly hot season had problems in maturing Pecorino cheese. The cheese had defects in the crust with visible cracks that prevented proper maturation of the cheese. One of the old shepherds, having seen many of his Pecorino cheeses with crust defects, began to experiment to try to recover these ruined cheeses. He took a cheese aged about 20–30 days with cracks in the crust, opened it and found that the interior of the cheese was good with no further defects; he cut it into slices and dipped it in hot whey (85°C–90°C), just as in the production of Ricotta cheese, with the objective of remaking the same type of pressed cheese.

While waiting for the whey to cool, the cheesemaker dipped his hands into the container to control the cheese, and to his surprise saw that the cheese mass began to stretch, and various pieces of cheese began to melt. In a spontaneous attempt to apply the technique of stretching the curd, he began to knead the dough to promote fusion of the slices of the curd to form a compact block of cheese, which subsequently divided into small forms of cheeses. In the following days, he invited other shepherds to taste the product of his 'discovery', without divulging the background. He received appreciation from all of them, for the pleasant taste was sweet, delicate and different from the typical smell and taste of the classic Pecorino; they thought it was a product of cow's milk, goat's milk or a mixture. Technically, what happened can be understood: the Pecorino of 20–30 days had undergone fermentation, which decreased the pH from about 6.3 to a pH suitable for stretching; this development, in combination with the hot whey and the accidental intuition of the cheesemaker, led to the production of a new cheese.

The name of the cheese, Vastedda della Valle del Belice, is derived from the birthplace of origin of the product, the Belice Valley (between the Sicilian provinces of Agrigento, Trapani and Palermo), and the term 'vastedda' as mentioned earlier comes from the successful attempt of the cheesemaker to recover cheeses with crust defects, which were 'vastati', 'ruined', which is the root of the term 'vastedda'. According to other Vastedda producers, the name comes from the shape of the container bowls of ceramic, called 'vastedde', where the cheese is placed after stretching to acquire the typical shape of flat or loaf bread.

The geographical area of sheep farming, milk production, processing and packaging of the cheese Vastedda della Valle del Belìce PDO is included as part of the territories of several municipalities in the provinces of Agrigento (including the town of Santa Margherita Belice, the cradle of origin of the product), Trapani and Palermo in Sicily.

8.9.2 Type

The Vasteddadella Valle del Belìce PDO is a sheep's stretched cheese that is consumed fresh. The FDM is not less than 35% and fat 18%, whereas the salt content does not exceed 2.7%.

8.9.3 Description and Sensory Characteristics

The cheese Vastedda della Valle del Belìce PDO is made with whole raw milk, with natural fermentation, and only from the native Valle del Belice sheep breed. The shape is the typical round loaf with slightly convex sides. The diameter of the pot used to form the cheeses should be between 15 and 17 cm and the heel height between 3 and 4 cm. Each cheese weighs between 500 and 700 g in relation to the size of the form. The surface is without rind, off-white, compact and smooth without spots or creases; a layer of pale straw colour may be present. The body is white, uniform and smooth, not grainy but with possible patterns caused by the traditional kneading; the holes must be absent or very few, as well as the exudation. The taste is typical of fresh sheep cheese, slightly tangy but never too sharp.

8.9.4 Method of Manufacture

Milk preparation: The natural pastures are the predominant diet of the indigenous sheep of the Valle del Belice breed for much of the year. In some periods, it is supplemented with hay, straw and stubble from farm production, and with natural feeds (GMO free). The use of concentrates cannot exceed 50% of the total dry ration. It is prohibited to feed animal products and plants or parts of plants (seeds) of fenugreek, tapioca and cassava. Green forage pastures, aromatic herbs and a thousand flowers of wild plants enrich the milk used to produce this cheese, imparting to it a rich bouquet of aromas typical of the product itself.

The milk from one or two milkings must be processed within 48 hr after the first milking. Fresh milk is filtered and chilled with special sieves and/or filters in cloth. It is traditionally heated in a copper tin-plated vat, up to a maximum temperature of 40°C with direct fire of wood or gas.

Coagulation: The heated milk is poured into a wooden vat for the coagulation phase, where at the temperature of 36°C–40°C lamb rennet paste (60–100 g/100 L milk) is added with a clotting time which varies from 40 to 50 min.

Cutting: The breaking of the curd begins only after a wooden stick, called the 'rotula', placed in the tina (wooden vat), remains in the vertical position, a sign of the right consistency of the coagulated curd. The shepherd–cheesemaker breaks the curd with the rotula through rotations of varying intensity until the cheese curds are as small as a grain of rice; spontaneous syneresis is favoured by the hot water added during the breaking of the curd. The lumps of curd settle at the bottom of the tina and are left to rest for about 5 min, so that they cohere.

Acidification: The cheese mass is taken from tina and filed in fuscelle (baskets of rushes) without any pressing of the paste. The curd is then left inside the baskets at room temperature for maturation (natural fermentation of the paste). The time required for the aging changes with the variation in the temperature of the environment (the cooler the environment, the more the time required). After 24 hr – but in the cold season after even 48 hr – the extent of curd acidification is assessed with a portable pH meter or by empirical stretching tests of the curd. After having reached a pH between 4.7 and 5.5, the cheesemaker will proceed to the stretching phase.

Stretching: The shepherd–cheesemaker cut the mature curd into slices in a wooden vessel, called the 'piddiaturi', and covers it with hot whey or water at a temperature of 80°C–90°C. The processing of the curd starts with bland movements, using a stick wood immersed in 'piddiaturi' in order to facilitate the fusion of the slices in a single stringy block, and with light pressure on the curd to stimulate syneresis of the whey. The process lasts 3–7 min, depending on the degree of maturation of the curd and the temperature it achieves during processing, which usually does not exceed 48°C–50°C. One big amalgamated mass of curd is extracted from the piaddiaturi to begin forming vastedde as soon as the paste has taken on a white-shiny surface.

Some old cheesemakers place it on a board for a few turnings that allow further syneresis and a firming of the same, whereas others start directly to detach from the mass portions of sphere-shaped curd, which are processed manually and closed at the point of detachment. Sealing is performed by holding the lips of the sphere, which initially showed flaking, between the thumb and forefinger. The spheres of curd are then placed with the closing point down in vastedde, and after being turned, will take the form of characteristic Vastedda (typical shape of flat or loaf bread).

Salting: Subsequently, when the forms of Vastedda cool and achieve the right consistency (usually after 6–12 hr from stretching), it is salted; this is carried out by immersing the forms in saturated brine at room temperature, for a period of time that varies from 30 min to 2 hr. This is followed by drying in a cool, moderately ventilated room, and after 12–48 hr, the cheese can be consumed.

8.9.5 Relevant Research

The Vastedda della Valle del Belice has been studied to characterise the quality of the product in different seasons of the year (summer, autumn and spring) (Todaro et al., 2014). Further studies present the molecular characterisation of the dominant bacterial population (Reale et al., 2007), and the characteristics of LAB isolated from Vastedda della Valle del Belice cheeses (Gaglio et al., 2014).

References

Akın, M. S. (2012). Accelerated ripening of Kashar cheese with encapsulated protease. *African Journal of Biotechnology*, 11 (66), 13007–13015.

Akın, M. B., Akın, M. S., Atasoy, A. F., Kırmacı, H. A. & Eren-Karahan, L. (2012a). Accelerated Kashar cheese ripening with encapsulated lipase and protease enzymes. *Italian Journal of Food Science*, 24 (4), 358–366.

Akın, M. S., Akın-Güler, M. B., Kırmacı, H. A., Atasoy, A. F. & Türkoğlu, H. (2012b). The effects of lipase-encapsulating carriers on the accelerated ripening of Kashar cheese. *International Journal of Dairy Technology*, 65 (2), 243–249.

Anastasiou, R., Georgalaki, M., Manolopoulou, E., Kandarakis, I., De Vuyst, L. & Tsakalidou, E. (2007) The performance of *Streptococcus macedonicus* ACA-DC 198 as starter culture in Kasseri cheese production. *International Dairy Journal*, 17, 208–217.

Anonymous (2015) *Turkish Food Codex, Cheese Coomunique, Communique No: 2015/6*, Published by the Ministry of Food, Agriculture and Livestock of Turkey.

Anonymous (1997). *Kaèkavalj – quality requirements*. SRPS E. C2.010. Institute for Standardization of Serbia, Belgrade.

Avşar, Y. K. (2009). [*Kars Kaşarının aroma profilinin belirlenmesi*. pp. 172. *II. Geleneksel Gıdalar Sempozyumu*, 27–29 Mayıs, Van, Turkey.]

Cappelli (1871). *Vocabolario di agricoltura di Canevazzi-Mancini*.

Carpino, S., Acree, T. E., Barbano, D. M., Licitra, G. & Siebert, K. J. (2002). Chemometric analysis of Ragusano cheese flavor. *Journal of Agricultural and Food Chemistry*, 50, 1143–1149.

Carpino, S., Horne, J., Melilli, C., Licitra, G., Barbano, D. M. & and Van Soest, P. J. (2004a). Contribution of native pasture to the sensory properties of Ragusano cheese. *Journal of Dairy Science*, 87, 308–315.

Carpino, S., Mallia, S., La Terra, S., Melilli, C., Licitra, G., Acree, T. E., Barbano, D. M. & Van Soest, P. J. (2004b). Composition and aroma compounds of Ragusano cheese: Native Pasture and Total Mixed Rations. *Journal of Dairy Science*, 87, 816–830.

Costanzo, N., Santoro, A. M. L., Sarno, E., Di Loria, A., Grembiale, R. D., Britti, D. & Capuano, F. (2015). *Use of sonoelastography to evaluate texture modifications of Mozzarella di bufala campana D.O.P. during storage at different temperatures. Italian Journal of Food Science*, in press.

EC (1996a). Commission Regulation (EC) No 1107/96 of 12 June 1996 on the registration of geographical indications and designations of origin under the procedure laid down in Article 17 of Council Regulation (EEC) No 2081/92. *Official Journal of the European Union*, L 148, 1–10.

EC (1996b). Commission Regulation (EC) No 1263/96 of 1 July 1996 supplementing the Annex to Regulation (EC) No 1107/96 on the registration of geographical indications and designations of origin under the procedure laid down in Article 17 of Regulation (EEC) No 2081/92. *Official Journal of the European Union*, L 163, 1–3.

EC (2000). European Commission Regulation (EC) No 1509/2000 of 12 July 2000 amending items in the specifications for several names listed in the Annex to Regulation (EC) No 1107/96 on the registration of geographical indications and designations of origin under the procedure laid down in Article 17 of Council Regulation (EEC) No 2081/92. *Official Journal of the European Union*, L 174, 7–10.

EC (2008). Commission Regulation (EC) No 656/2008 of 10 July 2008 registering certain names in the Register of protected designations of origin and protected geographical indications (Chamomilla Bohemica (PDO), Vlaams-Brabantse tafeldruif (PDO), Slovenská parenica (PGI), Cipollotto Nocerino (PDO)). *Official Journal of the European Union*, L 183, 15–16.

EC (2010). Commission Regulation (EC) No 971/2010 of 28 October 2010 entering a name in the register of protected designations of origin and protected geographical indications (Vastedda della valle del Belìce (PDO)). *Official Journal of the European Union,* L 283, 23–24.

Gaglio, R., Francesca, N., Di Gerlando, R., Cruciata, M., Guarcello, R., Portolano, B., Moschetti, G. & Settanni, L. (2014). Identification, typing and investigation of the dairy characteristics of lactic acid bacteria isolated from 'Vastedda della valle del Belìce' cheeses. *Dairy Science & Technology,* 94 (2), 157–180.

Gianferri, R., D'Aiuto, V., Curini, R., Delfini, M. & Brosio, E. (2007a). Proton NMR transverse relaxation measurements to study water dynamic states and age-related changes in Mozzarella di Bufala Campana cheese. *Food Chemistry,* 105, 720–726.

Gianferri, R., Maioli, M., Delfini, M. & Brosio, E. (2007b). A low-resolution and high-resolution nuclear magnetic resonance integrated approach to investigate the physical structure and metabolic profile of Mozzarella di Bufala Campana cheese. *International Dairy Journal,* 17 (2), 167–176.

Guinee, T. P., Pudja, P., Miočinović, J., Wiley, J. & Mullins, C. M. (2015). Textural, cooking properties and viscoelastic changes on heating and cooling of Balkan cheeses, *Journal of Dairy Science,* 98 (11), 7573–7586.

Hayaloğlu, A. A. (2009). Volatile composition and proteolysis in traditionally produced mature Kashar cheese. *International Journal of Food Science and Technology,* 44 (7), 1388–1394.

Kindstedt, P. S., Carić, M. & Milanović, S. (2004). Pasta-filata cheeses. In Fox, P. F., McSweeney, P. L. H., Cogan, T. M. & Guinee, T. P. (eds.), *Cheese: Chemistry, Physics and Microbiology, Volume 2: Major Cheese Groups,* 3rd edition. Elsevier Ltd., London, pp. 251–277.

Koca, N. & Metin, M. (2004). Textural, melting and sensory properties of low-fat fresh kashar cheeses produced by using fat replacers. *International Dairy Journal,* 14 (4), 365–373.

Kurultay, S., Yaşar, K. & Öksüz, O. (2004). The effect of different curd pH and stretching temperatures on some chemical properties of Kashar cheese. *Milchwissenschaft,* 59 (7–8), 386–388.

Licitra, G. (2010) World Wide traditional cheeses: Banned for business. *Dairy Science & Technology,* 90 (4), 357–374.

Licitra, G., Campo, P., Manenti, M., Portelli, G., Scuderi, S., Carpino, S. & Barbano, D. M. (2000). Composition of Ragusano cheese during aging. *Journal of Dairy Science,* 83, 404–411.

Licitra, G., Portelli, G., Campo, P., Longombardo, G., Farina, G., Carpino, S. & Barbano, D.M. (1998). Technology to produce Ragusano cheese: A survey. *Journal of Dairy Science,* 81, 3343–3349.

Manzi, P., Marconi, S., Di Costanzo, M. G. & Pizzoferrato, L. (2007). Composizione di formaggi DOP italiani. *La rivista di Scienza dell'Alimentazione, anno* 36, 9–22.

Manzi, P., Marconi, S., Gambelli, L., Santaroni, G. P. & Pizzoferrato, L. (2005). Chemical-nutritional characteristics of buffalo Mozzarella DOP cheese produced in the year 2004. *Rivista di Scienza dell'Alimentazione,* 1–2, 39–52.

MIPAAF (2005). *Quinta revisione dell'elenco nazionale dei prodotti agroalimentari tradizionali.* Suppl. Ord. n. 133 Gazz. Uff. n. 174 of 28 July 2005.

Moatsou, G., Kandarakis, I., Moschopoulou, E., Anifantakis, E. & Alichanidis, E. (2001). Effect of technological parameters on the characteristics of kasseri cheese made from raw or pasteurized ewes' milk. *International Journal of Dairy Technology,* 54 (2), 69–77.

Nega, A. & Moatsou, G. (2012). Proteolysis and related enzymatic activities in ten Greek cheese varieties. *Dairy Science and Technology,* 92 (1), 57–73.

Özer, B. H., Uzun, Y. S. & Kırmacı, H. A. (2008). Effect of Microencapsulation on viability of *Lactobacillus acidophilus* LA-5 and *Bifidobacterium bifidum* BB-12 during Kasar cheese ripening. *International Journal of Dairy Technology,* 61 (3), 237–244.

Pizzolongo, F., Quarto, M., Nasi, A., Ferranti, P., Addeo, F., Sacchi, R. & Chianese, L.(2007). Sensory profile of P.D.O. Mozzarella di Bufala Campana Cheese. *Italian Journal of Animal Science,* 6, 1136–1139.

Radulović, Z., Radin, D. & Obradović, D. (2006). Determinacija autohtonih bakterija mlečne kiseline iz kačkavalja. *Prehrambena Industrija*, 17 (3–4), 82–86.

Rapisarda, T., Pasta, C., Carpino, S., Caccamo, M., Ottaviano, M. & Licitra, G. (2014). Volatile profile differences between spontaneous and cultivated Hyblean pasture. *Animal Feed Science and Technology*, 191, 39–46.

Reale, S., Vitale, F., Scatassa, M. L., Caracappa, S., Curro, V. & Todaro, M. (2007). Molecular characterization of dominant bacterial population in 'Vastedda della Valle del Belice' cheese: Preliminary investigation. *Italian Journal of Animal Science*, 6, 595–597.

Romano, R., Borriello, I., Chianese, L. & Addeo, F. (2008). Quali-quantitative determination of triglicerydic, fatty acids and CLA in 'Mozzarella di Bufala Campana' by high resolution gascromatography (HRGC). *Progress in Nutrition*, 10 (1), 22–29.

Temizkan, R., Yaşar, K. & Hayaloğlu, A. A. (2014). Changes during ripening in chemical composition, proteolysis, volatile composition and texture in Kashar cheese made using raw bovine, ovine or caprine milk. *International Journal of Food Science and Technology*, 49 (12), 2643–2649.

Tenore, G. C., Ritieni, A., Campiglia, P., Stiuro, P., Di Mauro, S., Somella, E., Pepe, G., D'Urso, E. & Novellino, E. (2015). Antioxidant peptides from 'Mozzarella di Bufala Campana DOP' after simulated gastrointestinal digestion: In vitro intestinal protection, bioavailability, and anti-haemolytic capacity. *Journal of Functional Foods*, 15, 365–375.

Todaro, M., Bonanno, A. & Scatassa, M. L. (2014). The quality of Valle del Belice sheep's milk and cheese produced in the hot summer season in Sicily. *Dairy Science & Technology*, 94 (3), 225–239.

Yaşar, K. & Güzeler, N. (2011). Effects of coagulant type on the physicochemical and organoleptic properties of Kashar cheese. *International Journal of Dairy Technology*, 64 (3), 372–379.

Zygouris, M. P. (1952). *The Dairy Industry*, 2nd edition. Ministry of Agriculture, Athens, Greece.

9

Mould Surface-Ripened Cheeses

Katja Hartmann[1] *and Jean L. Maubois*[2]

[1] Anton Paar GmbH, Germany
[2] Dairy Research Laboratory, INRA, Rennes, France

9.1 Altenburger Goat Cheese PDO – Germany

<table>
<tr><td>Name: Altenburger goat cheese</td><td rowspan="3">
</td></tr>
<tr><td>Production area: Eastern region of Germany</td></tr>
<tr><td>Milk: Cow's (>15% goat's), pasteurised</td></tr>
</table>

9.1.1 Introduction

Altenburger goat cheese is listed as a standard cheese variety in the German regulation on cheese and has held a PDO label since January 1997 (EC, 1997). Altenburger goat cheese is a traditional cheese product which was popular in its production region in East Germany in the late nineteenth century. The production of Altenburger goat cheese was mentioned for the first time in writing in 1862 in the border region between the districts of Thuringia and Saxony in Germany, where it was only produced on small farms. Until today, it has been produced traditionally according to a protocol from 1897, and since about 1900 the

Global Cheesemaking Technology: Cheese Quality and Characteristics, First Edition.
Edited by Photis Papademas and Thomas Bintsis.

production moved to larger dairies. Today, Altenburger goat cheese is produced in two dairies in Germany (Feinkäserei Zimmermann, Saxony and Käserei Altenburger Land, Thuringia) (Käsewelten, 2015). The geographical regions of Altenburger goat cheese production include the districts Altenburg, Schmölln, Gera, Zeitz, Geithain, Grimma, Wurzen, Borna and the town of Gera located in the states of Thuringia, Saxony and Saxony-Anhalt in Germany (Deutsche Käseverordnung, 2013).

9.1.2 Type

Altenburger goat cheese is a soft cheese produced with 30% fat-in-dry-matter (FDM) and 38% dry matter (DM). At least 15% of goat's milk has to be added.

9.1.3 Description and Sensory Characteristics

Altenburger goat cheese is produced in a cylindrical form with a production weight of 250 g. The diameter is 11.5 cm, and the height is 2–3 cm. The surface in covered in white camembert mould (e.g. *Penicillium camemberti*), and a red smear may also develop. The added caraway is visible as small brown spots (Iburg, 2003; Landwirtschaft Sachsen, 2015). The cheese texture is of a pale yellow colour with very few gas openings. The aroma is very aromatic, piquant and strong.

9.1.4 Method of Manufacture

Altenburger goat cheese is produced from a mixture of cow's and goat's milk. Caraway is added before the curd is transferred into the moulds (Iburg, 2003; Mair-Waldburg, 1974).

Milk preparation: Altenburger goat cheese is manufactured from milk produced in the permitted cheese production regions. The cheese milk is pasteurised and fat-standardised (0.9%–2.9%) depending on the fat level of the final cheese.

Starter culture: Usually, a mixture of 1%–3% of *Lactococcus lactis* ssp. *lactis* and *Lactococcus lactis* ssp. *cremoris* is added to the cheesemilk at a temperature between 32°C and 40°C. Fermentation lasts approximately 20 min until the desired acidity level is reached (Mair-Waldburg, 1974).

Rennet: Rennet is added at temperatures between 32°C and 40°C, and coagulation lasts between 60 and 90 min.

Cutting: The curd is cut with a cheese harp vertically and horizontally, and while stirring, walnut-sized curd cubes are formed. Some whey is removed and retained.

Curd draining: The curd is transferred into moulds and caraway is added. The cheeses are subsequently turned for the first time and for a second time after 6 to 8 hr. The cheeses are stored the next day.

Salting: The day after production, the cheeses are dry-salted or transferred in a brine (16°Bé–18°Bé) for 60 to 90 min at 14°C–16°C (Mair-Waldburg 1974).

Maturation: After salting, the cheeses are stored for 3 to 4 days at 16°C in a dry room. When mould development is noticed on the cheese surface, the cheese is transferred to a ripening room with 85%–95% Relative Humidity (RH). After one week, a red smear starts to develop, which is enhanced by cheese treatment with fresh whey. The cheese ripens from the outside to the inside. The minimum cheese age is 14 days (Landwirtschaft Sachsen, 2015; Mair-Waldburg, 1974).

Storage: The ripened cheese is packed in aluminium foil or wooden chip boxes and stored at 10°C–12°C.

9.2 Camembert de Normandie PDO – France

<table>
<tr><td>Name: Camembert de Normandie PDO</td><td rowspan="3"></td></tr>
<tr><td>Production area: The Bocage (hedged farmland) of the Calvados, Manche, Orne and Eure départements in Normandy</td></tr>
<tr><td>Milk: Cow's, raw</td></tr>
</table>

9.2.1 Introduction

Camembert de Normandie PDO has been recognised at the European level since June 1996 (EC, 1996). It is a soft mould-ripened cheese made from raw cow's milk. Its annual production amounts to 4,500 tonnes. That represents only 3.8% of French Camembert production and slightly over 1% of the total French soft cheese group. This cheese variety was created in 1791 by a farmer named Marie Harel in the small Normandy village of Camembert (Beaumoncel castle) with the help of a 'refractory' priest born in Brie (another French area producing well-known soft moulded cheese. Production was then increasing but consumption was limited because of time required for transportation. In 1890, the typical wooden box was proposed by J. Charrel which allowed widespread diffusion of the cheese all over the country.

9.2.2 Milk

The milk must be from herds comprising at least 50% of cows belonging to the Normandy breed which graze (over a pasture area with a minimum of 0.33 ha per cow) for at least six months per year. Hay, beets and corn silage are allowed as feed for the rest of the year. As much as 80% of the feed ratio of the milk cattle, calculated on dry matter, must come from the fodder area of the farm cattle. Other components of the feeding are also precisely defined in the PDO rules. Milk collection must be done in less than 72 hr.

9.2.3 Description and Sensory Characteristics

Camembert de Normandie is a raw milk soft cheese having a cylindrical form with 10.5–11 cm diameter. Its weight at packaging time must be over 250 g, and it must contain at least 115 g of total solids. Its crust is thin and covered by white mould with possibly some red spots created by the growth of *Brevibacterium linens.*

9.2.4 Method of Manufacture

Milk preparation: Milk is first fat-standardised in order to obtain a ratio of fat/protein in milk as around 1.1, which allows a fat/total solids ratio of 0.45 to be achieved in the cheese. About 2.2 litres of milk is necessary to obtain one piece of Camembert. Mesophilic lactic starters and often a proper bacterial ecosystem containing different species such as *Hafnia alvei* are then added at temperatures ranging between 22°C and 38°C. Heat treatments (thermisation, pasteurisation), bactofugation and membrane ultrafiltration are strictly forbidden. Membrane 1.4 μm microfiltration was provisionally authorised in 2002 because of several *L. monocytogenes* outbreaks and definitively allowed in 2007 (JORF, 2007) with the requirement to mention 'microfiltrated raw milk' on the cheese package.

Rennet/coagulation: Milk is poured in specific 100 litre bowls. Addition of calf's rennet is done after a 'maturation time' of either a night at 10°C–12°C or 2 hr at around 30°C. The amount of rennet is less than 23 mL per 100 L at a temperature of 30°C–35°C. Coagulation lasts 10–20 min.

Cutting: For a duration corresponding to five times the coagulation time, the curd is vertically cut (with a slice thickness over 2.5 cm) and then taken either manually in spoons or mechanically by hemispheric equipment for deposition in hoops (11 cm diameter). Five spoons each for 40 min are required in each mould.

Curd draining: Whey draining occurs for 18 hr in hoops placed on draining tables covered with a draining mat. The double kinetics of acidification and of draining is one of the key points for obtaining typical Camembert de Normandie. Indeed, both kinetics determine the migration of colloidal calcium salts in the aqueous phase of cheese and consequently the structure of the curd network and its mineral content which will determine microbial ripening (Maubois, 2016).

The drained curds contained in hoops are turned once, and a stainless steel plate is placed on the surface in order to apply a little pressure and complete the drainage.

Salting: Only dry salt is used. If salting is done manually, salt is first added to the round edges and then successively to the two faces. Moulds (*P. camemberti* and *Geotrichum candidum)* in a liquid solution are sprayed on the surface. The cheeses are then placed on stainless steel string racks, where they are stored for one or two days in an air-conditioned room at 14°C with an RH of 85% for partial drying of the surface.

Maturation: The cheeses are then ripened in air-conditioned curing rooms named 'Hâloirs' which are equipped with a powerful air circulation system for at least 11 days at 10°C–13°C with an RH over 95%. The cheeses placed on stainless steel string racks are returned every couple of days. They could be placed on wooden shelves for either the last two or three days of ripening time or after packaging. Development of moulds and yeast takes place which initiates lactate consumption at the surface. That induces a calcium salt gradient which triggers crust formation. The enzymatic activity of a microbial added ecosystem induces production of typical flavour components. Hydrolysis of casein allows release of methionine and then formation of characteristic sulphur derivatives.

Packaging/storing: After wrapping cheese in paper that is permeable to humidity and gas, conditioning is done individually in a cylindrical box whose bottom is made of poplar wood. If the cheese is stored at 4°C–6°C, it can be consumed within two months.

References

Deutsche Käseverordnung, Established 1965, *Revision 1986, Amendment 2010, 2013*. German Federal Ministry for justice and consumer protection.

EC (1996). Commission Regulation (EC) No 1107/96 of 12 June 1996 on the registration of geographical indications and designations of origin under the procedure laid down in Article 17 of Council Regulation (EEC) No 2081/92 *Official Journal of the European Union*, L 148, 1–10.

Iburg, A. (2003). *DumontskleinesKäselexikon*. DuMont monte Verlag, Cologne, Germany.

JORF (2007). *Journal Officiel République Française*, Décret 2007–628 27 Avril 2007 relatif aux fromages et spécialités fromagères.

Käsewelten (2015). www.kaesewelten.info/kasesorten/altenburger-ziegenkase [Accessed 14 July 2015].

Landwirtschaft Sachsen (2015). http://www.landwirtschaft.sachsen.de/landwirtschaft/14129.htm [Accessed 1 July 2015].

Mair-Waldburg (1974). *Handbuch der Käse: Käse der Welt von A-Z: eine Enzyklopädie*, Volkswirtschaftlicher Verlag, Kempten, Germany.

Maubois, J.-L. (2016). *Le Fromage*. Ed. Lavoisier Paris, Paris.

10

Bacterial Surface-Ripened (Smear) Cheeses

Ylva Ardö[1], Françoise Berthier[2], Katja Hartmann[3], Elisabeth Eugster-Meier[4], Marie-Therese Fröhlich-Wyder[5], Ernst Jakob[5] and Daniel Wechsler[5]*

[1] *Department of Food Science, University of Copenhagen, Denmark*
[2] *Unité de Recherches en Technologie et Analyses Laitières Rue de Versailles, France*
[3] *Anton Paar GmbH, Germany*
[4] *Bern University of Applied Sciences, School of Agricultural, Forest and Food Sciences HAFL, Zollikofen, Switzerland*
[5] *Agroscope, Research Division Food Microbial Systems, Federal Department of Economic Affairs, Education and Research EAER, Bern, Switzerland*

10.1 Danbo – Denmark

Name: Danbo	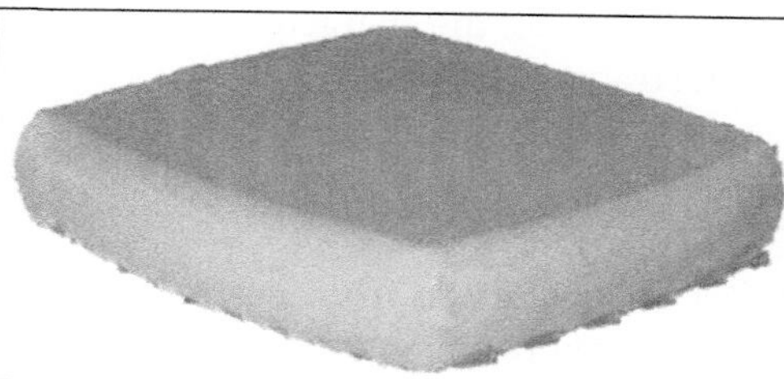
Production area: Denmark	
Milk: Cow's, pasteurised	

10.1.1 Introduction

Danbo is made in conformity with Danish Food Legislation (Danish Food Administration, 2013) and the General Standard for Cheese (Codex Alimentarius, 1978). Danbo is mainly produced industrially at several dairy plants, and it is the most popular cheese in Denmark. The name 'Danbo' was first documented by the Stresa convention and has been used since 1952 as a common name for several similar cheese varieties made at different dairies in Denmark that had been influenced mainly by Dutch, but also German, Polish and Russian cheeses. In 1897, Rasmus Nielsen, who was managing director of Kirkeby dairy in Denmark, travelled in eastern Prussia along the border with Russia to learn about their specific cheesemaking techniques as recommended by Professor Bernhard Bøggild of the Royal Veterinary and Agricultural University, Copenhagen. After studies in eastern Prussia, Rasmus Nielsen went to Holland, where he worked with cheese production in a number of different dairies. When he returned to Denmark, he used his experience and developed a new cheese variety.

Global Cheesemaking Technology: Cheese Quality and Characteristics, First Edition.
Edited by Photis Papademas and Thomas Bintsis.

A special feature of the new cheese was its square form, which was new for round-eyed cheeses, and a special smear surface-ripening technique that involved rubbing a suspension of bacterial and yeast cultures into the cheese rind. Two thirds of the Danbo produced today in Denmark is exported.

10.1.2 Type

Danbo is classified as a semi-hard cheese but at the soft end of the scale. Its fat in dry matter (FDM) is 20%, 30% or 45%, and its moisture in not-fat substance (MNFS) is 58%–60%. Interestingly, Danbo is made with 2% higher moisture content for the Danish home market, with which the really soft consistency that is highly appreciated is obtained, as Danbo is typically eaten sliced on bread (Danish Food Administration, 2013).

10.1.3 Description and Sensory Characteristics

Danbo is pale yellow and has a closed texture with small round eyes and a soft consistency that permits slicing with a thin wire. It is made in square or rectangular form with a height of about 7 cm and weight of 6–12 kg. It has a firm rind with smear, and the surface of the ripened cheese may be washed and coated by paraffin or aluminium folio. Danbo is smooth and soft. Long-time-ripened cheeses may be somewhat brittle. The body contains a few to a large number of equally distributed, evenly round, pea-sized eyes. The flavour is mild, salty, slightly acidic and with a pronounced piquant flavour that originates from the surface microflora. As the age increases, the characteristic smell and taste become more pronounced (Danish Food Administration, 2013).

10.1.4 Method of Manufacture

Milk: Danbo is made from pasteurised cow's milk, and sometimes bactofugation is performed to remove clostridia spores. The milk temperature is set to 30°C–32°C. Calcium chloride may be added at a concentration of up to 20 g per 100 L of cheesemilk to improve coagulation, and 20 g potassium nitrate/100 L milk may be added to prevent the late blowing caused by clostridia.

Starter culture/rennet: A mesophilic DL-starter (0.5%–1.0% outgrown culture) containing *Lactococcus lactis* ssp. *lactis* and *cremoris* with commonly less than 10% of gas-producing *Lc. lacis* ssp. *lactis* and *Leuconostoc* spp is used. Bovine rennet is employed.

Coagulation/cutting: The coagulation time is 30–45 min, and the curd is cut into cubes with a side of 6 mm.

Curd washing: After cutting, 30%–35% of the whey is replaced with 15%–25% water to obtain a sufficiently high minimum pH at about 5.2.

Cooking: Heating is performed to 37°C–38°C for full-fat Danbo and 34°C–37°C for reduced and low-fat Danbo.

Curd draining: The curd is pre-pressed under whey, cut into blocks and placed in moulds and further pressed.

Brine salting: This is done for 1.5–3 days at 12°C in brine of 20°Bé.

Maturation: A smear is applied to the cheese surfaces, and initially Danbo is ripened for 7–10 days at 18°C–20°C and 90%–95% RH to allow surface flora to develop. The smear may then be washed off and the surfaces dried and coated by paraffin. Further ripening is performed at 8°C–12°C and 85% RH, until sale. The characteristic taste and texture are achieved after maturation for 3–4 weeks at 12°C–20°C.

Storage: Cold (2°C–6°C).

10.1.5 Relevant Research

The lactose is completely converted to lactic acid by starter bacteria within a day or two, and thus a minimum pH at 5.2 is reached. During ripening, the pH increases slowly and reach 5.6–5.8 after 3–6 months as a result of the activities of the surface microflora. Yeast starts growing and by using lactic acid mediates an increase in pH at the surfaces, thereby creating a less hostile environment for smear bacteria, which contribute to a further increase in pH by performing oxidative deamination of amino acids. The fresh curd contains about 1.5–2 g/kg citrate at moulding, which is completely converted within about a week by the starter *Lactococcus lactis* ssp. *lactis* biovar. *diacetylactis* and *Leuconostoc* into diacetyl, acetoin and carbon dioxide. The gas produced is responsible for the formation of the typical small round eyes of Danbo. The intact casein is hydrolysed by rennet and plasmin, and the medium-sized peptides thus formed are further broken down by starter cell-bound proteases to obtain a size that makes it possible to be transported over the cell membrane and reach the intracellular peptidases, which release amino acids. Amino acids are catabolised into important cheese aroma compounds, including aldehydes and acids from leucine and characteristically for smear-ripened cheeses such as Danbo, sulphur compounds from methionine. Fat hydrolysis is limited because at least 95% of the milk lipase is inactivated by pasteurisation; however, lipolysis products may play a significant role in the flavour of long-time ripened Danbo (Madsen & Ardö, 2001; Waagner Nielsen, 1993). Starter LAB dominate the interior of Danbo during ripening together with non starter lactic acid bacteria (NSLAB) of mainly *Lb. casei/paracasei* but also other facultative heterofermentative lactobacilli such as *Lb. curvatus, Lb plantarum* and *Lb. rhamnosus*. The smear microflora containing yeast and bacteria such as *Debaromyces, Corynebacterium, Brevibacterium* and *Staphylococcus* contribute to ripening by increasing pH, performing oxidative processes and causing a migration of specific aroma compounds into the cheese body (Ardö, 2004; Antonsson, Molin & Ardö, 2003; Christiansen et al., 2005; Ryssel et al., 2015).

10.2 Epoisses PDO – France

<table>
<tr><td>Name: Epoisses PDO</td><td rowspan="3">
http://www.fromage-epoisses.com/img/bien_choisir_tome_epoisses.jpg</td></tr>
<tr><td>Production Area: Included in the Burgundy-Champagne region</td></tr>
<tr><td>Milk: Cow's milk, pasteurised or raw</td></tr>
</table>

10.2.1 Introduction

Epoisses PDO is one out of the 60 soft and smear-rind French cheeses which include seven PDO cheeses (Epoisses, Langres, Livarot, Maroilles, Mont-d'Or, Munster and Pont-L'Evêque). It is the only smear soft French PDO cheese which is acid-coagulated. This way of coagulation is frequent in Burgundy for goat's and cow's milk cheeses, which differentiate this region for the other French regions In 2014, it was ranked ninth among soft French PDO cheeses in respect of the volume produced, accounting for 4% of the total volume. Epoisses was recognised with AOC in 1991 and PDO in 1996 (EC, 1996). It is made and matured exclusively in a delimited area located in north-eastern France, which is an extension of its historical area; this area is part of a lowland region (250-600m) and is characterised by natural pastures on clay–limestone soils from the Lias geological period; this pasture leads to a specific mineral composition of milk that may facilitate proteolysis during cheese maturation (Delfosse and Letablier, 1999; Official website of Epoisses, 2010). Other non-PDO cheeses related to Epoisses (e.g. PGI Soumaintrain) are manufactured in this area. Actually, PDO Epoisses is manufactured at 3 dairy plants and 1 farm, from milk supplied by 44 farms producing 20,360,000 litres of milk (Official website for Epoisses, 2010).

Epoisses PDO is a soft cheese, acid coagulated and harbouring a smear rind. It is manufactured exclusively in the delimitated area from full-fat cow's milk, which is mostly pasteurised. Then, it is matured for a minimum of four weeks (usually six weeks, until eight weeks). After the maturation, cheeses from every plant are officially sampled four times per year for scoring their physico-chemical and sensory characteristics. When their global score is less than 28/40, their visual score less than 8/15 and their texture and flavour score less than 14/25, the cheeses are not labelled PDO.

Actual PDO Epoisses cheeses are derived from ancient cheeses that according to the oral tradition, may have been manufactured for the first time in the sixteenth century at an abbey located in a village of Burgundy named Epoisses (Delfosse & Letablier, 1999; Official website for Epoisses, 2010, Risoud, 2000). Then, it was manufactured by a farmer's wife according to specific know-how, as shown by specific rooms present in farms of the Epoisses region more than two centuries old. The Epoisses village was a place where cheeses manufactured at farms and reserved for the towns around it and Paris were exchanged. At the beginning of the twentieth century, Epoisses was produced at around 300 farms. Its production decreased progressively after the First World War, and then the Second World War, especially because the farmer's wife suddenly found herself farming alone and therefore had no time left for manufacturing a cheese requiring a lot of care and attention. Only two farms manufactured it in 1950, and manufacture halted some year later, until the efforts of some farmers, who used traditional manufacturing parameters for 25 years, progressively revived it, leading to its AOC denomination in 1991 with a production of 230 tonnes. In particular, these farmers reproduced the high humidity of the circulating air obtained in the traditional drying room by its northeast orientation, thus allowing a slow and progressive drying of the cheeses. The production of Epoisses increased almost regularly between 1992 and 2013 (4.7-fold); it decreased only in 1999 (by 70%), staying at low levels for three years after the death of two persons who consumed Epoisses contaminated by *Listeria*. Since this tragic event, most PDO Epoisses has been manufactured from pasteurised milk.

10.2.2 Description and Sensory Characteristics

Epoisses is a red-smear soft cheese obtained after milk acid coagulation. According to its specifications, it should have a minimum of 40% dry matter (DM) and 50% FDM. Its DM is 45%, fat content 23.8%, salt content 0.77%, protein content 16.5% and carbohydrate content 0.9%.

Epoisses's rind is smooth to slightly rippled, shiny and ivory-orange to brick-red as maturation progresses; it shows the counter-relief fingerprint of the maturation grid (parallel lines). Epoisses's body is creamy and has been described as balanced, well-marked with fruity aroma.

10.2.3 Method of Manufacture

Epoisses is manufactured according to rules applying from milk production to cheese maturation. All steps should be performed in the delimitated area.

Milk production: Milk has to be produced by Montbéliarde, Brune or French Simmental cows. The minimal surface area of pasture should be 20 m^2 per grazing cow. Grass from the beginning of grazing until at least 15 June is the main feed. Outside this period, a minimum of 30% of basic ration has to be dry feed. Every feed that leads to flavour or odour defects, coagulation defects or bacteriological risks is forbidden. Expressed as dry matter, 80% of the total ration per day, 85% of the annual total ration and 100% of the basic ration except farming by-products have to be produced within the delimited geographical area. Milk used for Epoisses manufacture has to be collected, stored and manufactured independently of milks for other purposes.

All additives other than salt, rennet and cultures of safe microbial strains are banned, as are milk concentration; storage of milk, curd or fresh cheese at a negative temperature and storage of fresh or maturing cheese under modified atmosphere.

Coagulation: It takes place with a low amount of rennet, and the acid is produced within 16 hr.

Owing to its specific technology (slow coagulation rate and acid coagulation), whey draining is a critical step, which requires a lot of care and experience in order to remove enough whey.

Cutting/moulding: The curd is cut roughly and moulded after 48 hr with two turnings.

Salting: Dry salting is carried out, and the pH is less than or equal to 4.5.

Maturation: The cheese is matured for four weeks or more. It is washed with brine or water first and then with added local grape spirit; the amount of local grape spirits, called Marc de Bourgogne, for the rind washing is gradually increased in the washing solution. Washing is done from one to three times per week until the end of maturation.

10.2.4 Relevant Research

The orange-red colour of its rind was shown to be related to the activity of different yellow bacteria, such as *Arthrobacter arilaitensis* and *Microbacterium gubbeenense,* rather than to the activity of *Brevibacterium linens* (Galaup et al., 2007). Pigments associated with this colour were the same as for PDO Maroilles and PDO Mont-d'Or, two other French red-smear soft cheeses, but their relative abundances were different between the three cheeses (Galaup et al., 2007). Recently, pigmentation by *Arthrobacter arilaitensis* was shown to be probably associated with the exponential growth phase of this species (Sutthiwong & Dufosse, 2014).

10.3 Esrom PGI – Denmark

Name: Esrom PGI	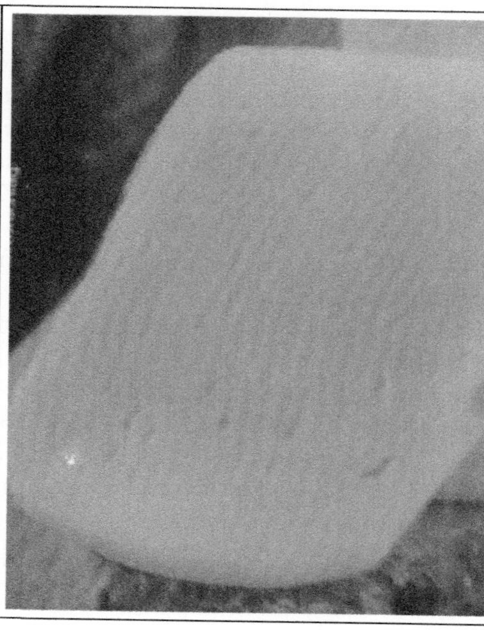
Production area: Denmark	
Milk: Cow's, pasteurised	

10.3.1 Introduction

Esrom was granted protected the geographical indication, PGI, by the EU (EC, 1996). The cheese was developed by the National Experimental Dairy, Hillerød, Denmark, in the 1930s, and because it resembled a cheese that had been traditionally made by Cistercian monks, it was named 'Esrom' after their head monastery in Denmark. Esrom is made from pasteurised cow's milk, and the main part of the produced cheeses is exported. A very strong flavour from the surface microflora, which was not always appreciated outside Denmark, has been considerably decreased over the years.

10.3.2 Type

Esrom may be made with varying fat content, which are 60%, 55%, 45% and 30% FDM, with corresponding moisture content of 43%, 46%, 50% and 54%, respectively (Danish Food Administration, 2004).

10.3.3 Description and Sensory Characteristics

The body of Esrom has a uniform yellowish to white colour. The cheese has a thin, supple, yellow to yellowy-orange edible rind with a clean, almost dry, thin and uniform yellowish brown to reddish brown outer skin. Older cheeses have a slightly greasy surface due to the breakdown of the outer skin. The cheese is made in a rectangular shape with a length that should be about twice the width. The height may vary from 3.5 to 7.0 cm, and the weight may be 0.2–0.5 kg or 1.3–2.0 kg. The body is soft and easily cut, and it has evenly distributed irregular holes about the size of rice grains. The flavour is mild, acidic and aromatic with clear notes of surface ripening that become more and more dominant as the cheese ages.

10.3.4 Method of Manufacture

Milk preparation: Raw cow's milk is standardised with regard to fat content and pasteurised (72°C/15 s). To facilitate coagulation and depending on the season, additions are made of up to 20 g of calcium chloride. To prevent growth of clostridia, 20 g potassium nitrate per 100 L of milk may be added before treatment with the starter and rennet.

Starter culture/rennet: Mesophilic DL-starter with undefined mixed strains of *Lc. lactis* ssp. *lactis* & *cremoris*, *Lc. lactis* ssp *lactis* biovar. *diacetylactis* and *Leuc. mesenteroides* ssp cremoris is used. Commercial bovine rennet is used for coagulation.

Cutting/curd washing: The curd is cut with a knife into grains with a diameter of 10 mm. About 20%–25% whey is removed, and 10%–20% water is added during stirring at 38°C–41°C for about 60 min.

Moulding/pressing: The curd is separated from the whey, poured into moulds and pressed lightly for 30–60 min.

Salting: The cheeses are chilled for 10–20 hr before they are placed in brine (12°C, 20–22°Bé) for 3–6 hr.

Maturation: The surfaces are treated with a smear culture, and the cheeses are stored for 1–2 weeks at high a RH (90%–95%) and 16°C–18°C to stimulate smear development on the surfaces. The cheeses are ripened for another 2–3 weeks at 13°C–14°C and 85% RH and finally at 8°C–12°C until it is sold. The cheeses are washed, dried and packed. They may be sold after only three weeks of maturation.

10.4 Hohenheim Trappisten – Germany

Name: Hohenheim Trappisten	
Production area: Hohenheim (Southern Germany)	
Milk: Cow's, pasteurised	

10.4.1 Introduction

Hohenheim Trappisten cheese is produced only in the Dairy for Research and Training at the University of Hohenheim in Stuttgart, Germany. It cannot be obtained on the open market, but must be ordered directly from the Dairy for Research and Training (Mair-Waldburg 1974). Hohenheim Trappisten cheese has been produced since the nineteenth century in the Dairy for Research and Training at the University of Hohenheim in Stuttgart. The university was constituted in 1818. Hohenheim Trappisten cheese does not bear a label or certification

10.4.2 Type

Hohenheim Trappisten cheese belongs to the group of semi-hard cheeses. The average composition of Trappisten cheese is as follows: FDM 45.8%, DM 57.1% and salt content 2.3%.

10.4.3 Description and Sensory Characteristics

The cheeses have the shape of a wheel with a diameter of 17–19 cm, a height of 6–7 cm and a weight of 1.7–1.8 kg. The cheese surface is treated with a red smear (Albrecht-Seidel & Mertz, 2006; Mair-Waldburg, 1974). The cheese body has a bright yellow colour. The texture is smooth and not short or crumbly. Evenly distributed slit holes are typical for Hohenheim Trappisten cheese. The flavour is mild and slightly sour. It becomes more intense with increasing maturation time.

10.4.4 Method of Manufacture

The manufacturing of Hohenheim Trappisten cheese is carried out according to a standardised protocol of the Dairy for Research and Training at the University of Hohenheim.

Milk preparation: Raw cow's milk is obtained from the Research Station Meiereihof (University of Hohenheim, Stuttgart, Germany). The milk is stored at 8°C–10°C and processed after a maximum of 12 hr. Furthermore, goat's milk may be used for cheese production. The milk is pasteurised (63°C/30 min) and standardised to a fat content of 3%. For pre-ripening, 0.1% of mesophilic culture is added at 10°C–12°C for 16 hr.

Starter culture/rennet: A mixture of starter cultures is added at 31°C which contains *Lc. lactis* ssp. *lactis, Lc. lactis* ssp. *cremoris, Lc. lactis* ssp. *lactis* biovar. *diacetylactis* and *Leuc. mesenteroides* ssp. *cremoris*. A total of 0.8%–1% of culture is added, and fermentation lasts approximately

30 min until a pH of 6.55 is reached. Calf's rennet is added (22 mL/100 mL cheesemilk) at 31°C, and coagulation lasts 50 min.

Cutting: The coagulated cheesemilk is cut with a cheese harp and an electronically operated stirrer until the curd particles exhibit a size of 5 mm.

Curd washing/cooking: The curd is washed by addition of 10%–15% water (30°C–35°C) and removal of 30% whey. Scalding takes place at a temperature of 39°C for 20 min (Albrecht-Seidel & Mertz, 2006; Hohenheim, 2015; Mair-Waldburg 1974).

Moulding: Further whey is removed (<30%), and the curd is transferred with a hose directly into the moulds. The cheese pH is 6.40, and they are stored for 2 hr at 20°C–24°C for further whey removal. Turning is performed regularly.

Salting: The cheeses are brined at 21°Bé and a temperature of 12°C–14°C for 30 hr. The brine has a pH value of 5.1–5.2.

Maturation: The cheeses are stored in a ripening room at 13°C–15°C und an RH of 85%–90% for three weeks. During ripening, the cheeses are treated with a red smear (10% NaCl and *B. linens*) and turned every other day.

Storage: The cheeses are coated with paraffin wax and stored at 6°C for another five weeks before they are ready for consumption. They are turned once a week.

10.5 Maroilles PDO – France

<table>
<tr><td>Name: Maroilles or Marolles, PDO
Three sub-denominations: Sorbais, Mignon and Quart</td><td rowspan="3">
http://sciencejunior.fr/wp-content/uploads/maroilles.jpg</td></tr>
<tr><td>Production area: Thiérache region (200–300 m)</td></tr>
<tr><td>Milk: Cow's, raw, thermised or pasteurised</td></tr>
</table>

10.5.1 Introduction

Maroilles PDO is one out of the 60 soft and washed-rind French cheeses which include 7 PDO cheeses (see Epoisses PDO earlier in this chapter). In 2014, it was ranked fifth among soft French PDO cheeses with respect to the volume produced, accounting for 11.4% of the total volume. Maroilles was recognised as AOC in 1976 and PDO in 1996 (EC, 1996). Maroilles is made and matured exclusively in a delimited area located in north France; this area of 200,000 m^2 is part of a lowland natural region sitting astride France (the Thiérache region) and Belgium (part of the Hainaut and Namur regions), characterised by its network of pasture and hedges. Other non-PDO cheeses related to Maroilles (e.g. Boulette d'Avesne) are manufactured in this area. Actually, Maroilles PDO is manufactured at 7 farms, 4 dairy plants, and 3 maturation plants while Maroilles's milk is produced at 455 farms (Official website for Maroilles, 2016).

Maroilles PDO is a soft cheese, rennet-acid coagulated and with a smear rind. It is manufactured exclusively in the delimitated area from cow's milk throughout the year. It is made in farms and from raw milk in very small amounts, that is, 6.7% of the production in 2014 (INAO-CNAOL, 2014), whereas most of the production takes place in dairy plants from thermised or pasteurised milk. It is then matured in one of the maturation plants, exclusively located in the delimited area, for 21–35 days (usually six weeks), depending on the cheese size. It retains its quality for 130 days for the two larger sizes, and 100 days for the smaller size (Dareville et al., 2003). After the minimum time of maturation for the cheeses are sampled for scoring their flavour, texture, odour and appearance by a trained panel on a 0–20 scale. A cheese that scores below 6 cannot carry the PDO label; if the score is less than 12, the manufacturer is informed (after three such notices, the PDO label is withdrawn for cheeses from this manufacturer).

Actual Maroilles cheeses are probably derived from ancient cheeses that were manufactured around Saint-Humbert de Maroilles, an abbey founded in the seventh century (Dareville et al., 2003; Official website Maroilles, 2016). These latter cheeses are cited as a specific product named 'Marolles' in a document dating back to 1723. Although its production area was very small, it was eaten by several French kings. The actual square form is derived from the form of moulds formerly used, because the latter could be made only from hard wood (associated with the local tree varieties). For centuries, Maroilles was only manufactured in north Thiérache, until cow breeding increased in Thiérache, two centuries ago, resulting in landscapes made up of pastures surrounded by hedges that determined the composition of the actual feeding ration. Until the First World War, at which time it disappeared, a local breed ('la Maroillaise') well adapted to the local conditions produced fat-rich milk adapted to Maroilles manufacture. Until 1970, Maroilles was manufactured only during the seasonal calving period. Since the nineteenth century, from a domestic product, the Maroilles cheese has progressively became a means of adding economic value to milk, taking advantage of favourable local conditions. Formerly, Maroilles was exclusively manufactured at farms. Every farm was equipped with a humid, aerated and tempered cellar, oriented southwest and generally made of bricks, which favoured the rind's red pigmentation. Since the end of the nineteenth century, cellars have been located in specific buildings, and maturation is supervised by specific persons. In the second part of the twentieth century, weakened by a trade war in favour of industrial production and by rigorous standards, cheesemaking at plants increased at the expense of cheesemaking at farms. Although the Maroilles volume manufactured in Thiérache was high after the Second World War, it then dropped to around 2,000 tonnes per year during period 1980–2007, probably because of dairy quotas, listeriosis cases and the retirement of aging cheesemakers. But this volume has consistently increased (twofold) during 2009–2014, since, in 2008, the film '*Bienvenue chez les Ch'tis*' enjoyed a huge success. The film showed the soaking of a piece of Maroilles in coffee cups, as is usual for the people of north France.

10.5.2 Description and Sensory Characteristics

Maroilles is a red-smear soft cheese obtained after milk is subjected to mixed coagulation (rennet and acid). According to the specifications, the cheese should have a minimum of 45% FDM and 50% DM. The mean (Dareville et al., 2003) DM is 53.8%; fat content 28.5%, including 18.1%

of saturated fat; salt content 1.05%; protein content 20.4% and carbohydrate content 0% (present only in traces).

Maroilles's rind is thin, regular, uniform, humid and orange; it shows the counter-relief fingerprint of the maturation grid (parallel lines). Maroilles's body is white to cream-coloured. Maroilles's texture is soft and smooth, and exhibits small mechanical and fermentative holes. The core texture is firmer and softens as maturation progresses. Maroilles has a slight odour of ammonia together with strong odours of mushroom, humid brick and cellar. Maroilles's flavour is lactic (acidified milk) at the end of its minimal period of maturation, and stronger and more persistent (sulphur aroma) when the maturation is longer.

10.5.3 Method of Manufacture

Maroilles is manufactured according to rules that apply from milk production to cheese maturation. All steps should be performed in the delimitated area.

Milk production: No specific cow variety is mandatory; most milk is actually produced by Holstein cows (Dareville et al., 2003). The minimal surface of the pasture should be 30 m^2/dairy cow. Every hectare of pasture has to include 90 linear metres of edges mainly made of lobed-leaved trees. Dairy cows should be fed fodder and concentrated feed specified in a list; grass in summer (for a minimum of 170 days) and by-products of local farming (i.e. beets and potatoes) in winter are the main feed. Four fifths of the fodder has to be produced within the delimited geographical area.

Starter culture and rennet: Milk is stored at 4°C and after heating at a temperature of 32°C–38°C is inoculated with mesophilic lactic bacteria at the beginning of its storage (Dareville et al., 2003). The time between two successive cheeemaking processes should not exceed 72 hr. In addition, adjunct cultures, with specific surface microorganisms, may be added. Commercial calf rennet (18–30 mL/100 L, with chymosine 520 mg/L) is added at the same temperature and at a pH of 6.35–6.65.

Cutting and moulding: The curd is cut and stirred and transferred into moulds.

Drainage: Drainage takes place at 16°C with more than three turnings.

Salting: The curd is either brined or dry-salted just after the mould is removed and left for two to three days (pre-maturation).

Maturation: The cheeses are matured in cellars with a temperature 9°C–16°C and an RH of more than 90%. Maturation cellars have to be built with unrecovered bricks, unrecovered bricks and stones or any inert material. The porous structure of bricks, the traditional material, allows maintaining a constant hygrometry. Dry brushing or washing with brine with or without maturation microbial strains is performed to prevent the mature Maroilles from developing a blue-coloured rind due to the activity of *P. camemberti* (syn. *album*) (Dareville et al., 2003).This colour develops after 12–15 days of maturation, indicating that lactose and lactic acid have been completely consumed. Mature Maroilles is wrapped in greaseproof paper that regulates the cheese humidity.

The cheeses are sold after maturation for at least 21 days for small cheeses and 35 days for large ones.

10.5.4 Relevant Research

Maroilles is closely related to Munster and Livarot by its cheesemaking parameters (Gauzere, 2013), but only to Munster also by the main yeast species present on its surface, namely, *Debaromyces hansenii* and *Geotrichum candidum*, at least for the pasteurised cheeses (Mounier et al., 2010).

10.6 Reblochon PDO – France

Name: Reblochon or Reblochon de Savoie PDO	
Production area: Mountains of Haute-Savoie and Val d'Arly in Savoie, two adjacent regions of the French Alps mountains (500–2500 m)	
Milk: Cow's, raw	http://www.reblochon-thones.com/wa_39_p/pa_40egpkhorxm5mi/th_IMG_3337_R.jpg?1lz2k81yh5agq2

10.6.1 Introduction

Reblochon PDO is ranked third among PDO French cheeses in respect of the volume marketed, accounting for 6.5% of the total French PDO cheese in 2014 (INAO-CNAOL, 2014). It is one out of the 187 pressed and uncooked French cheeses which include nine PDO cheeses. Reblochon is made and matured exclusively in a delimited area of 400,000 m^2 located north of the French Alps (high mountains located in Eastern France, near Switzerland and Italy). Reblochon is manufactured in 130 farms, 22 dairy plants and 11 maturation plants, while Reblochon's milk is produced in around 520 farms including 35,000 dairy cows in the 2000s (Official website for Reblochon, 2016). Based on an artisanal know-how developed over centuries in Haute-Savoie, Reblochon cheese is an uncooked and pressed cheese, rennet-coagulated, with a washed rind. It is made in the delimited area from full-fat raw cow's milk in farms in small quantities, that is, 13.4% of the production in 2014 (INAO-CNAOL, 2014), and most of the cheese is produced by dairy plants.

Reblochon cheese is consumed almost exclusively in France, with eastern central France accounting for one third of the consumption. Most of the consumption is in winter and less in spring, despite a peak in milk production during this latter period (Chesnay et al., 2011). Only a very small part of the cheese (about 3%) is exported, principally to three European countries (Germany, Switzerland and Belgium), because Reblochon's shelf life is very short.

Reblochon cheeses are derived from ancient cheeses that have probably been manufactured in the Alps since the thirteenth century, but their origin is still a matter of debate (Montigaud, 2016). According to the official version, Reblochon was first produced in the valley of Thônes, in the Haute-Savoie department. At that time, farmers had to pay an annual tax according to the quantity of milk they produced per day. In order to pay a lower tax, farmers did not complete milking until the quantity control was done. They then did a second milking, obtaining small amounts of a high-fat milk that they used for manufacturing small cheeses for their own consumption. The cheese was named according to this cheating using local word 'Re-blocher', which means to pinch the cow's udder once again. Reblochon has been probably manufactured legally since the seventeenth century. At the beginning of the twentieth century, only 40 tonnes or less of Reblochon was produced and consumed locally. Then, the development of tourism, the railway network and winter sports extended the distribution area to France, leading to

15,000 tonnes being marketed per year during 2011–2014. As a result of the increase in demand, the area of Reblochon's production extended rapidly beyond its original area, leading to the definitions of specific rules for production as early as 1953. Reblochon was registered as AOC since 1958 and as PDO in 1996 (EC, 1996). The total annual production of Reblochon cheese has been stable since the 2000s, after a steady increase between 1980 and 2000 (1.7-fold in 20 years) (PSDRRA, 2006).

10.6.2 Description and Sensory Characteristics

Reblochon is a red-smear, soft cheese. According to its specifications, the minimum value of both the FDM and DM should be 45%. The Official website for Reblochon (2016) states that the fat content is 27.4%, including 18.8% of saturated fat; salt content 1.3%; protein content 19.9% and carbohydrate content 0.4% including only traces of lactose.

Reblochon's rind is thin, regular, uniform, yellow to orange and entirely or partly covered with white foam. Reblochon's body is cream-coloured to light yellow (ivory). Reblochon's texture is soft, smooth and pliable and may exhibit a few small holes. Reblochon's flavour is creamy, characterised by a subtle nutty hazelnut aroma and is different depending on whether Reblochon is 'at farm' and 'at plant', being generally more pronounced and persistent for the former.

10.6.3 Method of Manufacture

Reblochon is manufactured according to rules that apply from milk production to cheese packaging. All the steps until the first packaging (included) should be performed in the delimited area. Some specifications are different, depending on whether Reblochon is 'at farm' or 'at plant'. Among them, for 'at farm', curd cutting, moulding with vegetal cloth and cheese turning one after the other have to be done manually.

Milk production: Milk has to be produced by Montbéliarde, Abondance or Tarentaise cows. The minimum surface area of the pasture should be 1.5 ha/dairy cow, and milk production should not exceed 500,000 L/year for a farm manufacturing Reblochon 'at farm'. Dairy cows are fed fodder, completed with non-GMO cereals in limited amounts; grass in summer (a minimum of 150 days) and hay in winter are the main feeds. Feeding of cows with any fermented foods is banned. The mineral fertilisation for grazing surfaces is limited. Fodder has to be produced within the delimited geographical area, except for 25% of the hay. Milking twice a day, in the morning and in the evening, with an interval of 8 hr or more in-between, is mandatory.

Starter culture and rennet: Milk is stored at 4°C and after heating at a temperature of 30°C–35°C, is inoculated with mesophilic starter culture and left for 30 min. Great importance is attached to the starter cultures inoculated. Only selected strains, specific for Reblochon, non-GMO and allowed by the Reblochon sector (listed) can be used. They are propagated in serum or milk, whether reconstituted or not. No more than 2% and 0.5% of the starter culture should be inoculated in vat milk 'at plant' and 'at farm', respectively. Then, commercial calf rennet (18–30 mL/100 L, with chymosine 520 mg/L) is added at the same temperature and is left for 10–20 min.

Cutting/moulding: The curd is cut at a size of a wheat to maize grain and is transferred into cylindrical moulds with 6–8 cm height and 14 cm diameter.

Pressing: The curd is pressed with a weight of 1.5–2 kg for approximately 1.5 hr. Each curd is pressed individually with a weight proportional to the curd weight. The curd should be turned once or twice during pressing. In addition, for 'at farm' cheeses, the curd should be turned at least once during the first 30 min after moulding, and should be pressed for a minimum of 6 hr (≥1.5 hr for 'at plant'). The resulting serum less than 24 hr old can be used as a drink for the dairy cows.

Salting: The curd is either brined or dry-salted just after the mould is removed and has a pH of 5.2.

Maturation: During pre-maturation, moulds are authorised only for 'at plant'. Pre-maturation for 'at farm' should last at least five days, including daily manual turning; cheeses should remain at the farm until at least the sixth day and have to be washed before leaving it. Then, the beginning of maturation for 'at farm' should include at least one damp treatment. It is then matured in one of the maturation cellars exclusively located in the delimited area, for a minimum of 15 days, but often more (three weeks or more). After the minimum time of maturation, the cheeses are sampled for assessing their compliance to their PDO specifications by scoring some of their sensory attributes and determining their gross chemical composition. Before leaving the delimited area, the cheeses should be individually packaged, except when they will be further used as a food ingredient. Each packaging should include a false wood bottom (120 mm and 70 mm for large and for small cheeses, respectively) which regulates humidity, and should preserve the cheese appearance. Cheese should be packaged until the sale to consumers.

10.6.4 Relevant Research

Since the 2000s, research has focussed on the surface microbiota, since Reblochon is a cheese whose maturation is mainly determined by the activity of this microbiota. Reblochon microbiota contains thermophilic LAB such as *S. thermophilus* and *Lb. delbrueckii* spp *bulgaricus* which are responsible for the curd acidification, but are still active during maturation, and present both in the core and surface, yeasts such as *G. candidum* and *D. hansenii* and aerobic bacteria such as *Brevibacterium* spp. active on the cheese surface during cheese maturation (Cogan et al., 2014; Monnet et al., 2016). A new species of gram-positive bacteria assigned to the family *Microbacteriaceae* (Actinobacteria class), namely, *Mycetocola reblochoni*, was first described from strains isolated from the Reblochon surface (Bora et al., 2008).

Geotrichum spp., responsible for the fuzzy appearance at the cheese surface, was especially investigated (Castellote et al., 2015; Marcellino et al., 2001). The red colour of the rind is probably linked to two families of pigments found also on three other red-smear soft cheeses, the first produced by the *B. linens* group and the second produced by different microorganisms, including *Micrococcus* spp., unidentified cocci and coryneform bacteria. Another pigment was only produced by Reblochon's bacteria (Galaup et al., 2005). One work described microbiota present on the wooden shelves used to achieve Reblochon maturation (Mariani et al., 2007); the biofilm was composed of the dominant populations present on the surface of the cheeses. It shows the stability of the biofilm present on the shelves depends on the season, geographical location and the age of the shelves.

10.7 Vacherin Mont-d'Or PDO – Switzerland

Name: Vacherin Mont-d'Or PDO	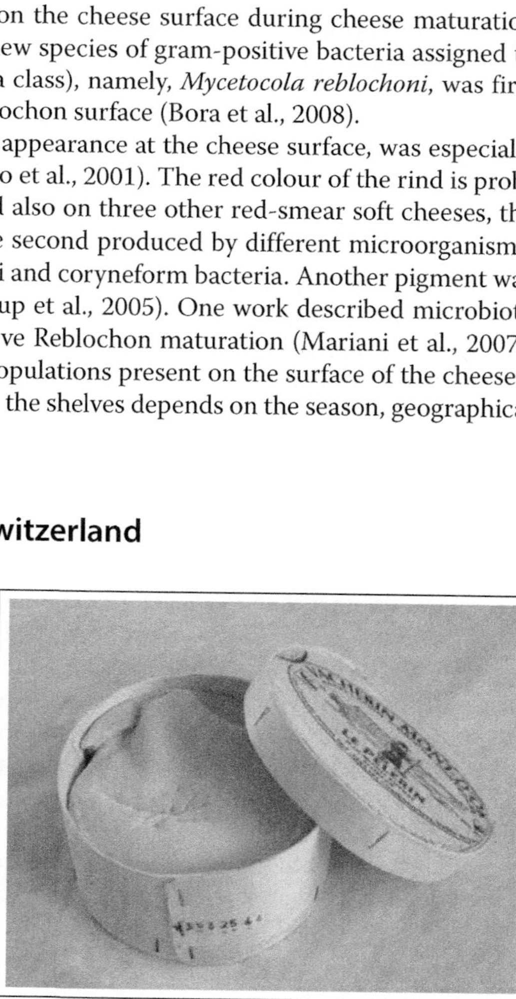
Production area: Several districts of the canton of Vaud; from Lake Geneva up to the Jura Mountains on the French border	
Milk: Cow's, thermised	

10.7.1 Introduction

Vacherin Mont-d'Or PDO is a soft, smear-ripened cheese made from thermised cow's milk. The cheese was named after a French mountain located on the Franco-Swiss border. A very similar cheese called Mont-d'Or, or Vacherin du Haut-Doubs PDO, is produced in the French region of Doubs bordering on the Swiss region of the cheese's production; however, the French version is made from raw milk. In Switzerland, the annual production of Vacherin Mont-d'Or PDO amounts to nearly 600 tonnes (TSM Treuhand GmbH, 2015). Vacherin Mont-d'Or was registered with PDO status in 2003 (FOAG, 2015).

According to legend, in 1870, during the Franco-Prussian War, a French soldier brought the secret of the production of Vacherin Mont-d'Or to Switzerland and disclosed it to his entourage after he had settled in Les Charbonnières, a small village in the Vallée de Joux near the French border. However, a law on road taxes, which dates from 1812, had already mentioned 'Vacherin' under the heading 'Cheese'. There is no question that Vacherin Mont-d'Or originates from France (Kulinarisches Erbe der Schweiz, 2008). Traditionally, the manufacture of Vacherin Mont-d'Or started late in summer when the milk yield declined and no longer allowed the production of larger cheeses like Le Gruyère. Even today, the period for manufacturing Vacherin Mont-d'Or is restricted. Production starts at the earliest on 15 August and ends on 31 March.

10.7.2 Type

Vacherin Mont-d'Or is a smear-ripened, full-fat soft cheese. According to the specifications, the cheese has a water content of 50%–57.5% (MNFS ≥ 65.0%) and an FDM of 49%–54.9%. The salt content has to be in the range of 1%–2%. The average composition of Vacherin Mont-d'Or PDO after a ripening period of three weeks is as follows: moisture content 54.8%, fat content 24.6%, salt content 1.7%, lactate 0.29 g/100 g and calcium 0.43 g/100 g (Goy & Jakob, 2009).

10.7.3 Description and Sensory Characteristics

Vacherin Mont-d'Or has a round shape and a soft rind with an amber to reddish brown colour. The cheeses are enclosed in hoops of spruce bark that come from the region of production and packaged in a wooden box. Vacherin Mont-d'Or is produced in three sizes. The cheese body is soft, smooth, and slightly runny and has an ivory colour. The taste is slightly salty and slightly sour. It is distinguished by an aroma of wood and spruce resin, which results from the ripening in hoops of spruce bark and on spruce shelves.

10.7.4 Method of Manufacture

The manufacture of Vacherin Mont-d'Or is carried out in small-scale dairies. The use of processing aids other than rennet, starter cultures and salt is forbidden.

Milk preparation: Dairy farms must be located within a maximum radius of 25 km from the cheese dairy. The cows whose milk is used in the production of Vacherin Mont-d'Or must not be fed with silage. At least 70% of the feed ratio for the dairy cattle (calculated on a dry matter basis) must come from the fodder area of the farm. At the farm, the milk must be filtered and cooled. If the milk is delivered less than twice a day, the milk temperature may not exceed 13°C. When the freshly collected evening milk is delivered to the cheese dairy, it must be kept until the next morning at a temperature of 4°C–15°C. Milk has to be processed no later than 20 hr after milking. Part of the milk may be centrifuged in order to standardise the fat content. The cheese milk is heat treated at 62°C for at least 15 s. However, the test for alkaline phosphates must still be positive after thermisation. After cooling, an amount of 5%–15% water (on the basis of the amount of milk) may be added.

Starter culture: The cultures are defined by the interprofessional organisation of Vacherin Mont-d'Or. Most often, yogurt cultures (*S. thermophilus* and *Lb. bulgaricus*) are used. However, mixed stock cultures containing *S. thermophilus, Lb. delbrueckii* ssp. *lactis* and *Lc. lactis* are used as well. The cheese factories prepare the bulk starter by inoculating skim milk every day. The bulk starter is added to the vat milk about 30 min before the addition of rennet.

Rennet: All kinds of 'natural' rennet are allowed; however, calf rennet is most commonly used. Coagulation is carried out at 32°C–38°C.

Cutting: The coagulum is allowed to become relatively firm. After cutting, the average size of the curd grains should be about 15 mm.

Curd washing: Immediately after cutting, part of the whey is removed (10%–25%) and replaced by the same volume of warm water in order to increase the temperature to ≤38°C and to ensure rapid acidification by the thermophilic starter culture. Moreover, lactose is diluted by addition of water in order to prevent an excessive drop in pH value.

Moulding: After the curd grains have attained the desired firmness, the curd is usually filled into tubular moulds with a diameter of 10–15 cm and a height of up to 60 cm. The cheese mass is allowed to set and the whey to drain off for 20–30 min while the moulds are turned at least once. If high tubular moulds are used, the pressed curd is cut into wheels that are 4–5 cm thick. Each wheel is then bound with a 3–4 cm wide strip of spruce bark that is fixed using a rubber loop or a wooden pin.

Cooking: Prior to use, the spruce bark strips are cooked in hot water (90°C–95°C) for at least 30 min.

Pressing: After strapping, multiple cheese wheels are covered with a wooden board and slightly pressed again for 1–2 hr in a warm room (25°C–35°C). The cheeses are then placed on a porous mat in order to drain off the residual whey until a pH value of about 5.4 is achieved (usually after another 1–2 hr).

Salting: Salting is done by immersing the cheeses in a brine bath with a concentration of 17%–19% NaCl (16°Bé –18°Bé) for 2–4 hr at a temperature of 14°C–20°C.

Maturation: The cheese is transferred onto red spruce shelves no later than 48 hr after the beginning of production. The ripening temperature may vary from 6°C to 16°C, and the RH is at least 85%. A surface culture containing *D. hansenii, B. linens* and *Arthrobacter* spp. is usually used in order to improve the quality of the product. The water used for washing can be slightly salty (≤5% NaCl). Initially, the cheeses are washed daily and turned over, but once sufficient smear is present on the rind, the cheeses are only washed every second day. During the final stage of ripening, the curing intervals are adapted to the dryness of the cheese rind, but curing is done at least once every three days. The cheeses are ripened for 17–25 days depending on the temperature of the cellar.

After ripening, the cheeses are packaged. For that purpose, the rubber band or the wooden pin that fixed the hoop of spruce bark during ripening is removed. The hoop is cut open, and the cheese is packaged into a wooden box made of spruce. The diameter of the box must be slightly smaller than that of the cheese in order to obtain the typical corrugation on the cheese surface. Until consumption, Vacherin Mont-d'Or is stored in cold rooms at a temperature between 0°C and 5°C.

10.7.5 Relevant Research

Bosset et al. (1997) investigated the volatile compounds (VCs) in Swiss Vacherin Mont-d'Or cheese and compared the obtained profile with that of a Swiss semi-hard, smear-ripened cheese called Vacherin Fribourgeois. A total of 66 components were identified. Vacherin Mont-d'Or cheeses contained some terpenoids, probably due to the direct contact of the cheese with the bark and the wood of spruce. The determination of four compounds (ethanol, α-pinene, ethyl caproate and α-terpineol) was sufficient to distinguish between the two cheese varieties. With the exception of the terpenoids, most VCs identified were present in both cheese varieties; however, the concentrations often differed.

References

Albrecht-Seidel, M. & Mertz, L. (2006. *Die Hofkäserei: Planung, Einrichtung, Produktion.* Grundrezepte, Ulmer Eugen Verlag, Stuttgart, Germany.

Antonsson, M., Molin, G. & Ardö, Y. (2003). *Lactobacillus* strains isolated from Danbo cheese and their function as adjunct cultures in a cheese model system. *International Journal of Food Microbiology*, 85, 159–169.

Ardö, Y. (2004). Semi-hard Scandinavian cheese made with mesophilic DL-starter. In *Handbook of Fermented Food and Beverages.* Marcel Dekker Inc., New York, pp. 277–290.

Bora, N., Vancanneyt, M., Gelsomino, R., Snauwaert, C., Swings, J., Jones, A. L., Ward, A. C., Chamba, J. F., Kroppenstedt, R. M., Schumann, P. & Goodfellow, M. (2008). Mycetocola reblochoni sp nov., isolated from the surface microbial flora of Reblochon cheese. *International Journal of Systematic and Evolutionary Microbiology*, 58, 2687–2693.

Bosset, J. O., Bütikofer, U., Berger, T. & Gauch, R. (1997). Study of the volatile compounds of the Vacherin Fribourgeois and (Swiss) Vacherin Mont-d'Or cheese varieties. *Mitteilungen aus dem Gebiete der Lebensmitteluntersuchung und Hygiene*, 88 (3), 233–258.

Castellote, J., Fraud, S., Irlinger, F., Swennen, D., Fer, F., Bonnarme, P. & Monnet, C. (2015). Investigation of Geotrichum candidum gene expression during the ripening of Reblochon-type cheese by reverse transcription-quantitative PCR. *International Journal of Food Microbiology*, 194, 54–61.

Chesnay, C., Laval, J. & Herbaux, C. (2011). *Etude filière laitière sous signe de qualité [Online]. Grand Genève, agglomération franco-valdo-genevoise, Cahier n°13-27a.* Available: www.arcdugenevois.fr/sites/.../cahiers.../cahier-13-27a_etude-filiere-lait_nov2011.pdf [Accessed 20 August 2016].

Christiansen, P., Petersen, M. H., Kask, S., Møller, P. L., Petersen, M., Vogensen, F. K., Nielsen, E. W. & Ardö, Y. (2005). Anticlostridial activity of *Lactobacillus* isolated from semi-hard cheeses. *International Dairy Journal*, 15, 901–909.

Codex Alimentarius (1978). *General Codex Standard for Cheese, CODEX STAN 283-1978.* Food and Agriculture Organization (FAO), Rome, Italy.

Cogan, T., Goerges, S., Gelsomino, R., Larpin, S., Hohenegger, M., Bora, N., Jamet, E., Rea, M., Mounier, J.V., Gueguen, M., Desmasures, N., Swings, J., Goodfellow, M., Ward, A., Sebastiani, H., Irlinger, F., Chamba, J., Beduhn, R. & Scherer, S. (2014). Biodiversity of the surface microbial consortia from Limburger, Reblochon, Livarot, Tilsit, and Gubbeen cheeses. In Donnelly, C. W. (ed.), *Cheese and Microbes.* Washington, DC: ASM Press.

Danish Food Administration (2004). *Bekendtgørelse om mælkeprodukter m.v.* [Announcement about dairy products etc.]. Notice No 335 of 10 May 2004.

Danish Food Administration (2013). *Bekendtgørelse om mælkeprodukter m.v.* [Announcement about dairy products etc.]. Notice No 2 of 04/01/2013.

Dareville, S., Perez, A., Dubois, M., Petitniot, V., Lesage, E. & Zakka, C. (2003). *Le Maroilles le plus fin des fromages forts* [Online]. Available: http://pfeda.univlille1.fr/iaal/docs/iaal2003/maro/maroilles.pdf [Accessed 31 August 2016].

Delfosse, C. & Letablier, M. (1999). *Comment renaissent les fromages ?' L'époisses, le rocroi, le soumaintrain. Carrières d'objets.* Editions de la maison des sciences de l'homme ed.

EC (1996). Commission Regulation (EC) No 1107/96 of 12 June 1996 on the registration of geographical indications and designations of origin under the procedure laid down in Article 17 of Council Regulation (EEC) No 2081/92. *Official Journal of the European Union*, 148, 1–10.

FOAG (2015). *Register of appellation of origin and geographical indications.* Federal Office for Agriculture. An Office of the Federal Department of Economy, Bern, Switzerland http://www.blw.admin.ch/ [Accessed 28 April 2015].

Galaup, P., Flamin, C., Carlet, E. & Dufosse, L. (2005). HPLC analysis of the pigments produced by the microfora isolated from the 'Protected Designation of Origin' French red-smear soft cheeses Munster, Epoisses, Reblochon and Livarot. *Food Research International*, 38, 855–860.

Galaup, P., Gautier, A., Piriou, Y., De Villeblanche, A., Valla, A. & Dufosse, L. (2007). First pigment fingerprints from the rind of French PDO red-smear ripened soft cheeses Epoisses, Mont d'Or and Maroilles. *Innovative Food Science & Emerging Technologies*, 8, 373–378.

Gauzere, Y. (2013). *Diaporama – L'affinage des pâtes molles [Online].* Available: http://idele.fr/rss/publication/idelesolr/recommends/perfectionnement-des-techniciens-produits-laitiers-fermiers-2013.html [Accessed 20 August 2015].

Goy D. and Jakob E. (2009). The composition of Vacherin Mont d'Or. Unpublished results. Agroscope Hohenheimer Protocol for manufacture of Hohenheim Trappisten cheese (2015). *Dairy for Research and Training.* Hohenheim, Germany.

INAO-CNAOL (2014). *Produits laitiers AOP – les chiffres clés, 2014.* [Online]. Available: https://www.google.fr/search?q=produits+laitiers+AOP++2014&ie=utf-8&oe=utf-8&client=firefox-b&gfe_rd=cr&ei=iI3jV6OgJ8nDaKWRn8AO [Accessed 16 July 2016].

Kulinarisches Erbe der Schweiz (2008). Lausanne, Switzerland. http://www.kulinarischeserbe.ch/ [Accessed 28 April 2015].

Madsen, J. S. & Ardö, Y. (2001). Exploratory study of proteolysis, rheology and sensory properties of Danbo cheese with different fat contents. *International Dairy Journal*, 11, 423–431.

Mair-Waldburg (1974). *Handbuch der Käse: Käse der Welt von A-Z: eine Enzyklopädie.* Volkswirtschaftlicher Verlag, Kempten, Germany.

Marcellino, N., Beuvier, E., Grappin, R., Gueguen, M. & Benson, D. R. (2001). Diversity of Geotrichum candidum strains isolated from traditional cheesemaking fabrications in France. *Applied and Environmental Microbiology*, 67, 4752–4759.

Mariani, C., Briandet, R., Chamba, J. F., Notz, E., Carnet-Pantiez, A., Eyoug, R. N. & Oulahal, N. (2007). Biofilm ecology of wooden shelves used in ripening the French raw milk smear cheese Reblochon de Savoie. *Journal of Dairy Science*, 90, 1653–1661

Monnet, C., Dugat-Bony, E., Swennen, D., Beckerich, J. M., Irlinger, F., Fraud, S. & Bonnarme, P. (2016) [Online]. Investigation of the activity of the microorganisms in a Reblochon-style cheese by metatranscriptomic analysis. *Frontiers in Microbiology*, 7, 536. https://www.ncbi.nlm.nih.gov/pmc/articles/PMC4837152/[Accessed 20 August 2016].

Montigaud, S. (2016). Le reblochon du Reposoir, du mythe à l'AOC [Online]. *Journal of Alpine Research* | Revue de géographie alpine. Available: http://rga.revues.org/3175 [Accessed 20 August 2016].

Mounier, J., Le Blay, G., Vasseur, V., Le Floch, G., Jany, J. L. & Barbier, G. (2010). Application of denaturing high-performance liquid chromatography (DHPLC) for yeasts identification in red smear cheese surfaces. *Letters in Applied Microbiology*, 51, 18–23.

Official website for Epoisses PDO (2010). *Décret n° 2010-1092 du 15 septembre 2010 relatif à l'appellation d'origine contrôlée « Epoisses » [Online].* Available: http://www.fromage-epoisses.com [Accessed 05 August 2016].

Official website for Maroilles PDO (2016). [Online]. Available: http://www.maroilles-infos.com [Accessed 31 August 2016].

Official website for Reblochon PDO (2016). [Online]. Available: http://www.reblochon.fr/ [Accessed 20 August 2016].

PSDRRA (2006). *Efficacité économique et dynamique agricole des filières fromagères IGP/AOP des Alpes du Nord [Online]*. Available: www.psdrra.fr/content/download/3389/.../1/.../4_pages_efficacite_economique.pdf [Accessed 20 August 2016].

Risoud, G. (2000). *Histoire du fromage d'Époisses, Éditions de l'Armançon, Précy-sous-Thil.*

Ryssel, M., Johansen, P. Al-Soud, W. A., Sørensen, S., Arneborg, N. & Jespersen, L. (2015). Microbial diversity and dynamics throughout manufacturing and ripening of surface ripened semi-hard Danish Danbo cheeses investigated by culture-independent techniques. *International Journal of Food Microbiology*, 215, 124–130.

Sutthiwong, N. & Dufosse, L. (2014). Production of carotenoids by Arthrobacter arilaitensis strains isolated from smear-ripened cheeses. *FEMS Microbiology Letters*, 360, 174–181.

TSM Treuhand GmbH (2015). *Mehrjahresvergleich ab 2004, Beilage zur Jahresstatiskik 20014.* Bern, Switzerland. http://www.tsmtreuhand.ch/ [Accessed 28 April 2015].

Waagner Nielsen, E. (1993). Danish cheese varieties. In Fox, P. F. (ed.), *Cheese: Chemistry, Physics and Microbiology. Vol. 2. Major Cheese Groups*, 2nd edition. Chapman & Hall, London, pp. 247–250.

11

Blue-Veined Cheeses

Maria Belén López Morales[1], *Ylva Ardö*[2], *Françoise Berthier*[3], *Kimon-Andreas G. Karatzas*[4] *and Thomas Bintsis*[5]

[1] *Food Science and Technology Department, International Excellence Campus for Higher Education and Research 'Campus Mare Nostrum', Veterinary Faculty, University of Murcia, Spain*
[2] *Department of Food Science, University of Copenhagen, Denmark*
[3] *Unité de Recherches en Technologie et Analyses Laitières Rue de Versailles, France*
[4] *Department of Food and Nutrition Sciences, The University of Reading, United Kingdom*
[5] *11 Parmenionos, 50200, Ptolemaida, Greece*

11.1 Cabrales PDO – Spain

Name: Cabrales PDO	
Production area: Asturias, Spain	
Milk: Cow's, sheep's and goat's (mixture), raw	

11.1.1 Introduction

Blue cheese is made from pasteurised mixtures of cow's, sheep's and goat's milks. Under the PDO regulation, 41 farmers and 30 dairy plants are registered. During 2013, production reached 428,189 kg, which represented a total of 4,280,000 euros. The cheese is mainly sold on the national market and represents 2.03% of the economic value of Spanish PDO cheeses (Ministry of Agriculture, Food and Environment, 2014). The PDO status of Cabrales

Global Cheesemaking Technology: Cheese Quality and Characteristics, First Edition.
Edited by Photis Papademas and Thomas Bintsis.

cheese was recognised by the Ministry of Agriculture, Fisheries and Food in 1990 and at the European level in 1996 (EC, 1996). In its area of production, the main activity includes the raising of cattle, sheep and goats, which are cared for by their owners, who stay with them in the corrals during the summer. These circumstances, and the distance to consumer centres, led to the transformation of milk into cheese by the shepherds themselves. Traditionally, the cheese was made for preserving milk for consumption throughout the year as part of the family diet. Only recently has it begun to be manufactured for retail (Ministry of Agriculture, Food and Environment, 2015). The production area comprises the Cabrales Municipality and some towns from the Peñamellera Alta Municipalities, all located in Asturias.

11.1.2 Type

Blue-veined, semi-hard cheese made essentially from whole, raw cow's milk or from a mixture of two or three types, cow's, sheep's and goat's milk. Ripening lasts for a minimum of two months and takes place in natural caves. The minimum fat in dry matter (FDM) is 45% with a minimum moisture content of 30% and a pH of 4.8–5.8.

11.1.3 Milk

The milk used is exclusively obtained from cows, sheep and goats registered under the PDO regulation (cow: Frisona or Pardo Alpina; sheep: Latxa and goat: Pirenaica breeds). There are no regulations concerning the proportions of each milk. The Alpine pastures of this region have traditionally been used for livestock raising according to traditional practices, encouraging and promoting the direct use of pastures in the area of production. The Regulatory Board is empowered to pass additional rules so that the milk used for the manufacture of Cabrales cheese meets their specific characteristics.

11.1.4 Description and Sensory Characteristics

The cheese has a cylindrical shape with a flat top and a bottom which, in some cases, may be slightly concave, with a maximum diameter and height of 19 and 10 cm, respectively, and a diameter/height ratio of between 1.5 and 2.2. The cheeses can weigh 300 g to 2.6 kg. The sides are straight or slightly convex. The rind is natural, soft, thin, creamy and orangey-brown or greyish, and there may be reddish or yellow patches caused by microbial growth. The texture is compact and creamy, with no eyes and medium-low stickiness. The ivory-white colour may vary depending on the type of milk used but is uniform throughout the body. There are greenish blue patches and veins caused by *Penicillium* spp. Moulds, which become darker in colour during ripening but must never be black. The *Penicillium* spp. should grow homogenously throughout the cheese. The smell of the Cabrales cheese is intense, clean, pleasant, slightly lactic and penetrating, and stronger when goat's milk is used. There is a note of spiciness and nuttiness reminiscent of hazelnut and almond in the curd cheese. The cheese flavour is balanced and intense, not overly salty, nor bitter nor intensely astringent. It has medium piquancy, especially when made from pure sheep's or goat's milk or mixtures. It has a pronounced, persistent aftertaste. The body is slightly rough, with a very fine, almost imperceptible graininess.

11.1.5 Method of Manufacture

Milk: Milk must be collected from the registered farms and meet the requirements laid down by legislation.

Coagulation: This stage is carried out by using natural rennet from kids' stomachs or powdered rennet. The milk must be kept at a temperature of 22°C–35°C for 60 min.

Cutting: The curd must be cut to a size of 1–2 cm in diameter.

Moulding/draining: After draining the curd in 'arnios' (cylindrical moulds), they are left for 2–4 days, turning them a couple of times so that an 'auto-pressing' is carried out instead of pressing.

Salting: They are then salted by sprinkling salt over the top, leaving them for 12 hr and then turning and sprinkling again.

Maturation: After a further 12 hr the cheeses are removed from the moulds, and left to air for approximately two weeks. Cheeses are ripened for two to five months in natural caves with a temperature of 8°C–12°C and a Relative Humidity (RH) of 90%. These conditions promote the natural development of *Penicillium* spp. without the need to add *Penicillium* spp. spores. The cheese should remain on wooden shelves and is turned and cleaned regularly. The different stages involved in producing the milk and making and maturing this type of cheese must take place in the defined geographical area. Although the use of sycamore maple leaves (*Acer pseudoplatanus*) is allowed to wrap the cheese, a special green aluminium foil bearing the legend 'PDO Cabrales' has now been established. As most of the cheesemakers have a small production quota, one label has been adopted for everyone, although in each label the type of milk used in the preparation – one, two or all three types – should be indicated.

11.1.6 Relevant Research

Lactococcus lactis ssp. *lactis* is dominant until the end of ripening, whereas populations of certain Lactobacillus species appear during ripening. Other, completely unknown lactococci have also been detected. The dominant eukaryotic populations from day 15 onwards are those of *Penicillium roqueforti* and *Geotrichum candidum* (Florez & Mayo, 2006; Mayo, Hardisson & Brana, 1990). During ripening, the soluble and insoluble volatile fatty acids indices of the fat decrease, as do the short-chain fatty acids (Nuñez, 1978). Different studies have been carried out to identify and characterise the different microorganisms involved in Cabrales ripening, including the phenotypic and molecular identification of yeasts (Alvarez-Martin et al., 2007; Florez et al., 2010; Nuñez et al., 1981), fungi (Fernandez-Bodega et al., 2009; Fernández-Salquero, 2004; Ferreira et al., 2012; Florez et al., 2007), LAB (Florez et al., 2006b; Mayo, Hardisson & Brana, 1990) and also the influence of *P. roqueforti* addition on the properties of Cabrales (Florez et al., 2006c). Different methods have also been studied to control *Acaraus farris* (Sanchez-Ramos and Castañera, 2009) in order to preserve cheese quality. Studies have also been carried out on the antimicrobial activity of LAB isolated from Cabrales cheese (Florez, Delgado & Mayo, 2005); the identification of ACE-inhibitory peptides (Gomez-Ruiz et al., 2006); the characterisation of kid rennet extracts (Florez et al., 2006a); the study of conjugated linoleic acid and isomer distribution during ripening (Luna, Juárez & Fuente, 2007) and the role of triacylglycerols for confirming milk authenticity (Fontecha et al., 2006). Finally, experimental models have been developed from isolated *Lactobacillus casei* strains to reduce biogenic amine concentration (Herrero-Fresno et al., 2012).

11.2 Danablu PGI – Denmark

Name: Danablu PGI	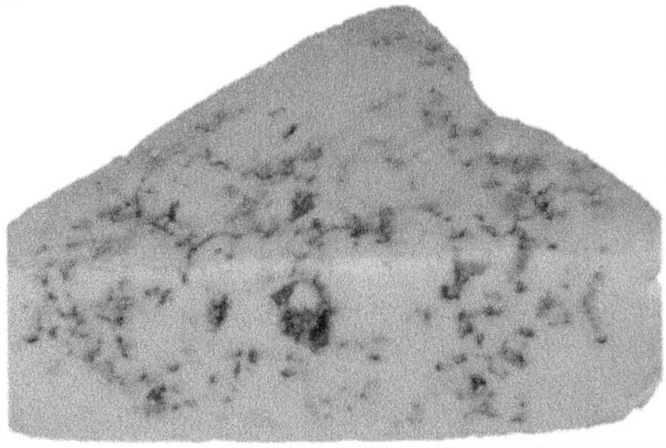
Production area: Denmark	
Milk: Cow's, thermised	

11.2.1 Introduction

Danablu is also known as Danish Blue. It has been granted PGI by the EU (EC, 1996), which dictates that the cheese can only be produced in Denmark. It is produced in medium-sized dairy plants with two different fat contents: Danablu 50+ and Danablu 60+ with 50% and 60% FDM, respectively. The production of blue cheese from cow's milk began around 1870 in Denmark at Havarti farm by the cheesemaking expert Hanne Nielsen, who had travelled around Europe to collect new recipes. Danablu was the result of further development by another cheesemaker, Marius Boel, who introduced homogenisation of the cream in 1927 as an attempt to make the cheese as white as Roquefort, which is made from ovine milk with less content of natural carotene than bovine milk. The fat breakdown in Danablu, which is important for the development of the typical blue cheese flavour, is accelerated by homogenisation. Further improvement of the flavour is achieved by replacing pasteurisation (72°C/15 s) with thermisation (61°C/15 s) of the milk to obtain a lower heat load and retain higher activity of the milk lipase that is responsible for a specific, highly desirable flavour in the cheese.

11.2.2 Type

Danablu is a blue-veined, semi-hard cheese with an FDM of 50% or 60 %. For optimum quality, the milk non-fat substance (MNFS) content should be 61%–63%; a value of up to 69% corresponding to a maximum of 47% moisture content in the cheese is allowed. The salt content is about 3%–4%.

11.2.3 Description and Sensory Characteristics

Danablu is made in cylinders that measure 20 cm in diameter and 9–10 cm in height, weighing 2.75–3.25 kg; or with a rectangular shape (30 cm × 12 cm) weighing about 4 kg. The surfaces are free of bacterial or mould growth and have a white to light yellowish or a light brownish colour. The cheeses may be slightly sunk in the centre. Openings of piercing channels are visible at the surface. The colour of the inner part is white to light yellow with regularly distributed pure marbling of blue-green mould veins in piercing channels and physical holes and slits. The marbling diminishes towards the cheese rind. Physical openings and piercing channels should be regularly distributed. The body is more closed towards the exterior of the cheese than in the centre. No eyes should be visible. The cheese is brine-salted, which results in a white rind that is quite dry. The texture of a young cheese is brittle, and it crumbles but becomes spreadable as it becomes more mature. The ripened body is loose and evenly soft and creamy, sliceable and spreadable and has a somewhat firmer and shorter consistency towards the cheese rind. Danablu has a round, piquant taste, is full of aromatic nuances and is strongly characterised by the growth of blue mould. The taste may be sharp and somewhat salty and acid with a slightly bitter note. Characteristic odour impressions originate from the methyl ketones introducing fruity, floral and musty and, specifically for 2-heptanone, spicy and blue-cheese-like notes.

11.2.4 Method of Manufacture

Milk: The cream is homogenised, the cheesemilk is thermised (61°C/15 s) and the temperature is set to 29°C–32°C. Calcium chloride may be added up to 20 g per 100 L of cheesemilk.

Starter culture: 0.5%–1.0% mesophilic culture of DL-starter and spores of *P. roqueforti* are added.

Rennet: The milk is coagulated with bovine rennet; coagulation is carried out for 50–70 min at 29°C–32°C.

Cutting: Grains with a diameter of 10–12 mm.

Scalding: There should be no stirring or very gentle stirring, and no heating for cheeses made with 50% FDM and heating up to 33°C for the cheeses with 60% FDM.

Curd washing: 20%–25% of the whey is removed, and 5% water is added to the variety with 60% FDM.

Moulding: Cheese curd is gently moved to cheese moulds and left to drain naturally (i.e. no pressure is applied) at 18°C for 20–24 hr.

Salting: Brine-salted at 18°C for 48 hr.

Piercing: Several needles are forced through the cheese to permit exchange between oxygen from the air and carbon dioxide from microbial activities in the interior of the cheese.

Maturation: Maturation takes 3–5 weeks at 8°C–11°C and 90%–95% RH and at 2°C–6°C and 80%–90% RH until retailed. The cheeses are ripened for at least five to six weeks and are carefully turned over about three times a day.

Storage: The cheese will last if stored properly at cold temperatures about 2°C–6°C for six to twelve weeks.

11.2.5 Relevant Research

Cantor et al. (2004) describes in detail the several biochemical, textural and microbiological changes of Danablu during maturation. The lactose in fresh Danablu is rapidly converted by the starter bacteria into lactic acid, which is used by *P. roqueforti*. Lipolysis and proteolysis in Danablu is extensive when compared to other cheeses. The free fatty acids, hexanoic and octanoic acids, are especially important flavour compounds. The specific characteristic flavour of Danablu also originates from a significant conversion of fatty acids to methyl ketones 2-heptanone and 2-nonanone, but 2-pentanone and 2-undecanone. Methyl ketones with one less carbon atom than the corresponding fatty acid are produced by β-oxidation. Methyl ketones can be reduced to secondary alcohols, under anaerobic conditions, and the most important are 2-heptanol, 2-nonanol and 2-pentanol. Casein is hydrolysed at more sites and at considerably higher rates in Danablu than in many other cheeses, and intact caseins or its primary breakdown products are mainly absent in the ripened cheeses. A larger number of different peptides are produced than in semi-hard cheeses and a high concentration of amino acids are released as a result of the peptidases, especially from mould and LAB working in concert. Other important aroma compounds in Danablu with fruity and floral notes are esters, which are formed by microbial esterification of free fatty acids with alcohols, and lactones, which are formed by inter-esterfication of hydroxyl fatty acids. The texture of a young cheese is brittle and crumbly, but heavy proteolysis and pH increase make it soft and spreadable as ripening advances.

11.3 Fourme d'Ambert PDO – France

11.4 Fourme de Montbrison PDO – France

<table>
<tr><td>Name : Fourme d'Ambert PDO,
Fourme de Montbrison PDO</td><td rowspan="3">Fourme d'Ambert Fourme de Montbrison

http://www.fourme-de-montbrison.fr/wp-content/uploads/2012/08/DSC_0057-671x348.jpg (above) and http://www.fourme-de-montbrison.fr/wp-content/uploads/2012/08/DSC_0043.jpg (below)</td></tr>
<tr><td>Production Area: Part of the north Massif Central Mountains (600–1600 m)</td></tr>
<tr><td>Milk: Cow's (raw or pasteurised)</td></tr>
</table>

11.4.1 Introduction

Fourme d'Ambert PDO and Fourme de Montbrison PDO are 2 out of the 40 French blue cheeses which include 6 PDO cheeses and represent 15.5% of the total French blue cheese produced in 2014. Fourme d'Ambert is one of the two most produced French PDO blue cow's cheese (INAO-CNAOL, 2014). In 2014, 5,972 tonnes of PDO Fourme was produced, mostly d'Ambert (90%). In 2015, 542 tonnes of Fourme de Montbrison was manufactured, among the lowest volumes for a cow's French PDO cheese. The two Fourme cheeses represented 15.5% of the total annual production of French blue cheeses made from cow's milk. The blue Fourme made from cow's milk accounts for two thirds of the volume of blue cheeses manufactured in

France (one third of it is made from ewe's milk, exclusively as PDO Roquefort (CNIEL, 2016). They have a mild aroma, compared to that of Roquefort, for example. The two Fourme cheeses were recognised as AOC in 1972 and as PDO in 2009 (Official website for Fourme d'Ambert, 2016; Official website for Fourme de Montbrison, 2016). Until the end of 2001, they were recognised AOC together as 'Fourme d'Ambert et de Montbrison'; since 2002, Fourme d'Ambert and Fourme de Montbrison have been recognised as two different PDO cheeses. Fourme is made and matured exclusively in three delimited areas covering 850,000 m^2 included in the Massif Central Mountains, low-sized mountains located in southcentral France; Fourme de Montbrison is made in the Rhone-Alps region, spread over an area of 70,000 m^2 adjacent to one of the two distinct areas dedicated to Fourme d'Ambert in the Auvergne region.

The Fourme denomination derives from the Greek *phormos,* then from the Latin *forma,* which means the container in which milk coagulates. The word *fromage,* which actually means 'cheese' in the French language, is derived from 'fourme' via the ancient words *fourmage* and *formage.* 'Fourme' is actually used for naming different French cheeses from the Auvergne region (that of Fourme d'Ambert), whether blue or not.

Based on an artisanal know-how developed over centuries in the Monts du Forez mountains, the blue Fourme is uncooked, unpressed and rennet-coagulated; it is made exclusively from raw or heated cow's milk and in the delimited area. As in all blue cheeses, its interior is aerated when it matures using a traditional practice, namely, piercing with long thin needles. A few of them – that is, 1% of Fourme d'Ambert in 2014 – are made at the farm (INAO-CNAOL, 2014). Then, it is matured for a minimum of 28 days (Ambert) or 32 days (Montbrison), but often more for several months. After the minimum maturation period, cheeses are randomly sampled for scoring their appearance, flavour and texture, and for assessing their gross physico-chemical composition. Some of them are disqualified as PDO when the flavour, appearance and/or texture scores are less than or equal to half the maximal scores (Chomette et al., 2004). The PDO blue Fourme is manufactured throughout the year. It is mainly consumed in France but is also exported to Europe. The blue Fourme is probably among the oldest French blue cheese. The two actual PDO blue Fourme cheeses are derived from ancient cheeses that have been manufactured in the Monts du Forez since at least the eighth or ninth centuries, as shown by sculptures in a feudal chapel. Strongly anchored in the Massif Central region, *P. roqueforti,* which has provided the blue colour, is a fungus present on rye, a grain that was very widely produced in the Auvergne region and was used for making bread. These ancient cheeses were produced daily exclusively at farms. Every herd was located at two different altitudes, depending on the season. A migration of livestock and farmers from villages to a higher altitude (1,000–1,6000 m) occurred from May to October. From collective and annually built units, highland farms at their peak (the end of the nineteenth century) became family-run and miniature farms called 'jasseries'. Thus, the herds reached pastures named 'estives', while hay for winter was harvested from the pastures located in the lowlands. In the jasserie, herd care and milking, as well as the cheese and butter manufacturing, were done exclusively by women (mothers and their daughters), who lived completely isolated for five months while hay was harvested by the men, who stayed in the villages throughout the year. In the eighteenth century, the blue Fourme was sold at Paris and Lyon, which was unusual for most cheeses. Since the mid-twentieth century, dairy plants collecting milk from several dairy farms have completely replaced the summer dairy farms. In the 1960s, the blue Fourme began to be manufactured outside its historical region at places located further west, where another blue cheese was traditionally made, the actual PDO Bleu d'Auvergne. In a few years, similar volumes were produced in the historical and recent production regions. As production area increased, practices for manufacturing

Fourme differed between locations, leading gradually to two distinct Fourme cheeses, both visually and after tasting or heating. The total annual production of the blue Fourme increased during the twentieth century (35-fold), making PDO Fourme actually the 10th French PDO cheese, while it has been stable since 2010 at around 6,000 tonnes (INAO-CNAOL, 2014).

11.4.2 Description and Sensory Characteristics

Fourme d'Ambert and Fourme de Montbrison are both blue-veined cheese. According to their specifications, they should have a minimum of 50% (Ambert) or 52% (Montbrison) of both fat in dry matter and dry matter; Fourme d'Ambert contains 1.3% of salt (0.8%–1.7%) (Chomette et al., 2004). Fourme's texture is soft, and firmer and smoother for Fourme de Montbrison. Fourme melts in the mouth. Fourme's body is white to light yellow (ivory), with Fourme de Montbrison' body being less white. Fourme d'Ambert is more blue-veined. Fourme's rind is thin, dry, bloomy, and orange (Montbrison) or grey (Ambert). Fourme's flavour is characterised by a subtle blue aroma, less pronounced for Montbrison, and a milky/fruity flavour. At 220°C, Fourme d'Ambert melts and spreads, while Fourme de Montbrison only melts (Anonymous, 2016).

11.4.3 Method of Manufacture

Fourme is manufactured according to rules that apply from milk production to cheese maturation (Official website for Fourme d'Ambert, 2016). All manufacturing steps including maturation should be performed in the delimited area.

Many differences between the two Fourme cheeses are actually linked to the curd cutting, draining and salting practices. Fourme d'Ambert's curd is less drained and salted only on the surface. In addition, a traditional draining practice is used for the Fourme de Montbrison draining; after demoulding, the cheeses are laid one behind another on an inclined gutter, on which they acquire their orange-coloured rind; twice a day, they are rotated a quarter turn and displaced one place higher at more and more dry places as the draining progresses (Guillermain, 2014).

Milk production: According to the specifications, dairy cows are fed fodder, completed with non-GMO feeds and additives in a limited amount; grass (a minimum of 150 days), fresh or dried fodder (a minimum of 3 kg daily per cow) and fermented feed are the main feeds. Fodder for the basic ration is produced in the delimited area.

Milk preparation: At farms, Fourme should be manufactured from one to two successive milkings; milk from the first milking is mandatorily stored at 4°C; in addition, they should be manufactured from full-fat and raw milk, and renneting should be performed before 16 hr after the first milking. The milk is heated to 29°C–34°C, selected strains of thermophilic (*Streptococcus thermophilus*) and mesophilic (*Lactococcus lactis*) LAB are inoculated as starter (Guillermain, 2014) and selected strains of *P. roqueforti* are inoculated as adjunct (Chomette et al., 2004).

Coagulation: The milk is coagulated at 29°C–34°C (Montbrison) or 30°C–35°C (Ambert).

Cutting: The curd is cut at the size of a wheat (Montbrison) or corn (Ambert) grain.

Moulding: The curd is moulded and drained at 18°C–22°C for six days (Montbrison) or at 18°C–25°C for 24–48 hr (Ambert).

Salting: The cheeses are brined or dry-salted.

Piercing: Aeration takes place by piercing at 6°C–15°C, for 10-day-old cheeses (Montbrison) or 4-day-old cheeses (Ambert).

Maturation: The cheeses are matured at 6°C–12°C and 90%–98% RH for 15 days or more. Then, they are stored at 0°C–6°C for more than seven days. Fourme cheeses are sold after 32 days (Montbrison) or 28 days (Ambert).

11.4.4 Relevant Research

Penicillium spp. was the subject of two recent studies. The two *Penicillium* spp. used in cheesemaking, *P. camemberti* and *P. roqueforti,* were shown to grow faster and be better competitors on the cheese surface than eight other *Penicillium* spp.; this was related to the presence in their genome of newly discovered horizontally transferred genome regions similar within strains of the two cheesemaking species, although these two species are not closely related (Ropars et al., 2015). These regions contain genes highly expressed in the early stage of cheese maturation.

Taxonomically, *P. roqueforti,* genus *Penicillium link,* subgenus *Penicillium* and species *roqueforti* are currently recognised as a single species, although substantial morphological differences have been reported among strains. This diversity has led to numerous distinct 'technological' species names such as *P. glaucum, P. stilton, P. gorgonzolae* or *P. aromaticum* (Gillot et al., 2015). The valid species name is currently *P. roqueforti* (Visagie et al., 2014). *P. roqueforti* was isolated from 120 individual blue cheeses coming from 18 different countries and affiliated to 12 cheese varieties that include Fourme d'Ambert. While morphological differences were observed among *P. roqueforti* strains, microsatellites did not reveal the existence of different species.

Four types of French blue-veined cheeses from the Massif Central region, including the two PDO blue Fourme, displayed significantly different properties after heating; specific sensory attributes could discriminate these cheese types (Bord, Guerinon & Lebecque, 2016). Fourme d'Ambert exhibited useful culinary properties: it presented good meltability and stretchability, and a weak oiling-off. In addition, taste attributes (salty, bitter and sour) could also distinguish between these heated blue cheeses.

11.5 Gamonedo PDO – Spain

Name: Gamonedo PDO	
Production area: Asturias, Spain	
Milk: Cow's, sheep's and goat's (mixture), raw	

11.5.1 Introduction

The PDO status of Gamonedo cheese was recognised by the Ministry of Agriculture, Fisheries and Food in 2003 and at the European level in 2008 (EC, 2008).

Gamonedo cheese is made from pasteurised mixtures of cow's, sheep's and goat's milks. Under the PDO regulation, 20 farmers and 20 dairy plants are registered. During 2013, production was 102,294 kg, for a total value of 1,530,000 euros, sold mainly on the national market, which represents 0.72% of the economic value of Spanish PDO cheeses (Ministry of Agriculture, Food and Environment, 2014). Gamonedo cheese continues the region's ancient traditions of integrated farming of herds and flocks. These flocks and herds were looked after by their owners on the pastureland where the animals grazed during the summer. These conditions and the remoteness from consumer markets, due to the transport difficulties, led to milk being processed into cheese by the shepherds themselves (Ministry of Agriculture, Food and Environment, 2015).

The geographical area is in the territory of the municipalities of Cangas de Onís and Onís in the Autonomous Community of Asturias.

11.5.2 Type

'Gamoneu' or 'Gamonedo' cheese is a full-fat, blue-veined semi-hard ripened cheese with a natural rind, made from raw cow's milk, sheep's milk or goat's milk or a mixture of two or three of these types of milk. The duration of the ripening is at least two months. It is lightly smoked and has greenish-bluish streaks of *Penicillium* spp. around the edges. Two types can be differentiated: (1) 'Gamoneu del Puerto' or 'Gamonedo del Puerto', made in the high mountain passes of the municipalities of Cangas de Onís and Onís (Picos de Europa) in small cheesemaking installations in the plains or summer pastures referred to in the specification, in the months of June to September; the cheese is made from the milk of the dairy herds and flocks grazing on these plains, using a mixture of the milk of at least two of the three species referred to; at least 10% of sheep's or goat's milk is used; and (2) 'Gamoneu del Valle' or 'Gamonedo del Valle', made in the lower-lying areas of the region covered by the designation of origin belonging to the municipalities of Cangas de Onís and Onís; production is not seasonal because the farming is semi-extensive, based on the pasture-farming system.

The minimum fat content and protein content in dry matter are 45% and 25%, respectively, with a maximum moisture content matter content of 45% and a pH value of 4.5–6.5.

11.5.3 Milk

This cheese is made from cow's milk (from the Friesian, Asturiana de los Valles, Alpine Brown breeds and their cross-breeds), sheep's milk (from the Latxa, Carranzana and Milschalfe breeds and their cross-breeds) and goat's milk (from the Alpine-Pyrenean, Picos de Europa, Murciano-Granadina and Saanen breeds and their cross-breeds) or from mixtures of two of these three types of milk. The pastures of this region have traditionally been used for livestock, and traditional practices are followed.

11.5.4 Description and Sensory Characteristics

The cheese is cylindrical in shape with a flat top and bottom. The diameter is between 10 and 30 cm with a height in the range 6–15 cm and a weight in the range 0.5–7 kg. The natural rind is thin, taking on a reddish brown colour during smoking, with red, green and blue tinges.

The colour of the cheese is white or yellowish white, with a light bloom of greenish blue *Penicillium* spp. close to the edges. The outside colour is unusual – the smoking process to which it is subjected gives the cheese a reddish brown colour, which then takes on red, green and blue tinges from the moulds that form during ripening in caves or cellars. The texture is loose, with holes. Eyes are small and unevenly distributed. The body has a hard or semi-hard consistency. It is firm but crumbles when cut. The main, but mild flavour, is of smoke, with a slight pungency. In the mouth it becomes buttery, with a lingering aftertaste of hazelnuts. The aroma is a little smoky, but clear and penetrating, becoming more intense in mature cheeses.

11.5.5 Method of Manufacture

Milk preparation: Milk must be collected from registered farms and meet the requirements laid down by legislation.

Coagulation: This is carried out by using rennet or enzymes. The milk must be kept at a temperature of 24°C–30°C for 60 min.

Cutting: The curd must be cut to the size of 0.5–1.5 cm in diameter.

Curd draining: The curd is placed in cylindrical moulds and left to drain for 90 min. Slight pressure may be applied to ensure that drainage is complete.

Salting: Dry salting is done by covering the top and the base of the cheese with sodium chloride after 24 and 48 hr, respectively.

After removal from the moulds, the cheeses undergo a smoking process whose duration and intensity vary depending on the environmental conditions. This process ensures proper drying and the formation of a dark-coloured rind with a texture allowing *Penicillium spp.* to penetrate the cheese during ripening. The material used for smoking is ash wood (*Fraxinus excelssior*), heather (*Erica* spp.), beech (*Fagus sylvatica*) or other native non-resinous wood.

Maturation: Ripening takes place in limestone caves or cellars with an average temperature of 10°C and an RH of 90%. This process takes at least two months. The cheese should remain on wooden shelves and is turned and cleaned regularly. The different stages involved in producing the milk and making and maturing this type of cheese must take place in the defined geographical area.

11.5.6 Relevant Research

Growth of LAB is greater in the interior of the cheese than on the surface, and among the lactic acid bacteria, lactobacilli predominate. Levels of enterococci and micrococci reach high numbers during the last stages of maturation, which suggests that these organisms may play a significant role in ripening. Of the LAB, *Lb. plantarum* is present at the highest levels, followed by *Lactobacillus casei* spp. *casei*. Low levels of lactic streptococci and leuconostocs can be detected. Levels of proteolysis and lipolysis are high. The degradation of $á_{s1}$- and â-caseins is almost complete at the end of ripening (Gonzalez de Llano et al., 1992).

A study has been carried out to identify and characterise the volatile fraction of artisanal Gamonedo blue cheeses; it was observed that methyl ketones and 2-alkanols increased during the first part of the ripening process, decreasing after 60 days (Gonzalez de Llano et al., 1990). Several studies have also analysed lipolysis as a pathway for flavour generation (Collins, McSweeney & Wilkinson, 2003). Recent studies have also been published on the influence of high-pressure processing in decelerating the lipolysis in blue-veined cheeses (Calzada et al., 2013) and one study on on-line control of the coagulation process by a fibre optic sensor (Abdelgawad, Guamis & Castillo, 2014).

11.6 Roquefort PDO – France

Name: Roquefort PDO	 *http://www.fromages-aop.com/wp-content/uploads/cartes/fromages/roquefortrt.png*
Production area: Part of the south Massif Central Mountains (300 m –1000 m)	
Milk: Sheep's, raw	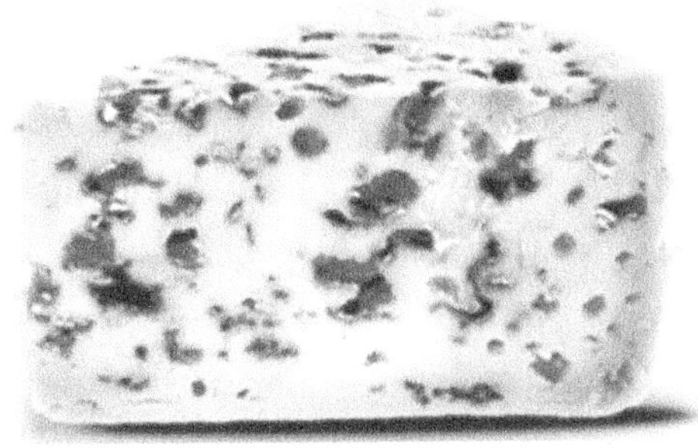 *http://www.soyez-roquefort.fr/wp-content/themes/bootstrap/img/bgd.jpg*

11.6.1 Introduction

Roquefort PDO is one of 40 French blue cheeses, which include 6 PDO cheeses. It represented 33% of total French sheep's milk cheese produced in 2014 (CNIEL, 2016). Roquefort is the second most commercialised French PDO cheese (behind Comté PDO), the most commercialised French PDO blue cheese and the most commercialised French PDO sheep's cheese (INAO-CNAOL, 2014). In 2014, 16,884 tonnes of Roquefort PDO was on the market. It has a strong aroma, compared to that of some blue cheeses, such as Fourme PDO, for example. Roquefort was the first cheese recognised as AOC, in 1925, and was recognised as PDO in 1996 (EC, 1996). Roquefort's milk is collected and Roquefort is made exclusively in a delimited area covering 1,500,000 hectares; it includes an old region named Rouergue and its adjacent regions; this area is located in the Massif Central Mountains, which are low-sized mountains located in southcentral France; it is part of the Larzac Plateau. Roquefort is matured exclusively in Roquefort-sur-Soulzon, a small village of this area, in cellars embedded in a limestone cliff overhanging the Roquefort-sur-Soulzon village (Official website for Roquefort, 2016). These cellars are built in natural cavities located under a scree 2 km long and 300 m wide at the bottom of the cliff. They are ventilated by natural chimneys ('fleurines') formed by rifts. Every cellar thus has its own relative humidity and temperature, depending on its location. Roquefort is manufactured (both cheesemaking and maturation) at 7 dairy plants (6 industrial and 1 cooperative), while Roquefort's milk (74,000,000 litres) is supplied by 1954 dairy farmers (Official website for Roquefort, 2016).

Roquefort is uncooked and unpressed, and rennet-coagulated; it is made exclusively from raw full-fat sheep's milk in the delimited area. It is manufactured exclusively at plants. Two thirds of the Roquefort volume is produced at one industrial plant. It is matured for a minimum of three months, but often more. After the minimum maturation period, cheeses are randomly sampled to score their appearance, flavour and texture, and for assessing their gross physico-chemical composition. The Roquefort PDO is manufactured from mi-November to August, when ewes lactate for 180–240 days.

Roquefort is probably among the oldest French blue cheese and is derived from ancient cheeses that have been manufactured for centuries in the delimited area. These cheeses are first mentioned in a document dated 1070 (Official website for Roquefort, 2016). Roquefort production really began in the nineteenth century, taking advantage of the popularity that Roquefort gained during the late eighteenth century, when Voltaire and Diderot, and people of the Enlightenment, liked it very much, owing to the specific know-how associated with its blue colour valued by the *Encyclopaedists.* (Vabre, 2016). Following this, Roquefort production rose continuously from 250 tonnes in 1810 to 850 tonnes in 1854 while its consumption increased remarkably in Paris compared to other types of cheese. This Parisian taste for Roquefort prompted and encouraged transformations in the cheesemaking and maturation processes as well as in the supply organisation; this is detailed in a recent thesis summarised in a book chapter (Vabre, 2016).

11.6.2 Description and Sensory Characteristics

PDO Roquefort is a blue-veined cheese. According to its specifications, it should have a minimum FDM of 52% and a minimum dry matter (DM) of 55%. On average, Roquefort contains 1.6% salt, 19.1% proteins, 1.8% carbohydrates, 32.1% lipids and 59.7% DM. Roquefort's rind is still wrapped up in aluminium foil when it is ready for consumption. Roquefort's texture is homogeneous and smooth. Roquefort melts in the mouth like butter. The Roquefort body is light yellow (ivory), exhibits small holes and uniformly distributed blue-green or blue-grey veins. Roquefort's odour is persistent and intense. Roquefort's rind is moist. Roquefort's flavour is pronounced, non-bitter, non-acid and not very salty. Nevertheless, its sensory characteristics change according to the microbial strains inoculated and the manufacturing practices used (Official website for Roquefort, 2016).

11.6.3 Method of Manufacture

Roquefort is manufactured according to rules that apply from milk production to cheese packaging (Official website for Roquefort, 2016). Milk production and cheesemaking should be performed in the delimited area, while all other manufacturing steps (from maturation to packaging for selling) are conducted exclusively in village Roquefort. About 12 litres of milk is needed to manufacture one cheese.

Milk production: Milk is produced exclusively by the Lacaune breed. In fact, there are actually two Lacaune breeds, one for milk production and one for meat production, which have different morphology and conformation. They are derived from both local (Larzac and Lauraguais) and non-local (Mérinos and Southdown) breeds. Dairy ewes are fed principally with grass (grazing is mandatory as soon as the weather conditions allow it), fodder and cereals produced for 75% of their dry matter in the delimited area. Feed purchase is limited to 200 kg per dairy ewe per year. In the sheepfold, every suckling ewe must have a minimum area of 1.5 m^2 (Anonymous, 2015b; EC, 2006).

The protein content of vat milk is not normalised. Any physical treatment of milk, except milk filtration to eliminate macroscopic impurities, is forbidden. The storage of milk, and curd under processing, at negative temperature is forbidden.

Starter culture: The milk is inoculated with mesophilic starter culture and spores of *P. roqueforti* at 28°C–34°C. Milk is inoculated with strains of *P. roqueforti* previously isolated in the Roquefort natural cellars. These strains differ by their textural, flavouring and colouring ability. They are propagated on a sterile medium still made on bread dough from wheat and rye flours.

Rennet: Coagulation takes place with rennet.

Cutting: The curd is cut and stirred.

Moulding/drainage: The curd is moulded and drained for 48 hr at 10°C–20°C. Then, after removing the moulds, the curds are frequently turned.

Curd labelling: The curds are labelled with hollow print.

Salting: Dry salting takes place for five days at 10°C.

Piercing: Aeration takes place by piercing at 6°C–15°C.

Maturation: The first maturation is for approximately two weeks (15–21 days) at 8°C–10°C and RH 95%. The second maturation is after wrapping in airtight foil at low temperatures, and the cheese is sold after at least three months of maturation. The storage of fresh or maturing cheeses under protective atmosphere is forbidden.

11.6.4 Relevant Research

Four blue cheeses including Roquefort (Blue Stilton, blue cheese with leaves and blue cheese spread) and Cheddar were analysed using an optoelectronic nose (easy-to-use chromogenic array) (Zaragozá et al., 2015). The array showed a distinct response to each cheese. By incorporating data measured at different times, it perfectly classifies all five cheeses in 5.5 hr. This suggests that this array can be a useful, rapid tool for discriminating between blue cheeses.

Penicillium spp. was the subject of two recent studies. The two *Penicillium* spp., namely, *P. camemberti* and *P. roqueforti*, were shown to grow faster and be better competitors on the cheese surface than eight other *Penicillium* spp.; this related to the presence in their genome of newly discovered horizontally transferred genome regions that are similar within strains of the two cheesemaking species, although these two species are not closely related (Ropars et al., 2015). These regions contain genes that are highly expressed in the early stage of cheese maturation. The second recent study was dedicated to the *P. roqueforti* genome for both taxonomic and diversity purposes (Gillot et al., 2015). Taxonomically, *P. roqueforti* is currently recognised as a single species, although substantial morphological differences have been reported among strains. This diversity has led to numerous distinct 'technological' species names such as *P. glaucum*, *P. stilton*, *P. gorgonzolae* or *P. aromaticum*. The valid species name is currently *P. roqueforti* (Visagie et al., 2014). *P. roqueforti* was isolated from 120 individual blue cheeses coming from 18 different countries and affiliated to 12 cheese varieties that include Roquefort; its microsatellites markers were studied. While morphological differences were observed among *P. roqueforti* strains, microsatellites did not reveal the existence of different species. Interestingly, at the intraspecific level, they revealed the existence of three highly differentiated populations, both genetically and morphologically, corresponding to a group of blue cheese varieties. In this respect, Roquefort was clustered together with two other PDO French blue cheeses, namely, Bleu des Causses and Bleu d'Auvergne, but far from nine other blue cheeses including Fourme d'Ambert PDO, Gorgonzola, Stilton and Blue Danish. This suggests that the *P. roqueforti* species has been shaped by different cheesemaking processes or that different *P. roqueforti* populations, either ecotypes or allopatric populations, were recruited for different cheese types.

Studies on the composition and evolution of Roquefort microbiota along manufacture were done in the 1960s, leading to a set of articles reminding readers that Roquefort was made from raw milk (Devoyod & Poullain, 1988b; Devoyod & Poullain, 1988a; Devoyod et al., 1972; Devoyod et al., 1971; Devoyod, Lapierre & Sponem, 1970; Devoyod, Desmazaud & Auriault, 1970; Devoyod, Bret & Auclair, 1968; Devoyod & Sponem, 1970; Devoyod & Lapierre, 1970; Devoyod & Lapierre, 1969; Devoyod, Lapierre & Ducelliez, 1969). Thus, besides the starter (*Lc. lactis* and *Leuc. (para)mesenteroides*) and adjunct (*P. roqueforti*) populations, which are inoculated in vat milk from cultures of selected strains, they identified and quantified other main microbial populations (such as yeasts, staphylococci, micrococci, lactobacilli and enterococci), all cohabiting and interacting during Roquefort manufacture. It would be interesting to describe this microbiota again using current microbial taxonomy and methods.

11.7 Stilton PDO – United Kingdom

Name: Stilton PDO	
Production area: Leicestershire, Derbyshire and Nottinghamshire	
Milk: Pasteurised cow's milk	

11.7.1 Introduction

Blue Stilton is well known as the 'King of English Cheeses'. Historical evidence has shown that a cream cheese was being made and sold in and around the village of Stilton possibly in the late seventeenth century and certainly in the early eighteenth century and was known as Stilton cheese. The cheese generally seems to have been matured for some time before being sold. A recipe for Stilton cheese was published in a newsletter by Richard Bradley in 1723, but no details were given regarding its size, shape or length of maturation. From the recipe, it appears that this would have been a hard cream cheese (it was pressed and boiled in its whey). In 1724, Daniel Defoe commented in his 'Tour through the villages of England & Wales' of Stilton being 'famous for cheese' and referred to the cheese as being the 'English Parmesan'. A later article by John Lawrence in 1726 suggested that the perfect Stilton should be ... 'about 7 inches in diameter, 8 inches in height and 18 lbs in weight'. Thus, it seems that some of the cheese being produced in the area was cylindrical and of a comparable size to that being made today. Lawrence also referred to the cheese as the 'recently famous Stilton' (Anonymous, 2015).

Stilton cheese has gained recognition as a PDO cheese (EC, 1996) and is made in three varieties: White Stilton, Blue Stilton, Mature Blue and/or Vintage Blue Stilton, which are made throughout the year in cheesemaking dairies within the county boundaries of Leicestershire, Derbyshire and Nottinghamshire. Each variety can only be made from pasteurised cow's milk.

11.7.2 Type

Stilton is an internally mould-ripened semi-soft blue cheese with a minimum FDM of 48%. Its gross composition is as follows: Moisture: 37%–41.6%, fat: 32%–35.2%, protein: 21%–28.7% and salt: 2.2%–2.7% (Cantor et al., 2004).

11.7.3 Milk

Pasteurised cow's milk.

11.7.4 Description and Sensory Characteristics

White Stilton: White Stilton is a white cheese made in a cylindrical form from full-cream pasteurised cow's milk (which can be standardised according to the season) produced by dairy herds from the three counties of Leicestershire, Derbyshire and Nottinghamshire (in times of shortage, milk may also be sourced from the surrounding counties of Cambridgeshire, Northamptonshire, Warwickshire, Staffordshire, Greater Manchester, Cheshire, Yorkshire and Lincolnshire), with no applied pressure and forming its own crust or coat. Internally, it has a uniform white colour, with a flaky or crumbly open moist and externally, a smooth, moist, white exterior rind, free from surface mould, blemishes or mites.

Blue Stilton: Blue Stilton is a blue-moulded cheese that is similar in form to White Stilton and is made from cow's milk produced in the designated area. Internally, it has a uniform creamy white colour except for blue/green veins radiating from the centre with a velvety or flaky open texture, free from gas holes and chalkiness and, externally, with a slightly wrinkled crust or rind with variable colouring.

Mature Blue and/or Vintage Blue Stilton: Mature Blue and/or Vintage Blue Stilton is a fully mature blue-moulded cheese.

Stilton has an open, slightly crumbly texture, while Mature Blue Stilton has a softer, creamy texture. As far as flavour is concerned, White Stilton has a fresh, clean, acidic flavour; Blue Stilton has a clean subtle, slightly sharp full flavour with a creamy taste and Mature Blue and/or Vintage Blue Stilton has a stronger, more complex flavour with a creamy background.

11.7.5 Method of Manufacture

Milk preparation: Fresh liquid cow's milk is collected from within the three counties of Leicestershire, Derbyshire and Nottinghamshire (in times of shortage, milk may also be sourced from the surrounding counties of Cambridgeshire, Northamptonshire, Warwickshire, Staffordshire, Greater Manchester, Cheshire, Yorkshire and Lincolnshire) and delivered regularly. The milk is then tested and may be standardised according to the season. The milk is then pasteurised and pumped into vats.

Starter culture: A mesophilic starter culture (e.g. *Lc. lactis*) is added at 21°C. In addition, blue mould – *P. roqueforti* – is added to the milk and/or sprayed onto the curds in order to generate the blue veins for Blue Stilton and Mature Blue and/or Vintage Blue Stilton only.

Coagulation: Coagulation is carried out with rennet at 30°C.

Cutting: The milk forms curds which are cut with knives to help release the whey.

Salting: Dry salting at the rate of 2.5% salt, and 2.5%–3% for white Stilton.

Moulding: The following day the curd is cut into small blocks, milled into walnut-sized pieces and salted, then tipped into cylindrical cheese moulds.

Ripening (Stage 1): White Stilton, Blue Stilton and Mature Blue and/or Vintage Blue Stilton go through the following stages. In a temperature- and humidity-controlled environment, the curds in their moulds are allowed to drain and settle evenly. After a short period, the moulds are turned top to tail to facilitate this draining. For up to seven days, they remain there and are turned regularly, to ensure an even distribution of moisture.

Ripening (Stage 2): Subsequently all three types of cheese are moved to the Second Stage. The delicate curds have now settled into the familiar Stilton drum shape. The mould is gently removed, and the surface is sealed. This helps prepare the surface of the cheese for the formation of the coat or rind and effectively seals it, preventing oxygen from entering the tiny airways which lie inside the cheese around the pieces of curd.

The process for White Stilton production stops here. After they have been wrapped and chilled, the cheeses are ripe and ready for selling after about seven days.

Ripening (Stage 3): However, the process for making Blue Stilton and Mature Blue and/or Vintage Blue Stilton continues as follows.

Blue Stilton cheeses may be placed in Stage 3 or may be moved straight to Stage 4. In Stage 3, cheeses are placed in a temperature- and humidity-controlled environment. Each cheese is turned on a regular basis to ensure even drainage, before moving on again.

Ripening (Stage 4) – pitching: Thereafter, the Blue Stilton and Mature Blue and/or Vintage Blue Stilton cheeses are turned less regularly and the ripening temperature and humidity carefully adjusted to give the optimal conditions for the final stage of maturation.

At about six weeks, the Blue Stilton and Mature Blue and/or Vintage Blue Stilton cheeses are pierced with stainless steel needles about the size of skewers. This allows air to enter the heart of the cheese, which activates the *P. roqueforti* blue mould present in the curd, and the blue veining begins to develop.

The cheeses are left to ripen, turned regularly and may then be pierced again.

Maturation and grading: Each batch of Blue and Mature Blue and/or Vintage Blue Stilton is graded individually so that customers, who specify the degree of ripeness and openness of texture they prefer, can have their Stilton packed and dispatched exactly to suit their needs.

11.7.6 Relevant Research

The acidification of Stilton is carried out by the starter culture (e.g. *Lc. lactis*), while its ripening is promoted by the development of moulds (*P. roqueforti*) as well as yeasts and other bacteria, so that the ripened product has a complex as well as typical microbial composition (Johnson, 2001). The complex microflora of Stilton cheese is responsible for cheese ripening as well as the typical aroma development. The microbial diversity occurring in Stilton cheese was reported to be dominated by *Lc. lactis, E. faecalis, Lb. plantarum, Lb. curvatus, Leuc. mesenteroides, Staphylococcus equorum* and *Staphylococcus* spp. (Ercolini, Hill & Dodd, 2003). Lactococci were found in the internal part of the veins as mixed colonies and as single colonies within the core. *Lb. plantarum* was detected only underneath the surface, while *Leuconostoc* spp. were homogeneously distributed in all the observed parts. It should be noted that *Leuc. mesenteroides* ssp. *cremoris* has been shown to stimulate the growth of *P. roqueforti*, the mould used in Stilton ripening (Hansen and Jacobsen, 1997). In addition, Gkatzionis et al. (2014) isolated certain strains of yeasts such as *Yarrowia lipolytica, D. hansenii* and *Trichosporon ovoides* and studied synergistic activities with the starter *Lc. lactis* and *P. roqueforti*. Furthermore, the synergy between *Y. lipolytica* and *P. roqueforti* (Gkatzionis et al., 2013) and between *Y. lipolytica* and *Kluyveromyces lactis* with *P. roqueforti* (Price et al., 2014) was reported to be responsible for the production of blue cheese odour.

Stilton cheese is made from pasteurised milk and is thus supposed to be free of undesirable bacteria; pasteurisation was introduced in 1989 as a result of a food poisoning outbreak in England associated with the consumption of Stilton cheese (Ercolini, Hill & Dodd, 2003; Maguire et al., 1991).

The maturation process is associated with changes in the cheese microenvironments, microorganisms that contribute to the process and various biochemical changes such as lipolysis and proteolysis which are important for aroma formation. In general, the microenvironments in the blue cheese are heterogeneous with pronounced gradients of pH, salt, water activity (aw) and redox potential (Cantor et al., 2004) depending on the handling of certain steps in the manufacturing process. Thus, the growth, interaction and biochemical activities of the various microorganisms present in different sites of the cheese are altered, and consequently affect the quality characteristics of the final product.

References

Abdelgawad, A. R., Guamis, B. & Castillo, M. (2014). Using a fiber optic sensor for cutting time prediction in cheese manufacture from a mixture of cow, sheep and goat milk. *Journal of Food Engineering*, 125, 157–168.

Alvarez-Martin, P., Florez, A. B., Lopez-Diaz, T. M. & Mayo, B. (2007). Phenotypic and molecular identification of yeast species associated with Spanish blue-veined Cabrales cheese. *International Dairy Journal*, 17, 961–967.

Anonymous (2015). History of Stilton. www.stiltoncheese.co.uk Assessed 25-07-2015.

Anonymous (2015b). Press pack, 'Be Roquefort' [Online]. Available: http://www.soyez-roquefort.fr/wp-content/uploads/2015/07/DP-ANGLAIS.pdf [Accessed 05 September 2016].

Anonymous (2016). Montbrison Fourme en forme. *Profession fromager*, 70, 18–25.

Bord, C., Guerinon, D. & Lebecque, A. (2016). Impact of heating on sensory properties of French Protected Designation of Origin (PDO) blue cheeses. Relationships with physicochemical parameters. *Food Science and Technology International*, 22, 377–388.

Calzada, J., Del Olmo, A., Picon, A., Gaya, P. & Nuñez, M. (2013). High-pressure processing decelerates lipolysis and formation of volatile compounds in ovine milk blue-veined cheese. *Journal of Dairy Science*, 96, 7500–7510.

Cantor, M. D., van den Tempel, T., Hansen, T. K. & Ardö, Y. (2004). Blue cheese. In Fox, P. F. et al. (eds.), *Cheese: Chemistry, Physics and Microbiology. Vol. 2. Major Cheese Groups*, 3rd edition. Chapman & Hall, London, pp. 175–198.

Chomette, P., Corale, S., De Broucker, C., Legrand, A. & Moulinot, M. (2004). La Fourme d'Ambert. Available: pfeda.univ-lille1.fr/iaal/docs/iaal2003/fourm/fourme.pdf [Accessed 25 August 2016].

CNIEL (2016). L'économie laitière en chiffre – édition 2016 [Online]. Available: http://www.calameo.com/read/002230051cdfa9ea35988 [Accessed 15 July 2016].

Collins, Y. F., McSweeney, P. L. H. & Wilkinson, M. G. (2003). Lipolysis and free fatty acid catabolism in cheese: A review of current knowledge. *International Dairy Journal*, 13, 841–866.

Devoyod, J. J. & Lapierre, M. (1969). La flore microbienne du fromage de Roquefort. IV. – Les entérocoques. *Le Lait*, 49, 637–650.

Devoyod, J. J. & Lapierre, M. (1970). La flore microbienne du fromage de Roquefort. V. – Les lactobacilles. *Le Lait*, 50, 277–284.

Devoyod, J. J. & Poullain, F. (1988a). Les Leuconostocs. Propriétés: leur rôle en technologie laitière. *Le Lait*, 68, 249–279.

Devoyod, J. J. & Poullain, F. (1988b). The leuconostocs – Characteristics – Their role in dairy technology. *Lait*, 68, 249–279.

Devoyod, J. J. & Sponem, D. (1970). La flore microbienne du fromage de Roquefort. VI. – Les levures. *Le Lait*, 50, 524–543.

Devoyod, J. J., Bret, G. & Auclair, J. (1968). La flore microbienne du Roquefort. I. I. Son évolution au cours de la fabrication et de l'affinage du fromage. *Le Lait*, 48, 613–629.

Devoyod, J. J., Desmazaud, M., Assenat, L. & Auclair, J. (1972). Les associations microbiennes dans le fromage de Roquefort. IV. – Action inhbitrice des microcoques caséolytiques sur l'ouverture du caillé. *Le Lait*, 52, 297–310.

Devoyod, J. J., Desmazaud, M., Sponem, D. & Auriault, C. (1970). Les associations microbiennes dans le fromage de Roquefort. I. – Action des entérocoques vis-à-vis des streptocoques lactiques et des leuconostoc. Nature des substances stimulantes produites par Streptococcus faecalis var. liquefaciens. *Le Lait*, 50, 374–390.

Devoyod, J. J., Desmazaud, M., Melcion, D., Degas, C. & Auriault, C. (1971). Les associations microbiennes dans le fromage de Roquefort. III. Action des entérocoques et des levures fermentant le lactose vis-à-vis des lactobacilles. *Le Lait*, 51, 399–415.

Devoyod, J. J., Lapierre, M. & Ducelliez, M. (1969). Flore microbienne du fromage de Roquefort. II. – Staphylocoques et microcoques. *Le Lait*, 49, 20–39.

Devoyod, J. J., Lapierre, M. & Sponem, D. (1970). Les associations microbiennes dans le fromage de Roquefort. II. Action de Penicillium roqueforti sur l'évolution de la flore microbienne. *Le Lait*, 50, 644–653.

EC (1996). Commission Regulation (EC) No 1107/96 of 12 June 1996 on the registration of geographical indications and designations of origin under the procedure laid down in Article 17 of Council Regulation (EEC) No 2081/92. *Official Journal of the European Union, L* 148, 1–10.

EC (2006). Cahier des charges consolidé de l'appellation d'origine protégée "Roquefort" [Online]. Available: http://ec.europa.eu/agriculture/quality/door/ [Accessed 15 September 2016].

EC (2008). Commission Regulation (EC) No 676/2008 of 16 July 2008 registering certain names in the Register of protected designations of origin and protected geographical indications (Ail de la Drôme (PGI), Všestarská cibule (PDO), Slovenská bryndza (PGI), Ajo Morado de Las Pedroñeras (PGI), Gamoneu or Gamonedo (PDO), Alheira de Vinhais (PGI), Presunto de Vinhais or Presunto Bísaro de Vinhais (PGI) published in the Official Journal of 13 November 1996. *Official Journal of the European Union*, L 189, 19–20.

Ercolini, D., Hill, P. J. & Dodd, C. E. R. (2003). Bacterial community structure and location in Stilton cheese. *Applied and Environmental Microbiology*, 69 (6), 3540–3548.

Fernandez-Bodega, M. A., Mauriz, E., Gomez, A., & Martin, J. F. (2009). Proteolytic activity, mycotoxins and andrastin A in Penicillium roqueforti strains isolated from Cabrales, Valdeon and Bejes-Tresviso local varieties of blue-veined cheeses. *International Journal of Food Microbiology*, 136, 18–25.

Fernandez-Salquero, J. (2004). Internal mould-ripened cheeses: Characteristics, composition and proteolysis of the main European blue vein varieties. *Italian Journal of Food Science*, 16, 437–445.

Ferreira, C. O., Alvarez-Martin, P., Florez, A. B., Diaz, M., & Mayo, B. (2012). Biochemical and technological properties of *Penicillium roqueforti* and *Geotrichum candidum* strains isolated from Cabrales, a Spanish traditional, blue-veined, starter-free cheese. *Italian Journal of Food Science*, 24, 115–124.

Florez, A. B. & Mayo, B. (2006). Microbial diversity and succession during the manufacture and ripening of traditional, Spanish, blue-veined Cabrales cheese, as determined by PCR-DGGE. *International Journal of Food Microbiology*, 110, 165–171.

Florez, A. B., Hernandez-Barranco, A. M., Marcos, I. & Mayo, B. (2006a). Biochemical and microbiological characterization of artisan kid rennet extracts used for Cabrales cheese manufacture. *LWT – Food Science and Technology*, 39, 605–612.

Florez, A., Alvarez-Martin, P., Lopez-Diaz, T. M. & Mayo, B. (2007). Morphotypic and molecular identification of filamentous fungi from Spanish blue-veined Cabrales cheese, and typing of *Penicillium roqueforti* and *Geotrichum candidum* isolates. *International Dairy Journal*, 17, 350–357.

Florez, A., Belloch, C., Alvarez-Martin, P., Querol, A. & Mayo, B. (2010). *Candida cabralensis* sp. nov., a yeast species isolated from traditional Spanish blue-veined Cabrales cheese. *International Journal of Systematic and Evolutionary Microbiology*, 60, 2671–2674.

Florez, A., Delgado, S. & Mayo, B. (2005). Antimicrobial susceptibility of lactic acid bacteria isolated from a cheese environment. *Canadian Journal of Microbiology*, 51, 51–58.

Florez, A., Lopez-Diaz, T. M., Alvarez-Martin, P. & Mayo, B. (2006b). Microbial characterisation of the traditional Spanish blue-veined Cabrales cheese: Identification of dominant lactic acid bacteria. *European Food Research and Technology*, 223, 503–508.

Florez, A., Ruas-Madiedo, P., Alonso, L. & Mayo, B. (2006c). Microbial, chemical and sensorial variables of the Spanish traditional blue-veined Cabrales cheese, as affected by inoculation with commercial *Penicillium roqueforti* spores. *European Food Research and Technology*, 222, 250–257.

Fontecha, J., Mayo, I., Toledano, G. & Juarez, M. (2006). Use of changes in triacylglycerols during ripening of cheeses with high lipolysis levels for detection of milk fat authenticity. *International Dairy Journal*, 16, 1498–1504.

Gillot, G., Jany, J. L., Coton, M., Le Floch, G., Debaets, S., Ropars, J., Lopez-Villavicencio, M., Dupont, J., Branca, A., Giraud, T. & Coton, E. (2015). Insights into Penicillium roqueforti morphological and genetic diversity. *PLoS One*, 10.

Gkatzionis, K., Hewson, L., Hollowood, L., Hort, J., Dodd, C. E. R. & Linforth, R. S. T. (2013). Effect of *Yarrowia lipolytica* on blue cheese odour development: Flash profile sensory evaluation of microbiological models and cheeses. *International Dairy Journal*, 30 (1), 8–13.

Gkatzionis, K., Yunita, D., Linforth, R. S. T., Dickinson, M. & Dodd, C. E. R. (2014). Diversity and activities of yeasts from different parts of a Stilton cheese. *International Journal of Food Microbiology*, 177, 109–116.

Gomez-Ruiz, J. A., Taborda, G., Amigo, L., Recio, I. & Ramos, M. (2006). Identification of ACE-inhibitory peptides in different Spanish cheeses by tandem mass spectrometry. *European Food Research and Technology*, 223, 595–601.

González de Llano, D., Ramos, M., Polo, C., Sanz, J. & Martínez-Castro, I. (1990). Evolution of the volatile components of an artisanal blue cheese during ripening. *Journal of Dairy Science*, 73, 1676–1683.

González de Llano, D., Ramos, M., Rodríguez, A., Montilla, A. & Juárez, M. (1992). Microbiological and physicochemical characteristics of Gamonedo blue cheese during ripening. *International Dairy Journal*, 2, 121–125.

Guillermain, H. (2014). La Fourme de Montbrison [Online]. Available: www.jeanyvesgriot.fr/wp-content/uploads/2014/07/Dossier-Fourme-Hervé.docx [Accessed 25 August 2016].

Hansen, T. K. & Jacobsen, M. (1997). Possible role of microbial interactions for growth and sporulation of *P. roqueforti* in Danablu. *Lait*, 77, 479–488.

Herrero-Fresno, A., Martinez, N., Sanchez-Llana, E., Diaz, M., Fernandez, M., Martin, M. C., Ladero, V. & Álvarez, M. A. (2012). *Lactobacillus casei* strains isolated from cheese reduce biogenic amine accumulation in an experimental model. *International Journal of Food Microbiology*, 157, 297–304.

INAO-CNAOL (2014). Produits laitiers AOP – les chiffres clés, 2014. [Online]. Available: https://www.google.fr/search?q=produits+laitiers+AOP++2014&ie=utf-8&oe=utf-8&client=firefox-b&gfe_rd=cr&ei=iI3jV6OgJ8nDaKWRn8AO [Accessed 16 July 2016].

Johnson, M. E. (2001). *Cheese Products*. In Marth, E. H. & Steele, J. L. (eds.), *Applied Dairy Microbiology*, 2nd edition. Marcel Dekker, Inc., New York, NY, pp. 345–384.

Luna, P., Juárez, M. & Fuente, M. A. de la (2007). Conjugated linoleic acid content and isomer distribution during ripening in three varieties of cheeses protected with designation of origin. *Food Chemistry*, 103, 1465–1472.

Maguire, H. C., Boyle, M., Lewis, M. J., Pankhurst, J., Wieneke, A. A., Jacob, M., Bruce, J. & O'Mahony, M. (1991). A large outbreak of food poisoning of unknown aetiology associated with Stilton cheese. *Epidemiology and Infection*, 106, 497–505.

Mayo, B., Hardisson, C. & Brana, A. F. (1990). Characterization of wild strains of *Lactococcus lactis* subsp. *lactis* isolated from Cabrales cheese. *Journal of Dairy Research*, 57, 125–134.

Ministry of Agriculture, Food and Environment (2014). Secretaría General Técnica Centro de Publicaciones. Datos de las Denominaciones de Origen Protegidas (D.O.P.) e Indicaciones Geográficas Protegidas (I.G.P.) de Productos Agroalimentarios, 3, 22–23.

Ministry of Agriculture, Food and Environment (2015). Electronic catalogue of Spanish cheeses http://www.alimentacion.es/es/conoce_lo_que_comes/bloc/queso/default/principal (Accessed 6 February 2015).

Nuñez, M. (1978). Microflora of Cabrales cheese: changes during maturation. *Journal of Dairy Research*, 45, 501–508.

Nuñez, M., Medina, M., Gaya, P. & Dias-Amado, C. (1981). Yeasts and moulds of Cabrales blue cheese. *Lait*, 61, 62–79.

Official website for Fourme d'Ambert (2016). [Online]. Available: http://www.fourme-ambert.com/en/ [Accessed 25 August 2016].

Official website for Fourme de Montbrison (2016). [Online]. Available: www.fourme-de-montbrison.fr/ [Accessed 25 August 2016].

Official website for PDO Roquefort [Online]. Available: http://www.soyez-roquefort.fr/ [Accessed 25 August 2016].

Price, E. J., Linforth, R. S. T., Dodd, C. E. R., Phillips, C. A., Hewson, L., Hort, J. & Gkatzionis, K. (2014). Study of the influence of yeast inoculum concentration (*Yarrowia lipolytica* and *Kluyveromyces lactis*) on blue cheese aroma development using microbiological models. *Food Chemistry*, 145, 464–472.

Ropars, J., De La Vega, R. C. R., Lopez-Villavicencio, M., Gouzy, J., Sallet, E., Dumas, E., Lacoste, S., Debuchy, R., Dupont, J., Branca, A. & Giraud, T. (2015). Adaptive horizontal gene transfers between multiple cheese-associated fungi. *Current Biology*, 25, 2562–2569.

Sanchez-Ramos, I. & Castañera, P. (2009). Chemical and physical methods for the control of the mite *Acarus farris* on Cabrales cheese. *Journal of Stored Products Research*, 45, 61–66.

Vabre, S. (2016). Rural Capitalism: The society des Caves de Roquefort, fromc. 1840 to 1914. In Oddy, D. J. & Drouard, A. (eds.), *The Food Industries of Europe in the Nineteenth and Twentieth Centuries*. Routledge, London and New-York.

Visagie, C. M., Houbraken, J., Frisvad, J. C., Hong, S. B., Klaassen, C. H. W., Perrone, G., Seifert, K. A., Varga, J., Yaguchi, T. & Samson, R. A. (2014). Identification and nomenclature of the genus Penicillium. *Studies in Mycology*, 343–371.

Zaragozá, P., Ros-Lis, J. V., Vivancos, J. L. & Martinez-Manez, R. (2015). Proof of concept of using chromogenic arrays as a tool to identify blue cheese varieties. *Food Chemistry*, 172, 823–830.

12

Acid-Coagulated Cheeses

Katja Hartmann[1,3]*, Françoise Berthier*[2] *and Giuseppe Licitra*[4]

[1] *Anton Paar GmbH, Germany*
[2] *Unité de Recherches en Technologie et Analyses Laitières Rue de Versailles, France*
[3] *Anton Paar, Germany*
[4] *Department of Agriculture and Food Science (DISPA), University of Catania, Italy*

12.1 Acid Curd (Harzer) – Germany

Name: Acid curd (Harzer)	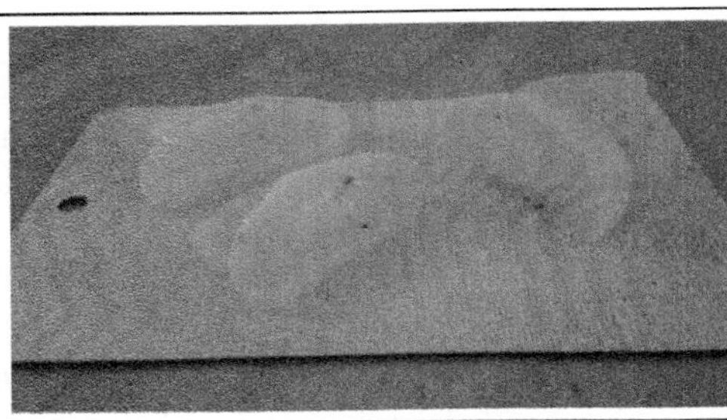
Production area: Central Germany	
Milk: Cow's, pasteurised	

12.1.1 Introduction

In Germany, the annual production of acid curd cheese is 28,000 tonnes, which accounts for 1.8% of the total. Over time, regional differences developed in the production of acid curd cheese, and today a large variety is available with diverse names depending on the manufacturing area (Kammerlehner, 2009). The cheese has no certification or label, but certain acid curd cheeses are listed as standard cheese varieties in the German regulation on cheese and/or bear a label of protected Designation of Origin (Deutsche Käseverordnung, 2013). The first historical account of acid curd cheese production dates back to 1785 in the eastern Harz region in Central Germany. During the last third of the nineteenth century, the production area increased to include the cities of Hannover, Mainz, Magdeburg, Halle and Leipzig in Central Germany. At the end of the nineteenth century, a cheese forming machine was developed, which further increased the production of acid curd cheese (Kammerlehner, 2009). Acid curd cheeses are mainly produced in Central Germany.

12.1.2 Type

Acid curd cheeses are produced from acid curd Quark and therefore constitute an extra cheese group. Quark is produced from acid-coagulated milk with or without the addition of rennet. It contains more than 30% biological high-grade protein with all the essential amino acids. Acid curd

Global Cheesemaking Technology: Cheese Quality and Characteristics, First Edition.
Edited by Photis Papademas and Thomas Bintsis.

Table 12.1 Acid-coagulated cheeses listed in the relevant German regulation, as cheese varieties, and information on production requirements.

Cheese variety	Manufacture
Harzer-Käse, Mainzerkäse	Ripens with yellow or red smear bacterial strains (type yellow cheese), caraway may be added, produced as low-fat cheese, production weight 25–125 g
Handkäse, Bauernhandkäse, Korbkäse, Stangenkäse, Spitzkäse	Produced as type yellow cheese and mould cheese (ripens with Camembert mould), caraway may be added, produced as low-fat cheese, production weight between 25 and 125 g
Olmützer Quargel	Produced only as yellow cheese, caraway may be added, produced as low-fat cheese, production weight 12 to 17 g

cheeses do not contain carbohydrates and have less than 1% fat. Furthermore, they contain approximately 1050 mg calcium and 0.3 mg riboflavin (B_2) per 100 g cheese. Different cheese types vary in the used cultures and added herbs. Table 12.1 lists acid curd cheeses that are described as standard cheese in the German regulation on cheese (Deutsche Käseverordnung, 2013).

12.1.3 Description and Sensory Characteristics

Acid curd cheese is produced with a weight between 1 and 13 kg. They are distinguished by their different properties into yellow cheese (German: *Gelbkäse*) with red smear and mould cheese with surface mould (German: *Edelschimmelkäse*). Yellow acid curd cheese has a smooth surface with a yellow to red/brownish colour. Mould acid curd cheese has an evenly mould-covered surface. The dry matter (DM) is between 27% and 40%. Acid curd Quark used for production of acid curd cheese has a whitish colour and a smooth but not greasy texture. The aroma is pure and slightly sour. Acid curd cheese exhibits a smooth and firm texture and a mild and piquant aroma which intensifies with increasing ripening time.

12.1.4 Method of Manufacture

Acid curd cheese is produced from Quark (see Section 12.3) and is consumed in a ripened state. Caraway seeds may be added. The complete manufacturing process is described in detail by Kammerlehner (2009).

Milk: Skimmed cow's milk is short-time pasteurised (72°C–75°C/15–30 s), and starter cultures are added. The acid curd Quark is processed into acid curd cheese after one day. Ripening salts (0.5%–1.5%) are added, which usually consist of sodium hydrogen carbonate, calcium carbonate or a mixture of both. Furthermore, 2.5%–4.5% NaCl – and caraway for some varieties – is added.

Starter culture: Usually, a mixture of cultures is used for acidification of the cheesemilk. Thermophilic cultures *Streptococcus thermophilus* and *Lactobacillus delbrueckii* ssp. *bulgaricus* are used, of which 2%–4% is added for the milk's fermentation at 40°C–42°C for 1.5–2.5 hr. For the cold fermentation (fermentation temperature 21°C–27°C), 1%–2% mesophilic cultures are used, and the fermentation time is 15–18 hr.

Rennet: Traditionally, no rennet is used for the production of Quark. For some cheeses, a mixture of acid and rennet coagulated Quark may be used.

Cutting: The acid gel is cut into cubes with approximately 10–20 cm edge length. The curd is stirred to induce a further decrease of the curd particles until an edge length of about 15–20 mm is reached.

Curd draining/pressing: After cutting and stirring, the curd sinks to the bottom of the cheese vat, and whey is removed. The resulting curd–whey mixture is skimmed and transferred into filter bags, which are stacked on top of each other. The resulting pressure enhances whey drainage until a DM content of 32% is reached. Pieces of approximately 2 kg are cut from the curd mass and moved into a cooling room.

Salting: The cheese is brined-salted. For some varieties, salt is added to the acid curd Quark.

Maturation: Yellow cheese ripens at temperatures between 22°C and 24°C and 90%–95% Relative Humidity (RH). Different yeasts start growing on the cheese surface, inducing a mild proteolysis. The yeast cultures can be sprayed on the cheese or arise from the ripening room flora. After 48–72 hr, a smooth rind has developed. The cheese ripens from the outside to the inside. A second ripening phase takes place at 16°C–18°C and 90%–95% RH. The cheeses are treated with a mixture of NaCl, red smear adjunct cultures, colouring agents (e.g. annatto) and sodium hydrogen carbonate. Mould acid curd cheese is sprayed with camembert cultures after formation of the warm cheese. The cheeses ripen at 16°C–19°C and 75%–85% RH for 10–16 days during which different mould and yeast cultures grow and cover the cheese surface.

Storage: The cheese is packed and stored in aluminium or cellophane foil. Sufficient gas exchange is necessary to avoid deterioration of quality characteristics.

12.2 Crottin de Chavignol PDO – France

<table>
<tr><td>Name: Chavignol or Crottin de Chavignol PDO</td><td rowspan="3">
© Isabelle TREVIDIC pour le syndicat du Crottin de Chavignol</td></tr>
<tr><td>Production area: Pays Fort, Sancerrois regions and their surroundings around the territory; all included in the Centre–Val de Loire region</td></tr>
<tr><td>Milk: Goat's, raw</td></tr>
</table>

12.2.1 Introduction

Chavignol PDO is one out of the 462 soft and mould-rind French cheeses which include 18 PDO cheeses and most goat's cheeses, one of the numerous French goat's milk cheeses, as well as one of the 14 French PDO goat's milk cheeses (INAO-CNAOL, 2014). Like most French goat's cheeses, including the PDO cheeses, it is an acid-coagulated cheese made exclusively from raw goat's milk. In 2014, it was ranked fourth among French PDO goat's cheeses in regard to the volume produced, accounting for 11.5% of the total volume (INAO-CNAOL, 2014). Chavignol was among the first French goat's cheeses recognised as AOC in 1976, and it was then recognised as PDO in 1996 (EC, 1996). Chavignol is made and matured exclusively in a delimited area of 550,000 m^2 which is its traditional production area, namely, the Pays Fort and Sancerrois regions and their surroundings (Official website for Chavignol, 2016). This area is characterised by pastures on clay-limestones or clay soils that are perfectly suited for goat feeding; it is included in the actual Centre–Val-de-Loire region and is also well known for its wine. Centre–Val-de-Loire is the first French region producing PDO goat's milk cheeses, with 1/3 of the cheeses (5/14) and 60% of the production located in it in 2014 (INAO-CNAOL, 2014). Chavignol PDO cheeses are derived from ancient cheeses that were probably manufactured when goat breeding was introduced in the Sancerrois region, namely, in the sixteenth century (Official website for Chavignol, 2016). Goats, along with fodder, grains and vineyard, were traditional in this poor agricultural region. The wine grower's wife generally took care of them and manufactured cheeses that were consumed by the farm staff. They used pre-draining, a practice that is still prevalent, which allowed delaying a time-consuming step, namely, moulding, so that other farming works could be done. This practice consists of removing part of the whey by pouring just-coagulated milk on cloth. Depending on the season and milk quantity, cheeses were with or without blue or white mould-rind or 'repassé' in winter, the last being stored in stoneware containers after their maturation. At the end of the nineteenth century, a phylloxera attack on a French vineyard freed up many plots for goat breeding, leading to an increase in goat's milk production, especially within and around the Sancerrois region. In the 1900s, some persons specialised in maturation and distribution of Chavignol made at farms; in addition, Chavignol reached Paris, taking advantage of a new railway line in the vicinity of its production area.

12.2.2 Description and Sensory Characteristics

Chavignol PDO is a soft cheese, acid-coagulated (a small amount of rennet may be added) and harbouring a mould-rind. According to its specifications, it should have a minimum fat in dry matter (FDM)of 45% and a minimum DM of 41%. Its DM is 49.2%, fat content 26.9%, salt 0.7%, protein 19.7% and carbohydrate 1.3% (Official website for Chavignol, 2016). It can be consumed at various stages of maturity from about two weeks old (minimum maturity; 'demi-sec') until about 15 weeks old ('repassé'). Chavignol's rind is thin and soft, and white or blue (Chavignol 'blue' and 'bleuté'), or brown and blue (Chavignol 'repassé'). Chavignol's body is white or light yellow (ivory). Although it has not been investigated, the blue colour of Chavignol's rind is probably associated with the production of spores by *Penicillium* spp. strains as reported for other cheeses for which it is a rind defect. Depending on its maturing conditions, Chavignol's texture can be creamy (especially Chavignol 'repassé'), thick and melting in the mouth (Chavignol 'mi-sec'), or brittle and dry (Chavignol 'sec'); and Chavignol's flavour can be predominantly lactic (Chavignol 'mi-sec'), or roasted with a dried (hazel) nutty aroma (Chavignol 'sec') or vegetal with humus and mushroom aroma (Chavignol 'blue').

12.2.3 Method of Manufacture

Chavignol is manufactured according to rules that apply from milk production to cheese maturation. All the steps should be performed in the delimited area. As goat's milk is seasonally produced with peak production in April/May and low production from November (in winter) whereas consumer's demand is regular throughout the year, the frozen storage of curd has been allowed.

Milk: Milk has to be produced by Alpine goats. By its colour and morphology, this breed is most closely related to the extinct traditional breed. Every goat should have a minimum straw surface area, which is different depending on whether they stay exclusively in a shed or not. The minimum surface area of pasture providing fodder should be not less than 1 hectare for 12 goats. Expressed as DM, no less than 75% of total rations per day are produced within the delimited geographical area. This ration has to include not less than 50% of fodder exclusively produced in the delimited area, supplemented by specified concentrated feeds. Specified methods of that extending the period of kid birth are allowed.

Starter culture/rennet: All additives other than salt, rennet, calcium chloride and cultures of safe microbial strains are banned; storage of curd at negative temperatures is allowed for less than 15 months. Commercial starters have to be cultivated on goat's milk from the delimited area before their inoculation in vat milk; fresh or frozen whey from previous cheesemaking can be used as the source of starter strains; and more than one starter strain has to be inoculated.

Coagulation: Frozen curd (after pre-draining) can be mixed with fresh curd until a ratio 1:1 (frozen:fresh) is reached. A cheese manufactured under this condition cannot be labelled 'at farm'. Any practice that accelerates draining before the pre-draining on cloth is not permitted.

Drainage: Transport of cheese to the maturation plant and its storage before salting should not exceed 72 hr, at a positive temperature less than 10°C. This period is not included in the maturation time.

Maturation: The cheeses are matured for a minimum of 10 days, up to 15 weeks. After the minimum time of maturation, the cheeses are randomly sampled to assess compliance to their PDO specifications by scoring some of their sensory attributes and determining their gross chemical composition.

Maturation in oil, alcohol, wine and herbs or spices, as well as under a film or any packaging modifying the cheese environment and ash on cheese are forbidden. Maturation without any air flow after the usual maturation is required for Chavignol 'repassés'.

12.2.4 Relevant Research

The species of *Penicillium* involved would have to be identified according to the standard working method for species description and identification recently published (Visagie et al., 2014), since the taxonomy of the *Penicillium* genus has been considerably revised. *P. camemberti* and *P. roqueforti*, were shown to grow faster and be better competitors on the cheese surface than eight other *Penicillium* species, owing to the presence in their genome of newly discovered horizontally transferred genome regions similar within strains of the two cheesemaking species, although these two species are not closely related (Ropars et al., 2015). These regions contain genes highly expressed in the early stage of cheese maturation. Salles et al. (2000) reported that minerals and lactic acid, but not amino acids and small peptides, were probably the direct taste-active components present in the water-soluble fraction of commercial Chavignol PDO. The maximal amount of frozen curd that can be mixed with fresh curd and the length of storage at negative temperatures of the curd before mixing have been studied to draw up the specifications for Chavignol PDO manufacture (Leroux, 2016). The sensory characteristics of experimental mature Chavignol were statistically similar when either 50% or 25% of frozen curd stored for either 10 months or 20 months were mixed with fresh curd. Nevertheless, storage of curd for 20 months instead of 10 months could lead to more acid and bitter cheeses, while mixing fresh curd with 50% instead of 25% of frozen curd could lead to more bitter cheeses that comply less with the Chavignol sensory specifications.

12.3 Quark – Germany

Name: Quark	
Production area: Germany	
Milk: Cow's; Goat's, Sheep's or Buffalo's, pasteurised	

12.3.1 Introduction

Quark does not bear a certification or label, but is listed as a standard cheese variety in the German regulation on cheese (Deutsche Käseverordnung, 2013). In 1999, a total of 748,350 tonnes of fresh cheese was produced in Germany, which is approximately half of the total fresh cheese production in Europe. Quark is very popular in Germany, and it is produced throughout the country; thus, different names depending on the region can be found (Fox & McSweeney, 2004; Mair-Waldburg, 1974; Statistica, 2017).

The production of Quark dates back to the Stone Age and was known by the ancient Germanic tribes north of the Alps. Furthermore, it is documented that the ancient Romans were intensively trading in Quark. Quark has been industrially produced since the nineteenth century in the German-speaking areas (Iburg, 2003; Kammerlehner, 2009). Quark is being produced in all regions in Germany.

12.3.2 Type

Quark belongs to the group of fresh cheese, which are ready for consumption directly after manufacture. Quark is a white, creamy fresh cheese, which is produced from whole, skimmed or semi-skimmed milk, cream or buttermilk. The whey protein content is limited to a maximum of 18.5%. According to the German regulations on cheese, Quark must exhibit a moisture content of >73%.

12.3.3 Milk

Mainly cow's milk is used, but goat's, sheep's or buffalo's milk are also occasionally utilised. The milk must exhibit good quality; that is, only milk from healthy cows may be used, and the composition, especially the protein content, must be within the normal values (Deutsche Käseverordnung, 2013; Mair-Waldburg, 1974).

12.3.4 Description and Sensory Characteristics

Coagulation is normally induced through isoelectric precipitation of casein micelles by lowering the pH initialised by lactic acid bacterial cultures. However today, in most dairies, rennet is also added for coagulation in order to reduce the slightly sour taste resulting from acid coagulation (Kammerlehner, 2009).

The colour is milk-white to a weak yellow, depending on the fat content. The texture is evenly smooth, paste-like and spreadable with no gas openings. When cream is added, it is evenly distributed within the Quark. The taste is slightly clean and mildly sour (Deutsche Käseverordnung, 2013; Kammerlehner, 2009).

12.3.5 Method of Manufacture

Quark is produced via acid coagulation supported by slight enzymatic (rennet) activity. Traditionally, the curd was hung in special cheesecloth for whey removal. Today, in industrial manufacture, separator technology is most often used, where the whey is removed from the Quark via centrifugal forces.

Milk preparation: The cheesemilk is pasteurised and fat-standardised; however, this method leads to rather high fat loss during whey removal. Thus, the final Quark is usually blended with cream in order to achieve the desired fat content. A high-temperature pasteurisation increases the yield as well as the shelf life due to inactivation of mould spores (Kammerlehner, 2009).

Starter culture: For the production of Quark, a warm (fermentation temperature 27°C–31°C) and cold (fermentation temperature 20°C–26°C) fermentation may be applied. The warm fermentation (lasts 1 to 3 hr) is applied for the most common separator technology, where 0.7%–4% of a mixture of cultures is added (e.g. *Lc. lactis* ssp. *cremoris*, *Lc. lactis* ssp. *lactis*, *Leuconostoc mesenteroides* ssp. *cremoris*) (Kammerlehner, 2009).

Rennet: Between 0.3 and 0.5 g per 100 L cheesemilk of rennet is added when a Soxhlet Henkel degree (°SH) of 9–10 is been reached. Coagulation lasts between 5 and 8 hr.

Cutting: The coagulated milk is cut into squares with edge lengths of 6–8 cm.

Curd draining: During the traditional manufacturing, the curd is transferred manually into special curd bags or on tables prepared with cloths in order to remove whey. In the separator technology, the curd is separated via centrifugal forces from the whey.

Maturation: The curd is cooled down to 5°C–6°C afterwards in order to terminate further acidifying processes.

Storage: Quark is stored and sold in small plastic containers with different weights (e.g. 62.5 g, 125 g, 250 g and 500 g). However, due to the fact that Quark is not ripened, the shelf life is limited.

12.4 Robiola di Roccaverano PDO – Italy

Name: Robiola di Roccaverano PDO
Production area: Asti and Alessandria province in the Piedmont Region
Milk: Goat's cow's, sheep's (in the mixtures, at least 50% of the milk must be goat's), raw

12.4.1 Introduction

The Robiola di Roccaverano is recognised by the Italian government as a product with the Designation of Origin in 1979. Its PDO was registered by the European Union in 1996 (EC, 1996).

The Robiola of Roccaverano is produced in the provinces of Asti (including the town of Roccaverano, 'cradle of origin' of the product) and Alessandria in the Region of Piedmont. Robiola of Roccaverano has ancient origins; testimonials date it back to Celtic-Ligurian times (second millennium BC). The cheese was later described by Plinio il Vecchio and Pantaleon, who appreciated the quality of the cheese and illustrated its production method.

A manuscript signed by the priest Pistone dated 1899 shows the history in the period 960–1860 of the Parish of Roccaverano. Among the historical information of political interest also emerges elements of economics that serve to highlight the importance of 'Robiola', such as the fact that five exhibitions per year are held in the Municipality of Roccaverano. On such occasions are sold for export 'excellent cheeses Robiole', especially in France. Its name recalls both the Latin *robium*, a reference to the reddish colour of the outer part of the crust, and the town of Roccaverano in the Asti province, where this cheese originated.

12.4.2 Type

Robiola of Roccaverano is a fatty, fresh, soft, acid-coagulated cheese. Its minimum fat, protein, and ash content are 40%, 34% and 3% per dry matter, respectively.

12.4.3 Milk

Robiola of Roccaverano PDO is an artisanal product using raw whole milk from goat breeds Roccaverano and Camosciata Alpina and their crosses, and ewes of the Langhe Sheep and cows of the Piemontese and Bruna Alpina breeds and their crosses. It is the only Italian PDO cheese which can be produced exclusively with goat's milk, or goat's and cow's milk or goat's and sheep's milk. In any case, at least 50% of the milk must be goat's, and the remaining quantity can be sheep's or cow's milk. The cheese processing time is between 24 and 48 hr after milking. The Robiola di Roccaverano PDO is mainly produced from spring to late autumn.

12.4.4 Description and Sensory Characteristics

Its shape is cylindrical, with a diameter of 9–14 cm, slightly convex, up to 2.5–4 cm. The weight varies from 0.25 to 0.40 kg.

The sensory characteristics differ for the fresh or aged product. The product is defined fresh from the fourth to the tenth day of maturation. In the early days of maturation, the crust can be in the form of a light natural bloom of mould or be non-existent. Externally, it is milky white, creamy white or straw. The inside is milky white. The texture is creamy and soft. The flavour and aroma are delicate, tasty and/or slightly sour. The product is defined as aged from the 11th day of production. The skin has a natural bloom of mould and is wrinkled. The colour is light straw or straw or light red to red. The paste is milky white. The structure is soft, slightly compact with the continued maturing. It can be creamy under the crust. The flavour and aroma are stronger compared to fresh products and become spicy with aging.

12.4.5 Method of Manufacture

Milk preparation: Animal feed is obtained by leaving the animals on pasture in the period between 1st March and 30th November, and with the use of green fodder and/or stored cereal grains, cereals, and legumes.. All feed must come from at least 80% of the area of origin. The animals use herbs and scrub from the hills of Piedmont; goats and sheep also graze in the woods. The natural flora (wild grass and shrub) give to the Rabiola the smells, flavours and fragrance that make it unique. The use of silage and of foods containing genetically modified organisms (GMOs) is not permitted. Further, the use of milk from farms without land is prohibited.

Acid coagulation: Raw milk is left to acidify for hours, and can be inoculated with natural and autochthonous LAB (milk and/or whey starter culture). In the early stage of acidification, a very small dose of calf rennet is added. The process of acidification (lactic) lasts from 8 to 36 hr depending on the climatic and environmental conditions of processing and the possible presence of starter cultures.

The slow acid coagulation process at a mild temperature, about 25°C, does not facilitate the syneresis of the whey, and therefore the fresh curd has a high degree of moisture. Before moulding, the curd can be left to drain in fine-mesh canvases.

Moulding: The next step is to gently transfer sour curd in moulds, 'fuscelle', equipped with perforated bottoms, for 48 hr with periodic turnings in order to promote further and slow dripping of the whey.

Salting: Dry salting must be carried out on the two faces of the product while turning the cheese in the mould or at the end of the forming process.

Maturation: The natural aging is carried out keeping the fresh Rabiola in special rooms for at least three days at temperatures from 15°C to 20°C. From the fourth day, the cheese can be sold or the maturation can be continued at the farm level or by affineurs. Furthermore, from the fourth day of moulding, vegetable flavouring can be used. Robiola of Roccaverano is considered refined since the tenth day of placement in the moulds. Some traditional producers may keep the Robiola for six months in glass jars with oil, or place the forms in straw.

12.4.6 Relevant Research

The chemical and nutritional parameters of Robiola di Roccaverano have been studied (Bonetta et al., 2008a; Coisson, Arlorio & Martelli, 2000; Manzi et al., 2007; Pattono et al., 2001). From Bonetta et al. (2008b), the variability of the product in relation to the period of production, spring, summer and winter is clear. The summer cheeses show much lower moisture in the aged cheeses, which will lead to harder cheeses, quite different from most common Rabiola. The results obtained highlight some product differences between the artisanal and industrial products. The use of Principal Component Analysis (PCA) allowed cheese samples to cluster on the basis of their age (fresh or ripened), the origin of production (artisanal and industrial) and even the season of production. Gross composition, microbiological parameters and gas chromatographic analyses of FAMEs provided the most important parameters for Robiola di Roccaverano cheese characterisation.

The microbiological quality of Rabiola di Roccaverano has been also studied using different techniques by Grassi et al. (2002) and Bonetta et al. (2008c), including the biogenic amine producer bacteria by the latter. Cocolin et al. (2007) explored the spoilage of the cheese by film-forming yeast.

References

Bonetta, S, Coïsson, J. D., Barile, D., Bonetta, S., Travaglia, T., Piana, G., Carraro, E. & Marco Arlorio, M. (2008b). Microbiological and chemical characterization of a typical Italian cheese: Robiola di Roccaverano. *Journal of Agricultural Food Chemistry*, 56, 7223–7230.

Bonetta, S., Bonetta, S., Carraro, E., Rantsiou, K. & Cocolin, L. (2008c). Microbiological characterisation of Robiola di Roccaverano cheese using PCR–DGGE. *Food Microbiology*, 25, 786–792.

Bonetta, S., Bonetta, S., Carraro, E.,Coïsson, J. D., Travaglia, F. & Arlorio, M. (2008a). Detection of biogenic amine producer bacteria in a typical Italian goat cheese. *Journal of Food Protection Volume*, 71 (1), 205–209.

Cocolin, L., Rantsiou, K., Dolci, P. & Zeppa, G. (2007). Spoilage of Robiola di Roccaverano DOP cheese by film-forming yeasts. *Industrie Alimentari*, 46 (470), 655–660.

Coïsson, J. D., Arlorio, M. & Martelli, A. (2000). Chemical characterization of Robiola di Roccaverano DOP (protected denomination of origin) cheese. *Scienza e Tecnica Lattiero-Casearia*, 51 (1), 38–49.

Deutsche Käseverordnung, Established 1965, Revision 1986, Amendment 2010, 2013. German Federal Ministry for justice and consumer protection EC (1996). Commission Regulation (EC) No 1263/96 of 1 July 1996 supplementing the Annex to Regulation (EC) No 1107/96 on the registration of geographical indications and designations of origin under the procedure laid down in Article 17 of Regulation (EEC) No 2081/92. *Official Journal of the European Union*, L 163, 19–21.

Fox, P. F. & McSweeney, P. L. H. (2004). Acid- and acid/rennet-curd cheeses Part A: Quark, cream cheese and related varieties. In *Cheese: Chemistry, Physics and Microbiology, Vol 1: General Aspects*. Academic Press, London, United Kingdom.

Grassi, M. A., Civera, T., Pattono, D. & Turi, R. M. (2002). Microbiological quality of "Robiola di Roccaverano", a typical Italian cheese. *Industrie Alimentari*, 41 (420), 1321–1327.

Iburg, A. (2003). *Dumonts kleines Käselexikon*. DuMont monte Verlag, Cologne, Germany.

INAO-CNAOL (2014). *Produits laitiers AOP – les chiffres clés, 2014*. [Online]. Available: https://www.google.fr/search?q=produits+laitiers+AOP++2014&ie=utf-8&oe=utf-8&client=firefox-b&gfe_rd=cr&ei=iI3jV6OgJ8nDaKWRn8AO [Accessed 16 July 2016].

Kammerlehner, J. (2009). *Cheese Technology*. Josef Kammerlehner Verlag, Freising, Germany.

Leroux, V. (2016). Le report de caillé en AOC Crottin de Chavignol: Incidences technologiques et sensorielles. *The use of frozen curd in AOC Crottin de Chavignol: Technological and sensorial effects*. Available: http://www.journees3r.fr/IMG/pdf/2006_10_qualite_13_Leroux.pdf.

Mair-Waldburg, H. (1974). *Handbuch der Käse: Käse d. Welt von A-Z: eine Enzyklopädie*, Volkswirtschaftlicher Verlag, Kempten, Germany.

Manzi, P., Marconi, S., Di Costanzo, M. G. & Pizzoferrato, L. (2007). Composizione di formaggi DOP italiani. *La rivista di Scienza dell'Alimentazione, anno* 36, 9–22.

Official website for Chavignol PDO (2016). [Online]. Available: https://crottindechavignol.fr/ [Accessed 25 August 2016].

Pattono, D., Grassi, M. A., Civera, T. & Turi, R. M. (2001). Characterisation of hand-crafted "Robiola di Roccaverano" goat milk cheese. *Industrie Alimentari*, 409, 1351–1355.

Ropars, J., De La Vega, R. C. R., Lopez-Villavicencio, M., Gouzy, J., Sallet, E., Dumas, E., Lacoste, S., Debuchy, R., Dupont, J., Branca, A. & Giraud, T. (2015). Adaptive horizontal gene transfers between multiple cheese-associated fungi. *Current Biology*, 25, 2562–2569.

Salles, C., Herve, C., Septier, C., Demaizieres, D., Lesschaeve, I., Issanchou, S. & Le Quere, J. L. (2000). Evaluation of taste compounds in water-soluble extract of goat cheeses. *Food Chemistry*, 68, 429–435.

Visagie, C. M., Houbraken, J., Frisvad, J. C., Hong, S. B., Klaassen, C. H. W., Perrone, G., Seifert, K. A., Varga, J., Yaguchi, T. & Samson, R. A. (2014). Identification and nomenclature of the genus Penicillium. *Studies in Mycology*, 343–371.

13

Whey Cheeses (Heat Coagulated)

Photis Papademas[1], Thomas Bintsis[2], Efstathios Alichanidis[3] and Ylva Ardö[4]

[1] Department of Agricultural Sciences, Biotechnology and Food Science, Cyprus University of Technology, Cyprus
[2] 11 Parmenionos, 50200, Ptolemaida, Greece
[3] Department of Food Science and Technology, School of Agriculture, Aristotle University of Thessaloniki, Thessaloniki, Greece
[4] Department of Food Science, University of Copenhagen, Frederiksberg, Denmark

13.1 Anari – Cyprus

Name: Anari	
Production area: Cyprus	
Milk: Sheep's and/or goat's and cow's whey (with the addition of sheep's, goat's and cow's milk)	

13.1.1 Introduction

Anari is a whey cheese produced during the manufacture of Halloumi cheese. Therefore, it dates back to Halloumi making, and it is a valuable by-product, both for its taste, nutritional characteristics and its market value.

Global Cheesemaking Technology: Cheese Quality and Characteristics, First Edition.
Edited by Photis Papademas and Thomas Bintsis.

13.1.2 Type

Anari has a cylindrical shape. It is approximately 10–12 cm in diameter and 20–25 cm in height. Fresh Anari has been reported to have a wide range of fat content, that is, 7.0%–39.5% and an average of 21.7%. The protein content also varies from 9% to 16.3%, with an average of 11%. The average moisture content of the fresh cheese is 65.4%. Dry Anari, on the other hand, has average content of 33.8% moisture, 21.7% protein and 34.9% fat (Cyprus State Laboratory, 2013). The addition of milk to the whey during the manufacture of Anari is not performed by all cheesemakers nor is the type of milk added standardised, therefore this might explain the wide range in fat and protein content (Papademas, 2016).

13.1.3 Description and Sensory Characteristics

Fresh Anari is a soft, high-moisture, fragile cheese, with a very mild nutty sweet flavour, and it can be sold in vacuum packs either slightly salted or not. Fresh unsalted Anari is a highly perishable product that has a very short shelf life (2–3 days once the packaging is opened). Salted Anari cheese is usually dried, under cold air streams, in specially designed drying rooms until it becomes hard and easy to grate. Salted Anari can be microbiologically safe for much longer periods (i.e. six months).

13.1.4 Method of Manufacture

Whey: The whey used for Anari cheese is a mixture of sheep's, goat's and cow's obtained from Halloumi cheese production. Sheep's and Goat's whey is also used for Anari.

Heating/scalding: The whey is gradually heated to 65°C–70°C, and a quantity of milk (sheep's, goat's, cow's or a mixture) is added (10% of the total whey). The heating is continued until the temperature reaches 90°C. The whey is cooked for about 30 min until the crumbly curds of 'anari' (mainly denaturated whey protein and fat) rise on the surface of the whey. Sometimes a quantity of citric acid is added in order to reduce the pH of the whey, demineralising the whey proteins to assist their denaturation.

Moulding: The curds are hand–ladled in moulds and left to drain overnight at refrigeration temperatures.

13.2 Anthotyros – Greece

<table>
<tr><td>Name: Anthotyros</td><td rowspan="3"></td></tr>
<tr><td>Production area: Greece</td></tr>
<tr><td>Milk: Sheep's and/or Goat's whey (with the addition of sheep's or goat's milk or cream)</td></tr>
</table>

13.2.1 Introduction

The name 'Anthotyros' was given by the cheesemakers of Crete and can be translated as 'blossom' cheese (*anthos* = 'flower'; *tyros* = 'cheese') because of the way the 'curds' rise to the surface when the whey is heated, before being separated and strained.

13.2.2 Type

It is a whey soft cheese. The gross composition of Anthotyros is as follows: moisture 65.0%–66.5%, fat 16.5%–16.6%, fat-in-dry-matter (FDM) 47.4%–49.3%, protein 9.6%–9.7%, salt 0.5%–0.6% and pH 6.3–6.4 (Tsiotsias et al., 2002).

13.2.3 Description and Sensory Characteristics

It is a whey soft cheese, with a cylindrical shape, 10–12 cm in diameter and 10–15 cm in height. Anthotyros cheese has a compact, closed body and a typical mild pleasant flavour.

13.2.4 Method of Manufacture

Whey and milk preparation: The whey from sheep's and goat's milk semi-hard and hard cheeses is filtered and cream and/or milk from sheep or goats are added.

Coagulation: The whey (pH 6.3–6.4) is preheated to 55°C. Salt (0.5%) and raw sheep's or goat's milk (~10%) is added at approximately 70°C. At the point of initiation of milk protein coagulation (about 80°C–83°C), heating is reduced and the cheese–whey system is left for 20 min to complete protein aggregation; during this period the temperature is increased up to 90°C.

Curd draining: The cheese curd from each treatment is then surface-collected by means of a perforated spatula and transferred to a stainless steel mould to drain. Draining of Anthotyros cheese is carried out for 3–4 h at room temperature, after which is cooled at 4°C overnight.

Salting: See above.

Maturation: The following day the cheeses are vacuum-packed (portions of 200–500 g) and can be consumed immediately. Also, it can be more salted and partially dried to about 40% moisture and used either as a table cheese or for grating.

13.2.5 Relevant Research

Kalogridou-Vasiliadou, Tzanetakis and Litopoulou-Tzanetaki (1994) studied the microbiological characteristics of Anthotyros and reported that the predominant microflora consisted of enterococci (44.3%), lactococci (*Lc. lactis* subsp. *lactis*, 13.5%), leuconostocs (27.9%) and heterofermentative lactobacilli (14.4%). Arvanitoyannis, Kargaki and Hadjichristodoulou (2011) studied the suitable modified atmosphere packaging (MAP) composition for effective prolonging of shelf life for Anthotyros with regard to the microbiological and sensory data. They found that a MAP with 60% CO_2/40% N_2 and a MAP with 50% CO_2/50% N_2 mixtures proved to be most effective for inhibiting total mesophilic microorganisms and *Escherichia coli*. Neither *Staphylococcus aureus* nor *Listeria monocytogenes* were detected over the duration (56 days) of the experiment. Tsiraki and Savvaidis (2013) studied the effect of packaging on the quality characteristics of Anthotyros and reported that the combined use of either vacuum packaging or MAP, and basil essential oil (0.4% v/w), can extend the shelf life of whey cheese and maintain the freshness and the sensorial characteristics of the product.

13.3 Manouri PDO – Greece

Name: Manouri, PDO	
Production area: Thessaly, Western and Central Macedonia, Greece	
Milk: Sheep's or sheep's and goat's whey (with the addition of sheep's and goat's milk or cream)	

13.3.1 Introduction

Manouri is a whey cheese which is manufactured in Thessaly, Western and Central Macedonia. Despite the fact that Manouri cheese is a whey cheese, it was the main product of the cheese-making process, and Batzos (see Part II, Chapter 7, Section 7.1) was the by-product.

13.3.2 Type

It is a soft, whey cheese with a maximum moisture of 60% and a minimum of FDM 70%. The PDO status for Manouri was recognised by the EC in 1996 (EC, 1996).

13.3.3 Whey

Whey with the addition of sheep's and goat's milk and/or cream from sheep.

13.3.4 Description and Sensory Characteristics

It is a whey soft cheese, with a cylindrical shape, 10–12 cm in diameter and 20–30 cm in height.

Manouri cheese has a compact, closed body and a rich flavour with creamy, fatty notes.

13.3.5 Method of Manufacture

Whey and milk preparation: The whey from sheep's and goat's milk semi-hard and hard cheeses is filtered, and cream and/or sheep's and goat's milk , so that the fat content is at least 2.5%.

Coagulation: Coagulation takes place with heating at 88°C–90°C for 40–45 min. When the temperature of the whey reaches 70°C–75°C, salt is added at 1% as well as whole milk or cream up to 25%. Quantities of whole milk and/or cream must be such as to ensure that the final product contains, at a minimum, 70% FDM. The flocculation of whey proteins starts at 80°C, and the heating is continued up to 88°C–90°C. At this temperature, the curd is left for 15–30 min.

Curd draining: The curd is transferred with cheesecloths into moulds and drained for 4–5 hr.

Salting: See above.

Maturation: This takes place in cold rooms (4°C–5°C).

13.3.6 Relevant Research

Lioliou et al. (2001) studied the microbiology of Manouri cheese throughout storage at 4°C for 20 days, and concluded that the counts of all the microbial groups increased throughout storage and reached higher levels in cheeses made in summer than in those made in spring. Moreover, the microorganisms developed better on the cheese surfaces than in the interiors, especially in summer. The pH (6.78–7.33) and salt-in-moisture content (2.53–3.72) of Manouri cheese did not seem to affect the growth of the bacteria and yeasts present. The isolates of *Enterobacteriaceae* spp. were mainly *Hafnia* spp. (68.75%), while the isolates from the Baird–Parker medium were mainly staphylococci. There was a great diversity of yeast species, but *Debaromyces hansenii* and *Pichia membranefasciens* predominated. Kaminarides, Nestoratos and Massouras, (2013) studied the effect of the addition of milk and cream on the physicochemical, rheological and volatile compounds of Greek whey cheeses and reported that the addition of milk to whey increased hardness, while the addition of cream reduced hardness. In addition, the whey cheese produced with 79% whey, 15% milk and 6% cream had the highest concentration of aromatic components.

13.4 Mesost and Messmör – Sweden

Name: Mesost, Messmör	
Production area: Northern Sweden, mainly in the counties Jämtland and Härjedalen	
Milk: Cow's whey	

13.4.1 Introduction

Mesost and Messmör are unripened whey cheeses that differ from each other only in water content. They are made by boiling whey obtained from cheesemaking to concentrate and preserve the nutrients. The soft, spreadable version – Messmör (soft whey butter) –contains more water than the sliceable Mesost. They are specialties of the northern part of Sweden and were originally made only during summer when the animals were on pasture in the mountains. Traditionally, whey from goat's milk cheese was used, because cow's milk was mainly used for making butter and fermented milk in this part of the world. Milder versions of the whey cheeses were later developed by replacing half of the volume of goat's whey with cow's whey. Most of the cheeses produced in Sweden today are made of whey of cow's cheesemilk. Mesost and Messmör contain milk sugar, lactose, healthy milk minerals, whey proteins and sometimes milk fat. In earlier days, the cheeses were made in iron pots, which increased the amount of iron, and today when the cheese is mainly made in stainless steel containers, iron is commonly added to increase the nutritional value. Cream may be added at the end of production. The flavour is very sweet with a slightly bitter aftertaste. This highly nutritious product is especially appreciated by children (Ränk, 1987).

13.4.2 Type

Mesost has a moisture content of about 20% and Messmör 30%–35%. The fat content varies between 2% and 17%; the low fat content is typically found in the softer version, Messmör.

13.4.3 Whey

Whey is obtained from cow's and/or goat's milk cheese production.

13.4.4 Description and Sensory Characteristics

Mesost is yellowish brown, sliceable and sold in small blocks ready to be cut and put on bread. Messmör is lighter brown, spreadable and sold in small plastic or paper boxes as butter and spreads or in tubes. The cheeses have a pronounced sweetness, a slightly burnt caramelised taste and a slightly bitter aftertaste.

13.4.5 Method of Manufacture

Whey and milk preparation: Milk is transformed into cheese, and the whey is used for the production of Mesost and Messmör.

Coagulation: The whey is boiled to evaporate water and obtain a moisture content of 20% (Mesost) or 30%–35% (Messmör). After concentration, heavy mixing is performed to avoid the build-up of large lactose crystals. Milk fat may be added at this stage. If Messmör is to be made, the concentrated mass is homogenised.

Moulding: The caramelised mass is poured into squared moulds and left to set. The block of Mesost is then cut into smaller pieces, wrapped up and sold to be eaten on bread. Messmör is poured into boxes and chilled.

Storage: A long shelf life is obtained if the cheese is kept cold (2°C–6°C). The product is practically free of bacteria.

References

Arvanitoyannis, I. S., Kargaki, G. K. & Hadjichristodoulou, C. (2011). Effect of three MAP compositions on the physical and microbiological properties of a low fat Greek cheese known as Anthotyros. *Anaerobe*, 17, 295–297.

Cyprus State Laboratory (2013). *Cyprus Food Composition Tables*. Press and Information Office (PIO), Nicosia, Cyprus.

EC (1996). European Commission Regulation (EC) No 1107/96 of 12 June 1996 on the registration of geographical indications and designations of origin under the procedure laid down in Article 17 of Council Regulation (EEC) No 2081/92. *Official Journal of the European Union*, L 148, 1–10.

Kalogridou-Vasiliadou, D., Tzanetakis, N. & Litopoulou-Tzanetaki, E. (1994). Microbiological and physicochemical characteristics of 'Anthotyro', a Greek traditional whey cheese. *Food Microbiology*, 11, 15–19.

Kaminarides, E. S., Nestoratos, K. & Massouras, T. (2013). Effect of added milk and cream on the physicochemical, rheological and volatile compounds of Greek whey cheeses. *Small Ruminant Research*, 113 (2–3), 446–453.

Lioliou, K., Litopoulou-Tzanetaki, E. Tzanetakis, N. & Robinson, R. K. (2001). Changes in the microflora of manouri, a traditional Greek whey cheese, during storage. *International Journal of Dairy Technology*, 54 (3), 100–106.

Papademas, P., (2016). Anari Cheese. In Donnelly, C. (eds.), *The Oxford Companion to Cheese*. Oxford University Press, Oxford, pp. 24–25.

Ränk, G. (1987). [*From milk to cheese*]. Nordiska Museets Handlingar 66. 2nd edition. Berlings, Arloev, Sweden. ISBN 91-7108-269-7.

Tsiotsias, A., Savvaidis, I., Vassila, A., Kontominas, M. & Kotzekidou, P. (2002). Control of *Listeria monocytogenes* by low-dose irradiation in combination with refrigeration in the soft whey cheese 'Anthotyros'. *Food Microbiology*, 19, 117–126.

Tsiraki, M. I. & Savvaidis, I. N. (2013). Effect of packaging and basil essential oil on the quality characteristics of whey cheese "Anthotyros". *Food and Bioprocess Technology*, 6 (1), 124–132.

Index

Global Cheesemaking Technology: Cheese Quality and Characteristics, First Edition.
Edited by Photis Papademas and Thomas Bintsis.

b

c

Printed and bound by CPI Group (UK) Ltd, Croydon, CR0 4YY
17/09/2026
14964299-0003